International Conference on Aerospace and Mechanical Engineering (ICAME'18)

Journal of Physics: Conference Series Volume 1355

Kollam, India
17 – 19 December 2018

Editors:

M. Jose Prakash
K.K. Abdul Rasheed
Mohammed Sadhikh

ISBN: 978-1-7138-1308-8
ISSN: 1742-6588

Printed from e-media with permission by:

Curran Associates, Inc.
57 Morehouse Lane
Red Hook, NY 12571

Some format issues inherent in the e-media version may also appear in this print version.

Institute of Physics
Dirac House, Temple Back
Bristol BS1 6BE UK

Phone: 44 1 17 929 7481
Fax: 44 1 17 920 0979

techtracking@iop.org

Additional copies of this publication are available from:

Curran Associates, Inc.
57 Morehouse Lane
Red Hook, NY 12571 USA
Phone: 845-758-0400
Fax: 845-758-2633
Email: curran@proceedings.com
Web: www.proceedings.com

TABLE OF CONTENTS

AEROSPACE AND FLUID FLOW (AF)

FLUID AND THERMAL ENGINEERING STREAM (FT)

DESIGN AND INDUSTRIAL ENGINEERING STREAM (DI)

MATERIALS AND MANUFACTURING STREAM (MM)

Journal of Physics: Conference Series **1355** (2019) 011001 doi:10.1088/1742-6596/1355/1/011001

Editorial

Proceedings of the 2nd International Conference on Aerospace and Mechanical Engineering (ICAME'18)

M. Jose Prakash, K.K. Abdul Rasheed, Mohammed Sadhikh
Department of Mechanical Engineering, TKMCE

Email: mdsadikh@gmail.com

Abstract: International Conference on Aerospace and Mechanical Engineering (ICAME'18) organized by the Mechanical Engineering Department was held at TKMCE from 17th to 19th December 2018. Previous conference was held on December 2015.

1. Introduction

The conference ICAME'18 is a biennial international conference in the field of Aerospace and Mechanical Engineering organised by the Department of Mechanical Engineering, Thangal Kunju Musaliar College of Engineering (TKMCE), Kerala, India. It is a notable event which brings together academia, researchers, engineers and students from various regions making the conference a perfect platform to share experience, foster collaborations across industry and academia, and evaluate emerging technologies across the globe.

2. Inaugural Session

The conference was inaugurated by Dr S Somanath, Director Vikram Sarabhai Space Centre, Indian Space Research Organisation(ISRO). The meeting was presided by Dr Shahal Hassan Musaliar, President TKM College Trust.
Dr Normah Binth Mohammed Ghazali, Professor, Malayasian Technologial University, Malaysia and Prof Kanchan Chouwdhary, Professor, IIT Kharagpur offered felicitations. Dr Mohammed Sadhikh, Organising Secretary presented conference rationale. Jb TK Jalaluddin Musaliar, Treasurer TKM College Trust, Dr S Ayoob, Principal, Dr K K Abdul Rasheed Head of Department, Dr S Jose Co Chair delivered complimentary speech.

3. Technical Session

There were 8 plenary talks delivered by the eminent scientists/technologists. After the inaugural function the keynote address was delivered by Dr.Normah Binti Mohammed Ghazali Professor, Malaysian Technological University, Malaysia on the topic Environmental friendly cooling with Thermal Acoustics - Theory and development. Afternoon session of the day started with the invited talk by Dr Kanchan Chowdhury Professor, IIT Kharagpur on Downstream LNG Technology.
The second day of the conference began with the expertise of the invited speaker Dr. Sher Afghan Khan, Professor International Islamic University, Malaysia about Noise pollution and control. The invited talk was followed by the "Space Technology workshop " by ISRO scientists K P Abhilash lecturing on the Topic: Launch Vehicle Programme and Integration, Lazar T Chittilappilli on the Topic : Propulsion Research, John Tharakan about the Topic : Thermal and flow studies in engines, Robin V T about the Topic : Technology Spin offs.
Afternoon session of the day started with the talk by Mr. T J Jacob DGM, Ship Manager, Mazagaon Dock Ltd. Mumbai on an enthralling Topic : Introduction to ship building in India; Intricacies in Launching of Ships.
The final day of the conference commenced with the invited talk by Prof. Kasthurirengan Srinivasan, Professor Emiretus, Indian Institute of Science, Bangalore about Cryocoolers – Technology for the future.

Other than the plenary talks there were oral presentations by researchers in different technical sessions.

4. Cultural Evening

As part of the cultural entertainment Ghazal Night was arranged in the college auditorium on the evening of the first day of the conference. Conference banquet followed the cultural programme.

5. Manuscripts

Based on the announcement of the conference we received about 75 manuscripts for the purpose of publication in IOP Conference series (Journal of Physics). To ensure the publication standards mandated by IOP every paper was reviewed by at least by two referee before it was accepted. Finally 50 manuscripts were accepted for publication based upon the comments of the referees and the final decision of the editorial team.

6. Acknowledgement

As editors of the conference proceedings we would like to express our sincere gratitude to all referees and members of the organising and steering committee involved in the evaluation of the papers for their critical evaluation and valuable suggestions. Our sincere thanks to the team from IOP Publication lead by Stephaine Gill from IOP Publication office for coordinating the publication. We would also like to thank the sponsors and all the committee members of ICAME'18 for their valuable support, guidance and cooperation in organizing and conducting the conference successfully.

K.K. Abdul Rasheed (Convenor)

M. Jose Prakash (Co-chair)

Mohammed Sadhikh (Secretary)

MEMORABLE MOMENTS OF **ICAME'18**

ICAME '18

ICAME '18
Environmentally Friendly Cooling with Thermoacoustics: Theory and Developments

ICAME '18

INTERNATIONAL CONFERENCE ON
AEROSPACE & MECHANICAL ENGINEERING
ICAME '18
FACING THE FUTURE
GROWTH THROUGH SUSTAINABILITY
Department of
Mechanical Engineering

DEPARTMENT OF
MECHANICAL ENGINEERING
WELCOMES YOU TO

INTERNATIONAL CONFERENCE ON
AEROSPACE & MECHANICAL ENGINEERING
ICAME '18
FACING THE FUTURE
GROWTH THROUGH SUSTAINABILITY
Department of
Mechanical Engineering
Thangal Kunju Musaliar
College of Engineering

DEPARTMENT OF
MECHANICAL ENGINEERING
WELCOMES YOU TO

INTERNATIONAL CONFERENCE ON
AEROSPACE & MECHANICAL ENGINEERING
ICAME '18
FACING THE FUTURE
GROWTH THROUGH SUSTAINABILITY
Department of
Mechanical Engineering
Thangal Kunju Musaliar
College of Engineering
INDIAN LAUNCH VEHICLE
PROGRAM & INTEGRATION

INTERNATIONAL CONFERENCE ON
AEROSPACE & MECHANICAL ENGINEERING
ICAME '18
Department of
Mechanical Engineering
Thangal Kunju Musaliar
College of Engineering
Thermal and Flow Studies in L
Rocket Engines

INTERNATIONAL CONFERENCE ON
AEROSPACE & MECHANICAL ENGINEERING
ICAME '18
FACING THE FUTURE
GROWTH THROUGH SUSTAINABILITY
Department of
Mechanical Engineering
Thangal Kunju Musaliar
College of Engineering
LAUNCHING OF

ICAME '18

ICAME '18

INTERNATIONAL CONFERENCE ON
AEROSPACE & MECHANICAL ENGINEERING
ICAME '18
Department of
Mechanical Engineering
Thangal Kunju Musaliar
College of Engineering
FEAST

MECHANICAL ENGINEERING
ICAME '18
SUSTAINABILITY

ME '18

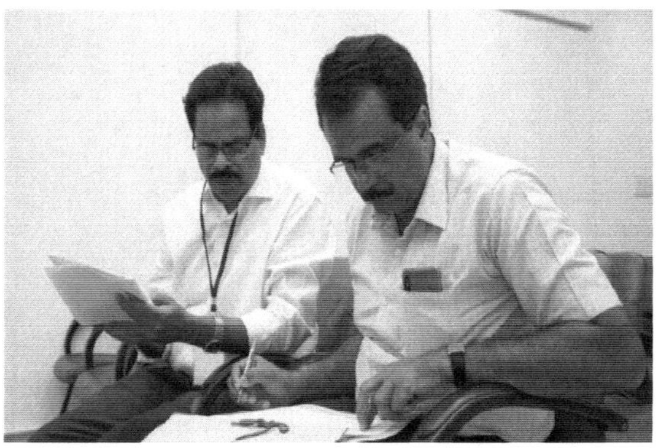

Cryocoolers :
Technology of the Future ?
S Kasthurirengan
Centre for Cryogenic Technology
IISc, Bangalore 560 012

Journal of Physics: Conference Series **1355** (2019) 011002 doi:10.1088/1742-6596/1355/1/011002

Peer review statement

All papers published in this volume of *Journal of Physics: Conference Series* have been peer reviewed through processes administered by the proceedings Editors. Reviews were conducted by expert referees to the professional and scientific standards expected of a proceedings journal published by IOP Publishing.

Numerical characterisation of the performance of flow rate on a non-contact vortex gripper

Anand Chandran C S[1], K S Sajikumar[2], K Jayaraj[3]
[1] M.Tech Student, Machine Design, College of Engineering Trivandrum, Kerala, India
[2] Associate Professor, Mechanical Engineering, College of Engineering Trivandrum.
[3] Professor, Mechanical Engineering, College of Engineering Trivandrum.

E-mail: anandchandrancs@gmail.com

Abstract. On manufacturing industries like semiconductor wafers, Manufacturing of medicine or Food processing industries require repeated handling of clean products such as wafers, Food products, etc. which are needs to be frequently handled and transported. This handling can be further ameliorated by Vortex Grippers. Vortex levitation can achieve Non-Contact handling by blowing air through a tangential nozzle into a vortex cup, thus swirling air flow forms which generates negative pressure. The suction force generated by the vortex is more than enough to keep the workpiece without contact. In the present study, a non-contact handling device called vortex gripper is modified by changing some design parameters. However, when a conical shaped vortex cup is used, a 60% increase in gripper lifting force is achieved with the same air consumption as reported early.

1. Introduction

The Production plants are becoming automated day by day, so repeated type of handling and manipulation is desirable which is mainly achieved by employing robots. The majority of the robotic grippers available are of contact type, but due to their unpredictable and unstable behaviour may cause damage to the object/workpiece. Such type of damages are very sensitive to certain objects like silicon material wafers, LCD or LED Screen panels, bakery products, manufacturing of biscuit, etc. The earliest study on the vortex gripper was proposed by Xin li, Kenji kawashima and Toshiharu kagawa in 2008. They demonstrated the vortex gripper mechanism of negative pressure generation and force development [1]. Mainly there are two kinds of pneumatic non-contact handling methods, and are Bernoulli levitation and vortex levitation. According to Vortex gripper, the suction force is developed by the constant supply of compressed air through the tangential nozzle attached with the vortex chamber which develops a swirling flow inside the cylindrical design of the cup creates a vacuum at the middle portion. The lifting force produced is due to the pressure difference in the upper surface of the workpiece and lower surface of the gripper. However, the pressure difference depends on the clearance between the device and the object, stated by Xin li et.al [1] Also, the dynamic characteristics and the dynamic modelling of the gripper are reported [2,3]. In addition, the main advantages of vortex levitation is that it requires low supply of air than the Bernoulli levitation, these characteristics are studied [2]. However, Stress generation and breakage, on handling and transportation of silicon material wafers clearly denotes the stress developed in the wafer region and also the crack size and location of breakage developed [4]. As reported in many previous studies for the high speed inspection procedure for LCD or LED glass screen panels, non-contact method is desirable [5]. Furthermore, for the reduction of food contamination as well as damages to the food products, non-contact type of grippers are desirable. Consequently, a rectangular shaped non-contact gripper for the handling of bakery products was developed by Rawal et.al [6] and Bernoulli gripper for handling of sliced fruits and vegetables was proposed by Davis et.al [7]. Furthermore, the best numerical method for predicting the flow characteristics of the vortex levitation is RNG K-Σ (Re-Normalisation group) [8]. Recently, the tangential nozzle are improved by the implementation of double nozzle and the pressure distribution is improved. Performance of the gripper is assessed by the maximum force, suspension – Region & stiffness. However, as the supply flow rate increases the relationship with force and clearance has an inclination towards the left side of the curve which decreases the suspension region [9]. Among these studies the vortex gripper is characterised by its gap height (Clearance) thus the levitation is stably suspended underneath as in Bernoulli effect gripper the suspension occurs an squeeze film flow which the workpiece performs a vibrational characteristics due to the damping force exerted into the workpiece which oscillates and finally makes it stable. Also, the outer diameter of Bernoulli gripper influence the suction force and the stiffness [10]. Based on a recent study the vortex chamber is slightly improved by the implementation of a diversion body, which improved the overall suction force of the gripper and the stability. Thus 25% larger suction force and the stability improves to 2.5 times greater than the already existing one [11].

The vortex gripper are continuously researched because of its high practical applications. This paper improves an early reported non-contact gripper. The simulation of the K-Σ model is used for the numerical purpose. The above said gripper is modelled and analysis is carried out to validate the model with the accepted gripper. The design is modified by changing the design parameters such as nozzle angle, nozzle shape, size of the nozzle and number of attached nozzles. The new design can generate higher lifting force than the early reported design with the same air supply which is discussed later.

2. Mechanism of vortex gripper

Figure.1 represents the schematic diagram of a newly developed vortex gripper. The table-1 shows different size of the gripper. The vortex gripper consists of a circular vortex chamber (Cup) and a cylindrical tangential nozzle. When high pressure air is blown through the tangential nozzle the air will be circulated cyclonically. By this action the center of the vortex cup will have less pressure. A vacuum is generated. The vacuum generated will levitate the object at bottom surface, but due to the flow of air the workpiece will not touch the gripper surface. The arrows inside the bottom view indicates the air flow direction.

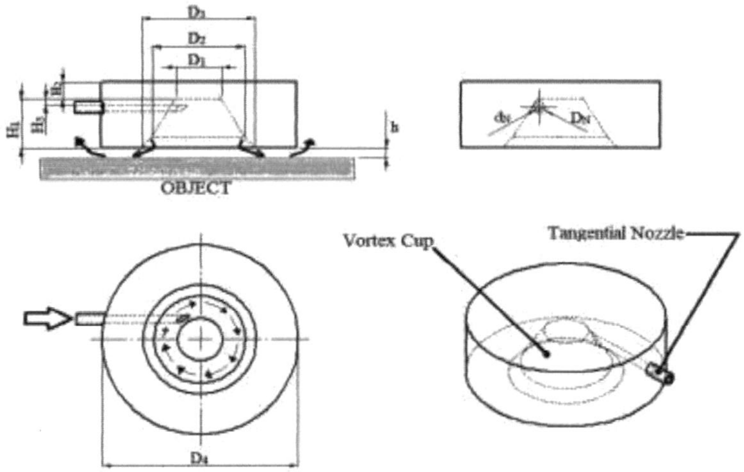

Figure 1. Schematic sketch of Vortex gripper (New model).

Table 1. Dimensions of vortex gripper (mm).

D_1	D_2	D_3	D_4	D_N	d_N	H_1	H_2	H_3	H
9	18.5	23	40	2.5	1.5	10.5	3.5	1.25	0.2 to 1.5

3. Numerical method

The newly developed model is numerically analysed to bring out the characteristics of the gripper. Ansys Fluent & CFX are the main numerical softwares used for the study. First an analysis is carried out for a vortex cup filled with one nozzle. The Navier-stoke equations are the governing equations for theses numerical study.

Continuity Equation - $\frac{d\rho}{dt} + \rho \frac{\partial U_i}{\partial x_i} = 0$ (1)

Momentum equation - $\underbrace{\rho \frac{\partial U_j}{\partial t}}_{I} + \underbrace{\rho U_i \frac{\partial U_j}{\partial x_i}}_{II} = -\underbrace{\frac{\partial P}{\partial x_j}}_{III} - \underbrace{\frac{\partial \tau_{ij}}{\partial x_i}}_{IV} + \underbrace{\rho g_j}_{V}$ (2)

Where $\tau_{ij} = -\mu\left(\frac{\partial U_j}{\partial x_i} + \frac{\partial U_t}{\partial x_j}\right) + \frac{2}{3}\delta_{ij}\mu\frac{\partial U_k}{\partial x_k}$ (3)

I: Local change with time
II: Momentum convection
III: Surface force
IV: Molecular-dependent momentum exchange (diffusion)
V: Mass force

3.1. Different Boundary Condition

The main boundary condition is the mass flow rate in the inlet and atmospheric pressure at the outlet. The cylindrical shaped geometries are the wall and the interiors. There is no heat transfer through the wall. The finite volume method were used for the Solution procedure with initial solution with first order discretization, latter it is changed into higher order discretization for improved accuracy of the final solution. The skewness is kept below 0.98. The boundary conditions used in the procedure was listed in the table 2. The RNG K-Σ method is used for the solution of viscous model.

Table. 2 Different boundary conditions that are used for the desired outlet

CONDITIONS	SETTINGS
Fluid	Ideal gas
Viscous	Turbulent
Inlet boundary	Mass flow rate (1.87x10-4 kg/s)
Outlet boundary	Pressure outlet (Atmospheric pressure)
Grid cell	1000835

4. Results and discussions

The Xin li et.al model is created and analysed to validate the technique used. The design dimensions are obtained from Xin li et.al [2]. The space (Clearance) between the gripper and work piece are modelled for analysis. The analysis is conducted for the clearance ranging from 0.2 to 0.85mm, using the ANSYS FLUENT software. The results are validated with the results of Xin li et.al [2]. Further, the technique is validated using the results of recently developed model using Xin li et.al [3]. This model consists of the tangential nozzle attached parallel with opposite faces of inlet nozzle. Figure. 2 shows the fluid domain model with 0.3mm clearance and the validation of the gripper is done for different clearance (h) from 0.15 to 1.5 mm. The comparison of the results with different clearance is shown in figure. 3. The results obtained are in good agreement with the results of Xin li et.al [3]. The maximum force occurs when the clearance is 0.3mm and the static pressure developed is about 680Pa.

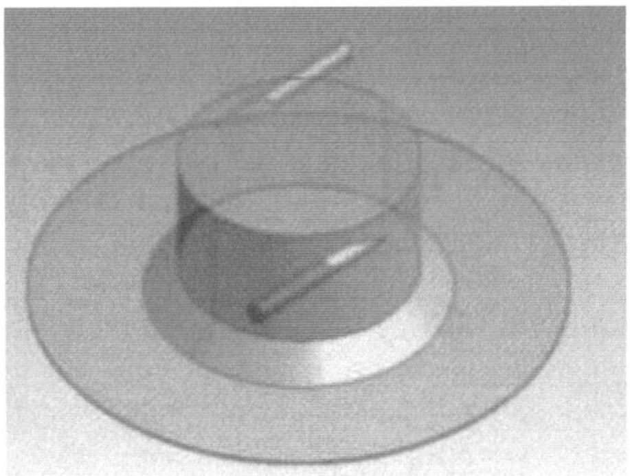

Figure. 2 Fluid domain model of recently developed two nozzle

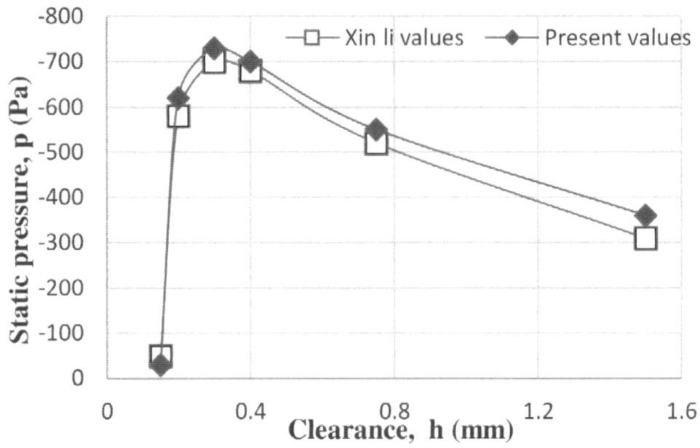

Figure. 3 Comparison between the two nozzle Vortex gripper with different clearance starting from 0.15 to 1.5mm

4.1 Optimisation

The optimisation of the vortex gripper is conducted by changing the design parameters such as the nozzle angle, nozzle shapes, number of nozzles and vortex cup design. Overall performance of the vortex gripper depends on the above parameters.

4.1.1 Optimisation of nozzle angles

The aim of optimisation of nozzle angle is to find the best angle for better pressure distribution and better stability of the work piece. The angles are varied from 5⁰ to 65⁰ with an interval of 5⁰ each. The figure. 4 shows the static pressure vs nozzle angle. From the figure it is understood that at the interval of 35⁰ to 40⁰ the static pressure is improved and also stable. These calculations were done for a clearance of 0.3mm. Figure. 3 shows that the maximum force obtained in the gripper is between 0.3 to 0.4mm clearances. However the selection of maximum force changes according with different flow rate of air and reduces the range for maximum force developed.

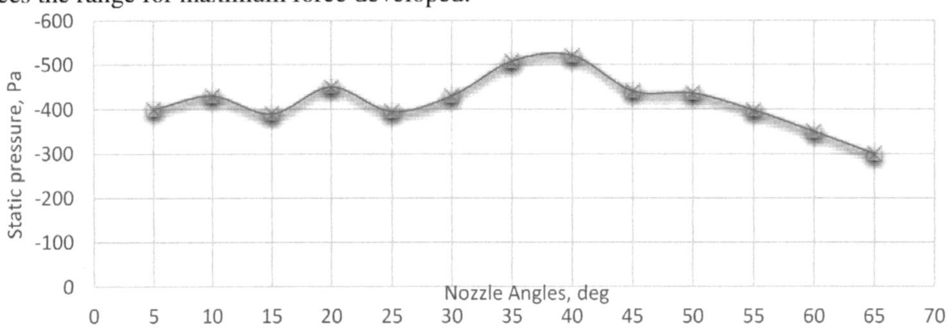

Figure. 4 Different angles and static pressure of vortex gripper

From the graph it is observed that at about 35⁰ – 40⁰ nozzle angle the static pressure developed was about 520Pa which is more than the value reported by Xin li et.al [2] and at this angle the workpiece is stable. When the angle is increased towards 40⁰ – 60⁰ the lifting force is slightly decreasing and also it was noted that there is an inclination of the work piece which may be due to the change of center of negative pressure slightly away from the central axis of the workpiece.

4.1.2 Optimisation of nozzle shape

The optimisation of nozzle shape is by changing the nozzle design by introducing three different types of nozzles namely Convergent, Divergent and combination of both convergent and divergent nozzles. The boundary conditions and the dimensions used were the same as proposed in the Xin li et.al [2]. A vortex gripper with convergent nozzle is developed with inlet diameter of 3mm and outer diameter of 1.5mm. Figure. 5 shows the model and Figure. 6 shows the pressure distribution, in this case the value obtained for the negative pressure is 635Pa, which is 1.5 times more than the Xin li design. And the velocity developed by the nozzle is also higher.

Figure. 5 Vortex gripper with Convergent nozzle

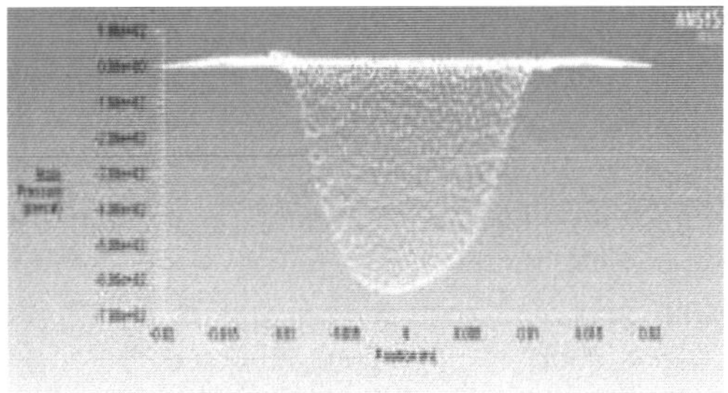

Figure. 6 Static pressure Distribution using convergent nozzle

For the divergent nozzle with inlet diameter of 0.75mm and outlet diameter of 1.5mm, the pressure developed is only 150Pa as compared to Xin li design which is about 450Pa. The convergent-divergent nozzle design, similar to that used in the rockets, reported by Biju kuttan et.al [12] is developed and analysed numerically.

4.1.3. Optimisation of Number of Nozzles

Model having two nozzle is already discussed and validated. Now vortex gripper with triple nozzle is developed with same diameter which are placed equally around the vortex cup having an angle of 120° separation each. The results were further analysed, the pressure distribution was not increased rather it decreased, the maximum negative pressure developed is of only 500Pa and the positive pressure increased rapidly. This may be due to the formation of turbulence from all the nozzle in spite of vortex formation. On increasing the number of nozzles from three to four nozzles the pressure needed for the levitation of the workpiece is increased but a high turbulent positive pressure is developing at the vortex cup which is highly unstable for the handing of the workpiece. In the four nozzle design the negative pressure developed is 600Pa and the positive pressure inside the vortex griper is 1200Pa. So it is evident that two nozzle is the optimum number of nozzle for vortex gripper.

4.1.4. New Design

From many designs developed, the one which gave best result is selected, which has high stability and lifting force than the Xin li's gripper. The model design consists of Cup having divergent shape. The inlet and other boundary conditions used are of same as that of the previous model. Figure. 1 consists of the schematic sketch of the design with the clearance of 0.3mm. The results of the analysis shows that the pressure distribution is 60% greater than the Xin li's gripper. The pressure developed inside the newly developed vortex gripper and the pressure generated is about 723Pa. In this case the gripper has a higher stability for the handling of the workpiece.

The design is further improved for higher lifting force by changing the number of nozzles to two. The results obtained are far better than the Xin Li's design and the stability of the design is enhanced. Figure. 7 provides the static pressure of the gripper with clearance of 0.3mm.

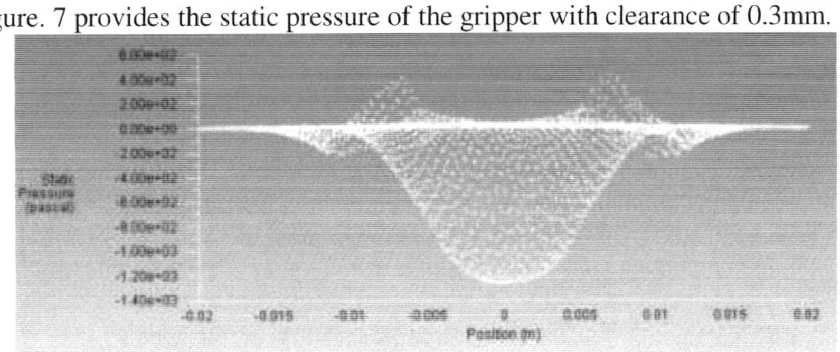

Figure. 7 Pressure distribution curve for 0.3mm clearance.

The pressure distribution curve has a high negative pressure of 1310Pa. The pressure distribution of the newly designed vortex gripper has about 2 times of the negative pressure than Xin Li's design for two nozzle gripper.

5. Conclusion

The non-contact gripper named vortex gripper can be worked with the help of swirling air flow to cause an upward lifting force and thus it can pick up a desired weight and hold the workpiece which is placed under the gripper without any contact. A vortex gripper is modelled and analysed using ANSYS CFX and Fluent software. The model is validated with pressure distribution of different position along the radius, with the results of Xin li et.al [2] & [3]. A detailed study of different models by changing the parameters such as nozzle shape, nozzle angle, Size of nozzle and number of attached nozzles. The effect of changing nozzle angle was analysed with different nozzle angle from 5^0 to 65^0. It was observed that, at about 35^0 - 40^0 nozzle angle the static pressure developed was about 520Pa for which the work piece is stable. The shape of the nozzle is varied using divergent, convergent & convergent-divergent nozzle and it is analysed. In divergent nozzle the static pressure decrease due to increase in nozzle diameter near the cup. In convergent nozzle static pressure was developed 1.5 times more than the existing tangential nozzle with the same amount of mass flow rate. In Convergent-divergent nozzle the static pressure observed was 5 times more than the existing tangential nozzle and the velocity augmented by the nozzle design structure. For three nozzle, the static pressure developed is 1.5 times higher than Xin li's designs, but the positive pressure of the vortex cup is higher. Further addition of nozzle showed a gradual decrease of static pressure because of high positive pressure inside the vortex chamber. From the above result it was found that the nozzle used need not to be increased beyond two nozzles.

The design with divergent vortex cup shows an improved result than Xin li gripper. The pressure distribution shows a 60% increase for both single and double nozzle design.

6. References

[1] Xin Li, Kenji Kawashima, and Toshiharu Kagawa, (2008a). "Analysis of vortex levitation." *Exp. Therm. Fluid. Sci.*, ***32(8)***, 1448–1454.

[2] Xin Li, Shouichiro Iio, Kenji Kawashima, and Toshiharu Kagawa (2011) "Computational Fluid Dynamics Study of a Noncontact Handling Device Using Air-Swirling Flow". *Journal of engineering mechanics © ASCE*, Page-(400-409).

[3] Jianghong Zhao, Xin Li(2016), "Effect of supply flow rate on performance of pneumatic non-contact gripper using vortex flow" *Experimental Thermal and Fluid Science* Page (91–100)

[4] X.F. Brun, S.N. Melkote(2009), Analysis of stresses and breakage of crystalline silicon wafers during handling and transport, *Sol. Energy Mater. Sol. Cells* **93** 1238–1247.

[5] Amano K, Yoshimoto S, Miyatake M(2011) "Basic investigation of noncontact transportation system for large TFT-LCD glass sheet used in CCD inspection". *Precis Eng*; **35**: 58–64

[6] Davis, S., Gray, J. O., and Caldwell, D. G. (2008). "An end effector based on the Bernoulli principle for handling sliced fruit and vegetables" *Robotics and computer-integrated manufacturing* **24**, 249-257.

[7] B.R. Rawal, V. Pare, K. Tripathi, (2008) "Development of noncontact end effector for handling of bakery products", *Int. J. Adv. Manuf. Technol.* **38** 524–528.

[8] Xin Li, Kenji Kawashima, and Toshiharu Kagawa (2008b). "Dynamic characteristics of vortex levitation." *Proc. of SICE Annual Conf., Tokyo,* Pages (1175-1180).

[9] X.F. Brun, S.N. Melkote(2009), Modeling and prediction of the flow, pressure, and holding force generated by a Bernoulli handling device, *J. Manuf. Sci. Eng. – Trans. ASME* **131**

[10] Xin Li, Kenji Kawashima, and Toshiharu Kagawa, (2008c). "Dynamic modeling of vortex levitation." *7th Int. Asia Simulation Conf., Beijing*, 218–224

[11] Q Wu, Q Ye, G X Meng, "Experimental and numerical study of vortex gripper with a diversion body" Proceedings of the Institution of Mechanical Engineers, Part C: *Journal of Mechanical Engineering Science* **Vol 226, Issue 6**, pp. 1526 – 1534

[12] Biju Kuttan P, Sajesh, "Optimization of Divergent Angle of a Rocket Engine Nozzle Using Computational Fluid Dynamics" 2014 , *The International Journal Of Engineering And Science (Ijes)* , **Volume2 issue 2** Pages 196-207

Fluid structure interaction simulations and experimental validation of a pipeline immersed in liquid

D Kumaresan[1], Lekshmi A[2], Kodati Srinivas[3], Suresh Mathew Thomas[4], Sruthy S[5]

[1]Scientist/Engineer, Structural Analysis Division, Mechanical Design and Analysis Entity, Liquid Propulsion Systems Centre, ISRO, Valiamala, Thiruvananthapuram-695 547, India.
[2]Graduate Student, Department of Civil Engineering, Amrita School of Engineering, Coimbatore, India.
[3]Head, Slosh Studies Division, Mechanical Design and Analysis Entity, Liquid Propulsion Systems Centre, ISRO, Valiamala, Thiruvananthapuram-695 547, India.
[4]Head, Structural Mechanics and Analysis Group, Mechanical Design and Analysis Entity, Liquid Propulsion Systems Centre, ISRO, Valiamala, Thiruvananthapuram-695 547, India.
[5]Assistant Professor, Department of Civil Engineering, Amrita School of Engineering, Coimbatore, India.

Abstract. In launch vehicles, there are number of fluid systems to cater to various requirements like propellant filling, draining to/from tanks, venting of ullage gases, pressurization of the tanks, feed systems to convey propellants to engine from tanks, lines to convey command gas to actuate various valves etc. These pipe lines are subjected to environmental loads i.e. vibration loads when the launch vehicle starts its course to place the satellite into its designated orbit. It is essential to study the dynamic characteristics of these pipe lines for the design of them for vibration loads. The response of these pipelines depends on fluid environment in which they are immersed. If a pipe line is immersed in liquid, its dynamic characteristics vary largely from that of a pipeline vibrating in ambient environment (air). If the pipe line immersed in fluid, natural frequency reduces and damping increases due to the added mass and fluid viscosity respectively. In order to study these effects, FSI studies are carried out on a pipe line immersed in water using ANSYS CFD and Mechanical software. Convergence studies with respect to time scale are carried out to benchmark the simulation procedure. This simulation procedure is validated by conducting the experiments.

1.Introduction

Launch vehicles generally consist of solid, liquid and cryogenic stages. Liquid and cryogenic stages use fuel and oxidizer for propulsion. Fuel and oxidizer are stored in two different propellant tanks. Gaseous medium required for actuation of valves and pressurization is stored in spherical gas bottles. There are number of fluid systems to cater to various requirements like propellant filling, draining to/from tanks, venting of ullage gases, pressurization of the tanks, feed line to convey propellants to engine from tanks, lines to convey command gas to actuate various valves etc.

A fluid system consists of a set of pipelines of different diameters, their supporting brackets and other elements like valves. In stages, four surrounding environment conditions arise. They are: (i) pipeline carrying ambient gas immersed in ambient air (ii) pipeline carrying ambient gas immersed in liquid (iii) pipeline carrying fluids immersed in fluid (iv) pipelines carrying fluid in ambient air. These pipelines are subjected to different environmental loads i.e. vibration loads when the launch vehicle starts its course to place the satellite into its designated orbit. It is essential to study the dynamic characteristics of these pipelines with the surrounding environment conditions for the design of them. The dynamic characteristics vary with respect to the conditions in which they were immersed. The response of a pipeline immersed in fluid environment varies largely from that of a pipeline vibrating in ambient air. This paper focuses on quantifying the effects of added mass and viscosity of fluid on natural frequency and damping of a pipeline immersed in water subjected to vibration through fluid structure interaction simulations. Simulation studies are carried out by sequential coupling of fluid and structural domains. Convergence studies with respect to time scale are done in order to bench mark the simulation procedure. Experimental studies are also carried out to validate the simulation procedure.

2. Literature Survey

Amin Zare et al used Euler Bernoulli beam theory for mathematical formulation for dynamic characterisation of pipelines subjected to fluid flow induced effects[1]. H. S. Simha et all analyzed fluid conveying pipes with simply supported, cantilever and fixed boundary conditions were analyzed[2]. Fluid structure interaction can be studied by two ways, one way coupling and sequential coupling, where later gives the more accurate results[3]. When a structure is immersed in fluid environment, fluid structure interaction takes place which alters the response of the structure[4]. Vipin Kumar et al discussed different methods for solving the dynamic equations. Ritz method was used to obtain dynamic equations for pipelines under different boundary conditions[5].

2.1. Added mass

Generally added mass is defined technically as a matrix which correlates the interaction of the mechanical structural elements through changes in fluid pressure. Expression for added mass for a perfect fluid at rest can be derived from the case of a simple harmonic oscillator. Added mass for a moving body inside an incompressible fluid does not depend on the viscosity [6]. Reduction in natural frequency of pipelines when immersed in an incompressible fluid is due to axial added mass coefficients caused by external fluid. Numerically the natural frequency can be obtained using Galerkin method[7]. The liquid layers surrounding a structure impose an asymmetry, which produces a difference in natural frequency in vertical and horizontal polarizations. The added mass coefficient is evaluated from the ratio between resultant force and acceleration acted on the wall [8].

2.2. Damping

There are various factors which govern damping of pipe lines immersed in fluid. Fluid viscosity and flow velocity are main factors which cause increase in damping of the system. The solution for this system is solved using spectral element method in frequency domain[9]. When fluid is flowing around a structure, it induces an additional damping effect as a result of its viscosity. Depending on an initial condition, added stiffness matrix and external force is together modeled to obtain the viscous force[10]. Transfer matrix is a matrix that represents the motion of a single pipeline section. The matrix incorporating the boundary conditions is called a point matrix. Both these matrices are combined to form overall transfer matrix[11]. To study the effects of added mass and viscosity, fluid structure interaction simulations of a pipeline made of polyimide immersed in water are carried out using ANSYS Mechanical and ANSYS CFD codes[12].

3. Numerical Methodology

Numerical simulations are done by CFD (computational fluid dynamics) simulations, which is the science of predicting fluid flow, heat transfer, mass transfer and chemical reactions. The pattern of fluid flow around a body immersed in liquid depends on the geometry of the body. For the FSI studies, fluid model and structure models are prepared. The interaction between the flow and the structure is based on sequential coupling of both.

3.1. Structural Domain

The structural domain represents the structure, its walls and boundary conditions. Basic equation of dynamic structural analysis for structural domain, static and transient analysis has to be carried out.

3.2. Fluid Domain

The fluid domain represents the fluid medium and its boundary conditions. Both the fluid and solid domain will have common boundaries where the fluid structure interaction takes place. Averaged Lagrangian Euler (ALE) method is used to study the dynamics of slowly varying waves in a moving medium. This method can be used for both linear and non linear systems. Momentum and continuity equations are also solved.

4. Numerical Implementation and Mesh Statistics

FSI studies are carried out on a polyimide pipeline immersed in water. Figure 1 gives the schematic of the pipeline (a), structural domain and fluid domain (c) used for FSI studies. Length and thickness normalized with the inner radius of the pipe line used for simulations are 16.95 and 0.034 respectively. Polyimide pipelines are fabricated by winding Fluorinated Ethylene Propylene (FEP) coated polyimide tape (Du Pont KAPTON film) and cured at 330°C. These polyimide pipe lines are used in cryogenic fluid systems as a flexible structural element to take care of thermal contraction requirements of propellant tanks. By virtue of their lower stiffness, it possesses flexibility to absorb differential contractions. Young's modulus and density of polyimide material are 2630 N/mm^2 and 1750 kg/m^3 respectively. Pipeline is discretized using three dimensional SOLID185 available in ANSYS Mechanical element library. Fluid domain is discretized as hexagonal grids with required boundaries.

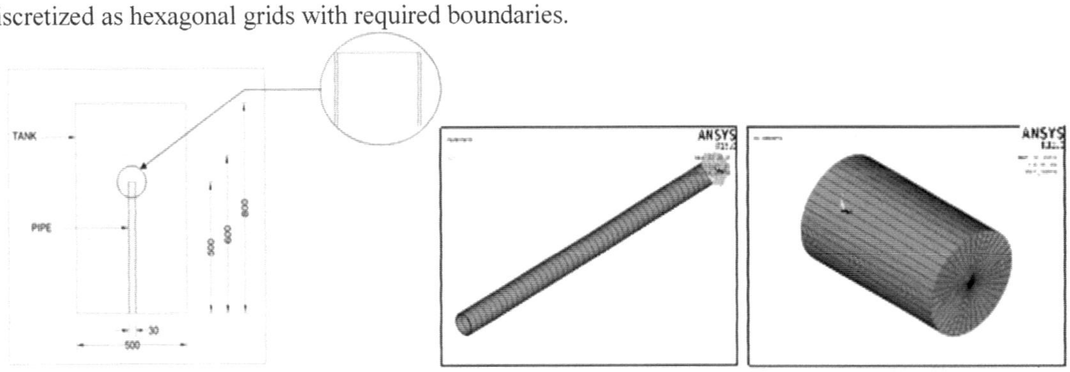

(a) (b) (c)

Figure 1. Schematic of the pipeline (a), structural domain (b) and fluid domain (c) used for FSI studies.

5. Analysis Procedure

Water and air contained in the tank are idealized as incompressible viscous fluids. Initial volume fractions have been set using step function as specified by free surface modeling of Volume of Fraction (VOF) method. Liquid height of fluid is chosen as 0.6m from the tank bottom. Coupled solver with appropriate convergence control parameters and different time step sizes are used for carrying out convergence studies. Homogeneous Eulerian-Eulerian multiphase model available in ANSYS/CFD is used for obtaining numerical solution in which time dependent momentum (Reynolds Averaged Navier Stokes – RANS) equations along with continuity and volume fraction equations are solved.

Initial pressure condition is $P_{initial} = P_{air} + \rho_{water}$ x g x (0.6 – z) x vf_{water}; set at opening boundary condition. Coupled field analyses are those analyses in which the input of one analysis is dependent on the results obtained from another analysis. There are two ways of coupling the results. For the present FSI studies, sequential coupling is employed. For solving sequentially coupled problem, ANSYS Multifield Solver (MFX solver) is used to obtain robust and accurate solutions. It is can also used to solve complex and larger models.

6. Results and Discussions

Convergence studies are carried out for different time scales (Δt) with structural damping (ξ) = 0%. The analysis was carried out for time steps of 0.005, 0.002, 0.001, 0.0002 and 0.0001 sec. Figure 2 give the plot of damping obtained with respect to different pipe line tip displacements. From the figure it is observed that damping increases with tip displacement. As time step reduces damping reduces and converges to a similar result for two simultaneous time steps.

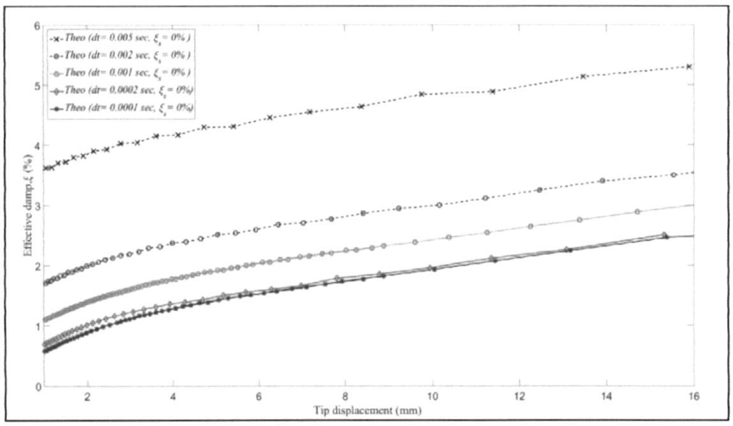

Figure 2. Damping obtained from the FSI simulations for different tip displacements.

From figure 2, it is observed that the result converges to a particular value for time steps of 0.0002 sec and 0.0001 sec. Damping of pipeline is estimated by using logarithmic decrement method. Experimental studies are also carried out for different tip displacements. Figure 3 gives the typical test data without and with filter (low pass band filter of 25 Hz) for a particular pipe line tip displacement.

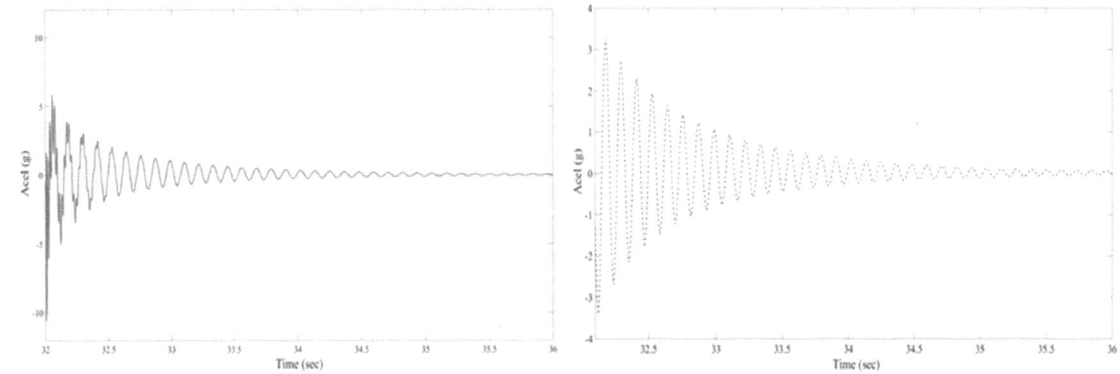

Figure 3. Typical test data without and with filter.[13]

Experiments are also carried out for the same pipe line in air in order to determine structural damping. From experimental tests, a damping ratio of 1% is obtained. This structural damping is included and FSI simulations are repeated for the time scale of 0.0002 sec. Results obtained are compared with the experimental values and given in Figure 4. From Figure 4, it is concluded that both experimental and numerical damping values match fairly well. Thus simulation procedure is validated. This can be taken as a benchmark criterion for obtaining convergence for simulation procedure.

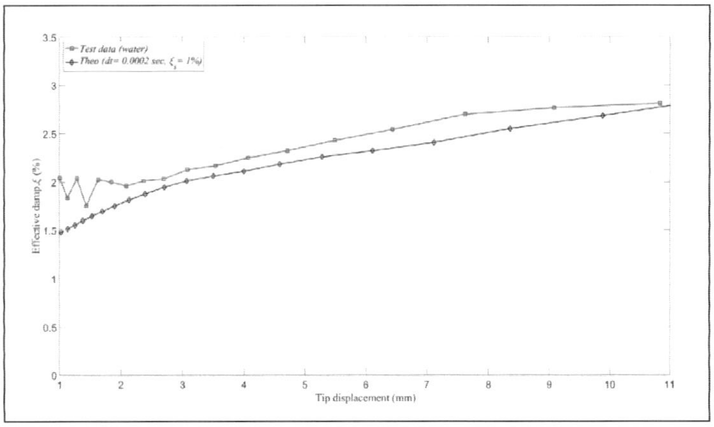

Figure 4. Comparison of experimental and FSI results.

7. Conclusion

Fluid structure interaction studies are carried out using ANSYS Multifield (MFX) solver for a polyimide pipe line immersed in water. It is found that damping ratio increases with increase in amplitude of test specimen. Convergence with respect to of time scale (Δt) is achieved for a time step of 0.0002 sec after several iterations without incorporating structural damping. FSI simulations are repeated for the converged time step after including the experimentally determined structural damping. Experiments are also carried out on similar pipe line immersed in water. Simulation results are compared with experimental results and they match fairly well. Thus the simulation procedure with respect to convergence criterion is validated/ benchmarked.

8. References

[1] Zare A, Eghtesal M and Daneshmand F., Numerical investigation and dynamic behaviour of pipes conveying fluid based on Iso-geometric analysis, 2017, *Ocean Engineering* **140**, 388- 400.

[2] Simha H S and Chellapila K., Vibration of fluid conveying pipes resting on two parameter foundation, 2008, Open Acoustics Journal, 1, 24-33.

[3] Badia S, Quaini A and Quarteroni A., Modular vs. non-modular pre-conditioners for fluid structure systems with large added mass effect, 2008, Computer Methods In Applied Mechanics and Engineering 197, 4216- 4232.

[4] Avcar M., Free vibration analysis of beams considering different geometric characteristics and boundary conditions, 2014, International Journal of Mechanics and Application, 4(3): 94-100.

[5] Kumar V and Singh K K., Analysis of natural frequencies for cantilever beams with I and T section using ANSYS, 2015, International Research Journal of Engineering and Technology, Volume: 02, Issue: 06.

[6] Forster C, Wall W A and Ramm E., Artificial added mass instabilities in sequential staggered coupling of nonlinear structures and incompressible viscous flow, 2007, Computer Methods In Applied Mechanics and Engineering 196, 1278- 1293.

[7] Liaghat T., Two way fluid structure coupling in vibration and damping analysis of an oscillating hydrofoil, April 2014, Universite De Montereal.

[8] Parameshwaran R and Dulipala S J., Fluid structure interaction and flow induced vibrations: A review, Journal on sound and vibrations, 2016, 12th International Conference on Vibration Problems, ICOVP2015, Procedia Engineering 144, 1286-1293.

[9] Hildago J A S, Gama A L and Moreira M., Natural vibration frequencies of horizontal tubes partially filled with liquid, 2017, Journal of sound and vibration 408, 31-42.

[10] Zou G P, Cherazhi N and Taheri F., Fluid induced vibration of composite natural gas pipelines, 2005, International journal of solids and structures 42, 1253-1268.

[11] Li Y and Liu G., Vibration analysis of liquid filled pipelines with elastic constraint, 2011, Journal of Sound and Vibration 330, 3166-3181.

[12] ANSYS Mechanical and CFD Software Manuals.

[13] Classified data of LPSC/ISRO.

Acknowledgement

Authors thank Dr. A. K. Asraff and Dr. R. Muthukumar, Mechanical Design and Analysis Entity of LPSC for their support for carrying out FSI simulations and experimental studies. Thanks are also due to Director, LPSC for permitting to publish this paper.

Numerical study on air-core vortex in a dual port cylindrical tank for lower speed of rotations

Sooraj PV, Joshy PJ
School of Engineering, Cochin University of Science & Technology, Cochin-22, Kerala

Abstract
Rotating cylindrical tank filled with liquid engendered an air core during draining. The extension of the air-core into the drain port reduces the flow area and consequently the flow rate. The ongoing study focuses on the numerical simulation results of such vortex formation in a rotating cylindrical tank with dual ports in which speed of rotation as the variable parameter here by keeping centre to centre distance fixed.

Nomenclature
C = Centre to Centre distance between the dual ports located at an equal distance from the base plate centre (mm)
d = Port diameter (mm)
D = Cylinder diameter (mm)
N = Rotational speed (rpm)

1. Introduction

Vortex is a rotating mass of fluid that forms a vacuum at the centre, draws the liquid into it. Vortex formation at intakes is a significant hydraulic engineering problem. The main reasons considered in the formation of vortex are the inertial distribution, fluid instability. Vortex extending to the free surface can lead to air entrainment that can cause mechanical damage and loss of pump performance. The phenomenon is of useful pertinence in the fuel feed systems in spacecrafts and rockets. The air core has a spiral shape, which perforates through the upper surface of the liquid to the base of the tank. Consequently, the effective cross-sectional area for the drain port becomes narrow because the area parting with the gas phase expands, reducing the draining flow rate remarkably. This paper focuses to assess the results of numerical simulations of the vortex formed during draining in a rotating cylindrical tank.

Numerous experimental investigations have been conducted on air core vortex formed during draining. More than one and a half centuries ago, Rankine *et al.* (1858) proposed a two-zone model to find out the tangential velocity within a liquid tank with a stable gas-core. Studies on vortexing in propellant tank were first carried out by Abramson *et al.* (1962) using a cylindrical tank [1]. Experimental investigations reported by Lubin and Springer (1967) have presented empirical relations for the dip formation during liquid draining [2]. The influence of various parameters like initial height of fluid, surface tension, viscosities on dip formation were analysed by them using a cylindrical tank with circular port. Granger *et al.* (1968) observed large surge velocities and rapid intensification of the air-core vortex only for particular ranges of drain port diameters when rotational flow currents were present in the liquid column in the tank. Ramamurthi and Tharakan (1995) found out the parameters affecting the gas-core formation and its growth an instantaneous suppression, but the suppressed vortex recovered after stopping the vibration [7]. Ramamurthi and Tharakan (1993) assessed effect of bell-mouthed, stepped and cylindrically shaped drain ports on suppression of air core vortex. Gowda *et al.* (1996) analysed the gas-core dynamics for liquids that are initially subjected to rigid body rotation. The effect of storage tank cross-section on vortex formed during draining was conducted by Gowda *et al.* (1996) and compared a storage tank with rectangular and square cross-section with cylindrical tank [6]. Influence of a dish or cup shaped suppressor positioned over a circular drain hole for vortex suppression was also studied. Gowda and Udhayakumar (2005) was able to prevent vortex formation during draining subjected to initial fluid rotation with the introduction of a vane type suppressor above drain port [10]. A variety of multiphase flow simulations, particularly related to gas-entrainment systems have been reported in the literature (Cogan *et al.* 2008, Li *et al.* 2008, Mahyari *et al.* 2010, Satpathy *et al.* 2013) [8][3]. Numerical studies of tank draining systems have mostly focused on cylindrical tanks with circular cross section and axi-symmetrically placed drain port. Mahyari *et al.* (2010) performed a CFD study to validate against their own experiments and analytical expressions for the development of effective vortex breakers during tank draining [4]. Robinson *et al.* (2010) have attempted the validation of their CFD predictions with the experimental work of Lubin and Springer (1967) using commercial CFD tools [9]. Park and Sohn (2011) have carried out systematic experimental and CFD study of gas-core vortex formation [5]. However, liquid draining through the cylindrical tanks is a function of parameters, such as pressurization, drain port size, shape, etc.

The current study focuses on the numerical simulations and analysis of air-core vortex using dual port in a rotating cylindrical tank keeping centre to centre distance same by taking drain ports diametrically opposite.

2. CFD Methodology

The commercial CFD code ANSYS FLUENT 15, 16 were used to simulate the transient draining of liquid from a cylindrical tank. FLUENT uses the finite volume method to break the problem domain into control volumes.

The time step is managed by Courant number, which was typically chosen between 0.1 and 0.4 depending on the mesh density. The time step sizes were varied accordingly. The CFD simulation involves tracking the gas-liquid interface using the VOF approach implicit scheme was used for time discretisation. The flow field for modeling this flow was assumed to be an axisymmetric swirling type, although the actual draining process accompanies a spiral motion on the free surface. So as to capture the gas vortex, whirling stream and turbulence viably, standard k– ω display for axisymmetric streams is applied, which uses second order differencing schemes for turbulent kinetic energy and dissipation rate. Navier Stokes equations are the governing equations, which can be represented in 3D cylindrical coordinates. A very common case is axisymmetric flow with the assumption of no tangential velocity ($u_\varphi = 0$) and the remaining quantities are independent of φ:

$$r: \quad \rho\left(\frac{\partial U_r}{\partial t} + u_r \frac{\partial U_r}{U_r} + u_z \frac{\partial U_r}{U_z}\right) = -\frac{\partial p}{\partial r} + \mu\left(\frac{1}{r}\frac{\partial}{\partial r}\left(r\frac{\partial U_r}{\partial r}\right) + \frac{\partial^2 U_r}{\partial z^2} - \frac{U_r}{r^2}\right) + \rho g_r$$

$$z: \quad \rho\left(\frac{\partial U_z}{\partial t} + u_r \frac{\partial U_z}{U_r} + u_z \frac{\partial U_z}{U_z}\right) = -\frac{\partial p}{\partial z} + \mu\left(\frac{1}{r}\frac{\partial}{\partial r}\left(r\frac{\partial U_z}{\partial r}\right) + \frac{\partial^2 U_z}{\partial z^2}\right) + \rho g_z.$$

$$\frac{1}{r}\frac{\partial}{\partial r}(r u_r) + \frac{\partial u_z}{\partial z} = 0.$$

3. Computational Mesh

A three dimensional tetrahedral structured grid was generated to represent test geometry. The number of cells was varied from 300000 to 700000 typically, for the meshes with configuration *d/D* ratio as 12/130, so as to achieve mesh independent solutions, which is assumed to be achieved when consecutive solutions show a change less than 1% in the outflow mass flow rate. The differencing schemes used were second order upwind for momentum and turbulent kinetic energy. Pressure was discretized using the PRESTO! Scheme, which is similar to the staggered-grid schemes used with structured meshes.

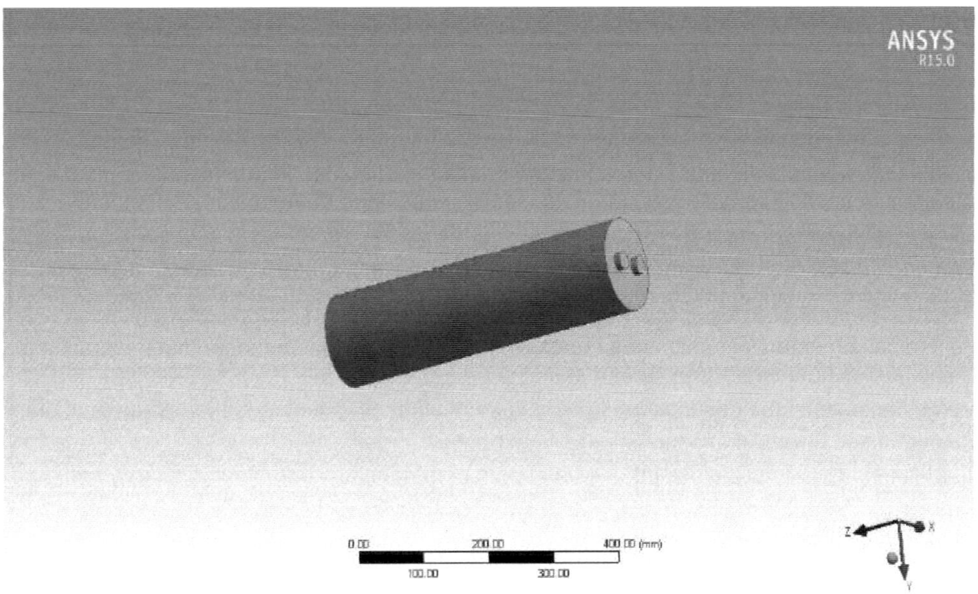

Fig. 1: Geometric representation of the model.

Boundary conditions: A large gas to liquid volume ratio was assumed so that the gas pressure above the liquid could be considered as constant. The boundary conditions for modelling the flow are given in Table.

Wall	Viscous / No Slip (u_Z; u_r= 0)
Inlet	Pressure Inlet (Gauge pressure = 0)
Outlet	Pressure Outlet (Gauge pressure = 0)
Interface	Stress free, Wall adhesion effects

Table 1: Boundary conditions

Model Geometry

To numerically investigate the tank draining problem, the models were particularly chosen so that boundary conditions should be properly satisfied. The cylindrical domain was designed with 130 mm diameter and 500 mm height. The drain port diameter was varied from 8mm to 20 mm where centre to centre distance was kept constant as 25 mm.

4 .Numerical results and discussion

VOF model has been shown to give a good account of draining characteristics in a qualitative and quantitative sense in the simulation of liquid draining from cylindrical tank. Generation of air core implies process of air entrainment in the cylindrical domain by park and sohn (2011). The dimple of the free surface at the early stage was due to the rigid body behaviour of the liquid in a tank before the draining was initiated. But sinking of the free surface was found near to the center area. Later air core was completely generated. The air core penetrated from the top face of the free surface to the drain port. The generated air core was sustained throughout draining. The effective exit section was reduced because the air occupied the center portion of the drain nozzle. Thus, the flow rate for draining decreased.

Vortex formation in dual port design

After setting the boundary conditions, respective mesh motion was achieved by setting the drain port diameter constant. Here the influence of speed of rotation on vortex formation was analysed.

Effect of speed of rotation

In the eccentric dual port system initially a dip forms and air entrains in the course of draining. When water was allowed to drain through the dual port rotating cylinder, the pressure just above the drain port was less than that of the atmospheric pressure because the velocity was sufficiently high. This draws the centre of the free surface down, there by forming a dip. The air pushes the vertical column of fluid over the drain port down, which results in air entrainment followed by air-core vortex.

Journal of Physics: Conference Series **1355** (2019) 012003 doi:10.1088/1742-6596/1355/1/012003

It is obvious that an increase in speed of rotation increases the turbulent velocity thus the vorticity magnitude.

Case 1: Rotational speed, N = 100 rpm

i) Vorticity Magnitude

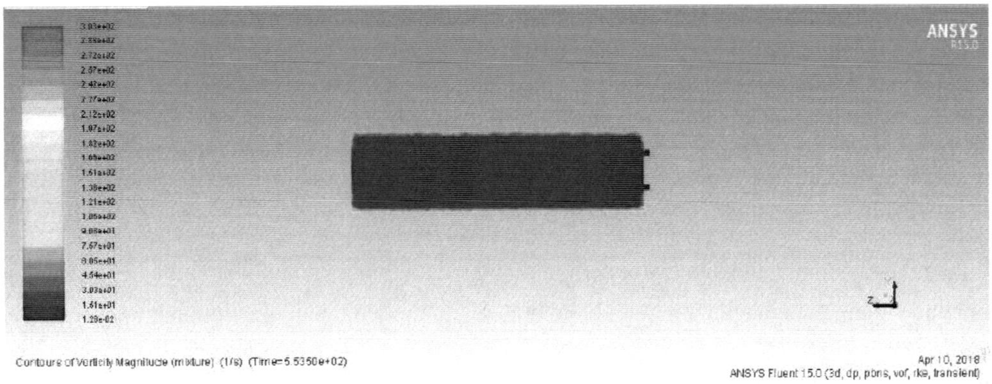

Fig. 2: Contours of vorticity at the dual port drain Outlet when given cylinder rotation, N = 100 rpm

The darker blue shades represent lowest vorticity downstream of the flow. So, it is inferred from the result that vorticity component is weak in the fluid domain. Hence there is no vortex formation. The vorticity magnitude and radial velocity profiles in eccentric dual port system are negligibly small compared to single port design. This is because; in the eccentric dualport system pressure gradient is not sufficient to enter the air core into the drain outlet.

ii) Radial Velocity

Fig. 3: Contours of radial velocity at the dual port drain Outlet when given cylinder rotation, N = 100 rpm

Radial velocity profiles are almost uniformly maintained throughout the whole liquid zone even if the liquid level decreased over time. As there is no significant speed of rotation, there are no flow-driving forces except gravity; therefore, the considerable radial and axial velocities near the drain port are entirely due to the sudden contraction of the flow passage.

However, significant velocity magnitude is observed on the free surface. According to the kinematic condition on a free surface, a particle on a free surface should dip in its position on the free surface. Thus, the low swirl case had small dip on the free surface, and the free surface dimples during nearly all of the draining time. The swirl velocity at the early stage was caused by the side-wall rotation before the draining. Thus, the magnitude was comparatively high near the side wall. As time goes on, the maximum swirl velocity zone moved to the bottom area because of the growth of the boundary layer on the side wall. Such a boundary layer also formed on the bottom wall region, is called the Ekman spiral layer.

Case 2: Rotational speed, N = 150 rpm

i) Turbulent intensity

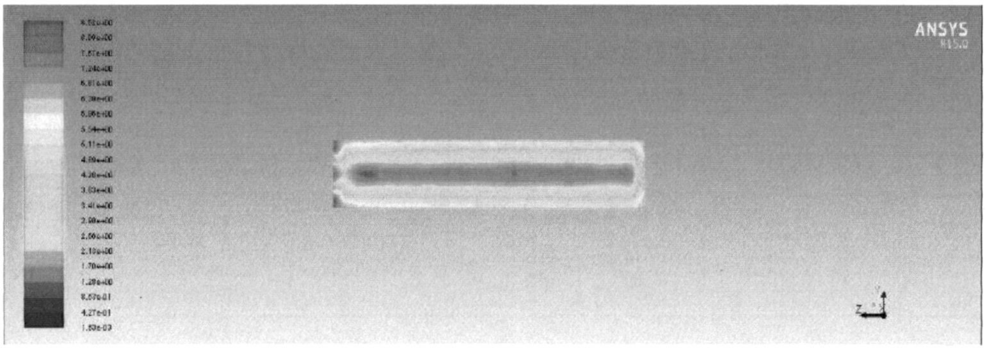

Fig. 4: Contours of turbulent intensity at the dual port drain outlet when given cylinder rotation, N = 150 rpm

When cylindrical rotation starts, the fluid particles in the middle of the volume of the tank doesn't move because none of the agencies which creates turbulence have yet come to play there. Next to the wall however the fluid does circulate because of the viscous stresses. These viscous stresses gradually decrease outwards from the wall and more and more fluid particles circulate. In the end, viscosity brings all the fluid in to a perfect solid body rotation. At this point paradoxically, the viscous forces vanish all together and there is nothing to force further changes in the turbulence.

ii) Velocity Magnitude

The fluid particles near the side wall attain the rotational energy due to no slip effect, so the magnitude of velocity will be at its maximum near the wall. But the velocities of the particles at the centre will be almost zero or negligible and after the passage of time the particles near the centre attains velocity, yet lower compared to the particles near the cylindrical wall. The dip shape of the free surface at the early stage was due to the rigid body behavior. The depression occurs on the free surface due to the pressure gradient existing between the free surface and the bottom of tank. The pressure force due to the rotation is minimum at the axis of rotation with its magnitude increasing radially outwards. Velocity profiles in eccentric dual port system are negligibly small compared to single port design. This is because; in the eccentric dual port system pressure gradient is not sufficient to enter the air core into the drain outlet.

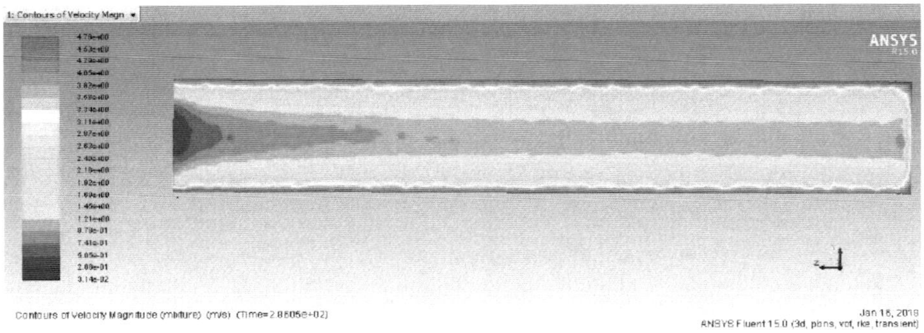

Fig. 5: Contours of velocity magnitude at the dual port drain Outlet when given cylinder rotation, N= 150 rpm

Case 3: Rotational speed, N = 250 rpm

i) Vorticity magnitude

As said, more and more vorticity is induced near the side wall of the cylindrical domain which decays before reaching the central axis of rotation. Hence, the dip formed due to the rigid body rotation fails to form the fully developed vortex and the pressure gradient will not be sufficient to generate the air-core.

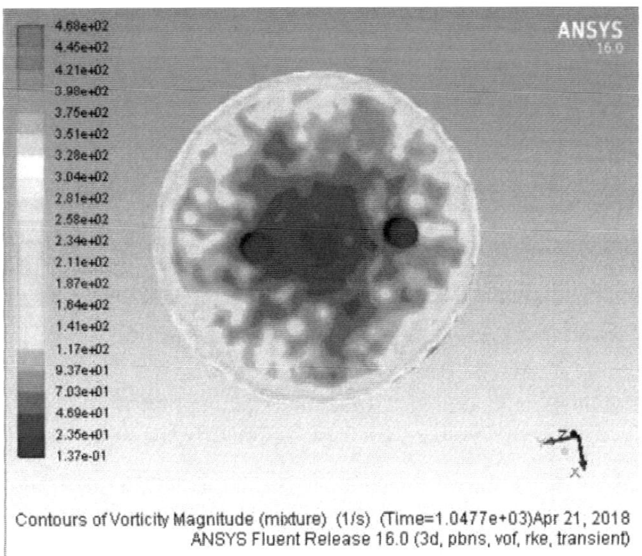

Fig. 6 Counters of vorticity magnitude at the dual port drain outlet when given cylinder rotation, N= 250 rpm

ii) Turbulent intensity

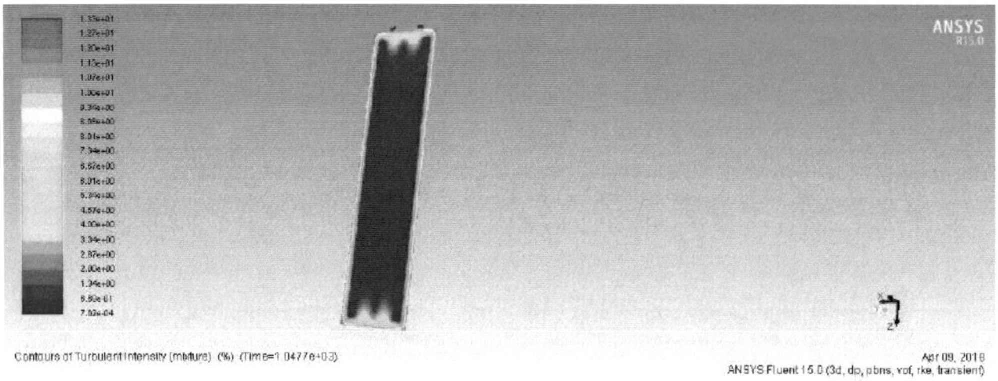

Fig. 7: Counters of turbulence intensity in the domain when given cylinder rotation, N= 250 rpm11

Turbulence intensity is also negligible at the drain port. Lower turbulence at the drain ports compared to adjacent walls. So there is no chance for vortex initiation or propagation. But as speed of rotation increases, there observed an increase in turbulent intensity compared to the other remaining cases.

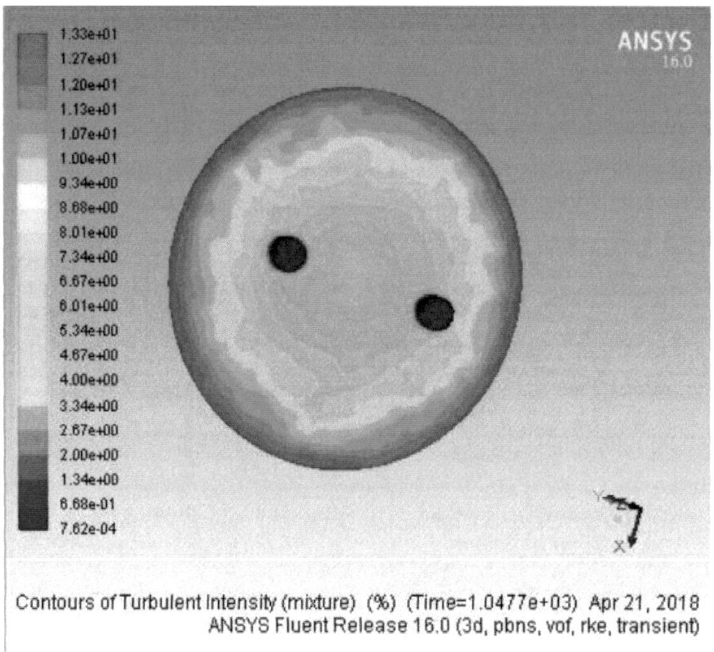

Fig 8 : Counters of turbulence intensity at the dual port drain outlet when given cylinder rotation, N= 250 rpm

5. Summary and Conclusions

Draining of liquid from a cylindrical tank was numerically investigated to study the gas-core vortex formation. VOF flow model was used to track the gas–liquid interface. The role of speed of rotation on the vortex formation was brought out with some numerical results. Speed of rotation is also a parameter which effects the formation of vortex. Preferably given low speed rotations, where vortex formation and its various characteristics are almost constant, which was elaborated with the help of the numerical results. However, for higher speed of rotations, the vortex formation was ineludible.

6. Reference

[1]. Abramson, H. N., and Chu, W. H. (1962)., Some studies of liquid rotation and vortexing in rocket propellant tanks (No. TN D 1212), *National Aeronautics And Space Administration Washington DC.*

[2]. Lubin, B. T., and Springer, G. S. (1967)., The formation of a dip on the surface of a liquid draining from a tank, *Journal of Fluid Mechanics*, 29(02), 385-390.

[3]. Satpathy K, Velusamy K, Patnaik B S V and Chellapandi P 2013 Numerical simulation of liquid fall induced gas entrainment and its mitigation *Int. J. Heat Mass Transfer* **60** 335–54

[4]. Mahyari M N, Karimi H, Naseh H and Mirshams M 2010 Numerical and experimental investigation of vortex breaker effectiveness on the improvement in launch vehicle ballistic parameters *J. Mech.Sci. Technol.* **24** 1997–2006

[5]. Park I S and Sohn C H 2011 Experimental and numerical study on air cores for cylindrical tank draining.*Int. Commun. Heat Mass Transfer* **38** 1044–9

[6]. Lakshmana Gowda, B. H., Joshy, P. J., and Swarnamani, S. (1996). Device to suppress vortexing during draining from cylindrical tanks. Journal of spacecraft and rockets, 33(4), 598-600.

[7]. K Ramamurthi and T John Tharakan. Intensification of a vortex during free draining. The Canadian Journal of Chemical Engineering, 73(3):292{299, 1995.

[8]. Cogan S J, Sheard G J and Ryan K 2008 Axisymmetric Hopf bifurcation in a free surface rotating cylinder flow *CTAC-2008: Proc. 14th Conf. on Biennial Computational Techniques and Applications* , G N Mercer and A J Roberts *ANZIAM J.* 50 251–65

[9]. Robinson A, Morvan H and Eastwick C 2010 Computational investigations into draining in an axisymmetric vessel *J. Fluids Eng.* **132** 1104–10

[10]. Gowda B H L and Udhayakumar H 2005 Vane-type suppressor to prevent vortexing during draining from cylindrical tanks *J. Spacecr. Rockets* **42** 381–3

Simulation of effect of bed temperature over producer gas composition using ASPEN PLUS, its experimental validation and emission characteristics of an IC engine

Vishnu A R[1], Saijal Kizhakke Kodakkattu[2], Syed Muhammed Fahd[3]

[1]Department of Mechanical Engineering, Baselios Mathews II College of Engineering, Kollam, Kcrala, India
[2]Department of Mechanical Engineering, Government Engineering College, Kozhikode, Kerala, India
[3]Department of Mechanical Engineering, TKM College of Engineering, Kollam

E-mail: vishnu.ar.1989@gmail.com, saijalkk@gmail.com, fahd@tkmce.ac.in

Abstract. Biomass is the source of energy that is generated from organic materials, which has the capacity to generate power in different means. Gasification is the technique that converts biomass materials into combustible gas by undergoing chemical reactions at high temperature with a controlled oxygen availability. The objective of this study is to determine the performance of downdraft gasifier and emission characteristics of IC engine using producer gases at various bed temperatures. The satisfactory performance of gasification, is theoretically attaining at a gasification temperature range of 500°C to 750°C. The temperature dependency of gasification process in the performance of downdraft gasifier is studied for different types of feed stocks. The downdraft gasifier is simulated using process simulation software ASPEN PLUS. It is found that at higher gasification temperatures the quality of producer gas increases considerably. The exhaust emission characteristics and efficiencies of engine with producer gas are comparable with normal gasoline exhaust characteristics.

Keywords: Gasifier, emission, IC engine

1. Introduction

Energy crisis and pollution are the main threats that we are facing now. Researches are going on at various parts of the world to develop an alternative energy source which has lower emission characteristics and better performance. Hydrogen which is considered as the future energy source due to its highest energy to weight ratio compared to any fuel. Gasification is considered to be as a better technology used to produce hydrogen from organic materials. By gasification process a considerable amount of organic waste materials can be converted into useful energy. Pyrolysis is the main chemical process which takes place in gasification process. It is the thermal degradation of biomass into combustible solid char, liquid bio oil by chemical processes with limited supply of oxygen.

Gasifiers are instruments used to develop gasification process. There are different types of gasifiers available, among them fluidized bed gasifiers and downdraft gasifiers are widely used based on nature of the feed. For solid feeds or for those feeds which can be solidified easily, can use downdraft gasifiers and those feeds which can powdered easily, can use fluidized bed gasifiers. Emission characteristics of the IC engines using different quality producer gases as fuels are to be studied, since producer gas is considered to be as an alternative fuel source of near future. If the emission characteristics are improved than conventional gasoline fuel, then producer gas can be used as alternative fuel, even there is a reduction in efficiencies or power outputs. Each feed gives different quality producer gases at different gasification temperatures. Current work deals with study of producer gas quality, performance and emission characteristics of solid feed coconut shell and fibrous feed at three different gasification temperatures using downdraft gasifier.

Rao *et al.* [1] carried out experiments on counter current fixed bed gasifier with different solid fuels to compare the efficiency of gasifier as well as suitability of dried fruit petals for gasification in comparison with gasification of wood chops and CSS using the very same gasifier under similar operating conditions.

Rajvanshi [2] mentioned about basics of gasification in the fourth chapter of his book 'Alternative Energy in Agriculture'. In the same, briefly described types of gasifiers, chemical reactions of gasification, different process zones, and yield products of gasification with its approximate percentage with different known feeds. The method of applying producer gas in SI and CI engines and the efficiencies that can be achieved from these engines.

Basu [3] in his text book named "Biomass gasification and pyrolysis practical design" explained in detail about gasification process and various related tests for the calculation of performance of the gasifier. He detailed about relevance and art of gasification. The practical approach of gasification and practical problems deals with gasification process.

Atnaw *et al.* [4] carried out simulation work on downdraft gasifier with oil palm fronds using aspen plus. He conducted the work on gasifier with different operating pressures, moisture contents etc. He simulated the gasification with various pressure ratios below 5 bar and optimised. The model study revealed the effect of moisture content, pressure ratio in gasification process.

Mathieu and Dubuisson [5] carried work on Performance analysis of a fluidized bed gasifier, using aspen plus process simulator. Used wood as feed material. Minimisation of the Gibbs free energy is the thermodynamic model used for process simulation.

Chittaranjan Panda [6] in his thesis simulated fluidized bed gasifier using aspen plus simulator for rice husk gasification. Studied the effect of reactor temperature, equivalence ratio and steam to biomass ratio on composition of product gas. The results showed hydrogen concentration in the producer gas increases with increase in temperature.

	Nomenclature	
CV	Calorific Value	kJ/kg
LHV	Lower Calorific Value	kJ/kg
P	Electrical load	W
CGE	Cold Gas Efficiency	%
V_i	Volume flow rate	m^3/s
m_i	Mass flow rate	kg/s
X_i	Weight percentage	%
x_i	Mass percentage	%
δ	Density	kg/m^3
BTE	Brake thermal efficiency	%

Forero-Núñez and Sierra-Vargas [7] conducted experiment on heat transfer analysis using thermal imaging. Heat transfer by convection and radiation is mentioned in this study by incorporating some dimensionless numbers. Coconut shell is used for the analysis. The biomass consumption was 8 kg/hr. which gives a thermal output of 3862 kW and an electrical efficiency of 20.71%. Cold Gas Efficiency depended on producer gas volumetric flow and heating value of producer gas.
Aspen plus is a typical user-friendly process simulation software, which can predict the nature of the process and its outcomes. Downdraft gasification process is simulated using Aspen plus, which is same as the experimental set up. Aspen plus simulation results are validated using experimental results. The emission characteristics of an IC engine with different quality of producer gases with feed stocks are studied and get it compared with conventional gasoline fuel.

2. Feed characteristics
The proximate and ultimate analysis of feeds are given in table 1.

Table 1. Proximate and ultimate analysis of feeds

Feed characteristics (wt. %)	Coconut shell	Sugarcane waste
Moisture	8.067	14.60
Fixed carbon	15.783	10.50
Volatile material	73.65	73.1
Ash	2.5	1.8
Carbon	45.26	43
Hydrogen	5.72	6.03
Nitrogen	0.13	0.77
Oxygen	48.89	50.08
Sulphur	traces	0.12

3. Experimental Setup
Figure 1 shows the downdraft gasifier used for experimental studies, in which the main part is its reactor itself. Then the producer gas flows through a cyclone, water scrubber, solid trap for removing solid wastes like ash, solid carbon .A chiller is provided at the exit for reducing the temperature of the producer gas below atmospheric temperature in order to connect it with the engine system. An orifice plate and u tube manometer is in the system to calculate producer gas consumed by the engine.

Figure 1. Downdraft gasifier

An exhaust gas analyzer is used to study emission characteristics and a calibrated producer gas analyzer is used to analyze the producer gas developed by gasification process.

4. Simulation Model

Figure.2 shows the process model of downdraft gasification process. The simulation consists of three reactors, cyclone, scrubber and a chiller. The gasification process is done in three reactors and the rest is arranged for removing the solid components present in the producer gas at exit. The Feed must be well dried before firing the gasifier. Even though it is dried it contains small amount of moisture, which is removed in drying zone which is simulated using the reactor R-stoic. All feeds in gasification process is considered as non-conventional by aspen plus, so feeds must be made conventional by using R-Yield reactor.

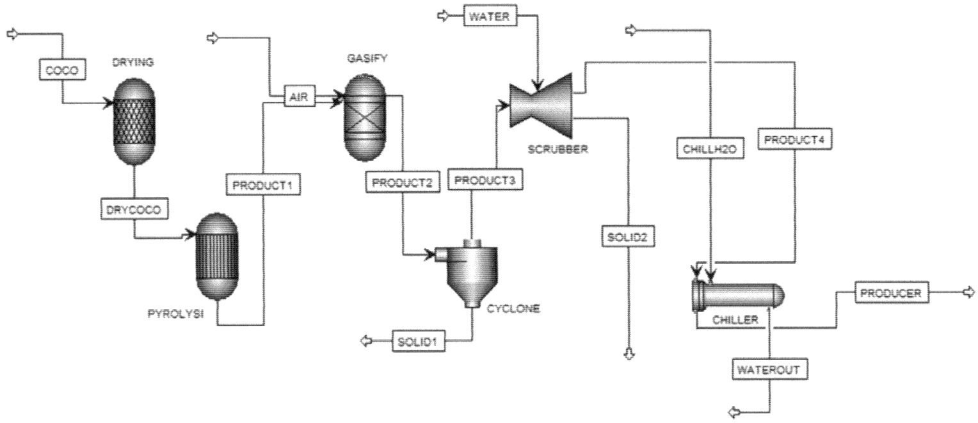

Figure 2. ASPENPLUS simulation model

The yield is done by simulation based on its ultimate and proximate analysis. The gasification is done by the reactor R-gibbs, which works on Gibbs free energy concept. The cyclone and water scrubber are used to remove solid components from producer gas and chiller is used to get producer gas below atmospheric temperature.

5. Results and discussions

5.1 Calorific values of biomasses

The calorific value or heating value of biomass can be calculated by using the equation. The calorific value of feeds is given in table 2

$CV= 0.3494X_C + 1.1783X_H + 0.1005X_S - 0.0151X_N - 0.1034X_O - 0.0211X_{ASH}$ [3][7].

Table 2. Calorific values of feeds

Biomass	Calorific value(MJ/kg)
Coconut shell	18.27
Sugarcane waste	15.34

5.2 Producer gas composition

From table 3 and table 4 the producer gas composition, both experimental and simulation can be compared, for coconut shell as well as sugarcane waste. The experimental and simulation results are compared for both the feeds by assuming gasification process as ideal in simulation. It can be seen that the simulation predicts the nature of producer gas.

As the gasification temperature increases, the mass flow rate of both hydrogen and carbon monoxide increases whereas mass flow rate of methane and carbon dioxide decreases for both the feeds. The results of simulation and experiment are comparable.

Table 3. Producer gas composition, both experimental and simulation of coconut shell feed

Temperature (°C)	625°C	675°C	725°C
Components (kg/hr.)		Simulation	
H_2	0.206	0.247	0.263
CO	2.382	2.654	3.251
CO_2	2.435	2.081	1.718
CH_4	0.267	0.185	0.151
		Experimental	
H_2	0.1516	0.1855	0.2485
CO	2.05	2.296	3.013
CO_2	2.344	2.242	1.833
CH_4	0.3411	0.262	0.179
		% Difference	
H_2	35	33	5.8
CO	16	15	8
CO_2	3	7	6
CH_4	21	29	15

Table 4. Producer gas composition, both experimental and simulation of sugarcane waste feed

Temperature (°C)	625°C	675°C	725°C
Components (kg/hr.)		Simulation	
H_2	0.177	0.199	0.203
CO	1.690	2.051	2.407
CO_2	2.212	1.909	1.726
CH_4	0.147	0.126	0.121
		Experimental	
H_2	0.1204	0.1547	0.1925
CO	1.54	1.89	2.227
CO_2	2.33	1.97	1.85
CH_4	0.232	0.2286	0.195
		% Difference	
H_2	47	28	5.4
CO	9	8	8
CO_2	5	3	6
CH_4	36	44	36

The composition of producer gas is predictable by using aspen plus process simulator. It is evident from the composition of producer gas that the quality is superior at higher gasification temperatures.

5.3 Density of producer gases

Density of producer gas is determined by hydrogen and carbon monoxide contents in the producer gas. As the amount of hydrogen and carbon monoxide increases, quality increases and density decreases.Table.5 and table 6 shows the density of producer gases at different gasification temperatures.

Table 5. Density of coconut feed producer gas

Temperature (°C)	Density (kg/m^3)		
	Simulation	Experiment	% Difference
625	0.97625	1.0349	5
675	0.94432	0.98598	4.2
725	0.8786	0.901332	2.5

Table 6. Density of sugarcane waste feed producer gas

Temperature (°C)	Density (kg/m^3)		
	Simulation	Experiment	% Difference
625	0.9942	1.12	11
675	0.9625	0.9999	3.7
725	0.9358	0.97	3.5

5.4 Calorific values of producer gas

Table 7 and table 8 shows calorific values of producer gas at different gasification temperature, for both the feeds. As temperature increases quality increases, calorific value increases. Equation used to find calorific value of producer gas is

$$LHV_{SYNGAS} = (x_{CO} \times LHV_{CO}) + (x_{H2} \times LHV_{H2}) + (x_{CO2} \times LHV_{CO2})$$ [2][3]

Table 7. Calorific Value of coconut shell producer gas

Temperature (°C)	Calorific Value (MJ/m^3)		
	Simulation	Experiment	% Difference
625	3.6	2.705	33
675	4.75	3.8	25
725	5.225	4.66	12

Table 8. Calorific Value of sugarcane waste feed producer gas

Temperature (°C)	Calorific Value (MJ/m^3)		
	Simulation	Experiment	% Difference
625	3.16	1.7	85
675	3.18	2.5	27
725	4.03	3.6	12

5.5 Cold gas efficiency (CGE)

Table 9 and table 10 shows cold gas efficiency of gasifier for different quality producer gases, both experimental and simulation. The maximum CGE attained by coconut shell feed producer gas is 71% experimentally and 77% by simulation. Similarly 49% and 60 % for sugarcane waste.

Table 9. CGE of coconut shell producer gas

Temperature (°C)	CGE (%)		
	Simulation	Experiment	% Difference
625	66	54	22
675	71	58	22
725	77	71	8

Table 10. CGE of sugarcane waste feed producer gas

Temperature (°C)	CGE (%)		
	Simulation	Experiment	% Difference
625	51	29	75
675	57	39	46
725	60	49	22

CGE = Flow of energy in gas /Energy contained in biomass

CGE = ($V_{syngas} \times LHV_{syngas}$)/ ($m_{biomass} \times CV_{biomass}$) [3][7]

5.6 Electrical efficiency of gasifier

Table 11 shows electrical efficiency of gasifier for different quality producer gases, experimentally. The maximum $\eta_{electrical}$ attained by coconut shell feed producer gas is 10%. Similarly 4.2% for sugarcane waste.

$\eta_{electrical}$ = Electrical power output /Energy contained in biomass

$\eta_{electrical}$ = P/ ($m_{biomass} \times CV_{biomass}$) [3][7]

Table 11. Electrical efficiency of gasifier, for both the feeds

Temperature (°C)	Coconut (%)	Sugarcane (%)
625	6.03	2.27
675	8.26	3.3
725	10	4.2

5.7 Brake thermal efficiency (BTE)

Table 12 shows BTE of IC engine with different quality producer gas, and with both the feeds. It is compared with the gasoline BTE over the same IC engine.

Table 12. Brake thermal efficiency (%)

Load (W)	Gasoline	Coconut			Sugarcane		
		(725°C)	(675°C)	(625°C)	(725°C)	(675°C)	(625°C)
0	0	0	0	0	0	0	-
200	5.51	3.9	3.7	3.2	3.1	3.1	-
400	10.11	6.9	6.5	6.2	5.8	5.62	-
600	12.11	10.4	9.8	9.3	8.2	-	-
800	15.18	12.51	12.52	-	-	-	-
1000	17.42	13.8	12.87	-	-	-	-
1200	17.65	14.5	-	-	-	-	-
1400	18.09	16.1	-	-	-	-	-
1600	16.56	-	-	-	-	-	-

The maximum BTE attained by gasoline fuel is 18.09% whereas the maximum BTE by producer gas is 16.1% for producer gas obtained from coconut feed at 725°C is used.

BTE = Output power of engine / heat input

Output power of engine = Output power measured / ($\eta_{coupling} \times \eta_{generator}$)
Where $\eta_{coupling}$ = 90% and $\eta_{generator}$ = 80%
Heat input = gas consumed × calorific value of producer gas

5.8 Emission characteristics

Table 13 to table 16 shows different emission characteristics of producer gas fuels, and compares it with gasoline fuel emission. By comparing CO_2 emission is high in the case of producer gas emission, whereas CO, NOx, HC emissions are very low for producer gas fuel, as compared to gasoline fuel emission.

Table 13. CO_2 Emission characteristics (% Volume)

Load (W)	Gasoline	Coconut			Sugarcane		
		(725°C)	(675°C)	(625°C)	(725°C)	(675°C)	(625°C)
0	4.2	9.72	8.78	8.8	6.58	6.3	-
200	4.3	9.99	9.58	9.6	7.82	7.4	-
400	4.5	10.3	10	9.61	8.42	8.12	-
600	4.7	10.59	10.6	11	9.82	8.67	-
800	4.8	11.86	11.7	11.2	-	-	-
1000	5.05	12.68	12.3	-	-	-	-
1200	5.15	12.97	12.82	-	-	-	-
1400	5.25	13.98	-	-	-	-	-

Table 14. CO Emission characteristics (% Volume)

Load (W)	Gasoline	Coconut			Sugarcane		
		(725°C)	(675°C)	(625°C)	(725°C)	(675°C)	(625°C)
0	4.22	0.92	0.94	0.96	0.94	0.9	-
200	4.24	0.47	0.53	0.63	0.7	0.84	-
400	4.29	0.29	0.36	0.45	0.76	0.84	-
600	4.34	0.18	0.21	0.33	0.72	0.81	-
800	4.38	0.18	0.19	0.21	-	-	-
1000	4.46	0.16	0.18	-	-	-	-
1200	4.61	0.14	o.15	-	-	-	-
1400	4.72	0.13	-	-	-	-	-

Table 15. HC Emission Characteristics (PPM)

Load (W)	Gasoline	Coconut			Sugarcane		
		(725°C)	(675°C)	(625°C)	(725°C)	(675°C)	(625°C)
0	162	78	118	125	92	127	-
200	161	71	114	121	78	106	-
400	156	35	53	106	69	88	-
600	141	29	33	71	54	82	-
800	135	20	29	50	-	-	-
1000	128	20	22	-	-	-	-
1200	131	18	20	-	-	-	-
1400	134	16	-	-	-	-	-

Table 16. NO Emission Characteristics (PPM)

Load (W)	Gasoline	Coconut			Sugarcane		
		(725°C)	(675°C)	(625°C)	(725°C)	(675°C)	(625°C)
0	29	14	7	5	8	4	-
200	33	15	7	7	8	5	-
400	36	15	8	7	9	5	-
600	41	16	8	8	9	5	-
800	46	16	9	8	-	-	-
1000	53	16	9	-	-	-	-
1200	60	19	10	-	-	-	-
1400	65	19	-	-	-	-	-

Lower BTE of producer gas fuel can be compensated with better emission properties of producer gas when compared with conventional gasoline fuels.

6. Conclusions

Using ASPEN PLUS process simulator, a model for biomass gasification with downdraft gasifier was simulated. Coconut shell and sugarcane waste were taken as feed materials. An experimental study of biomass gasification was conducted using a fixed bed downdraft gasifier using coconut shell and sugarcane waste as feeds. Experimental results are used to validate simulation results.

A series of experiments and simulations were performed to investigate the effect of gasifier bed temperature or the gasification temperature over the quality of producer gas. The mass flow of H_2, CO, CH_4 and CO_2 were calculated from experimental results and from simulation results. All other gases in the atmosphere is assumed to be present in lower concentration.

The results show that the hydrogen concentration in the producer gas increases rapidly with increase in temperature (600-750°C) at constant equivalence ratio. The calorific value of producer gas and cold gas efficiency of the gasifier shows good agreement between experimental and simulation results. The electrical efficiency of the gasifier is within the theoretical range. The brake thermal efficiency, exhaust gas characteristics of the engine is compared with gasoline fuel and producer gas. The results indicate that a good quality producer gas is a promising substitute for gasoline.

The coconut shell feed gave superior results with downdraft gasification rather than sugarcane waste feed, which effects the performance of gasifier as well as the engine. The variation is due to the fibrous nature of the sugarcane waste.

Acknowledgements

The authors are thankful to Government Engineering College, Kozhikode, and National Institute of Technology, Calicut, India, for providing necessary facilities for carrying out the experiments and various tests

References

[1] Rao M S, Singh S P, Sodha M S, Dubey A K and Shyam M 2004 *Biomass and Bioenergy* **27** 155-171
[2] Rajvanshi A K 1986 *Alternative Energy in Agriculture* **2** (4) 82-102
[3] Basu P 2010 Biomass gasification and pyrolysis: practical design and theory *Elsevier*
[4] Atnaw S M, Sulaiman S A and Yusup S 2011 *Journal of Applied Sciences* **11** (11) 1913-1920
[5] Mathieu P and Dubuisson R 2002 *Energy Conversion and Management* **43** 1291-1299
[6] Panda C *Aspen plus simulation and experimental studies on biomass gasification* 2012 Thesis, Chemical Engineering Department, NIT Rourkela, 2012
[7] Forero-Núñez C A and Sierra-Vargas F E 2016 *International Review of Mechanical Engineering* **10** (4) 239-246

Journal of Physics: Conference Series **1355** (2019) 012005 doi:10.1088/1742-6596/1355/1/012005

Numerical investigations on the heat transfer characteristics of tube in tube helical coil heat exchanger

Sheeba A[1], Mathew P[2] and Jose Prakash M[3]
[1]Associate Professor, Department of Mechanical Engineering, T K M College of Engineering, Kollam, Kerala
[2] Lecturer in Mechanical Engineering, Central Polytechnic College, Vattiyoorkavu, Thiruvananthapuram
[3]Professor, Department of Mechanical Engineering, T K M College of Engineering, Kollam, Kerala.

E-mail: sheebajameel100@gmail.com

Abstract. Heat transfer characteristics of a tube in tube helical heat exchanger in the laminar flow regime is analysed numerically. Numerical scheme is validated with the experimental results reported in literature. The effect of helix angle on the overall heat transfer coefficient is analysed. The overall heat transfer coefficient increases with Dean number and its rate of increase at higher Dean number is low. The rate of heat transfer is maximum at a helix angle of 27°. Simulation studies show that the secondary flows that originate in a coiled tube due to curvature and torsion produce an additional transport of the fluid over the cross section of the pipe. The overall heat transfer coefficient increases with increase in inner diameter.

1. Introduction

Energy and materials saving considerations, as well as economic incentives, have led to efforts to produce more efficient heat exchange equipment. Common thermo-hydraulic goals are to reduce the size of a heat exchanger required for a specified heat duty, to upgrade the capacity of an existing heat exchanger, to reduce the approach temperature difference for the process streams, or to reduce the pumping power. Due to their compact structure and high heat transfer coefficient, curved tubes have been introduced as one of the passive heat transfer enhancement techniques and are widely used in various industrial applications. Helical and spiral coils are well known types of curved tubes which have been used in a wide variety of applications, for example, heat recovery processes, air conditioning and refrigeration systems, chemical reactors, food and dairy processes. Due to the curvature of the tubes, as fluid flows through curved tubes, centrifugal force is generated. A secondary flow induced by the centrifugal force has significant ability to enhance the heat transfer rate as it moves fluid across the temperature gradient. Thus, there is an additional convective heat transfer mechanism, perpendicular to the axial flow, which does not exist in straight tube heat exchangers.

Dravid et al. [1] numerically investigated the effect of secondary flow on laminar flow heat transfer in helically coiled tubes both in the fully developed region and in the thermal entrance region. Patankar et al. [2] discussed the effect of the Dean number on friction factor and heat transfer in the developing and fully developed regions of helically coiled pipes. Yang et al. [3] presented a numerical model to study the fully developed laminar convective heat transfer in a helicoidal pipe having a finite

Journal of Physics: Conference Series **1355** (2019) 012005 doi:10.1088/1742-6596/1355/1/012005

pitch. The effects of the Dean number, torsion, and the Prandtl number on the laminar convective heat transfer were discussed. Rabin and Korin [4] developed a new simplified mathematical model for thermal analysis of a helical heat exchanger for long-term ground thermal energy storage in soil for use in arid zones. Acharya et al. [5] numerically studied the phenomenon of steady heat transfer enhancement in coiled-tube heat exchangers due to chaotic particle paths in steady, laminar flow with two different mixings. Zhang et al. [6] applied a control-volume finite difference method having second-order accuracy to solve the three-dimensional governing equations. Chen and Zhang studied the combined effects of rotation (coriolis force), curvature (centrifugal force), and heating/cooling (centrifugal-type buoyancy force) on the flow pattern, friction factor, temperature distribution, and Nusselt number.

The laminar forced convection and thermal radiation in a participating medium inside a helical pipe were analysed. Rennie and Raghavan [7] simulated the heat transfer characteristics in a two-turn tube- in-tube helical coil heat exchanger. Various tube-to-tube ratios and Dean numbers for laminar flow in both annulus and in-tube were examined. Prabhanjan et al. [8] compared the heat transfer rates between a helically coiled heat exchanger and a straight tube heat exchanger. Xin and Ebadian [9] considered the effects of the Prandtl number and geometric parameters on the local and average convective heat transfer characteristics in helical pipes. The heat transfer characteristics of a tube-in-tube helical geometry was studied by Timothy J Rennie and Vijaya G S Raghavan [14]. This is the relevant geometry of this analysis. Rennie and Raghavan conducted experimental and numerical studies 7eparately. They used Wilson [17] plots to determine the heat transfer coefficients in the experimental study. However, the numerical results showed significant divergence from the experimental values for smaller coils. They assumed that Wilson plots may be the reason behind this divergence. Most of the works reported in literature on heat transfer in helical coil tubes have been carried out considering wall boundary conditions of either constant heat flux or constant wall temperature. Physically realistic fluid to fluid forced convection heat transfer studies are to be done to get more insight to the problem.

Nomenclature

A	Cross sectional area, m^2	q	Heat transfer rate, W
L	Length of heat exchanger, m	U	Overall heat transfer coefficient, W/m^2K
D	Diameter of outer tube, m	ΔT	Temperature difference, K
d	Diameter of inner tube, m	k	Thermal conductivity, W/m K
R	Pitch circle radius of helix, m	h	Heat transfer coefficient, W/m^2K
p	Axial pitch of helix, mm	C	Specific heat, J/kg K
Q	Volume flow rate, ml/min	Re	Reynolds number
U	Velocity of flow, m/s	De	Dean number
$\dot{m}$	Mass flow rate, kg/s	Nu	Nusselt number
T	Temperature, K	Pr	Prandtl number

Suffixes:

i	Inner
o	Outer
h	Hot fluid
c	Cold fluid
1	Inlet
2	Outlet

2. Numerical Setup

2.1. Geometry modelling and meshing

The Geometric modelling of the heat exchanger is done using AutoCAD. Two helically coiled concentric tubes were created one over the another to form a double pipe helical heat exchanger. Both tubes have the wall thickness of 0.8 mm. All the dimensions are shown in Table 1. The mesh was created using meshing module of the ANSYS 14.5 WORKBENCH. Triangular elements were used in the Map type to mesh the faces with a spacing of 0.8. Tet/Hybrid meshing is applied to create volume meshes with a spacing of 0.8 in T Grid type. This resulted in 2,81,869 volume elements. The geometry for numerical modelling is shown in figure 1.

Table 1 Dimensions of coil used in the present study

Dimensions	Values
Outside diameter of inner tube	9.5 mm
Inside diameter of inner tube	7.9 mm
Wall thickness of both inner and outer tubes	0.8 mm
Outside diameter of outer tube	15.9 mm
Inside diameter of outer tube	14.3 mm
Radius of curvature of the helical coil	235.9 mm
Pitch of the helical coil	15.9 mm
Length of the coil	1.48 m

The geometry created and the mesh generated for the validation of the numerical scheme are shown in figure 1 and figure 2 respectively. Counter flow configuration was used. Both the hot and cold inlets were provided with mass flow inlet boundary condition. Both hot and cold fluid outlet boundaries were provided with pressure outlet boundary condition. The outer surface of the outer tube and the end concentric surfaces of the tubes were provided with wall boundary condition and the heat transfer is set to zero value.

Simulations were performed using five different mass flows in the inner tube and the annulus region. The flow rates used were 0.3lpm, 0.5lpm, 0.7lpm, 0.8lpm and 0.9lpm. This resulted in 25 simulations. Water and copper were selected in materials selection tab for the fluid and solid domains respectively. Temperature dependent properties like thermal conductivity (k), viscosity (μ), and specific heat (Cp) are given.

The realizable k-ϵ turbulence model with enhanced wall functions was used in the analysis. The SIMPLE algorithm is used for pressure velocity coupling with second order upwind scheme for the discretization of the momentum and energy equations. Boundary conditions are used according to the need of the model. The inlet and outlet conditions are defined as mass flow inlet and pressure outlet. . The outer surface of the outer tube and the end concentric surfaces of the tubes were provided with wall boundary condition and the heat transfer is set to zero value. Hybrid initialization method was used for initializing the flow variables which is found to have faster convergence as compared to standard initialization in this particular case.

A convergence criterion of 1.0e-04 was used for continuity k, ϵ and x, y and z velocities while for energy equation a convergence criterion of 1.0e-06 was found to be satisfactory. Grid independence study is conducted wherein the grid is refined successively using edge sizing and inflation growth rates. This is repeated until a tolerance of about 1% difference in two successive refinements. The results of grid independence study is shown in figure 3.

Journal of Physics: Conference Series **1355** (2019) 012005 doi:10.1088/1742-6596/1355/1/012005

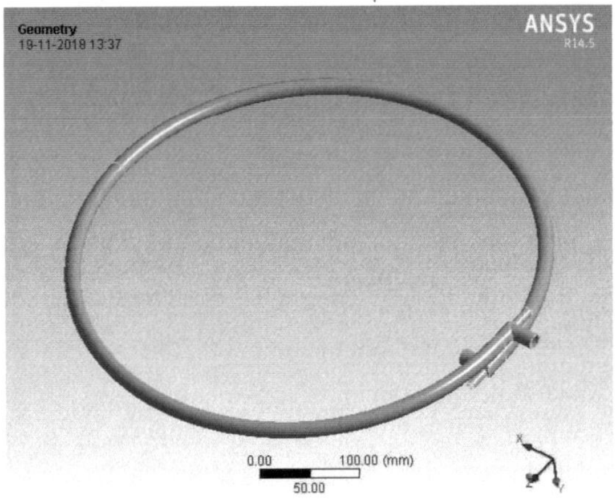

Figure 1. Geometry created

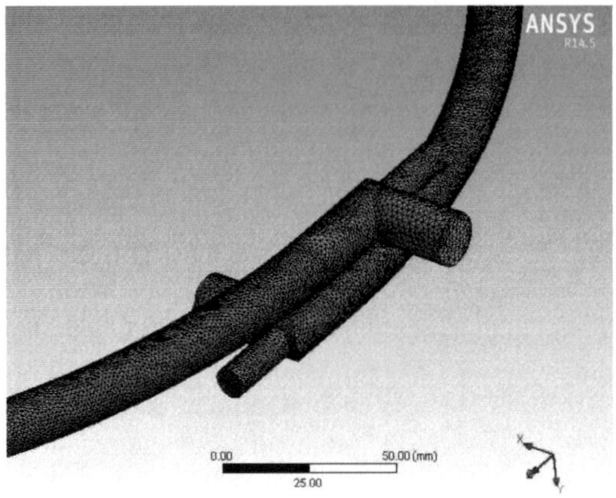

Figure 2. Meshed Geometry

2.2. *Data reduction*

Overall heat transfer coefficient, $U_{io} = \frac{Q}{A_{io} LMTD}$ (1)

Outside surface area of the inner tube, $A_{io} = \pi d_o L$ (2)

Where L is the length of the tube and d_O is the outer diameter of the inner tube.

Log mean temperature difference, $LMTD = \frac{(\Delta T_2 - \Delta T_1)}{\ln(\frac{\Delta T_2}{\Delta T_1})}$ (3)

Where, $\Delta T_1 = T_{h1} - T_{c2}$ and $\Delta T_2 = T_{h2} - T_{c1}$

Heat transfer coefficients for the inner and annulus were calculated using the average bulk fluid temperature and the average temperature of the coil.

3. Result and discussion

Numerical investigations have been carried out to study the effect of variations in Reynold's number. Dean number, helix angle and inner diameter on the heat transfer characteristics of the helical coil.

3.1 Grid independent study

Figure. 3 shows the grid independent study performed for the physical model to obtain the optimum number of elements. The number of elements is varied from 1.4 million to 3.9 million. The result from the calculation using 3.2 million cells agrees well with the case using 3.9 million cells. Thus, the mesh with 0.45 million cells has been chosen as the suitable mesh for further investigations.

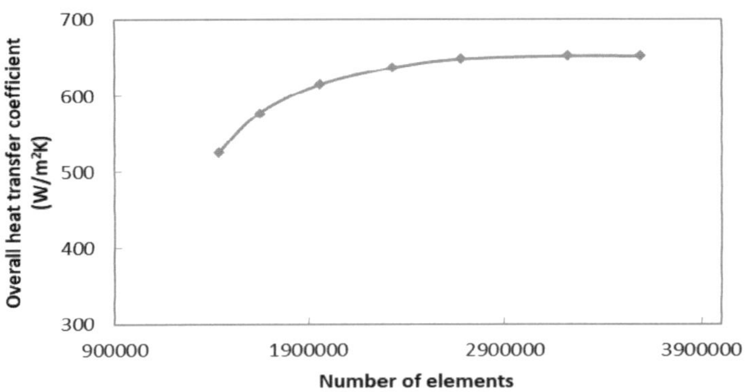

Figure 3. Grid independence study

3.1. Effect of Dean number on heat transfer

Figure 4 shows the effect of flow rate through the inner tube on overall heat transfer coefficient. The experimental results reported by Rennie et al. [14] is also shown for comparison. The overall heat transfer increases with increase in flow rate through the inner tube. Further, there is very good agreement between the results obtained from numerical studies and the experimental results reported.

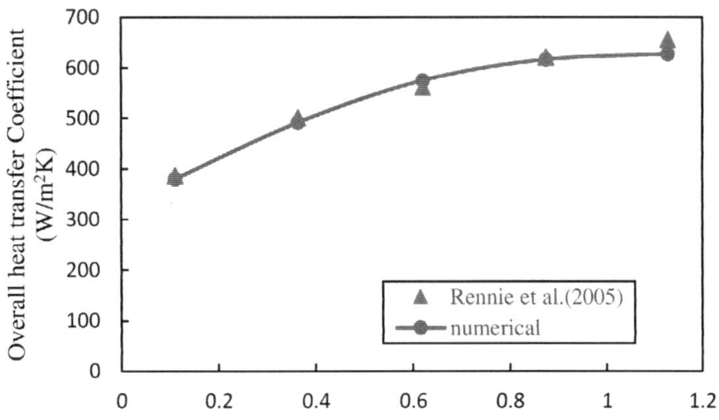

Figure.4 Variation of overall heat transfer coefficient with inner mass flow rate

Overall heat transfer coefficients for various annulus flow rates is presented in figure 5. The trends are typical for a fluid-to-fluid heat exchanger with the overall heat transfer coefficient increasing with both inner and annulus flows. However, the rate of increase of the overall heat transfer coefficient is less at higher inner Dean numbers. For a given annulus flow rate, increasing the inner flow rate results in an eventual asymptotic overall heat transfer coefficient.

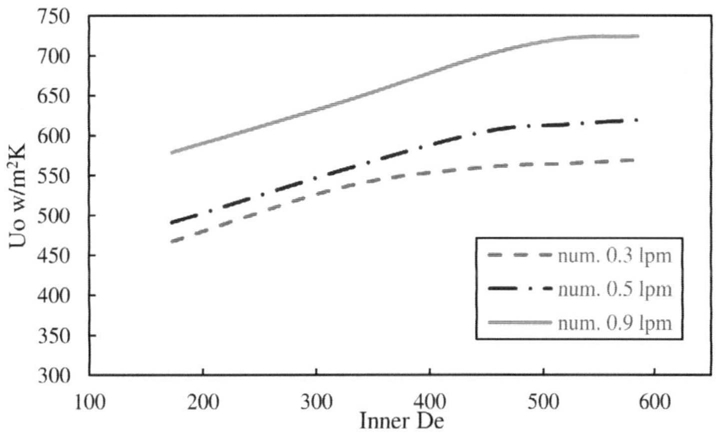

Figure.5 Variation of overall heat transfer coefficient with inner Dean number

Temperature contours at different outer flow rates are shown in figure 6. The secondary flows that originate in a coiled tube due to the curvature and torsion produce an additional transport of the fluid over the cross section of the pipe. From the temperature contours at different flow rates, it is clear that the high temperature contours are moved to the outer wall due to centrifugal force. As the flow rate increases the temperature contour symmetry distorts. The temperature gradient near the outer wall region increases and near the inner wall region temperature gradient decreases. As the Dean number increases, these variations in the temperature gradients in the outer and inner wall regions of the coil changes and which lead to high rates of heat transfer.

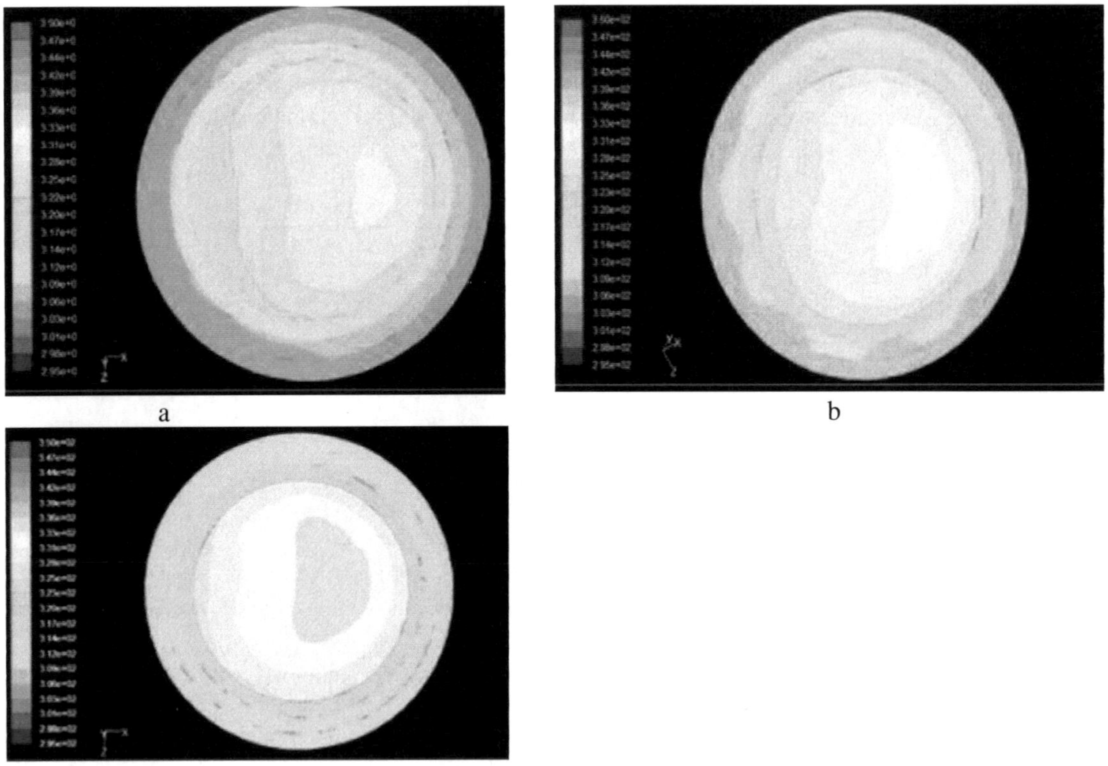

a b c

Figure.6 Temperature contour a) 0.3lpm, b) 0.7lpm and c) 0.9lpm

3.2. Effect of helix angle on heat transfer

Studies have been conducted to see the effect of helix angle on the heat transfer characteristics. The helix angle is varried from 5° to 45° in steps of 5°. Figure 7 indicates the variation of overall heat transfer coefficient with helix angle.

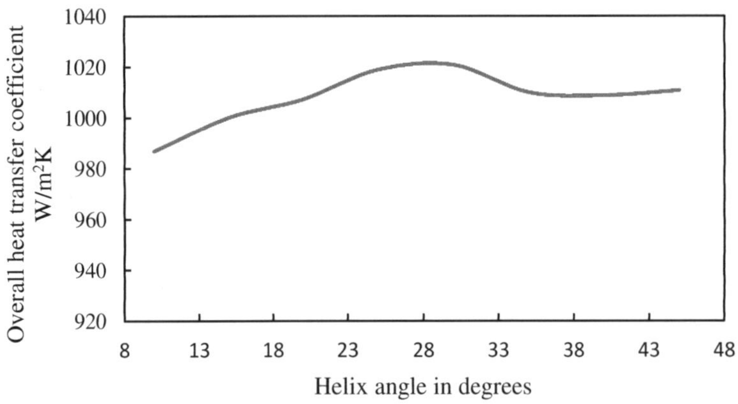

Figure 7. Variation of Overall heat transfer coefficient with helix angle

The heat transfer rate increases with helix angle, reaches a maximum value and then decreases. As the helix angle is increased, the torsion of the liquid also increases which may be the cause for the increase in heat transfer rate. It is currently accepted that the effect of coil curvature is to suppress the turbulent fluctuations arising in the flowing fluid, smoothing the emergence of turbulence and increasing the value of the Reynolds number required to attain a fully turbulent flow, with respect to a straight pipe. The above effect of turbulent fluctuations suppression enhances as the coil curvature increases. Torsion, on the other hand, is believed to destabilize the flow; reducing the Reynolds number at which turbulence emerges at values lower than the ones characteristic of straight pipe flow. The above destabilizing effect first increases, as torsion increases, reaches a maximum and then decreases with further increase in torsion. Variation of effectiveness with helix angle is shown in figure 8.

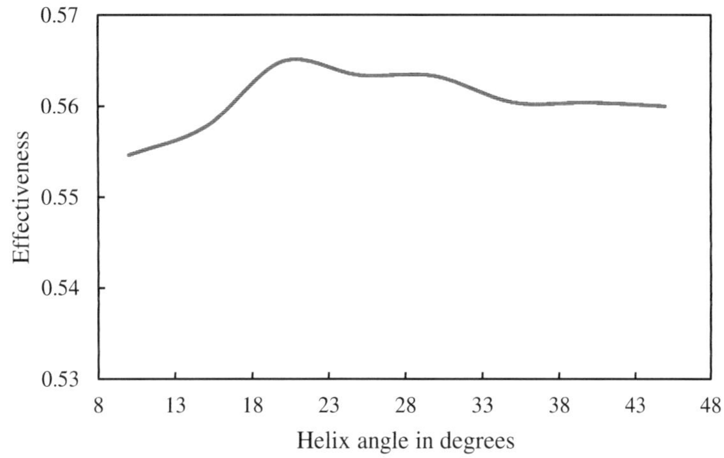

Figure 8. Variation of Effectiveness with helix angle

As the pitch of the coil increases, additional torsional flow patterns were observed. The torsional effect increases with increase in coil pitch. Therefore, an increase in heat transfer coefficient is expected. Maximum overall heat transfer coefficient is observed near a helix angle of 27°.

3.3. Effect of Inner tube diameter on heat transfer

In order to analyse the effect of inner tube diameter, simulations were performed for three different inner tube diameters. The dimensions of these three configurations are shown in Table 2.

Table 2. Dimensions for study of the effect of inner tube diameter

	do (mm)	di (mm)	t (mm)	Do (mm)	Di (mm)	R (mm)	p (mm)
First coil	12.7	11.1	0.8	15.9	14.3	235.9	15.9
Second coil	9.5	7.9	0.8	15.9	14.3	235.9	15.9
Third coil	6.35	4.75	0.8	15.9	14.3	235.9	15.9

Figure 9 and figure 10 indicate the effect of annulus flow rate on overall heat transfer coefficient and heat transfer rate respectively. Heat transfer rate and overall heat transfer coefficient are maximum for bigger tube since its surface area for heat transfer is the largest. Since the radius of curvature of the coils is the same, coil with larger inner diameter have larger curvature ratio which results in greater secondary flow effects and hence higher rate of heat transfer. Increase in tube size, increases the rate of heat transfer. When the annulus flow rate is high and the inner flow rate is low, the limiting heat transfer is in the inner tube. Hence any change in the inner flow rate will have significant effects on the overall heat transfer coefficient.

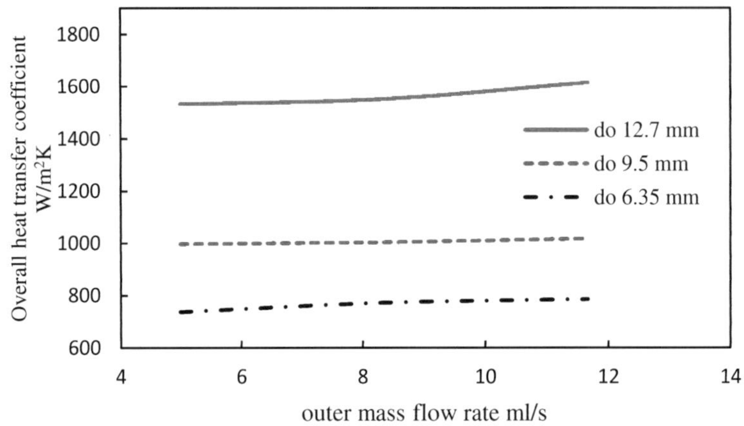

Figure.9 Variation of overall heat transfer coefficient with annulus mass flow rate

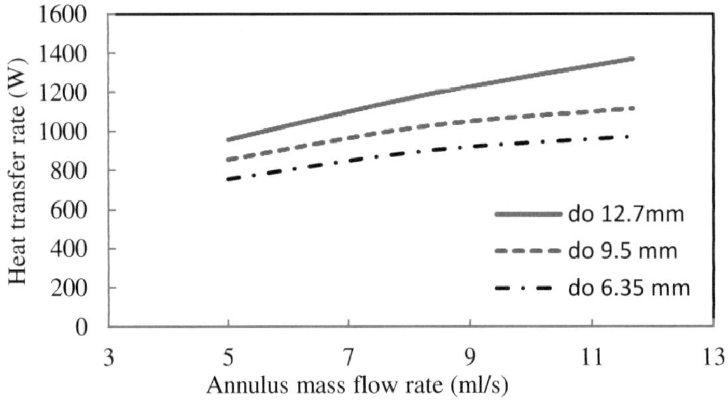

Figure.10 Effect of annulus mass flow rate on heat transfer rate

Increase in the inner tube diameter reduces the velocity of flow which resulted in lower Reynolds number. Increase in the inner tube diameter at the same radius of curvature increases the curvature ratio and increased secondary flow. This results in the decrease of inner Dean number and lower heat transfer coefficient.

4. Conclusion

Numerical study of a double-pipe helical coil heat exchanger was performed. A very good conformance with the experimental results reported in literature was observed. The following conclusions are derived out of this study. The Dean number and torsion have significant effect on the heat transfer characteristics. The overall heat transfer coefficient increases with inner Dean number. The rate of increase of the overall heat transfer coefficient is less at higher inner Dean numbers. For this heat exchanger the optimum value of helix angle is 27°. An inner tube to outer tube diameter ratio of about 0.88 leads to optimum performance of this helical heat exchanger.

References

[1] Dravid A N, Smith K A, Merrill E W, Brain PLT. 1971 Effect of secondary fluid on laminar flow heat transfer in helically coiled tubes AIChE J 17:1114–22.

[2] Patankar S V, Pratap V S, Spalding D B 1974 Prediction of laminar flow and heat transfer in helically coiled pipes J Fluid Mech 62:53–551.

[3] Yang G, Dong F, Ebadian M A 1995 Laminar forced convection in a helicoidal pipe with finite pitch Int J Heat Mass Transfer 38:853–62.
[4] Rabin Y, Korin E 1996 Thermal analysis of a helical heat exchanger for ground thermal energy storage in arid zones Int J Heat Mass Transfer 39:1051–65
[5] Acharya N, Sen M, Chang HC 2001 Analysis of heat transfer enhancement in coiled-tube heat exchangers Int J Heat Mass Transfer 44:3189–99.
[6] Chen H, Zhang B 2003 Fluid flow and mixed convection heat transfer in a rotating curved pipe Int J Thermal Sci 42:1047–59
[7] Timothy J Rennie, Vijaya G S Raghavan 2006 Numerical studies of a double-pipe helical heat exchanger Applied Thermal Engineering 26 1266–1273
[8] Prabhanjan D G, Raghavan G S V, Rennie T J 2002 Comparison of heat transfer rates between a straight tube heat exchanger and a helically coiled heat exchanger Int Commun Heat Mass Transfer 29:185–91.
[9] Xin R C, Ebadian M A 1997 The effects of Prandtl numbers on local and average convective heat transfer characteristics in helical pipes J Heat Transfer 119:463–7
[10] Andrea Cioncolini, Lorenzo Santini 2006 An experimental investigation regarding the laminar to turbulent flow transition in helically coiled pipes Experimental Thermal and Fluid Science 30 367-380.
[11] M Akiyama, K C Cheng 1972 Laminar forced convection heat transfer in curved pipes with uniform wall temperature International Journal of Heat and Mass Transfer 15 1426–1431.
[12] Lemenand T, Peerhossaini 2002 A thermal model for prediction of the Nusselt number in a pipe with chaotic flow Appl Therm Engg 22:1717–30.
[13] Rennie T J, Raghavan G S V July 2002 Laminar parallel flow in a tube-in-tube helical heat exchanger AIC Meeting CSAE/SCGR Program Saskatoon Saskatchwan 14-17;
[14] Timothy J Rennie, Vijaya G S Raghavan 2005 Experimental studies of a double-pipe helical heat exchanger, Experimental Thermal and Fluid Science 29 919–924
[15] Lin C X, Ebadian M A 1997 Developing turbulent convective heat transfer in helical pipes Int J Heat Mass Transfer 40:3861–73.
[16] M Akiyama, K C Cheng, 1974 Graetz problem in curved pipes with uniform wall heat flux, Applied Science Research 29 401-418
[17] J W Rose 2004 Heat-transfer coefficients, Wilson plots and accuracy of thermal measurements Experimental Thermal and Fluid Science 28 77-86.

CFD analysis of the performance of different airfoils in ground effect

Zubin Zaheer[1], Dr. K E Reby Roy[2], Gopakumar S Nair[3], Vineeth Ragipathi[4], Niranjan U V[5],
[1] Engineer, Tata Consultancy Services, Pioneer-ITPL, 560066, Bangalore, India
[2] Associate Professor, Dept. of Mechanical Engg. TKMCE, Kollam, India
[3] Address
[4] Dept. of Mech. Engg, BITS, Pilani, 333031, India
[5] PGDPM, NITIE, Mumbai, 400087, India

E-mail: zubinzaheer@gmail.com

Abstract. Ground effect is a phenomenon caused by the presence of a fixed boundary layer below the wing when it moves close to a fixed surface. This results in an effective increase in lift to drag ratio of the airfoil. In this paper, the performance of three airfoils used in general aviation is analysed in ground effect using ANSYS FLUENT at different operating conditions namely, the angle of attack and the height above the fixed surface. The main objectives are to are evaluate the performance of these airfoils in ground effect at different ground clearances and angles of attack and find the most suitable airfoil which can be used to exploit the advantages of ground effect. It was found that all the airfoils experienced an increase in lift and decrease in drag as distance from the ground reduced thus validating ground effect. The aerodynamic efficiency increases, reaches a maximum, and then decreased as the angle of attack increased due to the onset of flow separation at the top surface of the airfoil. It was also noted that the symmetric airfoil experienced negative lift at very low ground clearances which was attributed to the formation of a venturi effect between the airfoil and ground

1. Introduction

Ground effect is the phenomena when a fixed wing aircraft flies at close proximity (at or below the wingspan in case of a fixed wing aircraft or the rotor diameter in case of a helicopter) with a fixed surface such as the ground. Favourable aerodynamic conditions such as increased lift and decreased drag is observed when an aircraft flies in the ground effect region. This is attributed to the disruption of wind tip vortices which causes leakage from the high pressure bottom part of the airfoil to the top. Hence a high pressure "ram" effect is formed below the airfoil. Ground effect can be made use of to significantly reduce fuel costs, shorten the take-off distance and lower the stall speed.

2. Background

2.1 Literature Survey

The study of the influence of the ground on airplanes began as early as the 1920s by Raymond [1]. It was found out that the air trapped between the airfoil and the ground acts as cushion thus increasing the lift coefficient and causing the aircraft to skim over the runway for an abnormally long time before finally touching down. Over the years many aircrafts were developed to make use of ground effect called WIG (Wing in ground effect) crafts. They were developed both for civilian as well as military use. The most famous of these was the soviet "Ekranoplane". Ando [2] conducted a review of the WIG crafts used over the years. Sharma [3] conducted a comparative study between NACA airfoils and a new class of DHMTU airfoils which were specially designed for use in WIG aircrafts and obtained good results. In the research conducted by Razenbach and Barlow [4-6], it was found out that lift reaches a maximum at 0.08c (c denotes the chord length of the airfoil) and as the ground clearance reduces to zero, the boundary layers were seen to merge which explains the reduced lift force very close to the ground. Ahmed [7] studied the flow characteristics of symmetrical airfoils in ground effect at different angles of attack and found out significant high values of pressure coefficient at low ground clearances and this high-pressure coefficient extended over almost the full bottom section at high angles of attack. In further wind tunnel tests, Ahmed [8] found out that the airfoil geometry also played a role in the behaviour in ground effect. A strong suction effect was noticed at very small ground clearances of 0.05 h/c (h denotes the height of the airfoil from the ground) which was attributed to the formation of a convergent divergent passage between the airfoil and ground. Lot of research work is also ongoing for inverted WIG airfoils which are used as front wings and spoilers of race cars [9-10].

2.2 Airfoil Nomenclature
The following are the terms related to airfoil nomenclature.

a) The suction (or upper) surface is usually associated with low static pressure and high velocity.
b) The pressure (or lower) surface has a higher value of static pressure when compared to that of the suction surface.

The geometry of an airfoil is described using the following terms.

a) The point at the front of the airfoil with maximum curvature, called the leading edge.
b) The point at the rear of the airfoil with minimum curvature, called the trailing edge.
c) The straight line connecting the leading and the trailing edge, called the chord line. The length of the chord line is called the chord length.

The shape of the airfoil is defined by the locus of points, midway between the upper and lower surfaces of the airfoil, called the camber line.

2.3 NACA Series
The airfoil shapes developed by the National Advisory Committee for Aeronautics (NACA) are called NACA airfoils. The shape of the NACA airfoils is described using a series of digits following the word "NACA". The parameters in the numerical code can be entered into equations to precisely generate the cross-section of the airfoil and calculate its properties.

In Four-digit NACA series, the

a) First digit expresses camber in percent chord.
b) Second digit gives location of maximum camber located in tenth of chord.
c) Last two digits give thickness in percent chord.

For example, the NACA 2412 airfoil has a maximum camber of 2% located 40% (0.4 chords) from the leading edge with a maximum thickness of 12% of the chord. The NACA 0015 airfoil is symmetrical, the 00 indicating that it has no camber. The 15 indicates that the airfoil has a 15% thickness to chord length ratio: it is 15% as thick as it is long.

3. Objectives
A survey of the literature yields that ground effect is still not completely understood and more research is required at varied operating conditions and on different airfoil sections. This paper aims to

a) study the performance of three airfoil sections operating in ground effect during take-off and landing conditions.
b) compare and study the performance of the three airfoils at varying h/c ratios and varying angles of attack.

c) To obtain the aerodynamic coefficients at different ground clearances and angles of attack.
d) To find the most suitable wing airfoil section for WIG crafts from the three airfoils considered.

Here three different NACA airfoils are taken namely the NACA 4412, NACA 6412 and the NACA 0012. The NACA 0012 is a symmetrical airfoil whereas the other two are unsymmetrical.

4. Methodology

4.1. Airfoil and domain modelling

The airfoil coordinates are obtained by giving the particular NACA series number, angle of attack and the chord length from airfoiltools.com. The chord length was set as 1m and the airfoils were generated for AOA (Angle of Attack) 0, 4 and 8 degrees. The 2D models of the airfoils were then developed using the coordinate points in the Ansys design modeller.

A rectangular flow domain was created with distance of the airfoil from top surface of the domain wall taken as 6.6c [13] and the inlet and outlet distances set as 15c from the leading and trailing edges respectively as it was observed that this dimension does not affect the simluation. The airfoil boundary was then subtracted using Boolean and the rectangular flow domain is obtained.

4.2. Meshing

After naming the domain boundaries and the airfoil, the named edges and surface were selected and mesh controls were applied. Edge sizing is given to the ground, top, inlet, outlet and the airfoil. Inflation was given with the ground and airfoil as boundaries. Mesh independent study was also conducted and the grid sizes employed were 36782, 52343, 70502 and 172670 elements. The parameters monitored for convergence were C_L (Coefficient of Lift) and C_D (Coefficient of Drag). It was found that the mesh with 70502 elements provided sufficient convergence and this grid size was used for the simulation.

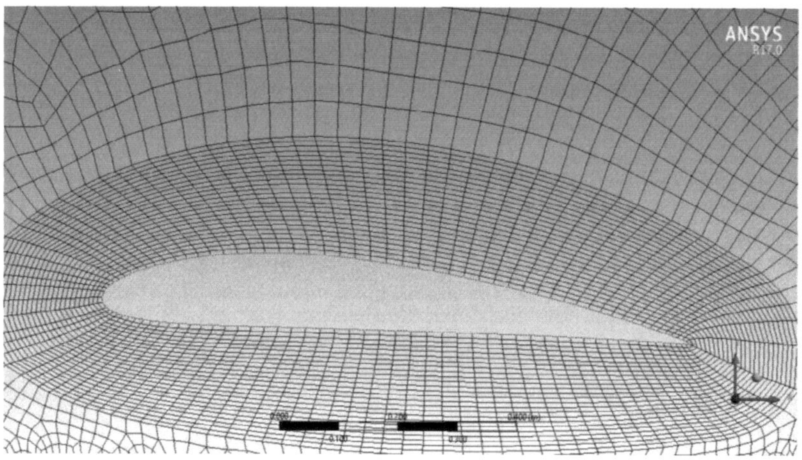

Figure1. A sample mesh shown

4.3 Solver settings

Ansys fluent was used for the simulations. The double precision solver is selected and the solver type is set to pressure based. The standard k-epsilon viscous model is selected and air is chosen as the flow material. Boundary conditions are given to make the airfoil stationary and the ground is considered as a moving wall and given a velocity of 30 m/s in the X- direction [13] to simulate ground effect. Slip wall boundary condition is given at the top boundary [13]. An inlet velocity of 30 m/s was applied

[13]. All the reference values were computed from inlet. The Simplec scheme is selected for the analysis and monitors were set up to monitor the values of C_L and C_D. The convergence criteria selected for continuity was set at 1x 10^{-6}. The initialization method is chosen as standard and the inlet values were chosen for initialization. Iterations were performed till convergence is obtained and pressure and velocity contours were plotted.

5. Results and discussion

Results are obtained for the three airfoils with a flow velocity of 30 m/s and at h/c ratios 0.2, 0.5 and 1 and taking different values of angles of attack(α) 0°, 4°and 8°. The values of C_L, C_D and the aerodynamic efficiency C_l/C_d are obtained. The pressure and velocity contours are also plotted.

It is observed in all cases that the C_L value decreases as the height from the ground increases due to the decrease in the ground effect. The drag is also observed to increase as the height from the ground increases due to the reduced ground effect.

5.1 NACA 6412

Table 1. Showing C_l, C_d values of NACA 6412 airfoil at various angles of attack (α) and h/c ratio

α°	h/c	C_l	C_d	C_l/C_d
0	0.2	0.67112	0.012566	53.41
0	0.5	0.65475	0.012614	51.66
0	1.0	0.65437	0.013279	49.28
4	0.2	1.1841	0.015862	74.65
4	0.5	1.0658	0.019398	54.94
4	1.0	1.0397	0.020825	49.93
8	0.2	1.5217	0.031304	48.61
8	0.5	1.3245	0.034725	38.14
8	1.0	1.3183	0.034992	37.67

The NACA 6412 airfoil has the maximum camber of all the three airfoils considered here. Hence at 0° angle of attack, the lift generated will be solely due to its curvature. At 4° angle of attack a substantial increase in C_L is observed[14]. This is primarily due to the increased lift as the angle of attack increases and also due to the wing shape. At 8°, it is seen that the lift increases due to increase in angle of attack but the drag almost doubles from that at 4°, leading to a decrease in aerodynamic efficiency. This increase in drag is due to the higher values of pressure drag forces which is because the flow over the airfoil surface is at the verge of separation.

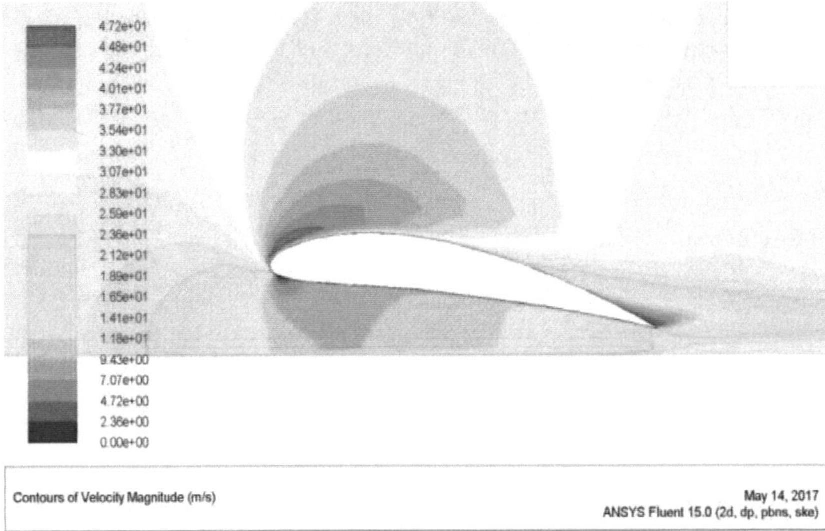

Figure 2. Velocity plot of NACA6412 at 8° AOA. Drag increases drastically at 8 AOA due to the onset of flow separation at the trailing edge on the upper surface as seen in the figure.

5.2 NACA 4412

Compared to the NACA 6412 airfoil, the value of maximum camber decreases from 6% to 4% for NACA 4412. Hence comparable results were obtained for both airfoils. Similar to NACA 6412, value of lift decreases and the value of drag increases as the height of the airfoil from the ground increases. At 4° angle of attack, an increase in the value of Cl is observed as expected. Its value decreases as h/c ratio increases. At 8° angle of attack too, the trend of lift and drag variation are similar to that of NACA 6412.

Table 2. Showing C_L, C_D values of NACA 4412 airfoil at various angles of attack (α) and h/c ratios

α°	h/c	C_l	C_d	C_l/C_d
0	0.2	0.447	0.0126	35.42
0	0.5	0.443	0.0136	32.57
0	1.0	0.442	0.0142	31.12
4	0.2	0.97783	0.01734	56.39
4	0.5	0.83693	0.018293	45.75
4	1.0	0.84758	0.0194439	43.60
8	0.2	1.2442	0.027928	44.55
8	0.5	1.1533	0.032597	35.38
8	1.0	0.65771	0.051955	12.66

5.3 NACA0012

The NACA0012 is a symmetric airfoil. Hence the lift will be generated solely due to the angle of attack. At 0^0 angle of attack, it produces no lift [14]. A suction effect is observed on the lower surface at 0^0 angle of attack. This phenomenon is due to the formation of a convergent–divergent passage between the airfoil and the ground, the velocity of flow increases between the ground and he lower surface leading to a decrease in pressure causing a local drop in lift force.

Figure 3. Here the pressure plot of NACA 0012 at 0 AOA is shown. A strong suction effected is observed on the lower surface due to the formation of a convegent divergent flow here.

Table 3. Showing C_l, C_d values of NACA 0012 airfoil at various angles of attack (α) and h/c ratio

α°	h/c	C_l	C_d	C_l/C_d
0	0.2	-0.27124	0.014698	-18.45
0	0.5	-0.06215	0.01223	-5.08
0	1.0	-0.0047938	0.011781	-0.4
4	0.2	0.45471	0.021619	21.03
4	0.5	0.42817	0.021712	19.75
4	1.0	0.39020	0.021493	18.15
8	0.2	1.0757	0.031047	34.65
8	0.5	0.85732	0.030974	27.68
8	1.0	0.79180	0.033440	23.68

6. Conclusions

A CFD Analysis of three airfoils were undertaken in ground effect. The effects at various operating conditions such as the height above the ground 'h', the angle of attack and also the effect of the airfoil geometry were studied and the following conclusions can be made.

1. All the three airfoils experience increased lift as the distance from the ground decreases. This increase in lift is due increased static pressure at the lower surface of the wing and due to the disruption of the formation of wind tip vortices by the ground.

2. The values of C_L and C_L/C_D are maximum at 4^0 AOA for NACA 4412 and NACA 6412. These values decrease when aoa is increased to 8 deg. This is due to the starting of flow separation on the upper surface at the trailing edge at high angles of attack. For NACA 0012, C_L/C_D increases till 8^0 AOA. This is because NACA 0012 has no camber and has less curvature of upper surface. Hence the flow separation at the trailing edge is delayed.

3. In the NACA 0012 airfoil, a suction effect is observed at 0^0 at all clearance ratios. This is due to the formation of a convergent- divergent section between the bottom surface of the airfoil and the ground. This creates a venturi flow below the wing leading to the formation of a low-pressure section, and hence results in negative lift.

4. NACA 6412 shows the highest aerodynamic efficiency of 74.65. This is seen at an angle of attack of 4^0 and at a h/c ratio of 0.2.

5. NACA 6412 shows the best performance of the three airfoils considered in ground effect. Compared to NACA 4412, NACA 6412 has more camber and hence more curvature of the top surface which can generate higher pressure drop. It also has a flatter bottom surface compared to NACA 0012 and hence the venturi effect which reduces lift is not observed at 0^0 angle of attack. Hence the best suited airfoil for WIG crafts from the three airfoils considered is NACA 6412.

7. References

[1] A.E. Raymond, 1921, Ground influence on airfoils, *NACA Technical Note* 67
[2] S. Ando, 1990, Critical review of design philosophies for recent transport WIG effect vehicles, Trans. Jpn Soc. Aeronaut. SpaceSci. 33 (99) (1990) 28–40.
[3]. S. SHARMA, 2014 CFD Simulation of the flow characteristics of NACA 0012, NACA 6409 and DHMTU airfoils in Ground Effect, *Proc. of the ASME 2014 4th Joint US-European Fluids Engineering Division Summer Meeting, FEDSM2014.*
[4] R. Ranzenbach, J.B. Barlow 1994 Two-dimensional airfoil in ground effect, an experimental and computational study, *Society of Automotive* Paper No. 94-2509
[5] R. Ranzenbach, J.B. Barlow 1995 Cambered aerofoil in ground effect: wind tunnel and road conditions *AIAA* Paper 95-1909
[6] R. Ranzenbach, J.B. Barlow 1996 Cambered aerofoil in ground effect, an experimental and computational study, *Society of Automotive* Paper No. 96-0909
[7] Ahmed, M. R. and Sharma, S. D. 2005 An Investigation on the Aerodynamics of a Symmetrical Airfoil in Ground Effect *Experimental Thermal and Fluid Science* Vol. 29, No. 6 pp. 633–647.
[8] M.R. Ahmed 2007 Aerodynamics of a NACA4412 airfoil in ground effect *AIAA Journal* Vol.45 No.1 2007
[9] J. Keogh 2014 The Aerodynamics of a Cornering Inverted Wing in Ground Effect *Applied Mechanics and Materials* Vol. 553 pp 205-210
[10] L.S.Roberts, J.Correia, M.V.Finnis, K.Knowles 2014 Investigation of forcing boundary layer transition on a single-element inverted wing in ground effect, *The Int. Vehicle Aerodynamics Conf.* 2014, pp 199-211
[11]. YILEI HE, QIULIN QU, AND RAMESH K. AGARWAL 2014 Shape Optimization of an Airfoil in Ground Effect for Application to WIG Craft *Hindawi Publishing Corporation, Journal of Aerodynamics, Volume 2014.*
[12]. NIKHIL S. PILLAI, ANIL T., ARAVIND R., RAHUL VINOD, SUDHEESH KUMAR E., ZAHIR UMMER ZAID, ANTONY J.K., MANOJKUMAR M., "Investigation on Airfoil Operating in Ground Effect Region", 2014, *Int. Journal of Engineering and Technology, 3 (4) pp 540-544.*
[13]. STEPHEN MAHON, XIN ZHANG 2005 Computational Analysis of Pressure and Wake Characteristics of an Aerofoil in Ground Effect *Journal of Fluids Engineering, Volume 127 pp 291-298.*
[14]. JOHN D ANDERSON JR. 2007 *Fundamentals of Aerodynamics* Tata McGraw-Hill Publishing Company Limited
[15] Hsiun, C. M., and Chen, C. K. 1996 Aerodynamic Characteristics of a Two- Dimensional Airfoil with Ground Effect *Journal of Aircraft*, Vol. 33, No. 2 pp. 386–392.

Memory Effect in Methane Hydrate Formation

Mohamed Iqbal Pallipurath[1*](Principal, iREZ Academy, India), Nandakishore Nagarajan[2], Yahiya Vahid[3] (Post graduates, TKM College of Engineering, India)

Email: mohamediqbalp@gmail.com

Abstract: Memory effect of formation water in methane hydrate formation and reformation is investigated in a high pressure reactor. Methane hydrate formation rates in fresh water and reacted water are measured at different pressures and temperatures. A significant reduction in formation time and incubation period is found with reused water. In the context of transportation of natural gas from distant gas fields as hydrate, this information can be utilized to reduce the transportation cost of natural gas.

1. Introduction

Methane hydrate is a clathrate. The word "clathrate" is derived from the Greek word "khlatron" meaning barrier, indicates crystalline inclusion compounds in which small guest atoms or molecules are physically trapped by three dimensionally shaped cavities formed by three dimensional assemblies of hydrogen atoms. [1] These compounds are called hydrates when the cage is formed from water molecules and gas hydrates when the enclosed molecules are gases. [2] The formation and decomposition of clathrate hydrates are first order phase transitions, not chemical reactions. Clathrate hydrates were first documented in 1810 by Sir Humphry Davy who found that water was a primary component of what was earlier thought to be solidified chlorine. [3]

According to size of the trapped molecule, three types of structures are commonly observed: cubic I, cubic II and hexagonal structure. These structures correspond to different arrangement of water molecules. [4] Structure formed by water molecules is determined by the size of guest molecules and composition of the mixtures. Common hydrate structures are shown in figure 1.

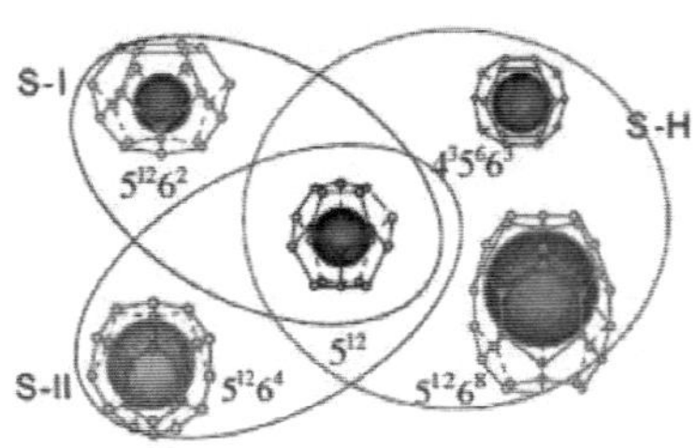

Figure 1. Molecular arrangements of different types of hydrate structures

The most common gas hydrate structures are SI and SII hydrates. SI has two types of cavity: a small pentagonal dodecahedral cavity consisting of 12 pentagonal rings of water molecules (20 water molecules) and a large tetrakaidecahedral cavity consisting of 12 pentagonal and two hexagonal rings of water molecules (24 water molecules).Structure II hydrate also has two cavity sizes, the pentagonal dodecahedral cavity and a larger hexakaidecahedral cavity consisting of 12 pentagonal and four hexagonal rings of water (28 water molecules). Unit cell of 7 (type H) consist of 34 water molecules, forming three types of cages, two small ones and one huge cage. In this case the unit cell consists of three small cages of 5^{12}; two small ones of type $4^3 5^6 6^3$ and one huge type of$5^{12}6^8$. The formation of type H requires the cooperation of two guest gases (large and small) to be stable.

Memory effect is a little studied phenomenon wherein water used for hydrate formation once appears to form hydrates faster when reused. The mechanism of memory effect is still imperfectly understood, some authors have suggested that the hydrate forming bonds may still be unbroken in the used water which facilitates the reformation later.

2. EXPERIMENTAL SETUP

2.1 MATERIALS

Methane gas with 99.5 % was used as guest molecule. Distilled water (host) was used to perform the experiments. Refrigeration test rig in the laboratory was modified to control the temperature inside the water bath. 99.5% pure methane gas was used in the experiments. The entire reactor was immersed in a constant temperature bath using a modified refrigeration setup as shown in figure 2.

Figure 2. Refrigeration rig modified to serve as the constant temperature bath.

2.2 The hydrate reactor

Stainless Steel 304 (SS-304) cylinder is the main part of the apparatus. SS 304 was selected because it can withstand high pressures and it would retain its properties even at low temperatures. [5] Monolithic piece of SS 304 was selected to avoid leakages due to weld and also to withstand high pressures. The reactor dimensions were such that it would have a total volume of 450 ml. Reactor's outer diameter was taken as 80mm, thickness is 10mm, and height of the cylinder was 90mm. The reactor is shown in the figure 3.

Figure 3. The high pressure hydrate reactor.

Figure 4. Pressure test with air compressor at IISC, Bangalore. Courtesy Prof. Kasturi Rangan

The reactor after fabrication by the authors was pressure tested successfully up to 20 Mpa at Indian Institute of Science, Bangalore with the help of Prof. Kasturi Rangan.

The reactor was immersed in the constant temperature bath as shown in figure 5.

Figure 5. Reactor in the constant temperature bath.

The reactor was rinsed with de-ionized water once. The amount of water injected into the apparatus was 200ml for every experiment. The solution is then introduced into the cell through the ball valve. The air was removed from the reactor cell by charging compressed CH_4 into the reactor while keeping ball valve open. The ball valve is then closed to pressurize the apparatus to required level. The ends of thermocouple were connected to the temperature indicator. The experiment was started by placing the cell in a NaCl bath. The time taken for each degree fall in temperature inside the apparatus was noted. The higher limit of temperature was taken as 20°C. The pressure inside the apparatus decreased with decrease in temperature. At the time of hydrate formation there was a rise in temperature due to exothermic nature of methane hydrate formation. [8]

The temperature became steady at the point of hydrate formation. When the pressure of the reactor reached a constant value the gas hydrate formation was completed. After the temperature and the pressure in the apparatus became constant, the temperature and the pressure were noted as equilibrium data. [9] The thermocouple readings corresponding to hydrate formation for the pressure inside the apparatus was noted. Total time required for the formation of hydrate was also noted. The experiment was repeated by varying the pressure. The hydrate crystals formed is shown in figure 6.

Figure 6. Methane hydrate formed in the reactor.

3. Investigation of Hydrate Dissociation

The apparatus was placed outside the temperature bath (normal atmospheric conditions). The temperature was then gradually increased inside the apparatus. Time required for each degree raise in temperature is noted. Time required for completion of hydrate dissociation was also noted. The procedure was repeated for a different pressure. Different levels of dissociation of hydrate are shown in figures 7a to 7c.

Figure 7a. Dissociation of hydrate. 10 minutes after exposure to atmospheric pressure.

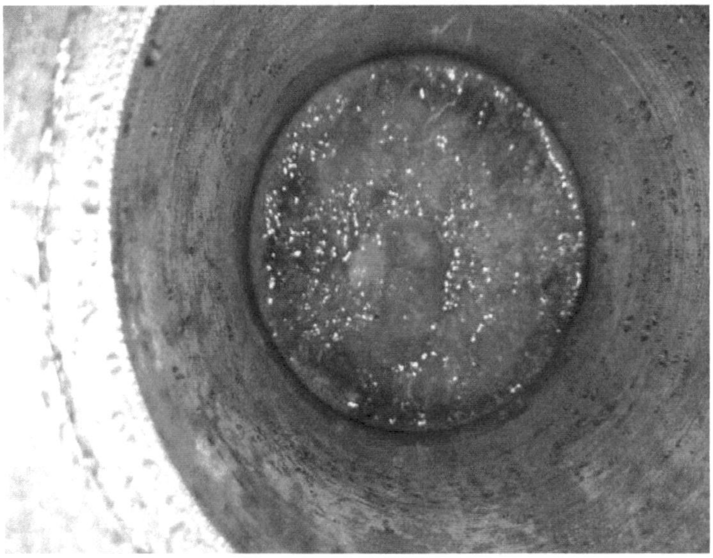

Figure 7b. Dissociation of hydrate. 15 mites after exposure.

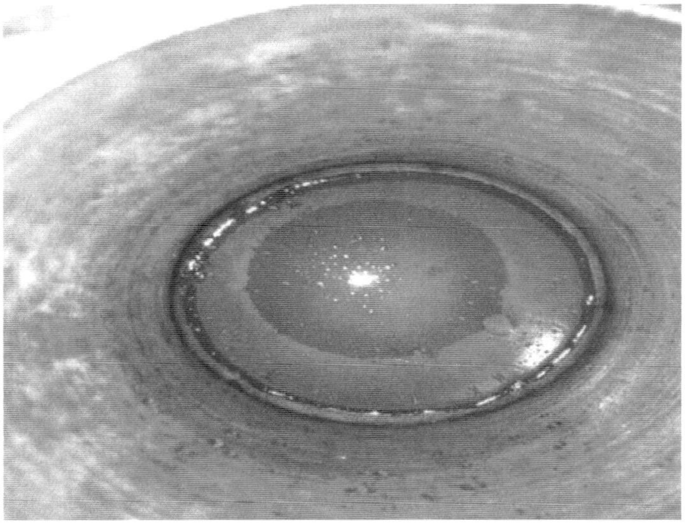

Figure 7c. Dissociation of hydrate. 20 minutes after opening pressure vessel.

4. Investigation of Memory Effect

Memory effect refers to the strikingly shorter time required for hydrate formation when the water used has been previously used to form hydrates. When hydrates are decomposed and the resulting water is used to form hydrates, the formation is faster. No conclusive explanation has been given for this phenomenon, although one theory proposes that the very tiny hydrate crystals which may persist in the used water may be responsible by acting as seeds for hydrate formation.

Fresh water which was used for methane hydrate formation was reintroduced into the apparatus after the completion of the hydrate dissociation. When the experiment was repeated, the pressure inside the apparatus was found to decrease with decrease in temperature. [4]

Methane hydrate formation has been investigated at different pressures, and primary focus was on induction time, number of moles of methane consumed per moles of water and then water to hydrate conversion. After completion of hydrate formation the dissociation was done and the rise in pressure was noted. Memory effect investigation was performed with once hydrated water.

5. Results and Discussion

5.1 Memory Effect in Methane Hydrate Formation

Induction time and hydrate formation temperature during dissociation and formation of methane hydrate with memory water (water already used to form hydrate) was investigated at different pressures. While using memory water hydrate formed at higher temperatures as compared to fresh water, but this trend was only observed at higher pressures. At lower pressure hydrate was formed at same temperature for memory water as that of fresh water. The time needed to form hydrate decreased while using memory water as compared to the time required for fresh water. The gas injected at high pressures remains at constant pressure for some time. Then with drop in temperature the pressure also falls. After the induction time there would be a sharp decrease in pressure for about 20-40 minutes. This is the period of methane hydrate formation. The sudden decrease in pressure shows the onset of methane hydrate formation. Then the pressure remains constant after the hydrate formation is completed. The time vs. pressure plot is shown in figure 8.

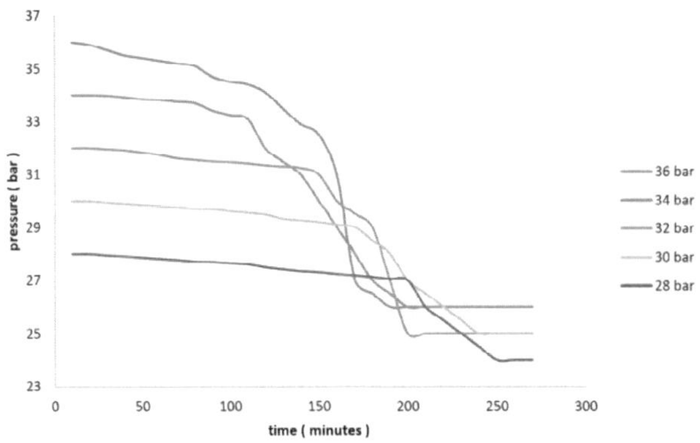

Figure 8. Time-pressure plot for Methane hydrate formation in memory water

Figure shown compares the time-pressure plots for methane hydrate formation at different pressures. The pressure drop is different at different conditions. Figure reveals that at high pressure the gas is easily soluble in water, so more number of methane moles get converted into hydrate, therefore pressure drop is high while at low pressure the solubility is low, so less number of methane moles get converted into hydrate, hence pressure drop is less. The time Vs temperature plot is shown in figure 9.

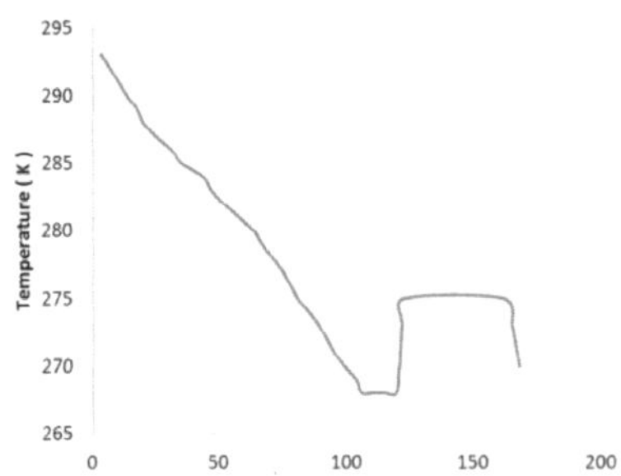

Figure 9. Time temperature plot of methane hydrate formation. X axis (Time) is in minutes.

Figure shows the Time-Temperature relation of methane hydrate formation at 36 bar. The temperature of the system is lowered using NaCl bath to 268 K. Temperature was reduced to 268 in order to keep the temperature well below the equilibrium point, which acts as a driving force for methane formation, known as under cooling. Therefore after the initial delay i.e. the induction time, temperature rises due to exothermic nature of methane hydrate formation. The experiment was continued till the temperature reaches steady state. This temperature was taken as methane hydrate formation temperature. The time Vs temperature plots at different pressures is shown in figure 10.

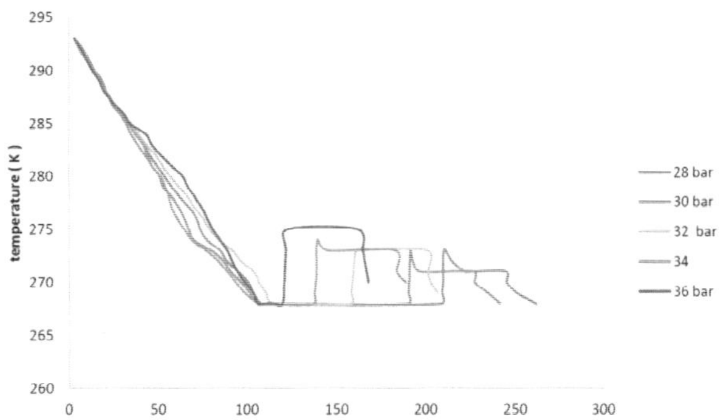

Figure 10. Time - temperature plot for different pressures. X axis is in minutes from introduction of methane.

Figure 10 shows time-temperature relation of methane hydrate formation at different pressures from 28 bar to 36 bar. It illustrates the variation in length of induction time with pressure and the exothermic nature of methane hydrate formation at different pressures.

The sudden rise in temperature, the plateau, and the subsequent drop in temperature are signatures of hydrate formation start and finish. The plot of number of moles of methane consumed Vs pressure is shown in figure 11.

5.2 Number of Moles Consumed

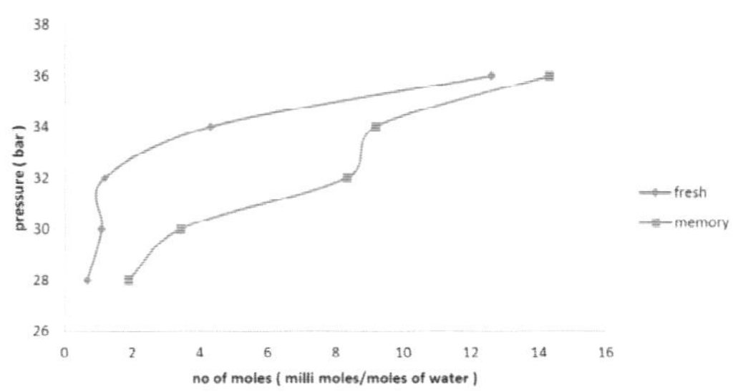

Figure 11. Number of moles of methane hydrate consumed vs pressure.

Journal of Physics: Conference Series **1355** (2019) 012007 doi:10.1088/1742-6596/1355/1/012007

Figure shows the comparison of number of moles consumed in memory water to moles consumed in fresh water. Usual trend is that number of moles consumed increases as pressure increases. But in memory water hydrate forms at higher temperature and the pressure drop is more as compared to the fresh water so number of moles consumed increases. Pressure Vs. temperature plots for both fresh and memory water are shown in figure 12.

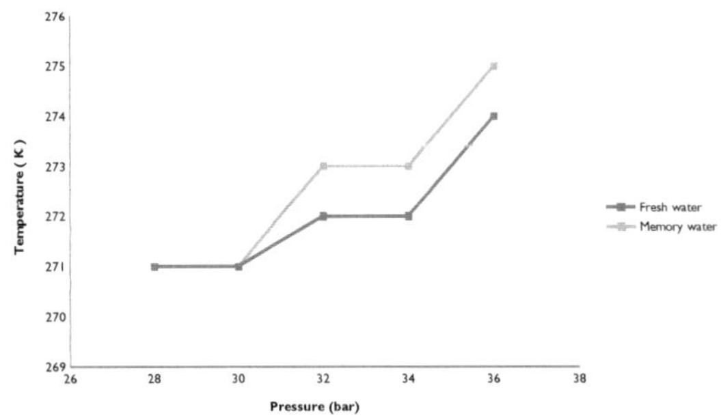

Figure 12: Pressure-Temperature relations for Methane Hydrate formation in memory water compared to pure water.

Results may be summarized as follows:

5.3 Hydrate Formation from Fresh water.

* As pressure increases the hydrate formation temperature increases.

* Induction time decreases with increase in pressure.

* As hydrate formation starts there would be a sudden decrease in pressure.

* Hydrate formation is exothermic in nature.

* Number of moles consumed increases as pressure increases.

* Water to hydrate conversion rate is higher in higher pressure range.

5.4 Hydrate Formation from Memory water.

* Hydrate formation shows an increase in formation temperature at higher pressure, but at lower pressure there is no discernible effect on formation temperature.

* Induction time reduces (12.9% of total time) in memory water as compared to fresh water (25%).

* The pressure drop is greater as compared to fresh water.

* Number of moles of CH_4 consumed in memory water is more as compared to fresh water.

* Water to hydrate conversion rate is also high.

5.5 Hydrate dissociation below 20°C.

* Hydrate dissociation temperatures are closer to hydrate formation temperatures.

* Hydrate formed at higher pressures can exist undissociated for more time in atmospheric conditions as compared to lower pressure conditions.

5.6 Hydrate dissociation above 25°C.

* Hydrate dissociated water when kept for long time at ambient temperature, its temperature would rise above 25°C and there is no significant memory effect for this water.

* Hydrate formed in this condition was similar to fresh water, temperature of hydrate formation was exactly similar, but time varied, showing that induction time for methane hydrate formation is stochastic.

6. CONCLUSION

Experimental investigation on methane hydrate formation in fresh water was conducted and it was compared with the formation in memory water. The induction time, temperature for hydrate formation, pressure drop during hydrate formation and exothermic nature of methane hydrate formation was studied.

It was found that the formation temperature is higher when forming hydrate with previously used (memory) water.

A significant drop in induction time was found to exist when forming hydrate with previously used (memory) water. (12.9% Vs 25%). This effect disappears when the used water is brought above 25°C.

The rate of hydrate formation is found to be 10 times higher in memory water as compared to fresh water. The number of moles of methane consumed is also found to be 10.5 times more as compared to fresh water.

Conflict of interest disclosure

The authors declare that there is no conflict of interest regarding the publication of this paper.

Acknowledgment

The authors would like to acknowledge Indian Institute of Science, Bangalore where pressure test of reactor was performed and Thangal Kunju Musaliar College of Engineering, Kollam, where the experiments were conducted.

REFERENCES

1. Vysniauskas and P. R. Bishnoi. A kinetic study of methane hydrate formation, Chemical Engineering Science 38 (1983), 1061-1972.
2. H. C. Kim, P. R. Bishnoi,T R. A. Heidemann and S. S. H. Rizvi. Kinetics of methane hydrate decomposition, Chemical Engineering Science 42 (1987), 1645-1653.
3. Sang-Yong Lee and. Gerald D. Holder. Methane hydrates potential as a future energy source, Fuel Processing Technology 71 (2001), 181–186.
4. Qiang Wu, Baoyong Zhang. Memory effect on the pressure-temperature condition and induction time of gas hydrate nucleation, Journal of Natural Gas Chemistry 19 (2010), 446-451.
5. Fedricco Rossi, Mikro Filipponi, Beatrice Castellani. Investigation on a novel reactor for gas hydrate production, Applied Energy 99 (2012), 167-172.
6. Garett C, Fitzgerald, Marco J, Castaldi, Yue Zhou. Large scale reactor details and results for the formation and decomposition of methane hydrates, Journal of Petroleum Science and Engineering (2012), 19-27
7. Dong-Liang Zhong, Shuang-Yi Heb, Dong-Jun Sunb, Chen Yangb. Comparison of methane hydrate formation in stirred reactor and porous media in the presence of SDS, Energy Procedia 61 (2014) ,1573 – 1576
8. Saheb Maghsoodloo Babakhani, Abdolmohammad Alamdari. Effect of maize starch on methane hydrate formation/dissociation rates and stability, Journal of Natural Gas Science and Engineering 26 (2015) , 1-5

9. Hesam Najibi, Morteza Mirzaee Shayegan, Hassan Heidary. Experimental investigation of methane hydrate formation in the presence of copper oxide nanoparticles and SDS, Journal of Natural Gas Science and Engineering 23 (2015), 315-323.
10. H. Fakharian, H. Ganji, A. Naderi Far and M. Kameli . Potato starch as methane hydrate promoter, Fuel 94 (2012) 356–360
11. M. I. Pallipurath. Effect of Bed Deformation on Natural Gas Production from Hydrates. Journal of Petroleum Engineering, Volume 2013, Article ID 942597, 9 pages, http://dx.doi.org/10.1155/2013/942597
12. M. I. Pallipurath. Dissociation and subsidence of hydrated sediment: coupled models. Energy Exploration & Exploitation · Volume 27 · Number 2 · 2009 pp. 105-131
13. M. I. Pallipurath. Dissociation of Hydrated Marine Sediment. Electronic scientific journal "Oil and Gas Business". 2007. №1. URL: http://ogbus.ru/eng/authors/Iqbal/Iqbal_1.pdf

Experimental and Numerical Investigations of Synthetic Jet Used for Electronic Cooling

Arun Jacob[1], Anandhu R[2], Leena R[3], Shafi K A[4]
T.K.M college of Engineering, Kollam, Kerala, India
E-mail: [1]arunjacobcool@gmail.com [2]anandhukeralapuram@gmail.com
[3]leenakalidas@gmail.com, [4]shafika.tkm@gmail.com

Abstract. Cooling of electronics and micro-electronics devices is an important task in our present world. Synthetic jet is a relatively new technique for electronic chip cooling which synthesizes stagnant air to form a jet resulted from periodic oscillations of a diaphragm in a cavity. Synthetic jet cooling increases the rate of heat transfer as compared to other cooling techniques. The impingement heat transfer characteristics of a synthetic jet is studying in this work. Synthetic jet is driven by a piston-cylinder arrangement a through circular nozzle for the impingement of jet on the heated surface. Air is considered as the cooling medium. Heat flux is taken as 8000W/m^2. Numerical simulations and experimental methods are conducted to study the effect of various distance between the orifice and the heated plated (Z). A circular orifice is used to study the characteristics of convective heat transfer. The results are verified by the time history of convective heat transfer characteristic and validated with experimental results. The model was simulated to investigate the dispersion of heat flow on the walls using a mathematical turbulent model of k- ω SST. The Reynolds number (Re) is in the range of 4000-8000 based on average velocity, while the normalized impinging distance varies between 2D to 10 D The results shows the significant influence of Z/D ratio and sinusoidal wave frequencies to the heat transfer rate obtained. Experimental and numerical investigations is carried out to study the variation of Nusselt number with jet velocity. It is found that the Z/D ratio at 6 gives the maximum amount of cooling for a flat plate.

Keywords: Jet impingement; Synthetic jet; Nusselt number

1. Introduction

Nowadays in the present world, there is a great demand for the electronic devices which are compact in size, occupying less space in system etc. This in turn poses a serious challenge in cooling the electronic devices due to the increase in heat generation coupled with the limited heat removal surface area. Thermal issues exist over a wide range of power dissipation levels from handheld devices that dissipate a few Watts to high-performance microprocessors dissipating over 100W. The engineers came up with some solutions like Heat sinks, forced air cooling (fans), heat pipes which are traditional cooling methods. But still they have some drawbacks, such as the heat sinks needs to have smooth and flat surface for better heat dissipation, which costs more for fabrication; forced air cooling (fan) requires more space in the system; heat pipes are hollow metal tubes which contains thermal liquid in it, which has got the chances of damaging the system. To overcome these drawbacks, synthetic jet can be potentially used for cooling applications. Impinging synthetic air jets can be used to transfer heat in diverse applications, which may vary from the cooling of a manufacturing process to the thermal management of electronics, microprocessors, graphic cards etc. While these steep cooling requirements have prompted the study and development of synthetic air jets. Synthetic jets are formed by periodic suction and ejection of fluid out of an orifice bounding a cavity by the time periodic motion of a diaphragm that is built into one of the walls of the cavity .During the ejection phase a coherent vortex, accompanied by a jet, is created and convected downstream from the jet exit. Once the vortex flow has propagated well downstream, ambient fluid from the vicinity of the orifice is the bulk of the high-speed air has moved away from the orifice, avoiding re-entrainment, while quiescent air from around the orifice is sucked into the orifice.

Smith & Glezer [1] investigated the evolution of synthetic air jet. This process of evolution of jet at jet plane is dominated by the time periodic formation and moving of these vertex pairs away from the orifice. These vertex pairs artificially are turbulent in nature then gradually undergoes transition becoming slow down and lose their coherence. B.L. Smith et al. [2] experimentally measured the flow visualization of synthetic jets and continuous jets with matched Reynolds numbers is conducted and they had found that they have the same profile shape, synthetic jets are wider and slower than matched continuous jets. synthetic jets grow more rapidly, both in terms of jet width and volume flux, than do continuous jets.

J.Z. Zhang [3] conducted experimental studies on excitation frequency forced on the actuator has a great effect on the synthetic jet flow field. There are two resonance frequencies at which the mean velocity and vorticity of the synthetic jet are maximized, especially at the higher resonance frequency. Clemens J.M. Lasance et al. [4] studied, they had an overview of the principles of heat transfer and acoustics related to a promising alternative for fans: synthetic jet cooling. It can be a compact, sensitive, and relatively efficient requiring a relatively low cone excursion, which makes the system very suitable for low-noise active cooling.

Harinaldi et al. [5], experimental investigation is carried out by varying impinging distance normalized to diameter of orifice (L/D). They got the best result at L/D=2 with excitation frequency=160Hz. Dae Hee Lee et al.[6]performed an experimental investigation to determine the effect of nozzle diameter on heat transfer and fluid flow for a round turbulent jet impinging on the surface of a flat plate for a dimensionless distance ranging from 2 to 14 between the nozzle and plate surface. The experiments are confined to Jet Reynolds numbers of 23,000. In their investigation, the local Nusselt numbers increased with increase of nozzle diameter in the stagnation region.

M. Arik [7] The experiments have been completed with synthetic jets, which are 12.5 mm in diameter and 2 mm thick with a square orifice of 1 mm. A synthetic jet has been driven at the resonance frequency of 4500 Hz, heat transfer enhancements over the specific heater sizes are presented, and it is found to be between 4 and 10 times of natural convection. Mangesh Chaudhary et al. [8] investigated the impingement of heat transfer characteristics of synthetic air jet. It was observed that heat transfer coefficient (h) increases rapidly up to axial distance of 48-50mm then decreases gradually with increase in distance for all frequency. Buddhika N. Hewakandamby[9] investigated numerically the effect of heat transfer coefficient with respect to the oscillation frequency, geometric parameters and the flows with low Reynolds number. He observed that the oscillatory flow jets achieved approximately 100 percentage improvement of heat transfer over conventional steady flow jets. Dipankar Sahoo and Sharif[10]developed a numerical model to predict the flow and heat transfer characteristics in the slot jet impingement cooling of a constant heat flux surface. They observed that the flow and thermal fields change with the Richardson number for a particular set of jet exit Reynolds number and domain aspect ratio. They also observed that the skin friction coefficient increases at high Richardson number due to decrease in aspect ratio.

2.Experimental Setup

Figure 1 shows the schematic diagram of the experimental setup. Synthetic jet is supplied by a piston cylinder arrangement which is mount on a stand. The thermal environment of the microprocessor is simulated using a copper plate of size 50 X 50 ×5 mm with a heater placed underneath. A thin heating coil having the same dimensions as that of the cross section of the target plate is used for ensuring uniform heating. The power input to the heater is varied using a dimmer stat and the voltage and current are measured using voltmeter and ammeter respectively. J- Type thermocouples were used to measure the temperature at various locations. The temperature data were acquired using a temperature scanner.

Power is supplied from DC power source through a voltage stabilizer, a variac and a current transformer. The voltage across the heater and the current are measured by digital meters whose ranges and the accuracies are of 0 to 20 ± 0.5% V and 0 to 400 ± 0.5% respectively. The oscillation of piston can be regulated by a RPM meter. The velocity of the jet exit air close to the nozzle was measured with a calibrated anemometer. In the experimental and numerical studies, the size of the orifice is 2 mm. Three circular holes, each with a diameter of 2 mm are drilled on the nozzle.

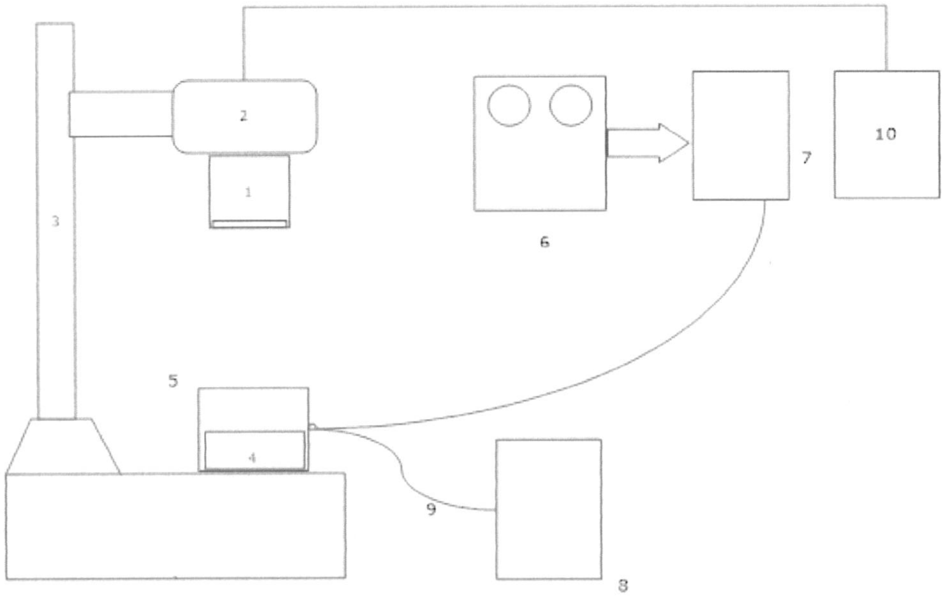

1.Piston cylinder 2. DC motor 3. Stand 4. Heater plate 5. Cabin 6. Power unit 7. Dimmerstat 8.temperature display unit 9. J type thermocouple 10. RPM controller.

Figure .1 Schematic layout of jet impingement system

3.Numerical Analysis

In this section, an overview on the numerical investigations are presented. A three-dimensional, incompressible, time-dependent scheme was used to compute the flow.

3.1. Boundary and Initial Conditions

Numerical analysis was conducted in 3D computational fluid dynamics field using impinging synthetic jet methods. An UDF (User Defined Function) was used to represent the motion of the membrane because the synthetic jet A computational study was performed to describe the membrane had two phase i.e. suction and blowing.

The flow is assumed to be steady, incompressible, three dimensional and turbulent. Heat transfer across the thickness of the plate, buoyancy and radiation heat transfer effects are neglected. Properties of the fluid such as density, viscosity and thermal conductivity are assumed to be constant. The impinging plate was kept at constant heat flux of 8000W/m^2. The three-dimensional Navier-stokes and energy equation with standard turbulent model are solved using Ansys Fluent software. The numerical simulation has been done using SST k-ω-turbulent model for jet impingement application. This model used to predict capture gradient effect and heat transfer distribution near the wall accurately with moderate computation cost. The schematic diagram 3-D computational domain is shown in Figure 2.

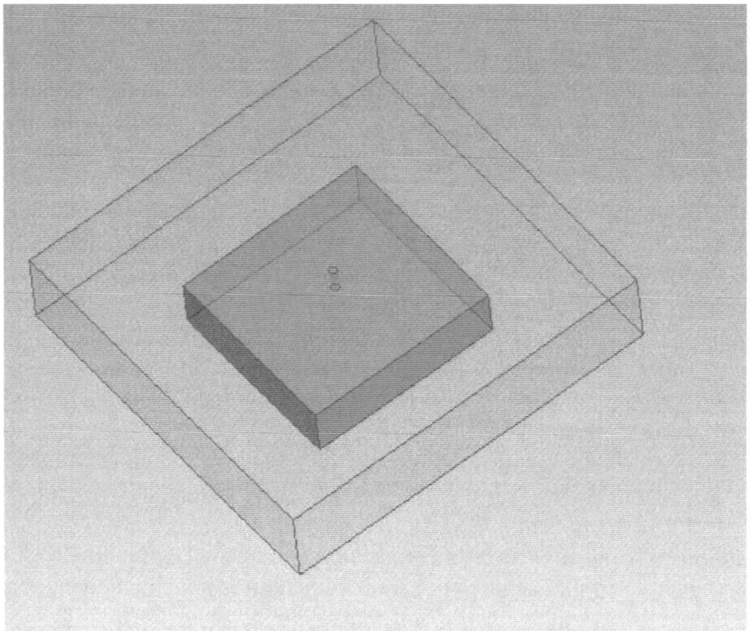

Figure 2. computational domain

3.2. Numerical grids independence study

Four different grid distributions are considered to ensure that the numerical results are grid independent, and the results obtained are shown in Fig. 3. The numbers of elements in the various cases are 3810, 25121, 196400, and 418200.By comparing with the third and fourth mesh configuration in terms of stagnation Nusselt number, the variation is relatively less and it shows that 6.6% variation. Therefore, for further analysis, a case with 418200 elements has been adopted because it ensured a good compromise between computational time and accuracy requirements.

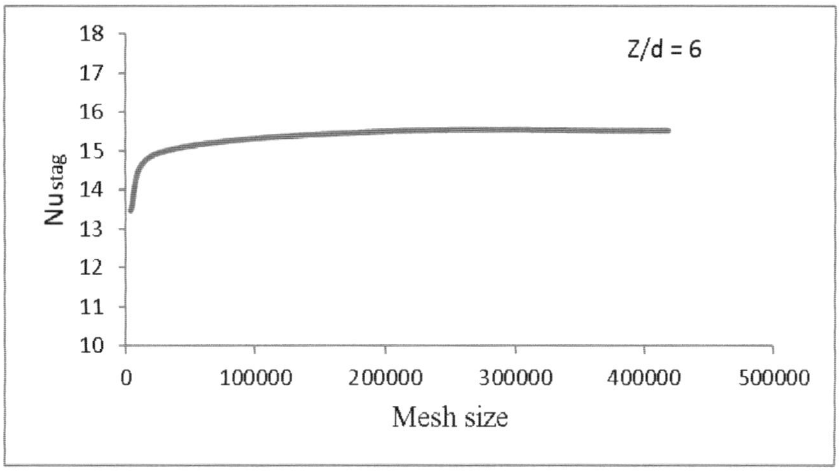

Figure 3 Grid independence study.

4. Numerical solver

The simulation was performed using Computational Fluid Dynamics (CFD) Ansys Fluent that solves equations of continuity, momentum end energy using the Reynolds-Averaged Navier-Stokes approach.

Fluid properties (Air):

- Density = 1.225 kg/m3
- Dynamic Viscosity = 1.846 e-5 Pa-s
- Thermal Conductivity = 0.024 W/ (m-K)

CFD model:

- Software: Ansys Fluent
- Transient analysis of single jet impingement on the flat plate
- Dimension of plate (50mmx50mmx5mm)
- Diameter of nozzle jet (2mm)

The computational domain is modelled as a 3D transient system in the present investigation. The turbulence model used is Shear Stress Transport (k-ε) model, which is found to be the best among available turbulence models for this type of flow configurations. Copper is used as the material for plate and air is used as the cooling medium. For an unsteady analysis, the pressure velocity coupling used and the velocity of the jet is kept unsteady by keeping it pulsating. A UDF has been introduced to generate a pulsating flow at the inlet of the nozzle.

5. Results and discussion

This section provides the analysis of the flow field behavior of sinusoidal jet. The heat transfer characteristics in terms of the Nusselt number is presented.

5.1 Validation of Experimental results.

Dimensional parameters in this studies as shown in Fig. 4 includes distance between the orifice and the heated surface (Z) or can be called as impinging distance from synthetic jet flow and diameter of circular orifice (D).Here, the frequency of the sine wave is 20Hz and diameter of jet is 2 mm. It is observed there is fair agreement between the results obtained from the experimental and numerical studies. From the studies, it is found that maximum heat transfer with single jet occurs when the *Z/D* ratio of 6. Stagnation temperature decreases up to 6 and then increases.

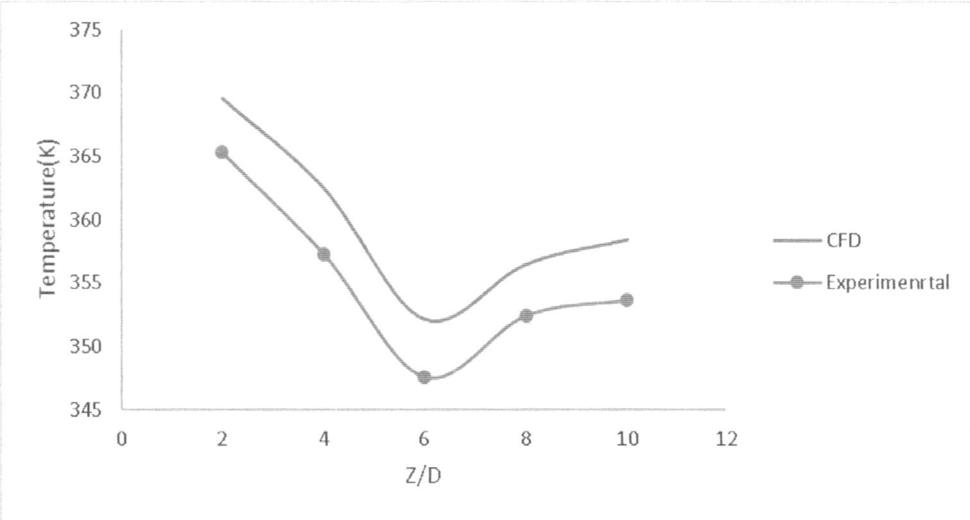

Figure 4 Variation of stagnation temperature with orifice-to-plate distance.

Figure 5 Temperature Contour on the surface of the heated plate.

5.2 Effect of heat transfer due to the frequency variation.

An increase in Nusselt number was found in fig. 7 when the pulsation frequency was increased from 10 Hz to 22 Hz. The increase in Nusselt number is due to the periodic destruction and reconstruction of thermal and hydrodynamic boundary layers, when unsteady impinging jet were used. The delay in flow velocity also allows the hot fluid after impingement to escape through the outlet. As a result, it does not affect the temperature of the incoming fluid. This would also enhance the heat transfer of the fluid. But due to the decay in kinetic energy of the jet results in low heat transfer at far end of the plate.

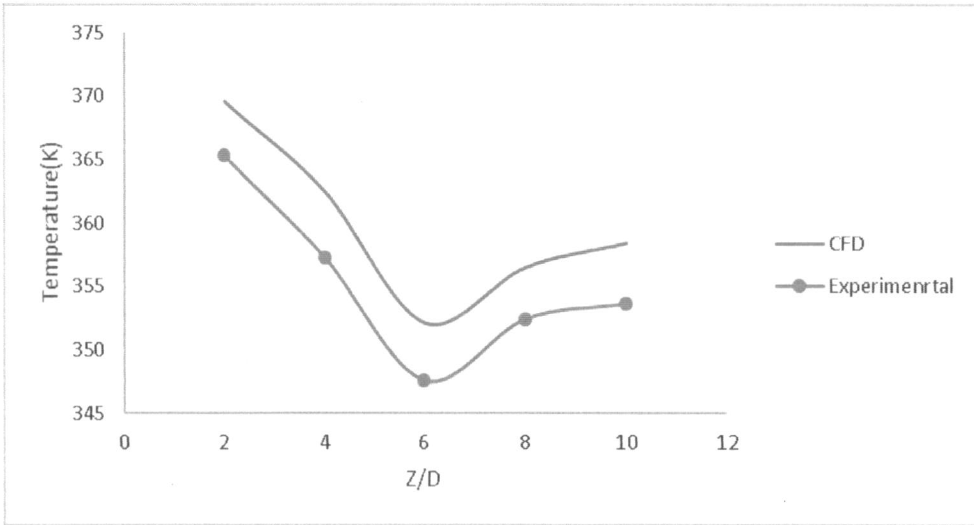

Figure 6 Streamwise Heat transfer distribution at Z/D = 6

5.3 Effect of stagnation Nusselt Number with Reynolds Number

The variation of the stagnation Nusselt number with the Reynolds number at Z/d =6 is shown in fig.7 As the graph indicates, the stagnation Nusselt number increases with the Reynolds number for the range of values tested. By increasing the Reynolds number, the axial momentum increases and turbulence produces in the flow direction. Hence heat transfer coefficient increases with the increase in the velocity of the fluid.

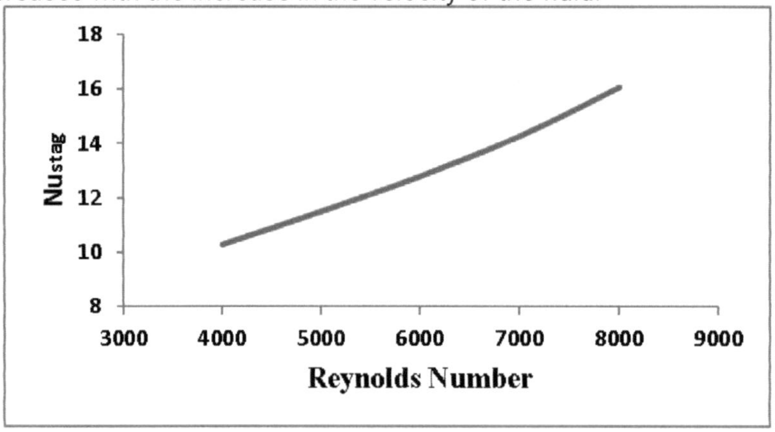

Figure 7 Variation of stagnation Nusselt Number with Reynolds Number

6.Conclusion

Experiments and numerical study are performed to evaluate the heat transfer coefficients by synthetic air jet from a circular orifice at Z/D=6. The numerical results of the steady jet agree satisfactorily with the experimental data. The salient parameters controlling the process are found to be orifice to plate spacing (Z/D), jet diameter (D), Reynolds number (Re) and frequency of the jet. The shortening and spreading of the synthetic jet also allow for higher turbulence levels and promotes flow instabilities.

7. References

[1] Smith B. L,Glezer A 1998 The formation and evolution of synthetic jets *Physics of fluids* 10pp 2281-2297
[2] B.L. Smith, G.W. Swift 2003 A comparison between synthetic jets and continuous jets

Experiments in Fluids *International Journal of Heat and Mass Transfer* 34 pp467–472
[3] J.Z. Zhang, X.M. Tan 2000 Experimental study on flow and heat transfer characteristics of synthetic jet driven piezoelectric actuator *Science in China Series E: Technological Sciences Sci China Ser E-Tech Sci*50pp221-229.
[4] Clemens J.M, Lasance, R.M. Aarts 2008 Synthetic jet cooling part-I: overview of heat transfer and acoustics, in: *IEEE Semi-Therm Symposium* pp 20-25.
[5] Harinaldi 2012 Effect of Impinging Distance for Convective Heat Transfer of Synthetic Jet *World Applied Sciences Journal* 20pp470-475
[6] Dae Hee Lee, Jeonghoon Song , Myeong Chan Jo 2004 The Effects of Nozzle Diameter on Impinging Jet Heat Transfer and Fluid Flow *International Journal of Heat and Mass Transfer* pp554–557
[7] M. Arik 2008 Local Heat Transfer coefficients of a High-Frequency Synthetic Jet during impingement cooling over Flat Surfaces *Heat Transfer Engineering* 29pp763–773
[8] Mangesh Chaudhari, A Puranik and A Agrawal 2010 Heat transfer characteristics of synthetic jet impingement cooling *International Journal of Heat and Mass Transfer* 53pp1057–1069
[9] Buddhika N Hewakandamby 2009 A numerical study of heat transfer performance of oscillatory impinging jets *International Journal of Heat and Mass Transfer* 52pp396–406
[10] Dipankar Sahoo, MAR Sharif 2010 Numerical modelling of slot-jet impingement cooling of a constant heat flux surface confined by a parallel wall *International Journal of Thermal Sciences* 43pp. 877–887

Numerical Analysis of Combustion and Film Cooling in LOx-Methane Rocket Engine

Arun M[1, 2, 3], Amith Raj R[3] and Jose Prakash M [3]

[1] Faculty of Engineering and Technology, University of Kerala, Thiruvananthapuram, Kerala-695034, India

[2] Department of Mechanical Engineering, Sree Chitra Thirunal College of Engineering, Thiruvananthapuram, Kerala-695018, India

[3] Department of Mechanical Engineering, TKM College of Engineering, Kollam, Kerala-691005, India

E-mail: arun.m@tkmce.ac.in

Abstract. Nowadays methane is considered as an alternative fuel for rocket engine application. It is abundant in the outer solar system that can be harvested from mars, Titan, Jupiter, many other planets and moon. The properties of cryogenic methane like higher density, higher vaporization temperature, less challenge in storage requirement compared to cryogenic hydrogen and its higher specific impulse, superior cooling capability and higher coking limit compared to kerosene created a renewed interest on space scientist in choosing methane as a propellant. Combustion and film cooling involving supercritical and transcritical regimes in a high pressure rocket engine combustion chamber is a challenging problem for the rocket engine designers. At supercritical and transcritical conditions, the thermophysical and transport properties exhibit tremendous variations which makes modeling combustion and cooling a demanding task. The present study is based on the oxygen methane thrust chamber model B at DLR P6.1 test bench. The finite volume based software ANSYS Fluent 15.0 is used for the numerical studies. The combustion is simulated using single step eddy dissipation approach. The real gas effects are incorporated using Soave Redlich Kwong cubic equation of state. The numerical scheme is validated by comparing the results obtained with the experimental results reported in literature. Investigations have been carried out to study the effect of film cooling on chamber wall temperature and wall heat flux for different oxidizer inlet temperature. The result revel that the hot gas temperature near the wall is higher in the transcritical combustion compared to supercritical combustion.

1. Introduction

Liquid propulsion rocket (LPR) engine uses liquid propellants to react and form products at high pressure and temperature. These products are then accelerated through a nozzle to get a high velocity jet. Liquid propellants are commonly used in higher stage engines due to many reasons. High density of Liquid propellants helps to reduce the volume of propellant tank which leads to have high thrust to mass ratio and high specific impulse. They can be started and stopped whenever required hence they give better overall control compared to solid propellant rockets. Recent advancements in rocket technology have introduced reusable rockets which also employ liquid propellants. Many of the launchers nowadays use Liquid Hydrogen (LH2) and liquid Oxygen (LOX) as propellants because of their high specific impulse. The combustion products are of low molecular weight and hence higher velocity for the jet. Also, the combustion products are non-toxic and eco-friendly. However, LH2 as a fuel has two significant disadvantages: (i) it is a very low density cryogenic fluid and therefore need larger tanks maintained at temperatures as low as 20K for storage (ii) the pumping power requirement is very high. There also exists a chance for resonance as the pumping frequency is close to the natural frequency. A proposed alternative propellant combination is methane and LOX. Methane being four times denser than LH2, requires smaller tanks for storage. Also the pumping effort required is significantly lower, compared to LH2. Methane does not require extremely low temperatures as LH2 for storage. Although specific impulse is slightly less, ease of handling and storage enables it to be used in future launchers. Methane (CH4) is abundant in the outer solar system. With fuel waiting at the destination, a rocket leaving Earth wouldn't have to carry so much propellant, reducing the cost of a mission. Also in comparison with kerosene, methane is easy to produce, having higher coking limit and better cooling capabilities. Since use of methane as propellant is an emerging technology, the different process parameters that control these systems has to be carefully investigated. Rocket engines uses different cooling method to cool the thrust chamber walls from the hot gases formed due to combustion. The primary cooling method used in LPR is regenerative cooling. Regenerative cooling is the method by which either cryogenic fuel or oxidizer is passed through the cooling channel around the thrust chamber. Film cooling is an auxiliary cooling method used to protect the chamber walls from localised heating. Even though a good number of research finding are there in the field of film cooling, a critical design and performance evaluation is required for thrust chambers with multi-element injectors functioning with LOX/CH4 propellant combination. The performance characteristics are deeply influenced by the chamber pressure, propellant inlet conditions and also configurations of the injectors. The near injector performance of the film coolant is deeply influenced by the combustion temperature, pressure and shape of the flame. Arnold et al. [1-4] has carried out a number of experimental studies on DLR P8 combustor B. Film cooling effects are tested with LOX-H2, and LOX-CH4 propellant combination. In their preliminary film cooling studies for a pressure range of 5-11.5MPa, a significant variation of circumferential wall temperature distribution and film cooling effectiveness was observed with GH2 as film coolant. Their results showed that tangential injection of film coolant in the vicinity of coaxial injector head has produced an effective reduction in wall temperature. They also tested the film cooling effectiveness in a nozzle segment of a subscale rocket combustion chamber and found that modified film cooling model is required to implement an additional correction factor for local Mach number. Their further investigation on DLR P8 combustor using LOx/GH2 showed that an increase of the film blowing rate results in a direct improvement of the local film cooling effectiveness. Their studies also confirmed the dominant influence of the film blowing rate on film cooling effectiveness. Suslov et al. [5] investigated the experimental results of film cooling performance in an Oxygen-Methane subscale combustion chamber. They explained the different complexities occurring due to film cooling like steep temperature gradients, turbulent mixing, and complexities in flow properties. Barbara Betti [6] studied numerically the effects of film cooling in hydrogen and methane thrust chambers. The capability of Reynolds Average Navier Stokes equation is verified using pseudo-injector approach. This approach well predicts the heat flux characterization, particularly at high chamber pressure. The data acquired and models developed during the initial period of In-Space Propulsion (ISP-1) programme were used by Ordonneau [7] to validate the models used in CFD simulations. Daimon et al. [8] experimentally and numerically investigated the film cooling aspects in a CH4/Oxygen subscale chamber with multi-injector elements. The effect of slot dimension on film cooling and the net heat flux reduction were analysed. Their results showed that mixing causes a low heat flux distribution near face plate along with high combustion efficiency.

In the present study, the film cooling effects of a multi-element injector thrust chamber are numerically investigated. The influence of film coolant on chamber wall temperature and wall heat flux are explored. Computationally cheap Eddy Dissipation Model (ED) along with single step reaction is used for numerical simulation.

2. Theoretical Formulation and Numerical Approach

In the present numerical study, the commercially available numerical tool ANSYS FLUENT 15.0 [9-10] is used. The conservation equations of mass, momentum, energy, and species equations are solved in fluid phase sequentially. The turbulence is modelled using k-ε standard. The thermodynamic properties of the fluids are taken from the National Institute of Standards and Technology (NIST) database [11].

2.1. Computational Domain

The present study was conducted on DLR Lampoldshausen test facility P 6.1 combustor 'B' [8]. The schematic of the thrust chamber is shown in figure 1. The thrust chamber consists of 6 segments, each 50 mm long. The first five segments constitute the combustion chamber, while the last segment performs as the nozzle. The diameter of the combustion chamber is 50 mm. The nozzle has a contraction ratio of 3.2 with throat diameter 28 mm. The chamber and the nozzle are cooled by water circulated in the cylindrical channels provided on the periphery of the segments.

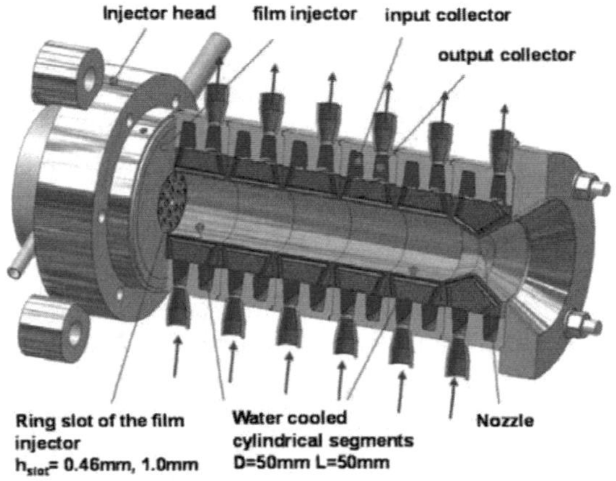

Figure 1. Schematic of the Thrust Chamber.

The injector plate has 15 coaxial injector elements. They are shown in figure 2. Oxygen is injected through the inner channel while fuel is injected through the outer ones. The injectors are placed in the face plate without recess. Among the 15 injectors, 5 injectors are placed on a pitch circle of diameter 19 mm and remaining are positioned on 38 mm pitch circle. The injectors are positioned uniformly on the pitch circle such that the elements form 5 identical triangles.

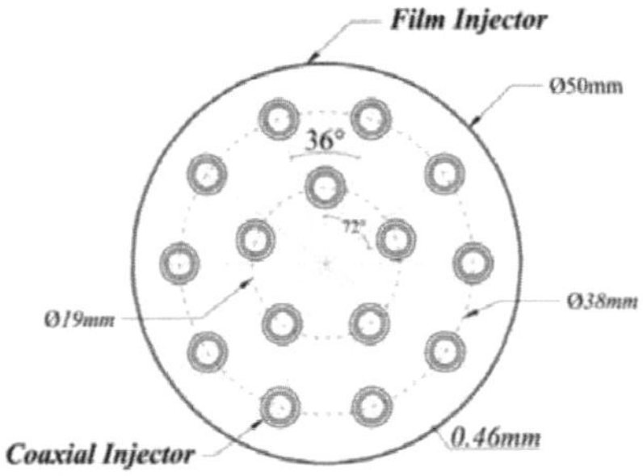

Figure 2. Injector face plate.

The film is injected through a continuous slot of 0.46 mm width. From the literature [5] it was found that the water flow rate was arranged such that a uniform wall temperature of 400 K is maintained on chamber lateral walls.

The computational domain was modelled using Design Modeler. The inlet for oxygen, fuel, and film are shown in figure 3. In order to reduce the size of the computational domain, only 72° of the thrust chamber was modelled. Periodic boundary condition was imposed on two sides of the domain. Periodic boundary condition was found most effective in modelling fluxes across the boundaries.

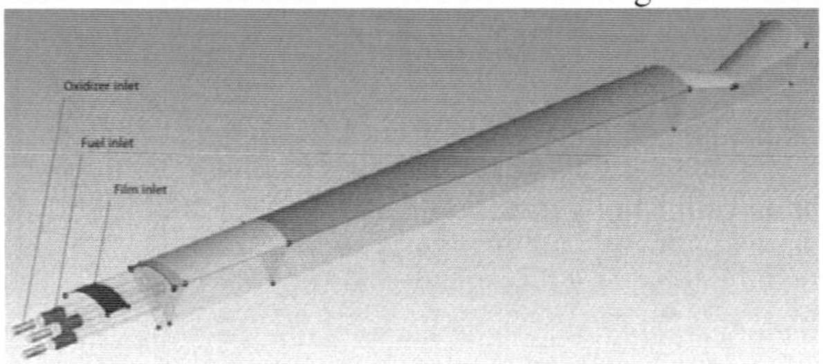

Figure 3. Computational Domain.

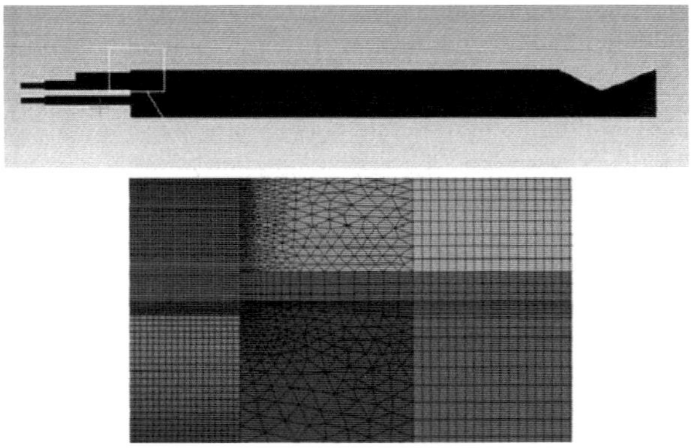

Figure 4. Generated Mesh.

The hybrid mesh was generated composing of hexahedral and tetrahedral elements. Fine elements were concentrated near the region of mixing. Tetrahedral elements were located up to a thickness of 5 mm from the faceplate. Inflation layers were created near the fluid-solid boundary for accuracy. A mesh having 1947112 elements is selected for the simulation after the grid independence study and the same is shown in figure 4.

3. Validation and Grid Independence Study

The results obtained from experimental studies by Daimon et al. [8] was chosen for validation of numerical scheme. The boundary conditions used for the validation study are shown in table 1.

Table 1. Boundary Conditions

Flow Parameters	Value
Fuel mass flow rate, kg/s	0.083
Fuel temperature, K	267
Oxidizer mass flow rate, kg/s	0.28
Oxidizer temperature, K	272
Film mass flow rate, kg/s	0.073
Film temperature, K	281
Velocity ratio (Vfilm/Vfuel)	1.08

Journal of Physics: Conference Series **1355** (2019) 012009 doi:10.1088/1742-6596/1355/1/012009

A mixture ratio of 3.4 is used for the simulation. The real gas effects are predicted by using Soave-Redlich-Kwong equation of state. Pressure based solver was used for the simulation. Computationally cheap Eddy dissipation (ED) was chosen as the reaction mechanism. The standard k-ε model was used for modelling turbulence. Constant wall temperature of 400 K was imposed on the outer wall of the thrust chamber.

The grid independence was checked with element sizes of 501273,593277, 1244477, 1947112, 2550000. The results of chamber pressure and wall heat flux were chosen to predict the grid independence. Among the tested grids a mesh with 1947112 elements showed good agreement with experiment results with maximum error of 18% in the case of wall heat flux. The details are illustrated in Figure 5.

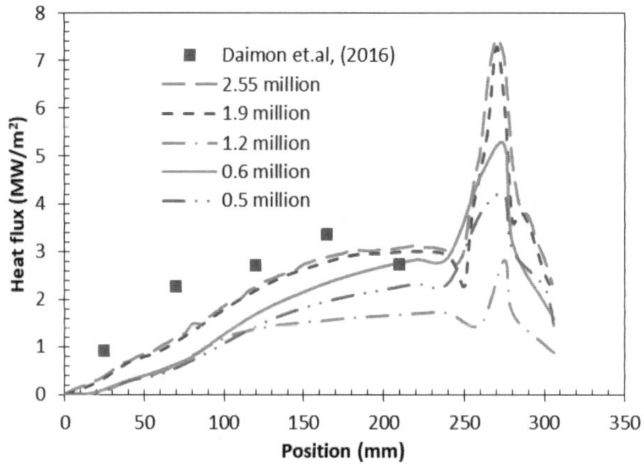

Figure 5. Details of Grid Independence Study

4. Results and Discussion

Most of the high-pressure rocket engines are provided with film cooling in addition to regenerative cooling to protect the wall surfaces from sudden heating. In the present study, efforts are taken to understand the influence of film coolant on wall heat flux and chamber temperature on PLR D6.1 subscale rocket thrust chamber at supercritical pressure.

The mass flow rates of oxygen and methane are 1.683 kg/s with 0.495 kg/s respectively. The film coolant methane is injected at a mass flow rate of 0.073 kg/s. Here the parametric studies were carried out by varying the oxygen inlet temperature. The chamber pressure lies in the near critical pressures of methane and oxygen where drastic variations in properties are observed. At 120K and 6 MPa density oxygen is 998 kg/m^3 and is approximately equal to that of water at 1 atmospheric pressure. Studies are conducted at three oxidizer inlet temperatures of 272K, 150K and 120K. In the second and third cases the injected oxygen is undergoing transcritical to supercritical transformation.

4.1. Effect of Inlet Temperature of Oxygen on Wall Temperature

Thrust chamber wall temperature increases gradually and shows a peak at the throat and then reduces due to thermodynamic expansion. In all the three cases the outer surface of the thrust chamber wall is imposed with a wall temperature of 400K assuming uniform regenerative cooling. In all the cases the wall temperatures are within safe limit. The maximum wall temperature obtained for 272 K is 509K and that for the 120K is 537K. This result indicates that regenerative cooling plays a vital role in maintaining the wall temperature under safer operating condition. When oxygen is injected at supercritical state, combustion occurs faster this is reflected in the graph. In the case of transcritical injection combustion occurs in the shear layer between the dense core of oxidizer and lighter fuel core. A slight delay in combustion is happening and due to this delayed combustion, the wall temperature increases slower in the case of transcritical combustion. The variation in wall temperature are illustrated in figure 6.

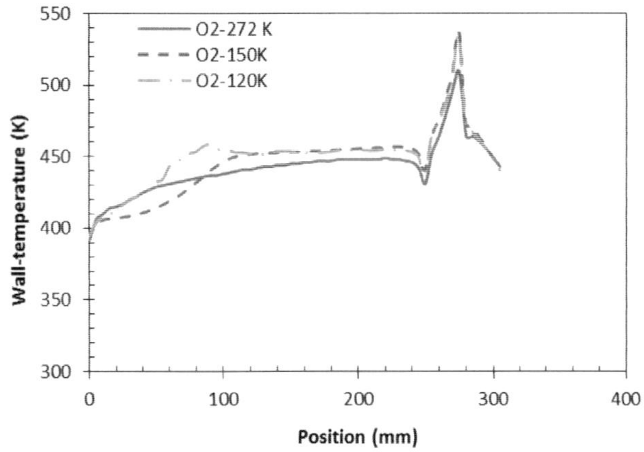

Figure 6. Variation in wall temperature for Isothermal wall Condition.

4.2. Effect of Inlet Temperature of Oxygen on Film Deterioration

Film cooling is an auxiliary cooling method used to protect the thrust chamber wall surfaces from localized heating caused due to sudden heat flux variations. Here the film coolant is injected at an inlet temperature of 281K. Figure 7 shows the consumption of film coolant methane along the axial direction measured 1mm below the thrust chamber. It can be clearly seen that in the case of cryogenic injection a stable film is maintained till the end of the thrust chamber. Primary recirculation occurring between the shear layers of propellants near the injector outlet, directs a part of the fuel towards the chamber wall. At the nozzle exit the mass fraction of methane is 0.1 for cryogenic injection cases; and for 272K injection case it is 0.03 only. Similar trend is obtained for mass fraction of methane at 2mm below chamber wall in figure 8.

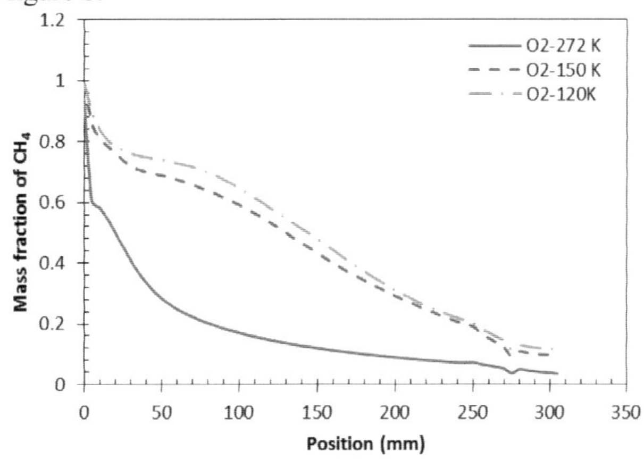

Figure 7. Consumption of methane in axial direction 1mm below chamber wall.

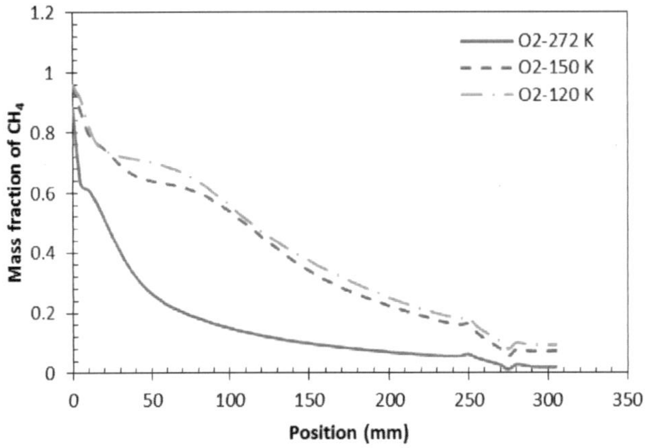

Figure 8. Consumption of methane in axial direction 2mm below chamber wall.

Studies indicate that due to the fuel injection pattern there is a layer of methane near the inner surface of the thrust chamber even without film cooling (figure 9). When oxygen is injected at 120 K and without film coolant, a significant amount of methane present near the wall is contributed by the upper injector elements. The mass fraction of methane near the wall increases with film cooling.

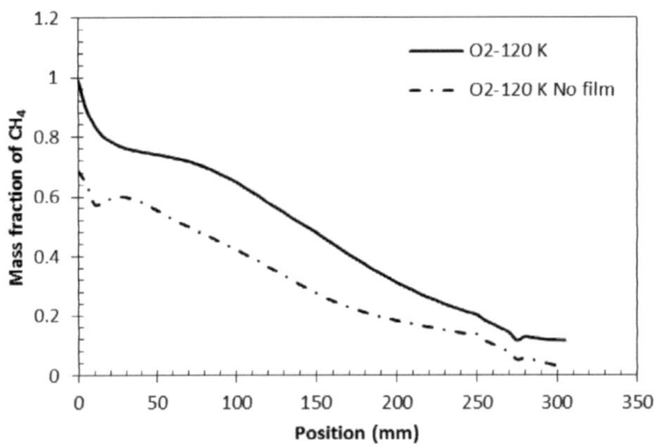

Figure 9. Comparison of methane mass fraction near wall with and without film cooling, oxygen temperature 120 K.

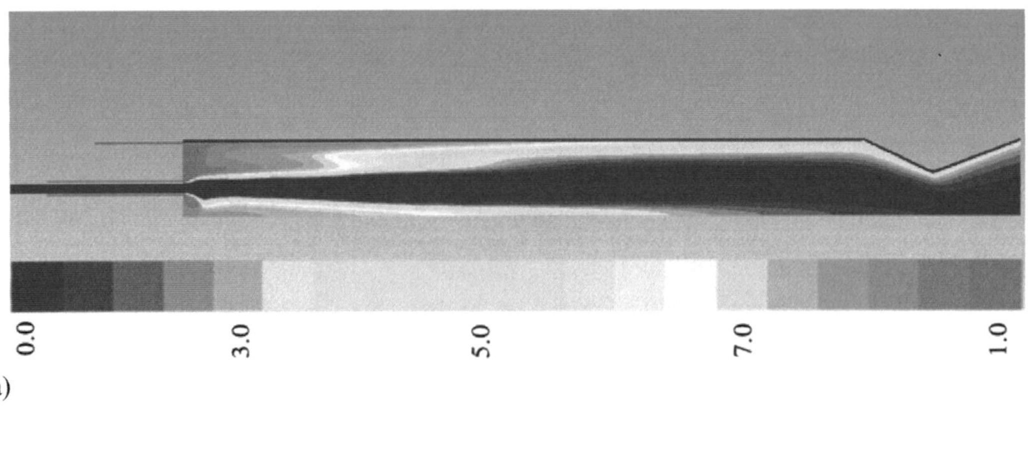

a)

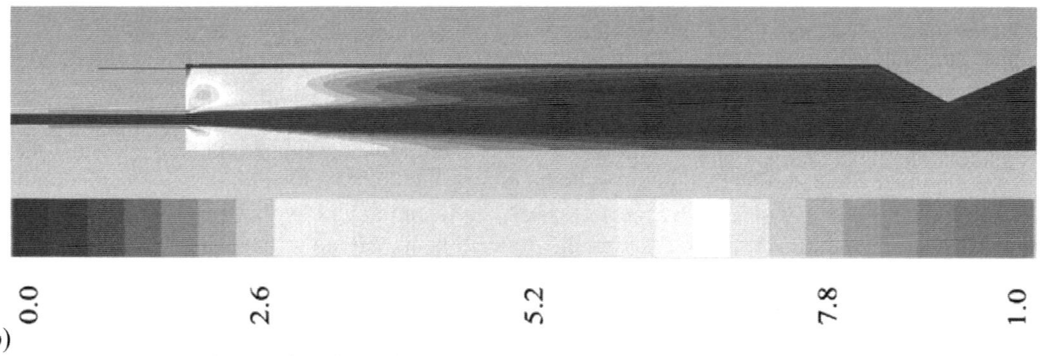

b)

Figure 10. Contour of mass fraction of CH4, (a) oxygen temperature 120 K (b) oxygen temperature 272 K.

The figure 10 shows the contour of mass fraction of CH4 on the plane passing through the centre of an inner injector element. In the case of cryogenic injection a significant amount of methane mass fraction is observed near the wall and this may affect the combustion efficiency.

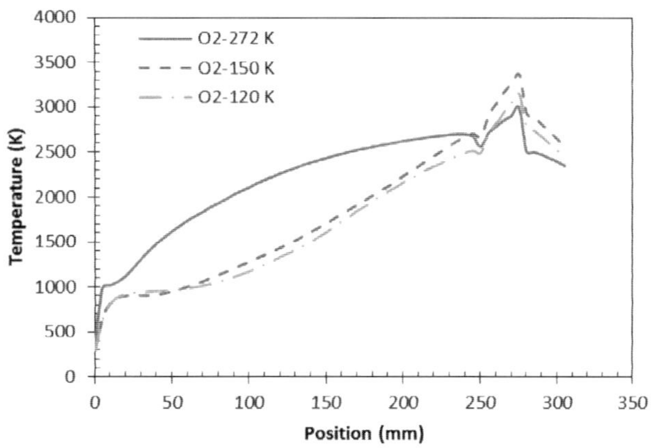

Figure 11. Variation of hot gas temperature in axial direction 1mm from the wall.

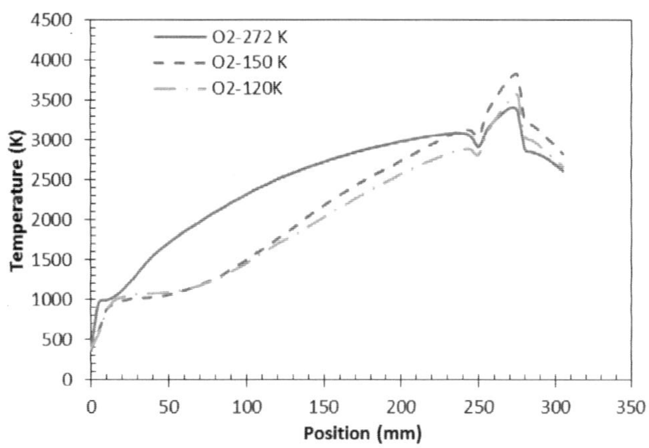

Figure 12. Variation of hot gas temperature in axial direction 2mm from the wall.

Hot gas temperature at distances of 1mm and 2mm from the wall is depicted in figure 11 and figure 12 respectively. In all the cases the hot gas temperature increases up to the throat then it decreases further downstream due to the thermodynamic expansion of gases. Hot gas temperature near the wall in the combustion chamber is high at supercritical combustion. In the case of transcritical injection due to delayed combustion, the hot gas temperature reaches the maximum value at a slower rate than supercritical injection.

4.3. Effect of Inlet Temperature of Oxygen on Wall Heat Flux

Once the oxygen is injected at 272K the wall heat flux increase gradually and reaches maximum at throat and decreases in the divergent section due to expansion. Figure 13 shows the average wall heat flux distribution along the axial direction. When oxidizer is injected at 272 K, the chamber pressure developed is 5.33MPa and the maximum value of heat flux at throat is 38 MW/m^2. In the cases of transcritical injection, upto an axial distance of 70mm the wall heat flux is lower compared to that at 272K. This reduction in wall heat flux is due to the influence of strong primary recirculation at cryogenic injection. For the inlet temperature of 120 K the density of oxygen is approximately 1000 kg/m^3 and due to this high density the combustion occurs in the shear layer between dense core of oxygen and methane. Once the propellant oxygen is injected at cryogenic state the combustion chamber pressure is raised to 6MPa. Similar wall heat flux conditions are observed for the injection temperature of 150K.

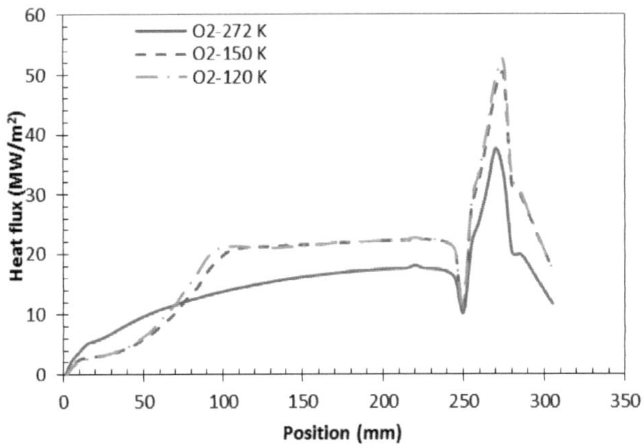

Figure 13. Variation of wall heat flux with oxidizer temperature.

5. Conclusions

Numerical studies on film cooled subscale thrust chamber was carried out to study the effect of inlet temperature of oxygen on wall temperature and wall heat flux. The numerical scheme is validated comparing the results with experimental results reported in literature. The combustion model used is single step eddy dissipation model and the turbulent model used is standard k-ε. The following conclusions are derived from the study.

For combustion where oxygen is undergoing transcritical to supercritical transformation, the methane concentration near the wall increases. A significant amount of the methane mass fraction near the chamber wall is contributed by the upper injector elements. The hot gas temperature near the wall is higher in the transcritical combustion compared to supercritical combustion. In the case of cryogenic injection a stable coolant film is maintained till the end of the thrust chamber. In the cases of transcritical injection, up to an axial distance of 70mm the wall heat flux is lower compared to that of supercritical injection.

References

[1] Arnold R, Suslov D and Haidn O J 2008 *proc. 44th AIAA/ASME/SAE/ASEE Joint Propulsion Conference & Exhibit* (Hartford) p 5242.
[2] Arnold R, Suslov D and Haidn O J 2009 *Journal of propulsion and power* **25** p 443.
[3] Arnold R, Suslov D I , and Haidn O J 2009 *Transactions of the Japan society for aeronautical and space sciences* (Japan) **7** p 81.
[4] Arnold R, Suslov D I, and Haidn O J 2010 *journal of propulsion and power* **26** p 428
[5] Suslov D, Betti B, Aichner T, Soller S, Nasuti F, Haidn O J 2012 *In Space Propulsion Conference.*
[6] Betti B, Onofri M, and Nasuti F 2012 *proc. 48th AIAA/ASME/SAE/ASEE Joint Propulsion Conference & Exhibit* (Atlanta: Georgia) p 3907.
[7] Ordonneau G 2013 *proc. 5th European conference for aeronautics and space sciences* p 11.
[8] Daimon Y, Negishi H, Koshi M, and Suslov D 2016 *Progress in Propulsion Physics* **8** p 129.
[9] Ansys Fluent 15.0 User's Guide 2013 (Ansys Inc: Canonsburg USA).
[10] Ansys Meshing 15.0 User's Guide 2013 (Ansys, Inc., Canonsburg USA).
[11] NIST Standard Reference Database 69.

Enhancement of the thermal performance of a loop heat pipe using silica-water nanofluid

Jobin Jose[1], Rajesh Baby[2*]

[1]Graduate Student, [2]Associate Professor
Department of Mechanical Engineering, St. Joseph's College of Engineering and Technology, Palai, Kerala, India.

*E-mail: rbaby55@gmail.com

Abstract. Loop heat pipes (LHPs) are heat transfer devices which are capable of transferring large amount of heat energy over long distances. This paper presents the results of numerical investigations carried out to determine the effect of silica-water nanofluid on the thermal performance of a LHP. Silica-water nanofluid at 2 % nanoparticle concentration and deionised water are considered as working fluids. The power input is varied from 40 to 320 W in steps of 40 W. The numerical results are validated with experimental data available in the literature. LHP using deionsed water is used as the baseline case. The results shows that silica-water nanofluid can significantly enhance the thermal performance of the LHP when compared with the baseline case. The evaporator temperature of LHP using nanofluid reduced by 27 % than the baseline case. The LHP attained steady state faster (150 s earlier) with the presence of silica nanoparticles due to enhanced thermal properties. Thus silica-water nanofluid based loop heat pipes proves its potential as a suitable working fluid for a wide range of applications.

Keywords : Loop heat pipe; nanofluid; thermal performance; thermal resistance; evaporator temperature

1. Introduction

The loop heat pipes (LHPs) operates on a multi phase fluid flow cycle which is maintained by a capillary medium in the evaporator. It is a heat transfer device which can transmit a large amount of heat over a long distance with negligible temperature difference. In steady state operation these heat pipes have high thermal conductance. Thus, LHPs are also known as super thermal conductors. Their advantages include electricity free operation, no moving parts and ability to work even with small temperature difference and transfer heat energy over long distances with negligible pressure drop [1].

Alekey *et al.* [2] numerically investigated 3D model of LHP evaporator to analyse the flow and conjugate heat and mass transfer. The simulation results shows that evaporation mainly occurs at the vapour groove corners close to the evaporator. To determine thermodynamic behaviour of LHP a steady state model was developed by Benjamin *et al.* [3]. They reported the major effect of evaporation coefficient and wick conductivity on the LHP evaporator temperature and its temperature field. Esarte *et al.* [4] experimentally investigated LHP to develop a new model for cooling diodes which emits high power. It was based on the steady state energy balance equations for each components. The effect of various parameters such as tube radius and length was also studied. Mariya *et al.* [5] numerically investigated the LHP to develop a 3D model of heat and mass transfer in the flat evaporator. The results were used for getting heat exchange dependences of heat load for various orientations of the evaporator. Jobin *et al.* [6] conducted a review on recent advances on loop heat pipes. It was concluded that investigations based on LHPs using nanofluids were scarce. The thermal performance of LHP using alumina nanoparticles were experimentally investigated by Akshay *et al.* [7]. It was found that the evaporator temperature can be reduced by 12% using nanofluids than deionised water. Riehl [8] studied the effect of nickel–water nanofluid in a LHP. Result showed that heat transfer coefficient was lower at the evaporator side and operating temperatures were higher throughout the LHP. Masahito *et al.* [9] also developed a mathematical model to investigate the heat transfer characteristics of evaporator in LHP with a small gap between case and wick. Also the effect of this gap on the thermal properties was discussed. It was found that microgap is an efficient approach for improving thermal performance.

Gunnasegaran *et al.* [10, 11] conducted experimental and numerical investigations on LHPs with alumina and silica as nanofluids. It was observed that the thermal resistance decreases when using silica-water nanofluid. The LHP charged with alumina-water nanofluid yields lower wall temperature difference between evaporator and condenser and attains steady state faster. Nookarajua *et al.* [12] experimentally and numerically investigated the thermal performance of LHP using copper sintered wick and compared with deionised water. From the results it can be concluded that as heat input increases the efficiency decreases.

The application of nanofluids in LHP is a challenging and fertile area of research, which is still in its nascent stages. Literature on LHPs are scarce and the use of nanofluids as the working medium in LHPs are scarcer. The present study deals with the numerical investigations on a LHP using silica-water nanofluid at 2% nanoparticle concentration. Heat input ranges from 40 to 320 W in steps of 40 W. The thermal performance of the LHPs using nanofluid is compared with that of deionised water (baseline case).

2. Numerical modelling

A three dimensional model (3D) of LHP is designed as shown in the figure 1. The detailed specifications of LHP is shown in table 1. The simulation is performed in the commercial software package ANSYS.

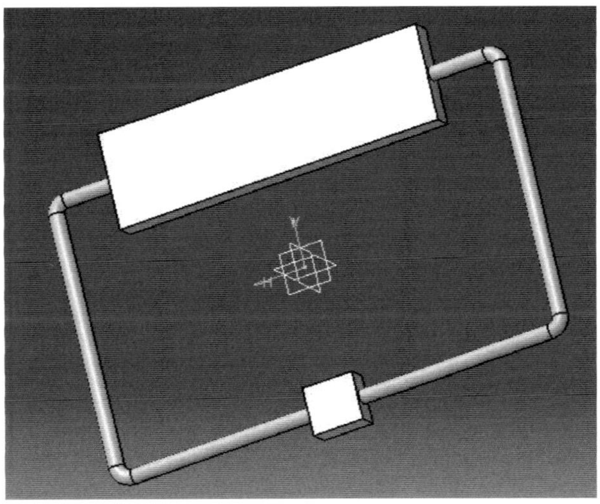

Figure 1. Model of LHP used in the present study

The 3D model of LHP is made in Catia V5 and is exported to ANSYS 14. The heat flux of 23000 W/m^2 is applied on the evaporator base. The initial temperature is taken as 22 ^{0}C. After modeling the LHP it is meshed with an element size of 2,40,514 as shown in figure 2.

Table 1. Specifications of LHP used for the numerical analysis

Specification	Dimension/material
Evaporator	
Dimension (mm)	L45 × W45 × H22
Compensation chamber	
Dimension (mm)	L45 × W38 × H22
Material	Copper
Condenser	
Dimension (mm)	L315 × W96 × H60
Material	Acrylic
Vapour line	
Outlet diameter (mm)	13
Inlet diameter (mm)	11
Length (mm)	490
Liquid line	
Outlet diameter (mm)	13
Inlet diameter (mm)	11
Length (mm)	490

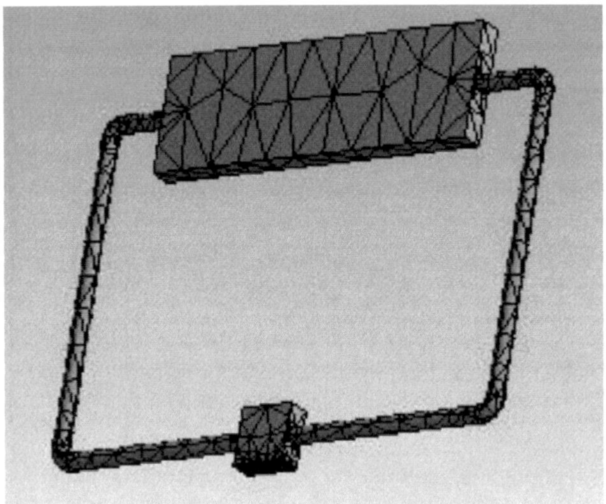

Figure 2. Meshed model of LHP

3. Equations used

During operation, heat input into the evaporator is absorbed by the coated surface of the evaporator and the coated surface of the evaporator will absorb the heat input provided. Part of the absorbed energy will be dispersed back to the ambient while the remaining energy will be absorbed by the heat pipe fluid. This process is expressed as:

$$Q_e = Q_i - Q_{loss,e} \tag{1}$$

Where Q_e = heat absorbed by evaporator and Q_i = heat input. Assuming the evaporator surface temperature to be T_e, the heat loss $Q_{loss,e}$ from the evaporator to the ambient is expressed as [13]:

$$Q_{loss,e} = \frac{T_e - T_{amb}}{R_{amb}} \tag{2}$$

$$R_{amb} = \left(\frac{1}{h_{amb}} + \frac{1}{\sigma\, \varepsilon_e (T_e + T_{amb})(T_e^2 + T_{amb}^2)}\right)\frac{1}{A_e N_e} \tag{3}$$

Where T_{amb} = ambient temperature, A_e = evaporator area, N_e = number/quantity, σ = Stefan Boltzmann constant, ε_e= emissivity of evaporator, h_{amb} = ambient convective heat transfer coefficient and R_{amb} = thermal resistance by ambient. The absorbed heat Q_e is instantaneously taken away by means of the heat pipes loop. This could be achieved by condensation of vapour on the condenser surface and evaporation of liquid in the evaporator section. This part of heat $Q_{e\text{-}c}$ is written as [13, 14]:

$$Q_{e\text{-}c} = \frac{T_e - T_c}{R_e + R_w + R_v + R_f + R_c} \tag{4}$$

Where R_e, R_w, R_v, R_f, and R_c represents thermal resistance of evaporator, wick, vapour line, and condenser respectively. Since the condenser temperature is high, certain amount of the heat which is transferred to the condenser surface will be dissipated to the ambient. The heat output from the condenser Q_o is expressed as:

$$Q_o = Q_c - Q_{loss,c} \tag{5}$$

Assuming the condenser surface with the temperature of T_c, $Q_{loss,c}$ which is the heat dispersion from the condenser to the ambient could be similarly calculated by using equation (2) .

4. Grid independence study and validation

Grid independence study was performed to analyse the influence of mesh density on the numerical results. The maximum evaporator temperature is measured at different mesh densities. Among them four grid sizes were selected considering the variation of obtained results from experimental results available in the literature [10]. The maximum evaporator temperature with different mesh densities are listed in table 2. The maximum evaporator temperature obtained from the present study is 98.15 °C for a mesh with 2,40,514 grid size and that reported in the experimental work is 101.24 °C. So, the grid size corresponding to 98.15 °C is used for further numerical investigations.

Table 2. Grid size used for grid independence check

Sl. No.	Grid size	Maximum evaporator temperature (°C)	Percentage deviation from the experimental result (%)
1	2,01,218	107.92	10.26
2	2,40,514	98.15	3.13
3	2,79,438	104.89	5.83
4	3,08,969	103.98	4.25

5. Results and discussions

5.1. Effect of heat input on evaporator temperature

Figure 3 represents the effect of heat input on evaporator temperature of LHP using deionised water and silica-water nanofluid at 2% nanoparticle concentration. The heat input is varied from 40-320 W in steps of 40 W. With the increase in heat input, the evaporator temperature of LHP increased in both cases. But it is lower for LHP using nanofluid. The maximum evaporator temperature obtained by LHP is 172.31 ^{0}Cand 138.25 ^{0}C for deionised water and silica-water nanofluid respectively. About 27% reduction in evaporator temperature can be obtained using nanofluid. It is due to better thermal properties of silica nanoparticles.

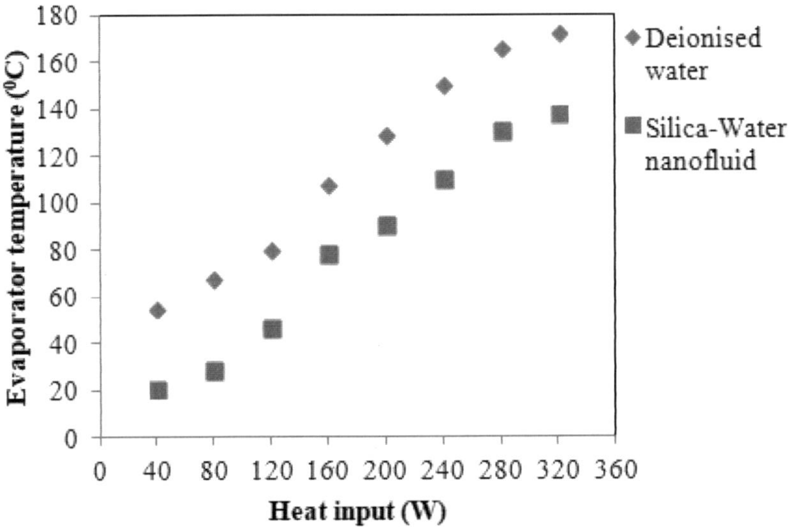

Figure 3. Heat input versus evaporator temperature of LHP

5.2. Effect of silica-water nanofluid on transient temperature distribution of LHP

Figure 4 shows the transient temperature distribution of LHP using deionised water and silica–water nanofluid at 2% nanoparticle concentration. It can be observed the LHP using deionised water is having the highest evaporator temperature. The temperature of all points diminished for LHP when nanofluid is used. This is due to the enhanced thermal properties of silica nanoparticle than deionised water. The LHP takes 800 s to attain steady state when deionised water is used whereas it takes only 650 s to attain steady state when silica-water nanofluid is used.

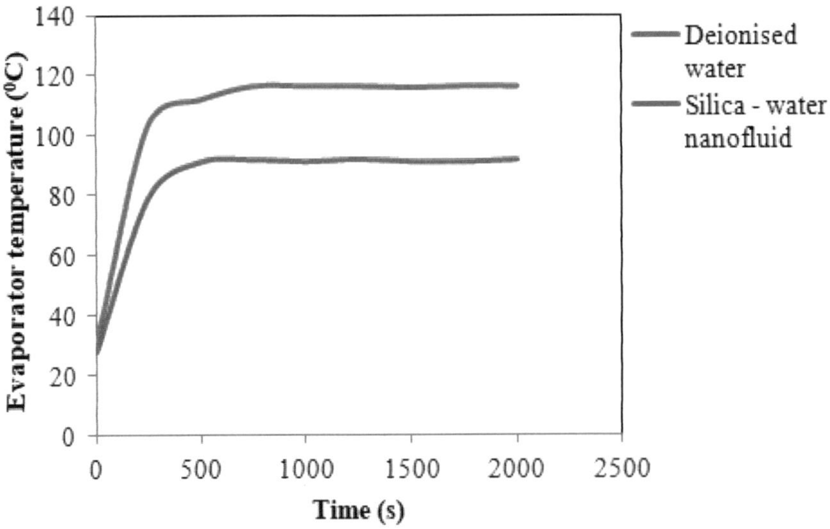

Figure 4. Transient temperature distribution of LHP

6. Conclusions

A loop heat pipe was numerically investigated to determine its thermal characteristics using silica-water nanofluid for a heat load of 40-320 W in steps of 40 W and nanoparticle concentration of 2%. The following conclusions are drawn from this work:

1. Evaporator temperature of LHP increases with increase in heat input. But it is lower for LHP using nanofluid than baseline case.
2. Silica-water nanofluid reduced the evaporator temperature of LHP significantly than deionised water. Nearly, 27 % reduction in evaporator temperature is obtained
3. LHP using silica–water nanofluid attained steady state faster than baseline case. The steady state is reached by LHP about 150 s earlier in comparison with the baseline.

References

[1] Adoni A A, Ambirajan A, Jasvanth V S, Kumar D and Dutta P 2010 Theoretical studies of hard filling in loop heat pipes *Journal of Thermophysics and Heat Transfer* **24** pp. 173–183.

[2] Aleksey A. Pozhilov, Dmitri K.Zaitsev, Evgueni M Smirnov and Aleksander A Smirnovsky 2017 Numerical simulation of heat and mass transfer in a 3D model of a loop heat pipe evaporator *St. Petersburg Polytechnical University Journal: Physics and Mathematics* **3** pp. 210-217.

[3] Benjamin Siedel, Valerie Sartre and Frederic Lefevre 2013 Numerical investigation of the thermohydraulic behaviour of a complete loop heat pipe *Applied Thermal Engineering* **61** pp. 541-553.

[4] Esarte J ,Bernardini A, Blanco J M and Sancibrian R 2016 Optimizing the design for a two-phase cooling loop heat pipe: Part A: Numerical model, validation and application to a case study *Applied Thermal Engineering* **99** pp. 892-904.

[5] Mariya A. Chernysheva and Yury F Maydanik 2012 3D - model for heat and mass transfer simulation in Flat evaporator of copper-water loop heat pipe *Applied Thermal Engineering* **34** pp. 124- 134.

[6] Jobin Jose and Rajesh Baby 2018 Recent advances in loop heat pipes: A review *IOP Conf. Series Material Science and Engineering* **396** pp. 1-8.

[7] Akshay J, Aswin J Joseph, Jobin Jose and Rajesh Baby 2018 Enhancement of the thermal performance of a loop heat pipe using alumina–water nanofluid: An experimental investigation *IOP Conf. Series Material Science and Engineering* **396** pp. 1-7.

[8] Riehl R R 2012 Heat transport behaviour of a miniature loop heat pipe using water–nickel nanofluid *Heat Pipe Science and Technology* **1** pp. 83–96.

[9] Masahito Nishikawara and Hosei Nagano 2016 Numerical simulation of capillary evaporator with microgap in a loop heat pipe *International Journal of Thermal Sciences* **102** pp. 39-46.

[10] Gunnasegaran P, Abdullah M Z and Shuaib N H 2013 Influence of nanofluid on heat transfer in a loop heat pipe *International Communications in Heat and Mass Transfer* **47** pp. 82–91.

[11] Gunnasegaran P, Abdullah and Yusoff M Z 2014 Effect of alumina-water nanofluid concentration on heat transfer in a loop heat pipe *Int. conf. on advances in Manufacturing and Materials Engineering* **5** pp. 137 – 146.

[12] Nookarajua B Ch, Kurma Rao P S V and Naga Saradac S 2017 Experimental and Numerical investigation on enhancement in thermal characteristics of Sintered Copper wick Heat pipe using deionized water as fluid Proceedings *5th Int. Con. of Materials Processing and Characterization* **4** pp. 1321–1331.

[13] Department of Energy 1992 *DOE fundamentals handbook: Thermodynamics, heat transfer and fluid flow* (U S)

[14] Riffat S B, Zhao X and Doherty P S 2005 Developing a theoretical model to investigate thermal performance of a thin membrane heat-pipe solar collector *Applied Thermal Engineering* **25** pp. 899-915.

Experimental study on heat transfer characteristics of liquid nitrogen through coated transfer line

Bindu S S[1], Aromal[2] and K E Reby Roy[3]
[1] Department of Mechanical Engineering, TKM college of Engineering, Kollam.
[2] PG Scholar, Department of Mechanical Engineering, TKM college of Engineering.
[3] Assistant Professor, TKM college of Engineering, Kollam

E-mail: bindusankar@tkmce.ac.in

Abstract. The effective and efficient utilisation of cryogen is made possible by optimised transfer lines which ensures reduction in chilldown time and cryogen usage. Cryogenic chill-down experiments were conducted on coated and regular copper tubes. Liquid nitrogen at different mass flux condition was used as the working fluid and Polyurethane coatings belonging to the family of Teflon was used as coating material. The performance of the regular and coated copper tubes at corresponding mass flux were compared. An increase in heat flux and heat transfer coefficient along with reduction in chilldown time was obtained by providing a low thermal conductive coating between the cryogen and wall material.

1. Introduction

Cryogen are widely being used in many industries for the purpose of food processing, material recycling, biological preservation etc. The cryogens will be stored in Dewar vessels and transported to the point of usage through transfer lines made of different materials. The process of Cryogen transport is so complicated since it incorporates intricate heat transfer between the wall material and cryogen, resulting in two phase flow and pressure surges. The identification of flow and heat transfer characteristics is badly at need in those industries which make of cryogens. Since the process involve two phase flow, the detailed heat transfer occurred during transport of cryogens are not well understood. Under chilldown conditions, a steady state will be attained between the transfer line walls and the cryogen being transported. A broad research on the flow structure and heat transfer characteristics are required to develop widespread understanding of this complicated phenomena. Earlier research works suggest reduction in time taken for chill down process during the cryogen transport across coated transfer lines[1]. Present study deals with heat transfer characteristics in coated transfer lines made of copper, coated with a low thermal conductive coating belonging to the family of Teflon.

2. Literature Review

According to Cowley et al.[2] , reduction in chilldown time can be attained by incorporating a low thermal conductive coating between the cryogen and pipe wall material. According to Allen et al. [3]response of cryogenic systems can be improved by providing such internal coatings on cryo-panels which in turn reduce the time taken for chilldown. Reed et al.[4] made a comparison between regular and internally coated propellant lines and identified faster cool down in coated lines.

In room temperature liquids, the modification in transition boiling could be attained by interposing thin insulating layers on transfer line wall [5]. Gaertner [6] witnessed reticence of Leidenfrost phenomenon in coated cryogen transfer lines. According to Leonard et al. [7] coated line chilldown is independent of cryogen flow rate, where shift to the nucleate boiling region is attained earlier. After a short period of time steady state will be attained there by indicating rapid cool down. Driester [8] conducted heat transfer enhancement using low thermal conductive materials which allows deep filling of the conduits with cryogens. This effectively shorten the time taken for chilldown to three times and will also significantly diminish the heat transfer surface of cryogenic liquid evaporators. The use of low conductive coatings permits to obtain the FB crisis at considerable temperature drops and its transition with more effective heat transfer. The use of low-conductive coatings on the internal tube surface resulted in thrice reduction of tube precooling time.

Hong Hu et al. [9] performed an experiment to evaluate the modification and enhancement on the quenching heat transfer by a nano porous heat transfer surface. The results indicated that the nano porous surface completely modified and enhanced the phase-change heat transfer in all three quenching regimes. Enhancements are mainly attributed to the super hydrophilic property and nano scale nucleation sites offered by the nonporous surface. Kim et al.[10] Studied the effect of surface roughness on pool boiling heat transfer coefficient and critical heat flux of copper surface having moderate wettability. It was confirmed that surface roughness enhances nucleate boiling heat transfer because of the increased number of active nucleation sites. CHF exhibited considerable augmentation with increasing surface roughness. In the same year Li-Wu et al. [11] conducted an experiment on super hydrophilic surface realized by enhancing the surface roughness with silica nano-particles on SS spheres following the spray coating method. The application of such surface to quenching in water was tested under both saturated and sub cooled conditions. It was shown that quenching is accelerated on the super hydrophilic surface by a factor of 60%, due to the marked boiling heat transfer enhancement with the critical heat flux being increased by 78%.

In 2017, Daisuke et al. [12] experimentally investigated a method for reducing the time and total mass of cryogenic fluid required for a chill down process for piping with Poly tetra fluoro ethylene coating. The results indicated that the temperature of the minimum heat flux point was higher for the pipe with insulating layer. This resulted in the decrease in chill down time and total mass of LN2 consumed in the chill down process. In the same year Ran Li and Zhongwei Huang [13]developed a new CHF model for saturated pool boiling on surfaces with micro-scale roughness, including micro-pillar, micro-ridge structures as well as random roughness made by emery paper or sand papers. The model accounted for the effects of roughness-augmented wettability and capillary wicking on CHF enhancement. In 2017, Shoji Moria and Yoshio Utaka [14] proposed numerous surface modifications to enhance the critical heat flux (CHF)in a saturated pool boiling. CHF enhancement is obtained as a result of effects of extended surface area, nucleation site density, wettability, capillary wicking, and wavelength decrease based on the modified Zuber hydrodynamic stability model.

Therefore, if the existence of a liquid meta-stability limit is presumed, a minimum in cool down time may be predicted to occur at an optimum coating thickness and is in qualitative agreement with observed behavior. These observations showed that it is essential to lower the coating surface temperature close to the meta-stability limit and to know more about time-averaged values of the vapor heat transfer coefficient. Therefore, further investigations have been undertaken with a technologically feasible coating material. Considering the theoretical aspects of the vapor film formation process, subsequently with the test specimens and experimental procedures, results and discussion has been made and arrived at conclusions concerning the unknown thermo hydrodynamic details of the cool down process.

3. Experiment work

Figure 1 shows the schematic of experimental set up which consists of Liquid Nitrogen Dewar vessel, Gaseous Nitrogen cylinder pressure regulator, valves, DAQ system, electric heater and flow meter. The test sections used for the experiment, was a normal copper tube [internal diameter 5/16 inches (0.0079m) OFHC Copper tubes (UNS C10100)] and second the copper tube with its interior coated with a thermally

Journal of Physics: Conference Series 1355 (2019) 012011 doi:10.1088/1742-6596/1355/1/012011

low conductive polyurethane coating. The pipe was coated with a simple method of fill-and-drain, creating a very thin coating. The tube was then placed vertically overnight to dry and solidify the coat. Thermocouples were attached at lengths 0.02 m, 0.07m, 0.12 m, and 0.17 m from the inlet side. At each point three thermocouples are attached at 120° apart. The heat in-leak to the test section was minimized using a thick layer of polyurethane foam insulation (thermal conductivity of 0.02 W/m K, density of 11 kg/m3) which is a type of expanded foam insulation. Test section was also covered by yarn and then by Nitrile rubber insulation.

Liquid Nitrogen was supplied to the test section through 1/2" SS 304 grade pipes and brass fittings from a 55 L Dewar made by IBP Co. Limited (TA-55). The liquid nitrogen was pumped from Dewar vessel by external pressurization using a gaseous nitrogen cylinder and having 47 L capacity. Flow rate of liquid Nitrogen was manually regulated by a pressure regulator mounted on external pressurization line. Initially the entire test section was purged with gaseous nitrogen. A by-pass line is introduced to ensure the entry of saturated LN2 into the test section. Temperature measurement of the test section was

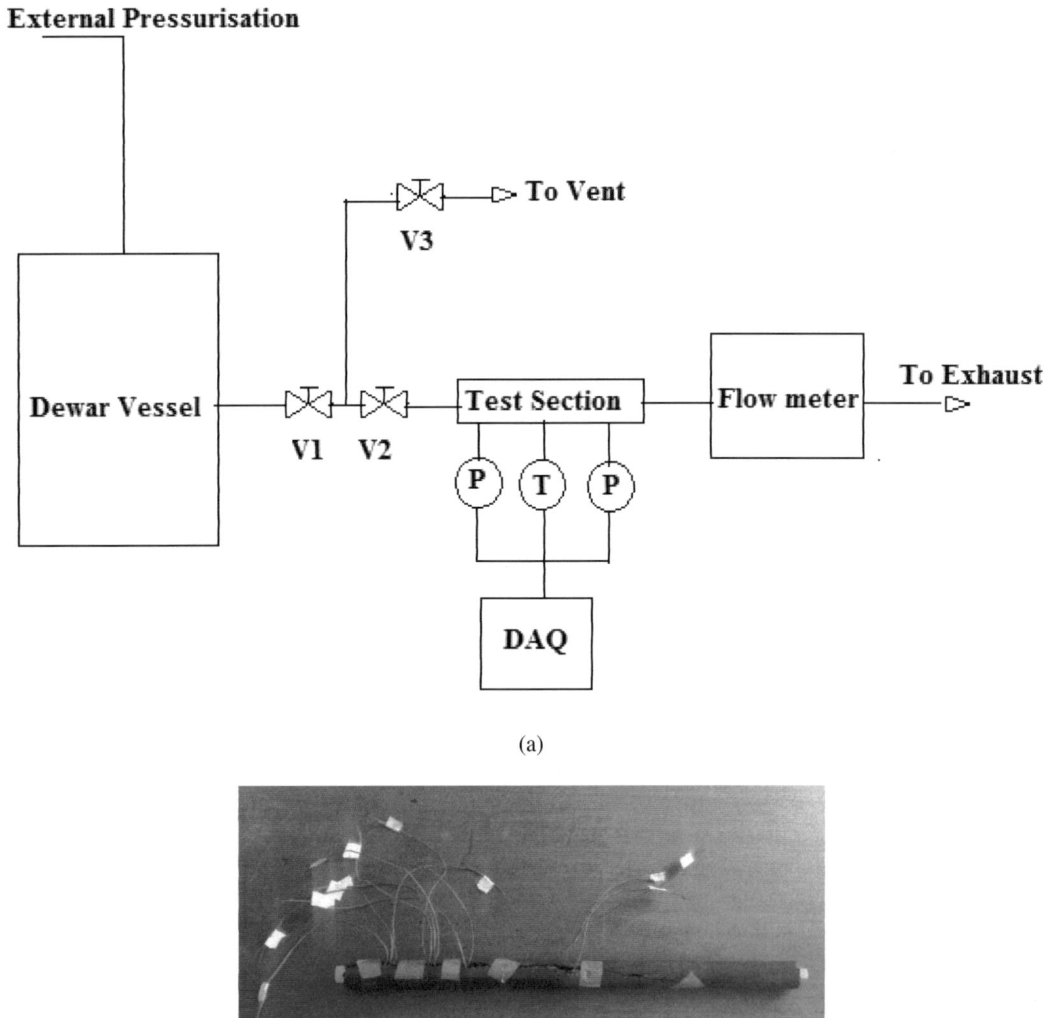

(a)

(b)

Figure 1. (a) Schematic of Experiment Set up. (b) Test section.

done using T-type thermocouples connected to Keysight 34972A data acquisition / data logger switch unit with scan frequency of 30 milliseconds which was also connected to the personal computer for data analysis. The average mass flux was measured using a Single phase volume flow meter having an accuracy of 0.05 mm3/s at the exit line from the test section. To ensure single phase gas flow entering the flow meter, the outlet line is placed in hot water bath heated by electric heaters. Pressure gauge is provided for measuring and regulating initial pressure. Different flow control valves are provided to control the flow.

4. Results and discussion

4.1 Time- Temperature comparison for different mass flow rates

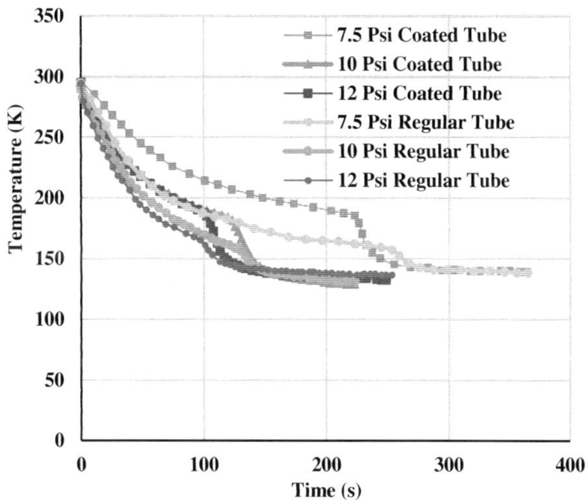

Figure 2. Temperature profile at different inlet pressures for coated and regular copper tube.

Figure 2 shows temperature profile for the coated and regular copper tubes at different mass flow rate conditions of 66 kg/ m^{2}s, 86 kg/ m^{2}s and 102 kg/ m^{2}s. The temperature profile is useful in identifying the chilldown effectiveness of coated and uncoated surfaces and it determines the rate at which the cryogen can cool down the tube.

Table 1. Percentage saving in film boiling time For copper tube.

Mass Flux (kg/m²s)	Test Section	Transition Temperature (K)	Transition Time (s)	% saving
66 kg/ m^{2}s	***Coated copper Tube***	180	227	13.2 %
	Regular copper Tube	154	257	
86 kg/ m^{2}s	***Coated copper Tube***	184	125	7.2 %
	Regular copper Tube	165	134	
102 kg/m^{2}s	***Coated copper Tube***	183	105	2.8 %
	Regular copper Tube	160	108	

At the mass flow rate of 66 kg/ m^{2}s, coated transfer line requires 225 seconds to cover the transition boiling regime, the temperature corresponds to this time is 170 K. The corresponding values for regular tubes were 253 sec and 158 K. At the mass flow rate of 86 kg/ m^{2}s, coated transfer line requires 119 seconds to cover the transition boiling regime, the temperature corresponds to this time is 183 K. The

Journal of Physics: Conference Series **1355** (2019) 012011 doi:10.1088/1742-6596/1355/1/012011

corresponding values for regular tubes were 130 sec and 157 K. At the mass flow rate of 102 kg/ m2s, coated transfer line requires 99 seconds to cover the transition boiling regime, the temperature corresponds to this time is 188 K. The corresponding values for regular tubes were 101 sec and 161 K. It can be seen that, with increase in mass flux, there is a substantial reduction in the time taken for transition.

For the regular tube, the solid-fluid thermal resistance forms the basis for the heat transfer at the interface. The vapor film developed immediately between the solid wall and fluid during the initial flow period causes an increase in the time required for chilldown as it takes time to break this boiling envelope.

In the case of coated surfaces, the low conductivity of the coating material results in a thermal gradient on its either side causing a reduction in temperature between fluid-fluid interfaces. Thus, resulting in reduction of chilldown time. The results presumed that a considerable reduction in chilldown time can be obtained by providing a coating of low conductivity on the inner walls of the conducting tubes and for any mass flow rate condition, the coated transfer line covers the film boiling regime at a higher wall temperature as compare to the regular tube. This results in a reduction in total chill down time since more than 75% of the total chill down time is covered for clearing the film boiling regime [15]. Table 1 shows the percentage saving in time for covering the film boiling regime.

4.2 Variation of Heat transfer coefficient

Inner wall temperature can be found by equation

$$T_i = T_0 + \left(\frac{r_0^2}{4\alpha}\left(\left(\frac{r_i}{r_0}\right)^2 - 1 - 2\ln\frac{r_i}{r_0}\right)\right)\frac{dT_0}{dT} + \left(\frac{1}{64\alpha^2}(r_i^4 - 5r_0^4)\ \frac{r_0^2 r_i^2}{8\alpha^2}\ln\frac{r_i}{r_0} - \frac{r_0^4}{16\alpha^2}\ln\frac{r_i}{r_0}\ \frac{r_0^2 r_i^2}{16\alpha^2}\right)\frac{d^2T_0}{dt^2} + .. \quad (1)$$

Inner wall surface heat can be calculated with the equation

$$q_i'' = \rho c\left(\frac{r_i^2 - r_0^2}{2r_i}\right)\frac{dT_0}{dt} + \left(\frac{(\rho c)^2}{k}\left(\frac{r_i^3}{16} - \frac{r_0^4}{16r_i} - \frac{r_0^2 r_i}{4}\ln\frac{r_i}{r_0}\right)\right)\frac{d^2T_0}{dt^2} + \frac{(\rho c)^3}{k^2}\left(\frac{r_i^5}{384} - \frac{3r_0^4 r_i}{128} + \frac{3r_0^2 r_i^3}{128} - \frac{r_0^6}{384r_i} - \frac{r_0^2 r_i^3}{32}\ln\frac{r_i}{r_0} - \frac{r_0^4 r_i}{32}\ln\frac{r_i}{r_0}\right)\frac{d^3T_0}{dt^3} + \quad (2)$$

Heat transfer coefficient can be found using the equation

$$h_i = \frac{q''}{(T_i - T_{sat})} \quad (3)$$

Table 2. Uncertainty analysis.

Measured parameters	Uncertainties
Length, L	±1 mm
Wall thickness	±0.01 mm
Temperature, K	±0.15 K
Pressure	±1.2 kPa
Volume flow rate	0.05 mm3/s
Maximum deviation in q''	12.8 %
Maximum deviation in h_i	10.7 %

Burggraf [16] in the year 1964 developed a method for the estimation of inner wall temperature from the measured outer wall temperature by using an inverse heat transfer technique. Table 2 shows the uncertainty analysis. The main source of uncertainty are from the temperature, pressure and volume flow rate measurements. Also the uncertainty in the measured values of heat flux and heat transfer coefficients are listed in table 2.

Figure 3(a) and (b) shows the variation of heat transfer coefficient for the un coated and coated transfer line. For a mass flow rate of 66 kg/ m^{2}s, the maximum heat transfer coefficient is obtained earlier for coated section (87 W/m^{2}K) compared to the uncoated section (27 W/ m^{2}K). For the mass flow rate of 86 kg/ m2s, the maximum heat transfer coefficient is obtained earlier for coated section (70 W/m^{2}K) compared to the uncoated section (59.7 W/ m^{2}K). For the mass flow rate of 102 kg/ m2s, the maximum heat transfer coefficient is obtained earlier for coated section (124 W/m^{2}K) compared to the uncoated section (57 W/ m^{2}K).

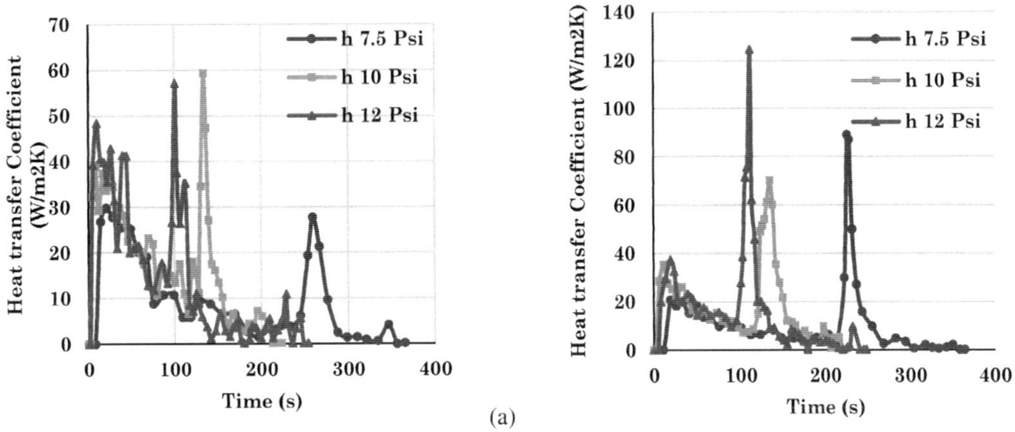

(a)

(b)

Figure 3. Variation of heat transfer coefficient with time for (a) un-coated Copper tube and (b) coated copper tube.

4.3 Variation of Heat flux:

Figure 4(a) and (b) shows the variation of heat flux for the un coated and coated copper tube. For a mass flow rate of 66 kg/ m^{2}s, the maximum heat flux attained is 9122 W/m^2 for the coated section at the same time the corresponding value for the uncoated section is 2091 W/m^2. For a mass flow rate of 86 kg/ m^{2}s, the maximum heat flux of 5745 W/m^2 is obtained for coated copper tube and the corresponding value

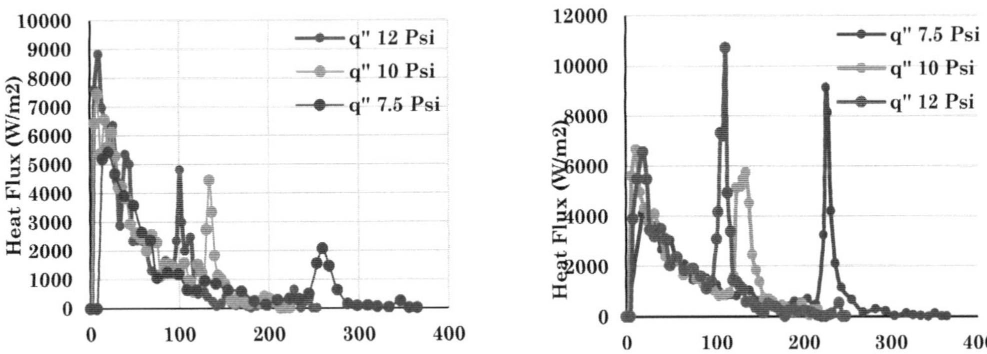

(a)

(b)

Figure 4. Variation of heat flux as a function of time for (a) Un coated copper tube and (b) Coated copper tube.

for uncoated copper tube is 4457 W/m^2. Also it can be inferred that the coated tube attains maximum heat flux earlier than the uncoated tube. For a mass flow rate of 102 kg/ m^{2}s, the maximum heat flux of 10698 W/m^2 is obtained for coated copper tube. Maximum heat flux of 4817 W/m^2 is obtained for uncoated copper tube. Also it can be inferred that the coated tube attains maximum heat flux earlier than the uncoated tube. Thus it can be concluded that for any mass flow rate condition, the coated tubes results in an earlier attainment of maximum heat flux.

Conclusion
The test results justifies that the use of Teflon coating in test section will enhance the heat dissipation characteristics with the reduction in cool down time. The data obtained indicates lower chill down time for various inlet pressures, due to the sudden increase in heat transfer after the transition. The obtained data will help in the optimization and design of transfer lines. This work shows the necessity for studying the effects of various coatings on the heat transfer enhancement and chill down time reduction in future. Future works can also employ different coating materials to cool down transfer lines made of stainless steel and other suitable materials. In depth understanding of this phenomena can be obtained by considering the heat transfer coefficient variation in the three flow regimes during chilldown process.

Acknowledgments
The authors would like to acknowledge the Space Technology laboratory of TKM college of Engineering and also the Kerala State Council for Science Technology and Environment (KSCSTE) for providing facilities for the successful completion of the project.

Nomenclature

d	Diameter, m
q"	Heat flux, W/m^2
r	Radius, mm
T	Temperature, K
t	Time, s
ΔT	Wall super heat, K
ρ	Density, kg/m^3
C	Specific heat, J/(kgK)
h_i	Heat transfer coefficient, W/m^{2}K
α	Thermal Diffusivity

Abbreviations

CHF	Critical heat flux
HTC	Heat transfer coefficient
FB	Film boiling
NB	Nucleate boiling
TB	Transition boiling

Subscripts

i	inner surface of wall
o	outer surface of wall
sat	saturation conditions

References

[1] S. S. Bindu, K. E. Reby Roy, and S. S. Kumar, "Experimental Study on the Effect of Low Conductivity Coating on Cryogenic Transfer Lines," *Proc. - 2018 2nd Int. Conf. Green Energy Appl. ICGEA 2018*, pp. 225–228, 2018.

[2] Cowley. C W, W. J. Timson., and J. A. Sawdye, "A Method For Improving Heat Transfer To A Cryogenic Fluid," *Adv. Cryog. Eng.*, vol. 7, 1962.

[3] L. D. Allen, "Amethod of Increasing Heat Transfer To Space Chamber Cryopanels," pp. 547–548, 1966.

[4] R. Reed, F. R. Fickett, L. T. Summers, and M. Stieg, "Advances in cryogenic engineering materials," *Adv. Cryog. Eng. Mater.*, vol. 40, pp. 528–532, 1994.

[5] B. J. Stock, "Observations on Transition Boiling Heat Transfer Phenomena," *Argonne Natl. Lab. Rept. ANL*, p. 6175, 1960.

[6] R. F. Gaertner, "Photographic Study of Nucleate Pool Boiling on a Horizontal Surface," *J. Heat Transfer*, vol. 87, no. 1965, p. 17, 1965.

[7] Leonard K.E Getty, "A comparison of cooldown time between internally coated and uncoated propellant line," in *Advances in Cryogenic Engineering*, 1967, pp. 331–339.

[8] G. A. Dreitser, "Heat transfer enhancement in channels for film boiling of cryogenic liquids," in *Applied Thermal Engineering*, 2005.

[9] H. Hu and T. K. Wijeratne, "Two-Phase Flow and Heat Transfer During Chilldown of a Simulated Flexible Metal Hose Using Liquid Nitrogen," *J Low Temp Phys*, vol. 174, pp. 247–268, 2014.

[10] J. Kim, S. Jun, R. Laksnarain, and S. M. You, "Effect of surface roughness on pool boiling heat transfer at a heated surface having moderate wettability," *Int. J. Heat Mass Transf.*, vol. 101, pp. 992–1002, 2016.

[11] L. W. Fan, J. Q. Li, L. Zhang, Z. T. Yu, and K. F. Cen, "Pool boiling heat transfer on a nanoscale roughness-enhanced superhydrophilic surface for accelerated quenching in water," *Appl. Therm. Eng.*, vol. 109, pp. 630–639, 2016.

[12] D. Takeda, K. Fukiba, and H. Kobayashi, "Improvement in pipe chilldown process using low thermal conductive layer," *Int. J. Heat Mass Transf.*, vol. 111, pp. 115–122, 2017.

[13] R. Li and Z. Huang, "A new CHF model for enhanced pool boiling heat transfer on surfaces with micro-scale roughness," *Int. J. Heat Mass Transf.*, vol. 109, pp. 1084–1093, 2017.

[14] S. Mori and Y. Utaka, "Critical heat flux enhancement by surface modification in a saturated pool boiling: A review," *International Journal of Heat and Mass Transfer*. 2017.

[15] R. Shaeffer, H. Hu, and J. N. Chung, "International Journal of Heat and Mass Transfer An experimental study on liquid nitrogen pipe chilldown and heat transfer with pulse flows," *Int. J. Heat Mass Transf.*, vol. 67, pp. 955–966, 2013.

[16] O. R. Burggraf, "An Exact Solution of the Inverse Problem in Heat Conduction Theory and Applications," *J. Heat Transfer*, vol. 86, no. 3, p. 373, 1964.

Design Optimization of Nozzle and Second Throat Diffuser System for High Altitude Test using CFD

Naveen Kumar K*, Vijayanandh R*, Ramesh M

Department of Aeronautical Engineering, Kumaraguru College of Technology, Coimbatore Tamil Nadu, India – 641049

* Corresponding author E-mail: naveenkumar.k.aeu@kct.ac.in, vijayanandh.raja@gmail.com

Abstract. Nozzles are used to produce thrust in airplanes, rockets, and even satellites. Different designs are employed for various applications. For propelling satellites in space, High Area Ratio supersonic nozzles (A.R. > 100) are used. During ground tests of these nozzles, it is found that a shock wave is formed and flow tends to separate in divergent section. A second throat ejector diffuser system (STED) is employed to simulate ambient vacuum conditions at high altitudes, in sea level, around the nozzle. This vacuum level is maintained throughout the test times and is used for smooth expansion of the gas across the nozzle without flow separation. The Shockwaves formed in the diffuser help in maintaining the desired low-pressure levels and reduce the flow velocity and increase the pressure. A High Area Ratio nozzle (A.R.= 114) and a Second Throat Diffuser are designed and modeled together. The flow analysis is carried out using CFD and the pressure, velocity and temperature variations in the nozzle and diffuser are studied. The diffuser design is then optimized, by varying different parameters of the diffuser, and analysis is carried out for each design in order to obtain full flow in the nozzle and the diffuser. The design of the diffuser must be such that the flow in the nozzle must be stable and should provide a suitable pressure recovery with the reduction in flow velocity. After the diffuser design is finalized, performance analysis of the diffuser is carried out, along with the nozzle, in order to achieve the working conditions of the diffuser.

Keywords: Aerodynamics; Nozzle; Area Ratio; Mach Number; Optimization; Rocket;

1. Design of Nozzle and Second Throat Diffuser

1.1. Design of Nozzle

1.1.1. High A.R. nozzles

A nozzle is used to produce thrust by converting pressure energy into kinetic energy. The high-pressure gases from the combustion chamber expand in the nozzle providing the required thrust for any flying body such as aircraft, rockets or missiles. As the pressure decreases the flow velocity inside the nozzle increases. The nozzle accelerates the gas through its expansion section. There are different types of the nozzle that are used depending on the exit velocity the nozzle shape varies. For subsonic operation, the nozzles have a convergent portion only from the combustion chamber. For supersonic exit velocities, a convergent-divergent nozzle is used.

High Area Ratio nozzles are nozzles having an A.R. above 100, which employed as thrusters for satellites in space. In space, the satellite needs to be carried to the desired altitude and due to atmospheric disturbances, the satellites tend to shift from its desired orbit. For propelling the satellite and rectifying any shift in its orbit, small thrusters are designed for operation at high altitudes. The thrust provided by the thrusters are very small and used for Attitude and Orbit Control of Satellites (AOCS), and form a crucial system for control of satellites for correcting its orbit. The thrusters are located in different positions of the satellite and are used only in space. In which there are two types of high A.R nozzles used, which are the main engine and steering engine nozzle. The Main engine nozzle, having A.R. 114, is considered for analysis in this paper [2]. The High A.R. nozzles need to be tested in the ground level before it can be used on satellites. However, while performing a ground level test, it is found that a shock wave is formed and complete expansion of the gases is not obtained in the divergent portion due to the high A.R. and high back pressures at the exit in sea level. To obtain full flow in the entire nozzle low-pressure atmosphere should be created at the nozzle exit while testing it. This low-pressure atmosphere replicates the vacuum conditions around the nozzle in space and also enables complete flow expansion of the gases inside the nozzle [3].

1.1.2. Design methodologies used

Bell nozzles are the commonly used nozzles as they are more advantageous than the other type of nozzles, which provides smooth expansion, maximum thrust along the axis and gives the required Mach number at a length less than that of the conical nozzle. Length is of major importance in the Aerospace industry so the bell nozzle is the optimum shape preferred for the supersonic applications. Unlike a conical nozzle where the deflection angle is 150 and the line connecting the throat and the exit cross sections, an arc connects the throat and exit area of the bell nozzle and the deflection angle varies at each point of the bell nozzle.

G.V. Rao method is employed for designing minimum length nozzles. In this method, the divergence angle at the inflection point and at the exit are determined using Prandtl Meyer flow relations and compressible flow equations. As a preliminary step the nozzle exit Mach number should be calculated using the Area-Mach number relation. The length of the bell nozzle is within 60% to 80% of that of the conical nozzle depending on the amount of expansion needed.

After the length is determined the arc must be drawn from the throat to the exit, which is completed by determining the throat radius and drawing a curve to the inflection point. The initial curve is plotted such that at the throat the slope is zero and increases gradually till the inflection point. At the inflection point, there is a sharp increase in the slope which is equal to the initial diverging angle. From this point, a series of points are plotted with each point corresponding to a bigger A.R. corresponding to a higher Mach number. The series of points are plotted until the exit point is reached which should be the same as the required exit radius. There are no formulae to calculate the curve in the converging section. Hence a suitable curve with a gradual decrease in slope is considered [6].

1.1.3. Nozzle Design Conditions and Calculated Values

The conditions that are used for designing the high A.R. main engine nozzle are:

Type	=	Bell/Contoured nozzle
A.R.	=	114
Chamber Pressure	=	33.86 bar
Chamber Temperature	=	3472 K
Mass Flow rate	=	8.7 kg/s
Length of Chamber	=	0.111 m

Initial expansion angle = 32°
Final expansion angle = 9°

From the above conditions [7] the other nozzle parameters are calculated using compressible flow relations: The contour points of the designed nozzle have been plotted using commercial software package ICEM-CFD, which is used for designing and meshing. An arc is drawn connecting all the points. The subsonic portion is drawn using straight lines. The nozzle has meshed and the boundary conditions are specified as Pressure Inlet, Pressure Outlet, Axis, and Wall. Type of Mesh used here is Quadrilateral and the number of cells is 7200, a typical discretized view of the nozzle has been shown in figure 1.

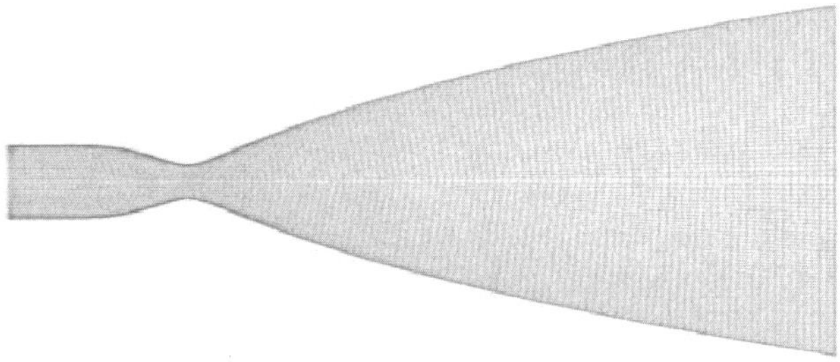

Figure 1. Nozzle contour after meshing

1.2. Diffuser Design Conditions and Calculated values

The location of the diffuser is aft of the nozzle so the inlet parameters are considered as the exit conditions of the nozzle. The throat of the nozzle is considered as the first throat and the throat of the diffuser is considered as the second throat. The diffuser design conditions [8] are as follows:

Inlet Pressure = 10 mbar
Minimum Exit Pressure = 40 mbar
Inlet Temperature = 481 K
Net mass flow rate = 8.7 kg/s
Minimum Throat Diameter = 0.8 m
Minimum Throat Length = 80 m
Converging section angle = 4°
Diverging section angle = 7°

The different sections of the diffuser are comprised of the straight section, convergent section, throat section, and divergent section. A weak final shock is formed at the end of the throat in the diverging section which causes a considerable increase in the pressure. The location of the final shock should be inside the diverging section for smooth operation of the diffuser. Hence all the components of the diffuser, the convergent section, throat section, and the divergent section play an important role in giving pressure recovery and thereby reducing the flow velocity from high Mach number to subsonic speeds [9].

The designed nozzle and diffuser system has meshed and the boundary conditions are applied as pressure inlet, pressure outlet, axis, and wall. Type of Mesh used in this paper is quadrilateral and the number of cells is 14050, a typical discretized view of the diffuser has been shown in figure 2.

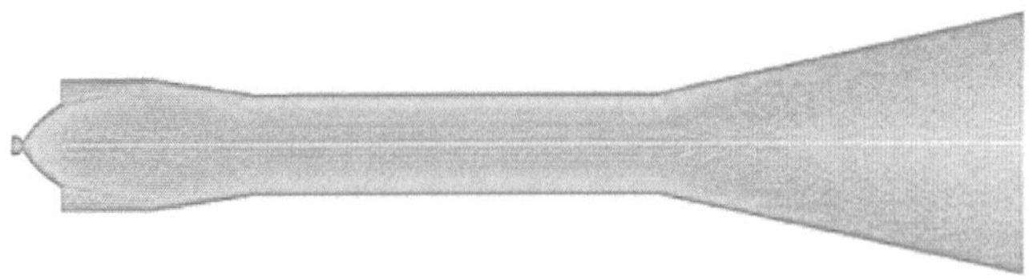

Figure 2. Nozzle and Diffuser mesh

2. Result Analysis of Nozzle and Second Throat Diffuser

2.1. Nozzle and Second Throat Diffuser System Analysis

The designed nozzle and second throat diffuser system which is plotted and meshed using ICEM-CFD are then imported into Ansys Fluent. The Axis-Symmetric option is selected to perform the analysis for 3D flow through the nozzle and diffuser system. The flow analysis is continuous as the results for complete expansion as well as pressure recovery are needed so the steady state option is selected. To obtain the flow solution faster and for quicker convergence, the implicit solver has been selected. The flow is compressible and the density is an ideal gas, hence the energy equation is selected and the Spalart-Allmaras has been selected as the Viscous Model, which is widely used in aeronautical applications also used to specify that the flow has turbulent viscosity. The major advantage of this model is that it assumes a thin boundary layer for the flow along the wall and there is no requirement to feed a user-defined function for the boundary layer. Water vapor has been selected as working fluid and the density is set to ideal gas which is required for the compressible flow of gases. Operating condition is set to zero Pascal because the initial pressure and when the flow enters the nozzle and the second throat diffuser system it is in vacuum condition. The pressure inlet is used to specify the total pressure and total temperature and the Initial Starting Pressure at the nozzle inlet. The pressure far field is used to specify the pressure outlet of the diffuser such that the nozzle flow is not affected by it [12, 13, and 14].

2.2. Nozzle Analysis

The nozzle is first analyzed in order to check the flow expansion through it, without the diffuser. The analysis is also used to check the pressure and temperature at the nozzle exit. This is to verify the exit Mach number, exit pressure, and exit temperature to ensure that at the specified total pressure there are no shock formations inside the nozzle. Here the nozzle exit is specified as the pressure outlet boundary condition. The Mach distribution over the nozzle is shown in figure 3.

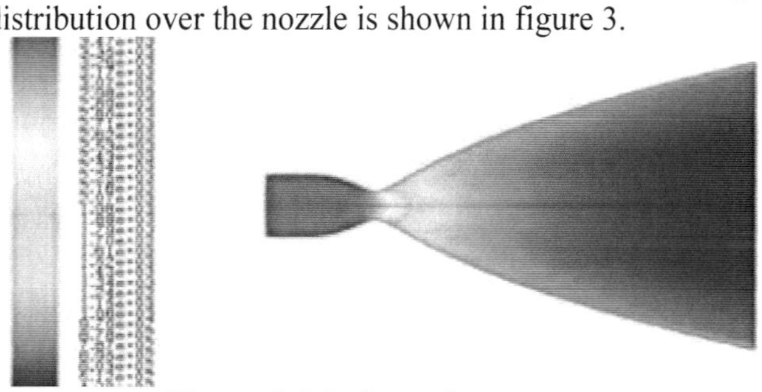

Figure 3. Mach number contour

Figure 4 shows the pressure variation along the nozzle

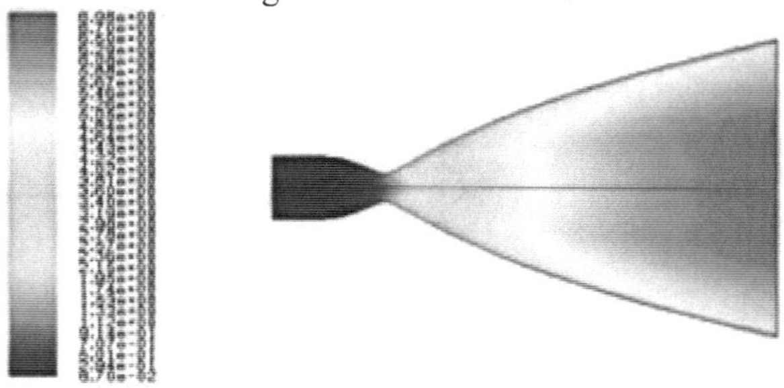

Figure 4. Pressure contour

The nozzle provides an exit Mach number of near 6.8, exit pressure of 250 Pa and an exit temperature of 1000 K approximately. These results, for exit Mach number and temperature, are similar to the flow results which are calculated using isentropic flow relations. The nozzle gives a smooth expansion and accelerates the flow to the required Mach number. Hence this nozzle can be used for the analysis along with the second throat diffuser [15].

2.3. Second Throat Diffuser Analysis
The initial design of diffuser, which is analyzed for different back pressures:

2.3.1. Case – 1 At Back Pressure, Pb = 200 m bar
The flow enters the nozzle, in which there is no complete expansion of the flow inside the nozzle. A shock is created within the nozzle and there is a drop in the velocity from supersonic to subsonic. This is diffuser unstarted condition, which means that there are no shocks created in the diffuser and the diffuser does not do any work in reducing the velocity or increasing the pressure of the gas (figure 5).

2.3.2. Case – 2 At Back Pressure, Pb = 40 m bar
This is the minimum pressure that can be provided by the diffuser in full flow condition, i.e. when both the diffusers are in operation. The flow expands through the nozzle but oblique and normal shocks are created within the straight portion of the diffuser. Hence the flow velocity is decreased and the diffuser does not start, which means the diffuser is in unstarted condition and the diffuser does not do any changes to the flow velocity or pressure (figure 6). As the back pressure value is reduced the maximum Mach number increased from 6.8 to 7.5. This means more expansion occurs inside the nozzle as the back pressure decreases, causing the flow velocity to increase.

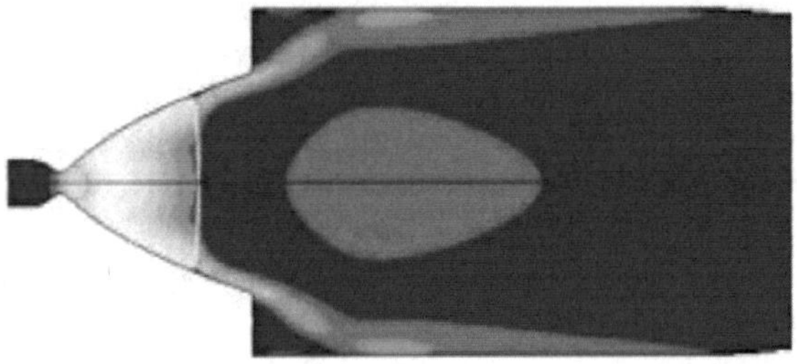

Figure 5. Diffuser at Back Pressure, Pb = 200 m bar.

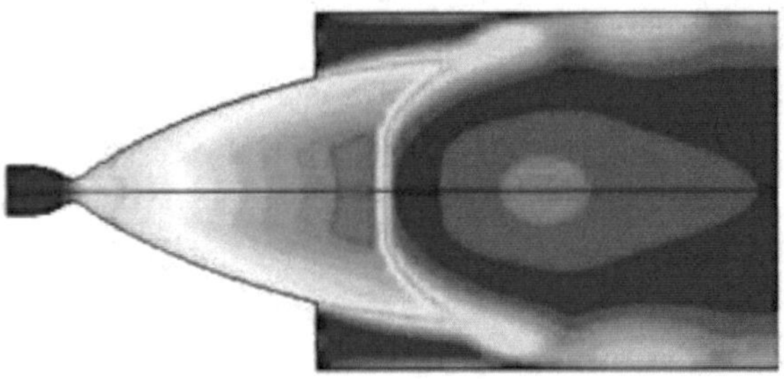

Figure 6. Diffuser at Back Pressure, Pb = 40 m bar

3. Optimization of the Diffuser

3.1. Need for Optimization
It is clear from the flow analysis of the second throat diffuser is not suitable for the nozzle, as the diffuser remains in unstarted condition and there is no pressure recovery or reduction of velocity required. If tests are conducted on this diffuser it would yield inaccurate and unsatisfactory results. The design of the diffuser needs to be optimized to obtain the desired flow within the nozzle and the second throat diffuser.

Journal of Physics: Conference Series **1355** (2019) 012012 doi:10.1088/1742-6596/1355/1/012012

The method is used for optimizing the diffuser is by fixing the backpressure value at 200 m bar and varying various design parameters of the diffuser and carrying out the analysis using the same conditions each time. Complete expansion of the gases inside the nozzle and the diffuser should be started are primary criteria for the design optimization. The flow stability in the nozzle and diffuser is the next vital criteria used for finalizing a particular diffuser design.

3.2. The methodology used for Optimization

The major factor that is used in optimizing the diffuser design is to obtain a complete expansion of flow within the nozzle. Initially, the straight section length (z) is adjusted as it is the primary reason for the flow to not enter into the diffuser. When the flow touches the convergent section it gets deflected into the throat immediately and causes a series of oblique shocks to be formed inside the throat. The straight length is reduced and the analysis is carried out. Few of the lengths that are analyzed (z = 0.5; 1; 1.7) are found to give complete flow inside the diffuser. These lengths are used to vary other parameters in order to obtain a stable flow inside the diffuser and provide a suitable pressure recovery.

The convergent angle is kept constant and the throat radius is varied. Strong shocks are also observed when the straight section length is 0.5 m. The throat section is adjusted so that the convergent angle remains at 40. Using A.R. and shock relations the diameter of the throat is calculated as 1.15 m. The divergent angle is set at 70. This angle is increased to 100 and then 140 for reducing the pressure as much as possible. The angle is not decreased since the flow velocity does not decrease and also it is not increased further because the flow will become unstable.

3.3. Unstable Flow with Pressure Recovery

A diffuser design having a throat diameter of 1.15 m and length 12 m, exit diameter 3 m and divergent angle of 140 gave sufficient pressure recovery. However, the flow is unstable as the final shock is formed within the throat of the diffuser. The exit pressure after analysis is found to be near 225 m bar. In this design, any flow disturbance will cause a shock to be formed within the nozzle affecting the nozzle test results. Hence this design can also not be used for testing the high A.R. nozzle. Figures from 7 and 8 shows the Mach number and pressure variation in unstable flow.

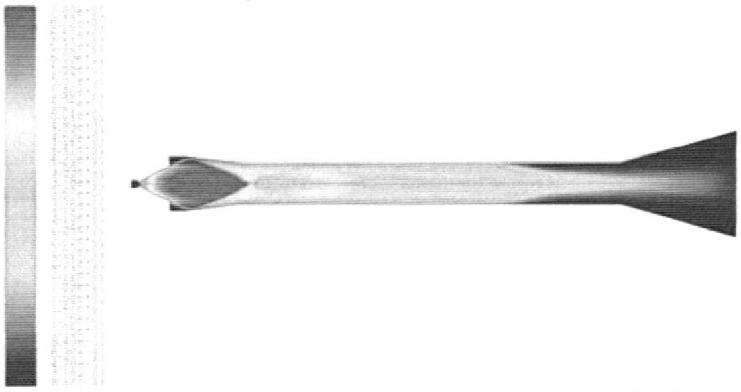

Figure 7. Mach number variation in unstable flow

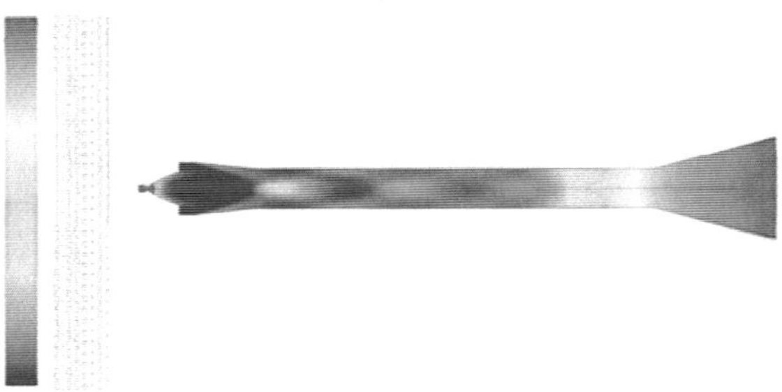

Figure 8. Pressure variation in unstable flow

3.4. Stable Flow with Pressure Recovery

The optimized diffuser design that gave a stable flow and a suitable pressure recovery is a diffuser of throat diameter 1.2 m, straight section length of 1 m, throat length of 12 m, exit diameter of 3 m and divergent angle of 70. The flow in this diffuser is stable since the final shock wave is created at the divergent section of the diffuser. This second throat diffuser also gives a suitable pressure recovery near 200 m bar at the exit and 190 m bar along the wall. This increase in pressure occurs in the divergent section hence any variation in the exit pressure will not cause any disturbance to the flow inside the nozzle. Figures 9 to 12 show the Mach number, pressure and temperature variation of the diffuser in stable flow operation.

At the throat, the flow velocity decreases sharply and a series of oblique shocks are created. The oblique shocks also cause the pressure to increase gradually. In the divergent section, there is a slight increase in Mach number and the final shock is created in the divergent section. After the slight increase in Mach number, the flow velocity decrease to sonic speed. The final shock also causes an increase in the downstream pressure of the flow and the pressure rises to 200 m bar along the axis and 190 m bar along the wall. This flow obtained is stable and the designed diffuser can be used to test the high A.R. nozzle. The complete expansion is obtained in the nozzle and the shock in the divergent section will assist in testing the nozzle for steady full flow condition and obtain accurate test results.

Figure 9. Mach number contours for Stable flow in the diffuser

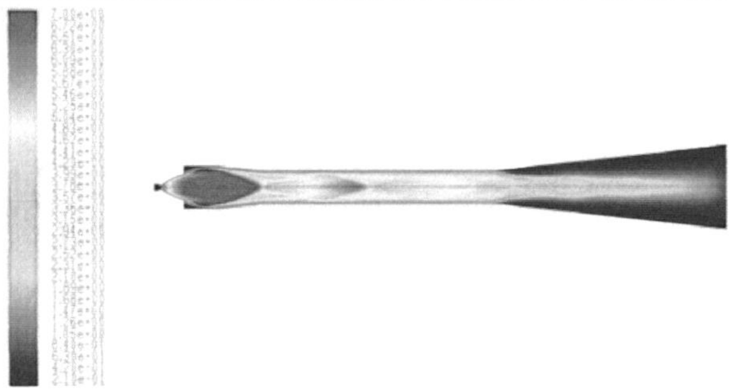

Figure 10. Pressure contours for Stable flow in the diffuser

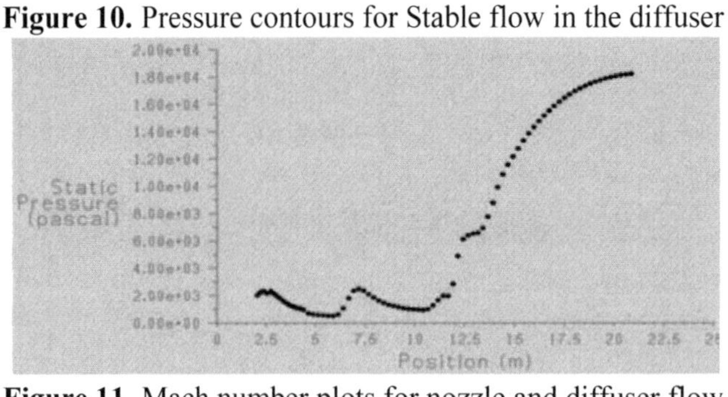

Figure 11. Mach number plots for nozzle and diffuser flow

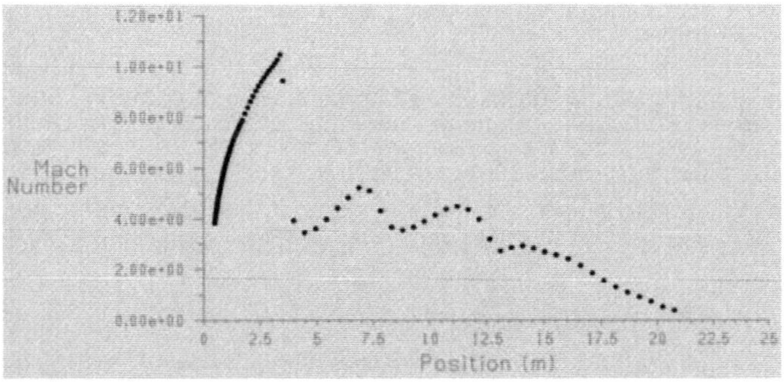

Figure 12. Pressure plots for nozzle and diffuser flow

4. Performance Analysis and Optimization of Diffuser

The diffuser design that is optimized must be analyzed for performance at different exit pressures. This is to verify the stability of the flow inside the nozzle when a test is conducted on it. It is seen that the diffuser is steady at an exit pressure of near 200 m bar. Above this pressure, the flow is unstable and will yield inaccurate results during testing of the nozzle. The pressure value below which stable flow is attained is the stable flow limit. Below this recovery Pressure value, the flow is stable and there is complete expansion inside the nozzle. This is required for testing the High A.R. nozzle and will give accurate expansion results during testing of the nozzle

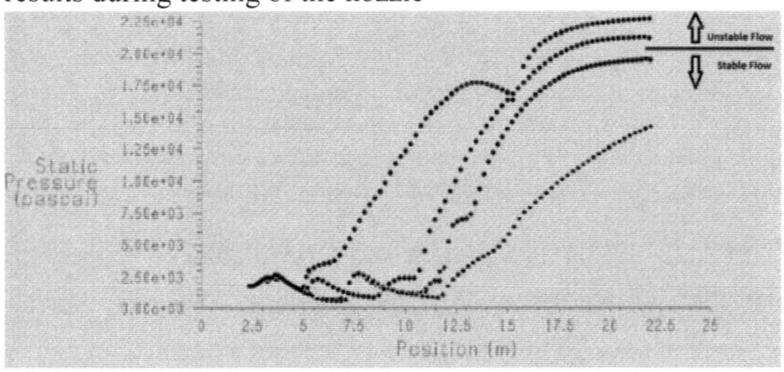

Figure 13. Pressure plot for different recovery pressures

The analysis that is carried out on the optimized nozzle and the second throat diffuser gives the various flow conditions at different recovery pressures. The unstable flow is obtained above 200 m bar and below this pressure value, the flow is stable. Hence 200 m bar can be considered as the stable pressure limit. If the exit pressure is maintained below 200 m bar and the nozzle is tested the expansion is stable and the flow results can be considered as accurate. When both the ejectors are in running condition, the minimum pressure value that can be obtained is 40 m bar.

4.1. Discussion

The normal nozzle without second throat diffuser provides an exit Mach number of near 6.8; exit pressure of 250 Pa. Since the nozzle gives a smooth expansion and accelerates the flow to the required Mach number, it is used for the analysis along with the second throat diffuser. The analysis carried out with the initial design of second throat diffuser shows that the diffuser remained in Unstart condition and the expansion that occurs in the nozzle is unstable. As the back pressure value is reduced the maximum Mach number increased from 6.8 to 7.5, which indicates more expansion occurs inside the nozzle as the back pressure decreases, causing the flow velocity to increase thus the need for optimization. The optimization is done by only changing the second throat diffuser wall angles and dimensions. The optimized second throat diffuser gives complete flow expansion inside the nozzle and suitable pressure recovery near 200 m bar at the exit and 190 m bar along the wall thus allows the nozzle to work at maximum efficiency.

5. Conclusion

The second throat diffuser must provide a suitable pressure recovery and at the same time, the flow in the nozzle should be stable. Optimization of the diffuser is done to create a complete expansion of the gases inside the nozzle. Various sections of the diffuser modified in such a way to move the final normal shock wave from the throat section to the diffuser section to attain the maximum efficiency. Even if there are any flow disturbances causing variation in the diffuser exit pressure the flow inside the nozzle will be stable because of the oblique shocks that maintain the vacuum level inside the nozzle. The shock formation inside the diffuser throat will not allow the variations in exit pressure to affect the flow in the convergent section and the nozzle. Thus the expansion inside the nozzle will not be affected and the flow results can be considered as accurate. Hence as long as the exit pressure is maintained below 200 m bar, the designed diffuser will give accurate test results in the nozzle.

References

[1] Alan Vincent E V, CFD Analysis of Supersonic Exhaust Diffuser System for Higher Altitude Simulation, International Journal of Engineering Studies and Technical Approach, Volume 01, No.6, June 2015, pp 1 - 8.

[2] Bruce Ralphin Rose J et al, A numerical optimization of high altitude testing facility for wind tunnel experiments, Chinese Journal of Aeronautics, (2015), 28(3), pp 636–648

[3] Dominic Xavier Fernando, Design and Analysis of Second Throat Diffuser System at Different Back Pressures for High Altitude Test Journal of Basic and Applied Engineering Research, Print ISSN: 2350-0077; Volume 1, Number 5; October 2014, pp. 67-70

[4] Ch V K N S N Moorthy, Computational Analysis Of A C-D Nozzle With 'Sed' For A Rocket Air Ejector In Space Applications, International Journal of Mechanical and Production Engineering Research and Development (IJMPERD), ISSN(P): 2249-6890; ISSN(E): 2249-8001, Vol. 7, Issue 1, Feb 2017, pp 53-60

[5] Thomas M. Berens, Numerical and Experimental Investigations on Subsonic Air Intakes with Serpentine Ducts for UAV Configurations, Challenges in Europian AeroSpace 5th air and space conference, 2015 paper no. 074, pp 1 - 22.

[6] Chen, F., Liu, C.F., Yang, J.Y., 1994. Supersonic flow in the second-throat ejector diffuser system, Journal of Spacecraft and Rockets

[7] Kim H., Lee Y., Setoguchi T., and Yu S., 1999. Numerical Simulation of the Supersonic Flows in the Second Throat Ejector- Diffuser Systems, Journal of Thermal Science 8

[8] Manikanda Kumaran R, Vivekanand P K, Sundararajan T, BalaSubramanian S, Raja Manohar D., August 2009. Analysis of Diffuser and Ejector Performance in a High Altitude Test Facility, 45th AIAA/ASME/SAE/ASEE Joint Propulsion Conference & Exhibit, AIAA 2009- 5008, Colorado, USA.

[9] Liepmann, H.W. and Roshko, A. (2007), Elements of Gas Dynamics, First South Asian Edition, Dover Publications, Inc., New Delhi. India

[10] Park, B.H., Lim, J., Park, S., Lee, J.H. and Yoon, W. (2012), "Design and analysis of a second-throat exhaust diffuser for high altitude simulation", J. Propuls. Pow., 28 (5), 1091-1104.

[11] Cooper GK, Phares CG. CFD applications in an aerospace engine test facility. Report No.: AIAA-1990-2003. 1990.

[12] Manikanda Kumaran R, Sundrarajan T, Raja Manohar D. Performance evaluation of second-throat diffuser for a high altitude testing facility. J Propul Power 2010;26(2):248–58.

[13] Mon KO, Lee C. Optimal design of supersonic nozzle contour for altitude test facility. J Mech Sci Technol 2012; 26(8), pp 2589–94.

[14] Pugazenthi RS, Mclntosh AC. Design and performance of a supersonic diffuser for plasma wind tunnel. Word Acad Sci Technol 2011;5(8), pp 1254–72.

[15] Rathakrishnan E. Gas dynamics. 5th ed. New Delhi: Prentice- Hall of India Learning Private Limited; 2011, pp 110–35.

[16] Bruce Ralphin Rose J, Veni Grace J. Performance analysis of lobed nozzle ejectors for high altitude simulation of rocket engines. International Journal Model Simul Sci Comput 2014; 5(4):1450019.

[17] Senthil Kumar M, Vijayanandh R, Rahul Srinivas, Arun Karthik D, Tamil Mani M, Acoustic Analysis and Comparison of Chevron Nozzle Using Numerical Simulation, International Journal of Mechanical and Production Engineering Research and Development (IJMPERD), ISSN(E): 2249-8001, Vol. 8, Special Issue 7, Oct 2018, 1089-1103.

Thermodynamic modelling of hybrid adsorption system

P Abhishek[1] and V Baiju[1]

[1] Research scholar, Department of Mechanical Engineering, T.K.M College of Engineering, Kollam, A P J Abdul Kalam Technological University, Thiruvananthapuram, Kerala, India.
[2] Assistant Professor, Department of Mechanical Engineering, T.K.M College of Engineering, Kollam,Kerala, India.

E-mail: abhishekp8999@gmail.com
baij84@yahoo.co.in

Abstract. The ability of adsorption based desalination system to cogenerate distilled fresh water and cooling effect provides an increasing attention in the field of cooling and desalination technologies. Eco-friendly adsorption cycles are capable of producing the hybrid effects (desalination& cooling) with low temperature heat source having temperature range from 50°C to 90°C. Adsorption characteristics (isotherm) of silica gel-water and zeolite-water have been presented in this study. Dubinin-Astakov (D-A) and Dubinin-Radushkevich (D-R) adsorption equilibrium equations are used for comparing the performance of silica gel and zeolite. Results show that maximum water production rate in the case of silica gel adsorbent is stabilised at lower regeneration temperature, ranges from 80°C to 120°C, where as in zeolite it is in the higher range from 150°C to 250°C. Maximum COP of the system obtained is 0.65 for silica gel-water at a regeneration temperature of 80°C where as 0.35 for zeolite–water at 110°C. Study concluded that the performance of the system using silica gel-water adsorbent is comparatively better than zeolite-water pair.

Keywords - Adsorption, desalination, cooling, silica gel, zeolite

1. Introduction

Today the water crisis affects billions around the world. Clean fresh water is an essential ingredient for a healthy human life. There is the same amount of fresh water on earth as there always has been, but the population has been exploded, leaving the world's water resources in crisis. In this scenario the need to tap potable drinking water is a great challenge. One of the practical solutions to overcome this crisis is desalination which aims in separating dissolved salts and other minerals from brackish water. The process of desalination can be done in various ways, like thermally activated systems or pressure activated systems or chemically activated systems. Multi stage flash (MSF), multi effect desalination (MED), solar distillation, reverse osmosis and electro dialysis[1] are the major desalination methods. But difficulties of maintenance, relatively high energy consumption and high cost are became the major shortcoming of these technologies. Therefore there is a need for low energy consumption desalination with low running cost. One such method is adsorption desalination [2,3] process capable of producing potable water and cooling effect.

Adsorption desalination system is an emerging process of thermal desalination capable of utilizing low grade heat easily obtainable from solar collectors or waste heat or biomass energy. Moreover it has a range of advantages including; the use of low temperature heat (waste heat, biomass, solar), very few moving parts leading to reduced maintenance cost, fouling & corrosion is reduced due to low operating temperatures, and confinement of brackish water to a fraction of the total system. Adsorption desalination is a thermodynamic cycle that consists of Evaporation/Adsorption and Desorption/Condensation processes where the adsorbent material adsorbs pure water from the boiling sea water in the evaporator. This evaporation process is maintained by the adsorbing action of the adsorbent .The desorbed water vapour is condensed in the condenser to produce potable water which is collected and pumped out of the condenser. Cooling effect is produced during the evaporation process which occurs at temperature ranging from 10°C to 30°C. As the adsorption desalination cycle utilizes only low temperature waste heat or solar energy [4-5], it has the potential to be the most efficient method of converting the sea or brackish water to potable water as the only parasitic electric power needed are those used for powering the valves and water pumps[6]. The paper first outlines about the working principle of single bed adsorption desalination system. The mathematical modelling of adsorbent is also presented in this study. The effect of regeneration temperature and cooling water temperature on water production, COP and energy requirement for water production is also investigated.

2. Single bed adsorption desalination cycle – Working principle

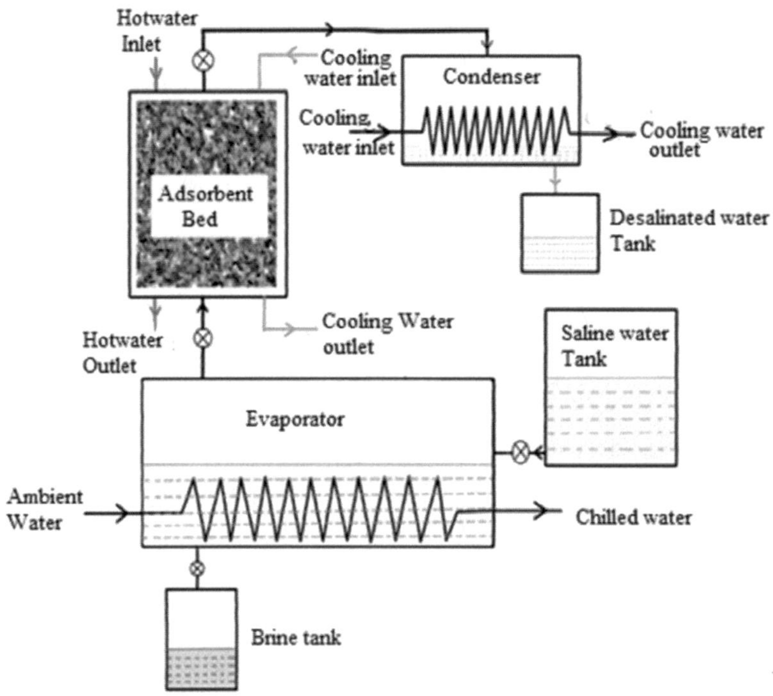

Figure 1: Single bed adsorption cooling and desalination system

Adsorption cycle utilizes adsorption assisted evaporation and desorption activated condensation processes to produce cooling in evaporator and fresh water at the condenser. These useful effects are achieved in an adsorption system with the help of low temperature heat source that can be extracted from industrial waste heat or solar energy [7-9]. Schematic diagram of single bed adsorption cooling and desalination system is shown in figure 1. Evaporator, adsorbent bed and condenser are the main parts of a single bed adsorption system. Firstly the saline or brackish water is charged into the evaporator. In the evaporator, the water is flash evaporated at the low pressure. Ambient water which is circulated in the evaporator be used for cooling purposes. The concentrated brine is discharged periodically from the evaporator to regulate the sea water level and salt concentration. When the valve between the adsorbent bed and the evaporator is opened, water evaporates and travels from the evaporator into adsorbent bed where it is adsorbed by the silica gel. During adsorption process, the heat of adsorption is removed by the cooling water, circulating in adsorbent bed heat exchangers. The adsorption continues until the adsorbent saturated with water vapour and closing the valve between the adsorbent bed and evaporator. The saturated silica gel is regenerated using hot water coming from the solar collector and this process is known as desorption process. The valve between the adsorbent bed and condenser is in open position, the hot water drives off the regenerated water vapour to the condenser. In the condenser, the heat of condensation is rejected into cooling water circulating through the condenser and the condensate is collected as pure water. Then the desalinated water is collected in the collection tank. This thermodynamic cyclic process is represented in figure 2.

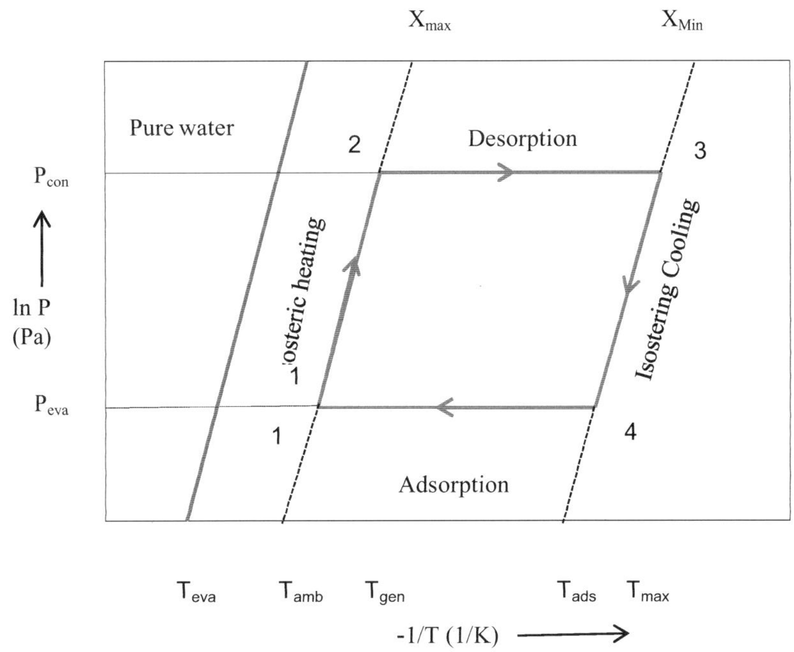

Figure 2: Adsorption desalination cycle

- Process 1→2 – Isosteric heating
- Process 2→3 – Desorption (Isobaric heating)
- Process 3→4 - Isosteric Cooling
- Process 4→1 – Adsorption (Isobaric cooling)

The P–T–X diagram on lnP vs. −1/T coordinates (where X is the amount of adsorbate adsorbed by the adsorbent at equilibrium conditions, kg of adsorbate/kg of adsorbent), is a convenient way to describe the thermodynamic cycle of an adsorption desalination system. Where X_{max} and X_{min} represent maximum and minimum amount of water adsorbed. T_{gen} and T_{ads} are the temperature at which desorption and adsorption process begins.

3. Mathematical modelling of single bed adsorption desalination and cooling system

Mathematical modelling is based on adsorption isotherm and energy balance of components of the system.
To simplify the models, the following assumptions are made:
(1) The temperature is uniform across the adsorbent (silica gel) layer.
(2) Adsorption of the water vapour by the silica gel inside the adsorber bed is uniform.
(3) The Adsorbent and the adsorbate gas phases are in equilibrium condition.
(4) Heat loses from the adsorbent bed, evaporator, condenser are neglected.
(5) The specific heat capacities of water, adsorbent, metallic wall are constant.

Adsorption isotherms represent the amount of adsorbate that can be adsorbed per unit mass of dry material at a specific vapor pressure. These are helpful to assess the maximum amount that can be adsorbed/desorbed under certain conditions. Different isotherm models [10] have been developed for predicting the adsorbent material performance. In this model two different adsorbent are used silica gel and zeolite. The uptake of water vapour by the silica gel at a temperature (T) and pressure (P) can be estimated using Dubinin–Astakhov (D-A) [11,12] equation as

$$X = X_0 \exp[-(\frac{RT}{E}\ln\left(\left(\frac{p}{p_o}\right)\right))^n] \quad (1)$$

Where X_0 (kg/kg of adsorbent) is the maximum adsorbed amount, E (kJ/mole) is the characteristic energy and n is the D-A constant, R (J/mol.K) is gas constant, P (Pa) is the adsorption pressure and P_0 is the equilibrium pressure. The unknown values in Dubinin-Astakhov equation for type RD silica gel are available in literature[12-14]. The values of parameters used are given in Table 1. The mathematical modelling equations of the adsorption desalination cycle are solved using MATLAB 2015.

Table 1. Properties of silica gel

Parameter	Value
X_0	0.592 (kg/kg of silica gel)
E	3.105 (kJ/mole)
n	1.1
R	8.314 (J/mol. K)

The adsorption behaviour of the zeolite–water pair for adsorption and desorption processes is modelled by Dubinin-Radushkevich Isotherm [13].

$$X = X_0 \exp\left[-\left(K\ln\left(\frac{T}{T_{sat}}-1\right)^n\right)\right] \quad (2)$$

Table 2. Properties of zeolite

Parameter	Value
X_0	0.124 (kg/kg of zeolite)
K	5.012
n	1.4

Energy balance is applied for all the processes of the adsorption desalination cycle for finding out COP and energy requirement.

$$Q_{heating} = Q^{1\text{-}2} + Q^{2\text{-}3} \quad (3)$$

Where $Q^{1\text{-}2}$ is the total sensible heat input during this process 1-2 and $Q^{2\text{-}3}$ is the heat of desorption

Energy required per kilogram of water produced is given by:

$$E_w = \frac{Q_{heating}}{N_{water}} \quad (4)$$

The cold production takes place in the evaporator is given by:

$$Q_{ev} = N_{bed}(X_1 - X_4)h_{fg_{cw}} \quad (5)$$

Where $h_{fg_{cw}}$ is the latent heat of vaporization of saline water at the evaporator temperature.

Coefficient of Performance,

$$COP = \frac{Q_{ev}}{Q_{1\text{-}2}+Q_{2\text{-}3}} \quad (6)$$

4. Results and Discussion

Performance of single bed adsorption system has been compared for two different adsorbents namely silica gel-water and zeolite-water. Using adsorption equilibrium isotherms and energy balance equation the system has been thermodynamically modelled for two adsorbents. Amount of water produced, COP of the system and energy required for desorption and adsorption have been discussed and presented.

4.1. Effect of regeneration temperature on water production

Figure 3 represents the variation of amount of water production during the cycle with respect to the varying inlet hot water temperature (regeneration temperature) which is circulated through the bed. Amount of water produced using silica gel and zeolite as adsorbent are presented in figure 3.(a)&(b). In both cases the water production rate increases with increase in bed temperature and gradually flat out at higher temperatures. As hot water inlet temperature increases desorption also increases and hence the quantity of water desorbed is also increased and reaches a maximum value when adsorbed quantity completely removed from the adsorbent. It is clear from these comparison that the heat source for desorption process using zeolite as adsorbent must have comparatively higher temperature range with silica gel as adsorbent. Hence solar energy can effectively be uttilized for complete desorption process using silica gel.

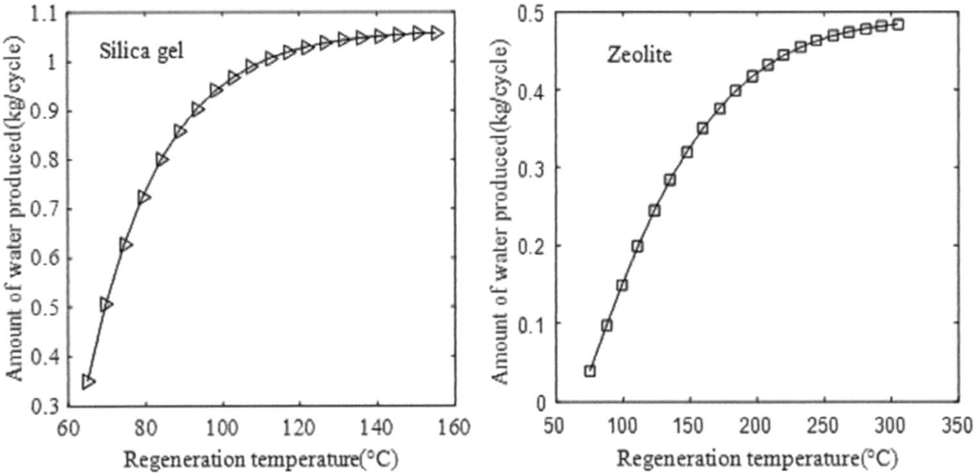

(a) (b)

Figure 3. Effect of regeneration temperature on water production for silica gel and zeolite

4.2 Effect of regeneration temperature on energy required per kg of water production

Variation of energy required per kg of water produced with hot water temperature is presented in figure 4. Energy required per kg of water produced is the ratio between the heating requirement of the system and the amount of water produced during a cycle. The energy consumption per kg of water production firstly decreases up to a minimum value and then increases. The decreasing tendency is due to the increased rate of water production. In the case of silica gel the energy consumption per kg of water production reduces to its minimum value at 80°C while in the case of zeolite is 150°C.

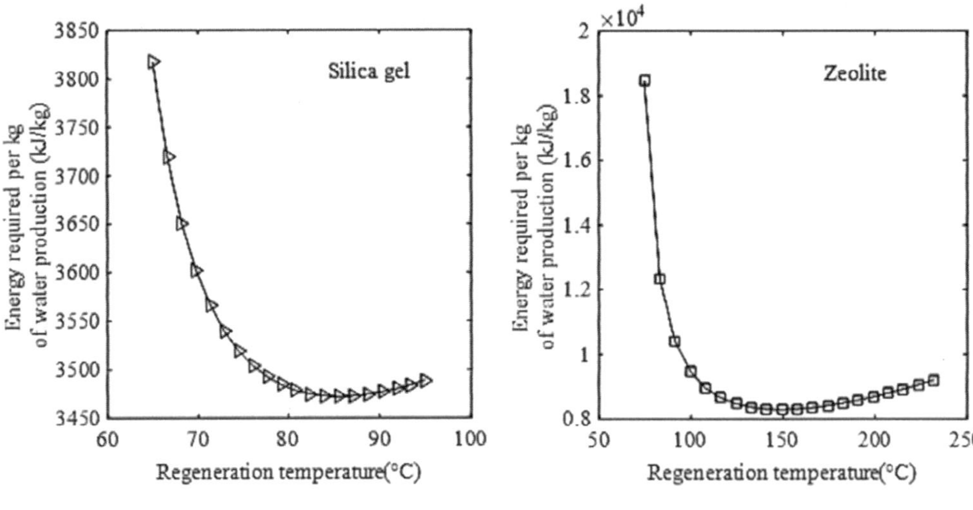

(a) (b)

4.3 1 **Figure 4.** Effect of regeneration temperature on energy required per kg of water production for silica gel and zeolite

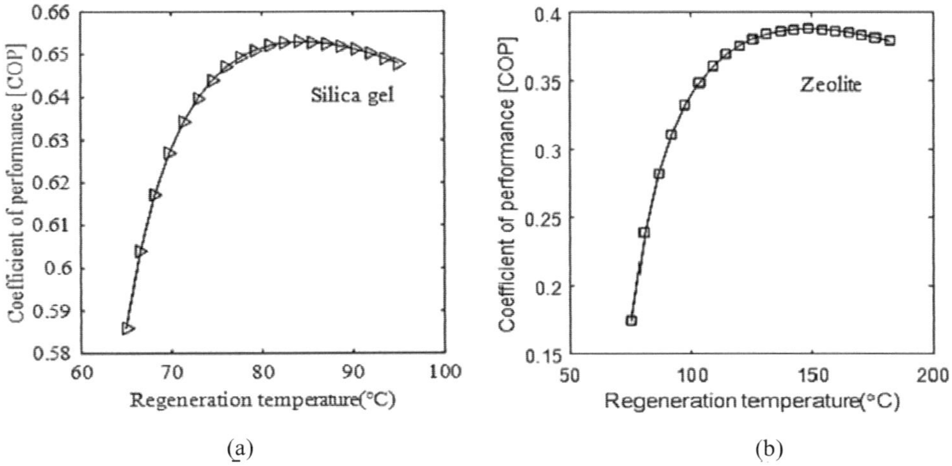

Figure 5. Effect of regeneration temperature on coefficient of performance

Figure 5 represents the variation of COP based on regeneration temperature. There is a certain temperature at which there exists a maximum COP for a fixed cooling water and condenser temperature. COP increases rapidly with the bed temperature and attains a maximum value then decreases slowly. It is because initially, the increase of refrigeration effect is more than the increase of heat input. COP is maximum at 80°C and 150°C in the case of silica gel and zeolite respectively. After these temperatures the heat input is very high, therefore COP decreases.

4.4 Effect of varying cooling water temperature on water production, energy required and cop.

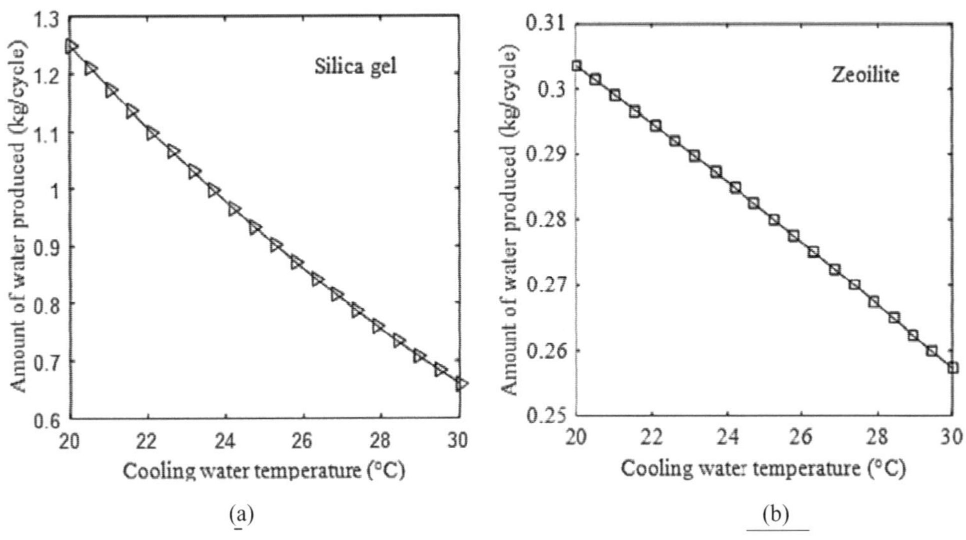

Figure 6. Effect of varying cooling water temperature on water production for silica gel and zeolite

The effect of cooling water temperature on water production is presented in figure 6. When cooling water temperature is low, the adsorbent bed temperature is also low and the adsorbent will adsorb more amount of water at low temperature. It will increase the amount of water production during the desorption stage. The increased adsorbed amount of water vapour at low adsorbent temperature increases the refrigeration effect. Due to decrease in cooling water temperature, more water production occurs and hence energy consumption per kg of water production decreases.

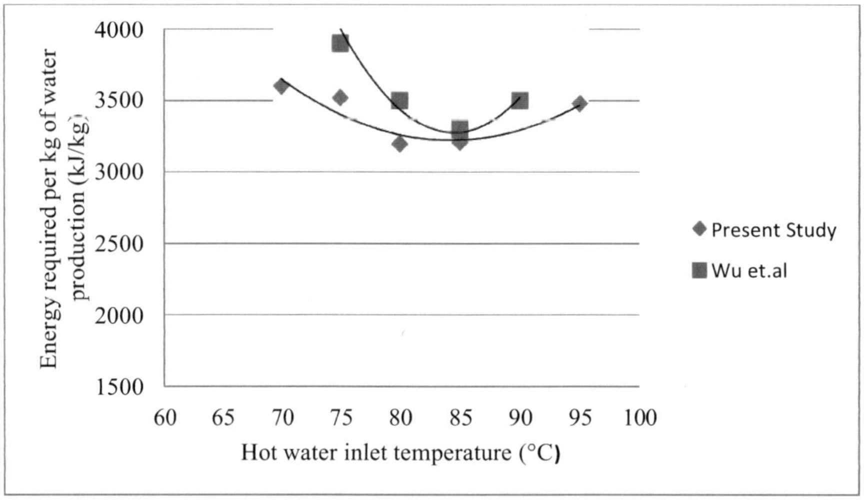

Figure 7. Comparison of energy required (kJ/kg) vs hot water inlet temperature

For validating the result, energy required for water production with hot water temperature is compared with result obtained by Wu et al. [14] ,who had conducted a thermodynamic analysis on two bed adsorption system using silica gel. As can be observed from figure 7, the simulated result of present study agree well with reported data. Both study indicates that the energy consumption per kg of water firstly decreases to a minimum value and then increases. The decreasing tendency is due to increased rate of water production.

5. Conclusion

In this study working principle of single bed adsorption desalination system is described and a mathematical model is developed using Dubinin-ashtakov isotherm. The performance characteristics of two working pair namely silica gel-water and zeolite-water are discussed. Results shows silica gel-water pair have better adsorption characters and zeolite-water pair requires comparatively high temperature heat source for stable desorption process. The obtained simulation results show that the hot water inlet temperature of the adsorbent and cooling water temperature have an impact on the water production,energy consumption and COP of the system. The better performance of the system can be obtained for low cooling water temperature and high hot water inlet temperatures of the adsorbent bed. From the study it can be concluded that siliga gel is a better performer than zeolite for this hybrid system.

ACKNOWLEDGEMENTS
Author's gratefully acknowledged Centre for Engineering Research and Development (CERD), A P J Abdul Kalam Technological University, Trivandrum, for sponsorship under CERD seed money scheme.

6. References

[1] Karagiannis, I C, & Soldatos P G 2008. Water desalination cost literature: review and assessment. *Desalination*, *223*(1–3), 448–456

[2] Ng, K C, Saha, B B, Chakraborty A and Koyama S 2008 Adsorption desalination quenches global thirst. *Heat Transfer Engineering*, *29*(10), 845–848

[3] Ng K C, Thu K, Kim Y, Chakraborty A and Amy G 2013 Adsorption desalination: An emerging low-cost thermal desalination method. Desalination, 308, 161–179

[4] Aly N H and El-Figi A K 2003 Thermal performance of seawater desalination systems. *Desalination*, 158(1-3), 127-142

[5] Li C, Goswami Y and Stefanakos E 2013 Solar assisted sea water desalination: A review *Renewable and Sustainable Energy Reviews*, 19, 136-163.

[6] Wu J W, Hu E J and Biggs M J 2011 Thermodynamic analysis of an adsorption-based desalination cycle (part II): Effect of evaporator temperature on performance. *Chemical Engineering Research and Design*, 89(10), 2168–2175

[7] Kim Choon Ng, Kyaw Thu, Anutosh Chakraborty, Bidyut Baran Saha and Won Gee Chun 2009 Solar-assisted dual-effect adsorption cycle for the production of cooling effect and potable water. *International Journal of Low-Carbon Technologies* 4; 61–67.

[8] Ahmed S Alsaman, Ahmed A Askalany, K Harby, Mahmoud S Ahmed 2016. A state of the art of hybrid adsorption desalination–cooling systems. *Renewable and Sustainable Energy Reviews* 58; 692–703.

[9] Kim Choon Ng, Kyaw Thu, Bidyut Baran Saha, Anutosh Chakraborty 2012. Study on a waste heat-driven adsorption cooling cum desalination cycle. *International journal of refrigeration* 35; 685-693.

[10] Nimibofa Ayawei, Augustus Newton Ebelegi, and Donbebe Wankasi 2017. Modelling and Interpretation of Adsorption Isotherms. *Journal of Chemistry Volume* , Article ID 3039817.

[11] K C Ng, H T Chua, C Y Chung, C H Loke, T Kashiwagi, A Akisawa, and B B Saha 2001 Experimental investigation of the silica gel–water adsorption isotherm characteristics, *Appl. Therm. Eng.* 21 (16) 1631–1642.

[12] Ali S M,and Chakraborty A 2016 Adsorption assisted double stage cooling and desalination employing silica gel + water and AQSOA-Z02 + water systems. *Energy conversion and management*, 117, 193–205

[13] Solmuş I, Yamali C, Kaftanoğlu B, Baker D and Çağlar A 2010 Adsorption properties of a natural zeolite-water pair for use in adsorption cooling cycles. *Applied Energy,* 87(6), 2062–2067

[14] Ali S M,and Chakraborty A 2016 Adsorption assisted double stage cooling and desalination employing silica gel + water and AQSOA-Z02 + water systems. *Energy conversion and management*, 117, 193–205

Journal of Physics: Conference Series **1355** (2019) 012014 doi:10.1088/1742-6596/1355/1/012014

Simulation of Stratified Two-phase Flow Regime using Air-Water Model in *ANSYS Fluent*®

Binet Monachan, Rijo J. Thomas, Deepak Steaphen, Mathew Skaria, Shafi K.A.
Department of Mechanical Engineering, T K M College of Engineering, India.

E-mail: binetmonachan007@gmail.com

Abstract. Two-phase flow has widely grown in importance in various fields of science and engineering system such as nuclear reactors, heat exchangers, transport system, chemical processing plants etc. One of the major issues faced by these sectors is the development of an accurate and reliable multiphase flow measurement system. At present there is no single system which could measure the two phase flow parameters irrespective of its incoming conditions. Characterization of the single phase flow parameter is easier when compared with two phase flow as the velocity distribution, mass flow rate and void fraction of each component phases are not easily measurable. The other factors that add to these difficulties are the prediction of flow regimes and the effect of pipe or channel orientation on the flow properties in the two phase flow. The hold-up of the individual phases and their relative velocity along with pressure drop needs to be taken care while dealing with these flows. Various researches have clearly shown that no correlation can be used for predicting the flow regimes in a two phase flow satisfactorily. By using volume of fluid multiphase flow model air water two phase flow was simulated for inlet velocities of water and air as 0.121 m/s and 6.56 m/s respectively. Stratified laminar flow was obtained from the CFD simulation. Variation in volume fraction of both air and water, mass flow rate and pressure drop with respect to change in time are investigated using the simulation.

1. Introduction

Liquid-gas or gas-liquid two-phase flows are common phenomena in various engineering fields like chemical, nuclear and petroleum industries. These two phase flow are most common in all applications wherever cryogens are being used as coolants which include aerospace, superconductivity applications etc. The characteristic study of these two phase flow is complex mainly because of the different flow patterns encountered during the flow. On comparing with the single phase fluid flow, its concept is well understood but its complex behaviour limits its application. As a result of this many experimental, simulation and modelling studies are being carried out to characterize the two phase flow behaviour with the aim of simplifying the complex concept behind it.

Many different flow regimes are possible for two phase flow whereas in single phase flow there exists only laminar or turbulent flow along with a transition region in between the two. In two phase flow it may be turbulent in liquid phase and laminar in vapour phase or any of the four different combinations possible. Some sources vary on the descriptions of the flow and the exact number of variations. This is because certain flow patterns are either very similar or are transitional regimes, so they are grouped together as single pattern. While the flow regime cannot be determined from the void fraction or flow quality alone, the flow does typically change with respect to the increase or decrease of either of these values [1].A brief description of the flow patterns is given, arranged with respect to increasing flow quality:

• Bubbly flow: Gas bubbles of various sizes flow at approximately the same velocity as the liquid. At lower liquid flow rate, the bubbles tend to travel at the top of the channel. At higher liquid flow rates, the bubbles are seen more evenly dispersed in the channel.

• Plug flow: As the gas flow rate increases, the small bubbles begin to combine to form 'plugs' of gas in the channel.

• Stratified flow: Gravitational forces separate the liquid and gas phases. The liquid gas interface is smooth. This is the rarest type of two-phase flow because it occurs most often in larger diameter pipes and laminar flow.

• Wavy flow: Increasing mass flow rates creates disturbances at the phase boundary. The amplitude of these waves increases as the mass flow rates continue to increase. At the higher gas flow rates, semi-slug flow begins to appear and is sometimes considered a separate flow regime.

• Slug flow: Wave amplitude is high enough that the liquid touches the top of the channel. The vapour is completely separated by 'slugs' of liquid traveling in the channel.

• Annular flow: At very high vapour flow rates, the liquid is forced to the walls of the channel and the gas travels in the center. At higher gas flow rates, liquid droplets are found in the vapour flow.

• Dispersed flow: The vapour flow rate is so much higher than the liquid flow rate that the liquid is broken into small droplets and carried in the vapour stream. As the vapour flow rate continues to increase, the droplets continue to get smaller and eventually become froth flow. Figure 1 shows common flow patterns seen in horizontal two-phase flow.

These flows were mostly been investigated only experimentally due to their complex nature and all the empirical relations derived on the basis of these experimental results were valid only in a limited range of operating condition [1]. The growth in computer technology provided flexibility for using and constructing large-scale computational models for analysis of these complex two-phase flow types [2]. Determination of the geometry of the flow patterns or flow regimes is considered as one of the major difficulties in the modeling of two-phase gas/vapor–liquid flow [3].

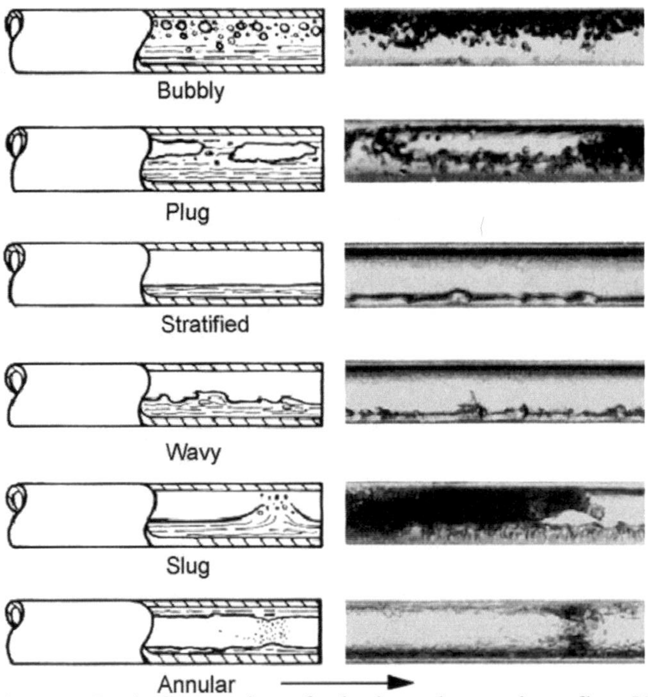

Figure 1: Basic flow regimes for horizontal two-phase flow [1]

2. Two-Phase Flow Theory

The mathematical model for two-phase flow can vary significantly depending on the flow regime that is in question. There are, however, some common terms that should be recognized when dealing with two-phase flows. The key factors and common terms that determine and describe two-phase flow are void fraction (α), flow quality (χ), and slip (S). In all the equations given below suffix *v* and *l* denote the vapor and liquid phase in two phase flow. Void fraction is the ratio of the vapor cross sectional area to the total area. This is one of the most basic commonly mentioned terms in two-phase flow and can be written as

$$\alpha = \frac{A_v}{(A_v + A_l)} \tag{1}$$

The flow quality is the ratio of vapor mass flow rate to the total mass flow rate and is given as

$$\chi = \frac{m_v}{(m_v + m_l)} \tag{2}$$

Slip refers to the ratio of the vapor velocity to the liquid velocity which helps to determine how the liquid gas phase boundary behaves and is given by the relation

$$S = u_v / u_l \tag{3}$$

The slip can be related to the other parameters by the following relation

$$S = \frac{\chi}{(1-\chi)} \frac{(1-\alpha)}{\alpha} \left(\frac{\rho_l}{\rho_v}\right) \tag{4}$$

When the slip ratio is unity, the void fraction and flow quality are related simply by the different liquid and vapor densities. When the slip ratio is not known, the void fraction cannot be easily determined from the flow quality and vice versa. Due to this, the flow regimes have been identified by numerous maps developed for different channel configurations. Baker map, later modified by Scott has been used by several authors for predicting flow characteristics in horizontal flow. Attempts have been made to simplify the problem of determining the flow regime in a quantitative way. There have been numerous maps developed for different channel orientations. Some of the maps are relatively simple and use only two axes to distinguish the flow patterns, while others involve different axes for different flow patterns.

3. Flow Pattern Maps

While liquid-gas two-phase flow is complicated, the flow patterns are distinct and can be consistently categorized. The primary factor which governs the flow behavior in a two phase flow is the vapor-liquid phase interaction that varies as the flow velocity of the phases change. Along with phase interaction, fluid characteristics like viscosity and surface tension plays a prominent role in determining the behaviour of the flow. The flow characteristics depends on the flow channel orientation whether it is horizontal, vertical or inclined at an angle and also depends on the gravitational force acting on it. Various researches have shown that no correlation or single theory can be used for predicting the pressure gradient or liquid hold-up over all flow regimes encountered in two phase flow in pipes satisfactorily [4]. Both empirical correlation methods and mechanistic models are still widely used to predict accurately what flow pattern will occur for given input flow rates, pipe size, and fluid properties [5]. Then only proper flow model can be selected for the two phase flow analysis. Various literatures have presented different methods for this purpose, usually in the form of two-dimensional maps in which the locations of the boundaries between flow pattern regions are based on empirical observations. Many of these maps covers a rather limited range of fluid properties and pipe diameters. Consequently, large discrepancies are often observed between a predicted flow regime and that actually observed in a subsequent test.

Different flow pattern maps developed by various researchers are briefly described below in chronological order. One of the first flow pattern maps was suggested by Bergelin and Gazley which

was based on the air water system in a 1 inch pipe. It uses the liquid and gas mass flow rates m_l and $m_g$as the coordinates [5]. A similar flow pattern map based on air water data in 0.87 inch pipe was proposed by Johnson and Abou-Sabe [6]. Alves [7] suggested a flow pattern map based on data for air water and air--oil mixtures in a 1 inch pipe utilizing the superficial liquid and gas velocities, v_{sl} and v_{sg} as the coordinates. He represented both of the systems on a single map.
Figure 2 shows the Baker map, the most commonly used chart for predicting flow characteristics in horizontal flow. The Baker flow regime map shows the boundaries of the various flow pattern regions as functions of the superficial mass velocity of the gas phase and the ratio of superficial mass velocities of the liquid and gas phase. It has been used as a rough guide in prediction of the two phase flow regime in horizontal pipes. The dimensionless parameters λ and ψ were added so that the chart could be used for any gas/vapor–liquid combination rather than the standard combination, at which both the parameters λ and ψ will be equal to one. The standard combination is water and air flow under atmospheric pressure and at room temperature. Appropriate values for λ and ψ will help in proper prediction of the geometry of two-phase flows with any gas/vapor and any liquid at other pressures and temperatures. Although the transition from one flow pattern region to another is presented as a line, in reality these lines are rather broad transition zones. Each researcher probably observed the transition at slightly different superficial flow combinations during the experimental work [1]. Baker chart and was later modified by Scott and it does not have clearcut transition boundaries but instead shows relatively wide bands depicting regions of transition from one flow pattern to another [4].

In this map the vertical axes and horizontal axes denotes G_G/λ and $(\lambda\, G_L\, \psi)/G_G$respectively. G_G and G_L are the mass flux of the gas phase and liquid phase respectively. Expression for λ and ψ are given in the below equations.

$$\lambda = [(\rho_G/\rho_a)\ (\rho_L/\rho_W)]^{\frac{1}{2}} \quad (5)$$

$$\psi = (\rho_W/\rho_L)\ [(\mu_L/\mu_W)(\rho_W/\rho_L)^2]^{\frac{1}{3}} \quad (6)$$

There has also been a large amount of work done in predicting the pressure drop in a two-phase flow system. These mathematical models of the pressure drop vary significantly from one to the other and usually require at least the void fraction and flow quality to be known. These models also usually only work for very specific flow conditions and have strict sets of assumptions that must be met for them to apply.

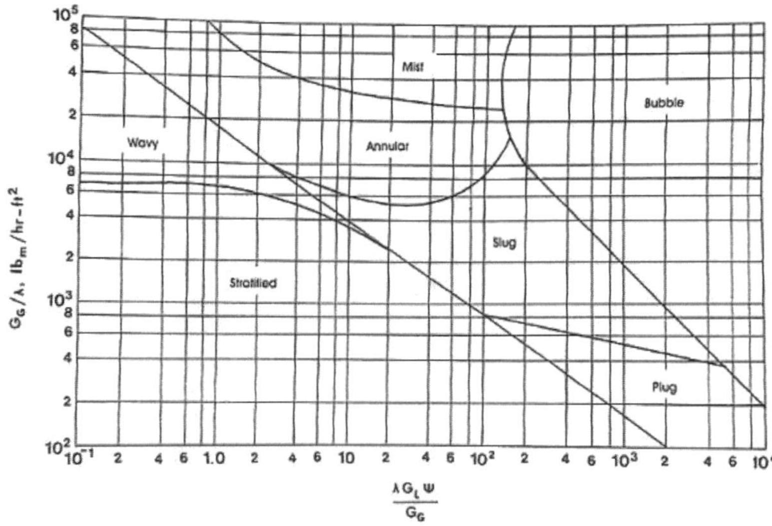

Figure 2: Baker map for two-phase flow in horizontal tubes [8]

4. Experimental Techniques for Flow regime identification

In consideration with the importance and complexity of two phase flow, various technique was developed for the identification of different flow regimes. The characteristics of two phase flow regime depend on the length, diameter, orientation of the tube, superficial velocities of two phases and properties of the fluids. Since two phase flow patterns play a vital role in two phase flow pressure drop, void fraction, heat and mass transfer characteristics there is a need to precisely identify the flow regimes. Several experimental techniques were adopted in the past to study the two phase flow patterns. Guilizzoni [9] studied the flow pattern of two phase flows from the change in density of the phases sampled using impedance probe. But the impedance probes interrupts and changes the characteristics of the fluid flow. Later it was found that nonintrusive techniques such as gamma ray or X-ray based techniques do not disturb the flow. Kumar et al. [10] and Ikeda et al. [11] used Gamma ray tomography to measure the void fraction in two phase flows and X-rays to characterize the two phase flows respectively. The amount of the received Gamma ray was used to characterize the flow pattern. The change in dielectric constant of two phase medium is used in determining the void fraction using the capacitance probe to determine the flow regime. Jaworek et al. [12] developed a capacitance sensor operating at 80MHz to measure the void fraction in air water two phase flow. Image analysis and Particle Imaging velocimetry techniques were used to study the gas-liquid slug flow along vertical pipes and to characterize the two phase flows.

Very few works have been done earlier using Infrared sensing technique to study the two phase flow patterns. One such work is by Ruixi et al. [13] where the voltage output from the sensor is used to characterize the two phase flow pattern. In the experiments a lower voltage around 0.46 indicates a water flow and a higher voltage indicates air. The elongated portion at higher voltages indicates a slug flow. However the fluctuations obtained at the lower voltages as shown in figure 3 were observed in the experimental results. Arunkumar et al. [14] designed a pair of parallel copper electrodes on the two sides of a glass tube acts as a dielectric sensor for two phase flow regime identification. The principle involved in the measurement is that the changes in effective permittivity of the two phase fluid mixture. The effective permittivity of the medium changes as the void fraction in the glass tube changes and causes variation in the capacitance value across the electrodes. Adhavan et al. [15] used an infrared sensor for identifying the two phase flow regimes. The current flowing across the receiver is used in characterizing two phase flow and three flow regimes were identified using this technique. The basic principle behind measurement was that the current changes with intensity of IR rays falling on receiver and on the basis of current output the flow pattern was characterized. However experimental results in figure 4 shows comparatively lesser fluctuations because of refined acquisition speed and good ambient conditions (dark room environment) maintained throughout the experiment.

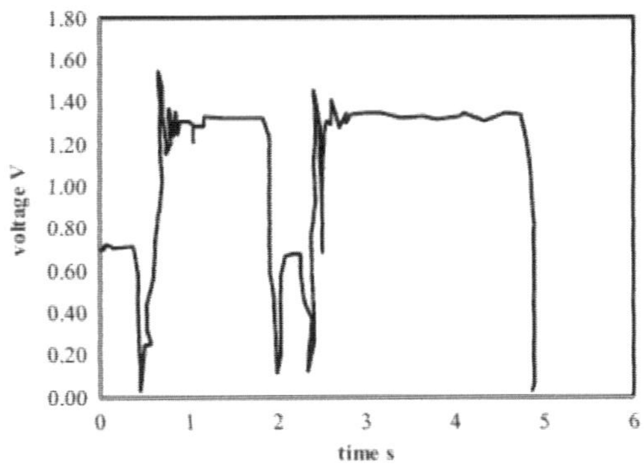

Figure 3: Slug flow pattern characterization [13]

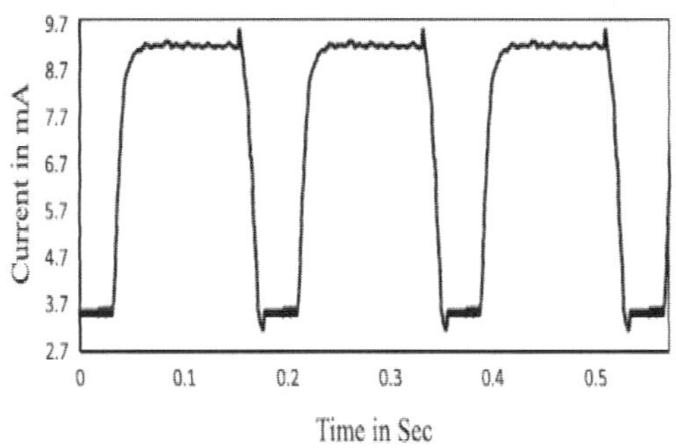

Figure 4: Slug flow pattern characterization [15]

Void fractions can be monitored by capacitance probes by measurements between the two electrodes or multiple pairs of electrodes. The change in void fraction leads to change in capacitance in view of the changes in the dielectric constant changes between the vapour and liquid. The capacitance between two flat plates is given by the relation

$$C = \varepsilon A/l \quad (5)$$

While that of a cylindrical capacitor with two concentric cylinders having outer and inner radius as r_o and r_i respectively, with a given space between them is given by

$$C = 2\pi\varepsilon L/(\ln(r_o/r_i)) \quad (6)$$

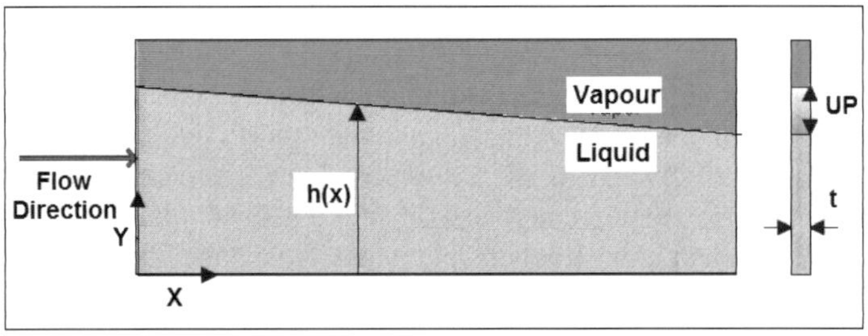

Figure 5: Laminar stratified flow of fluid

Depending on the orientation of the capacitor plates with respect to the flowing fluid, the capacitances get arranged either in series or parallel configurations. The laminar two-phase flow meter proposed to be developed behaves similar to that of an open-channel. This means that the fluid flows in the channel, but the liquid surface is also exposed the vapour zone, similar that of a canal or river. The idea of the flow meter is that there is laminar, two-phase flow which passes through a number of tall, narrow channels. In this laminar flow, the liquid and gas phases get separated into a stratified flow. Such a flow leads to the development of a slope down the channel length due to the difference in viscosity between the liquid and gas and the friction with the narrow channels as shown in figure 5. The viscous force between the liquid and the wall is more than that between the gas and the wall and this leads to a negative slope of the liquid level as it travels. Figure 6 shows the schematic of the proposed typical laminar flow stratified flow meter.

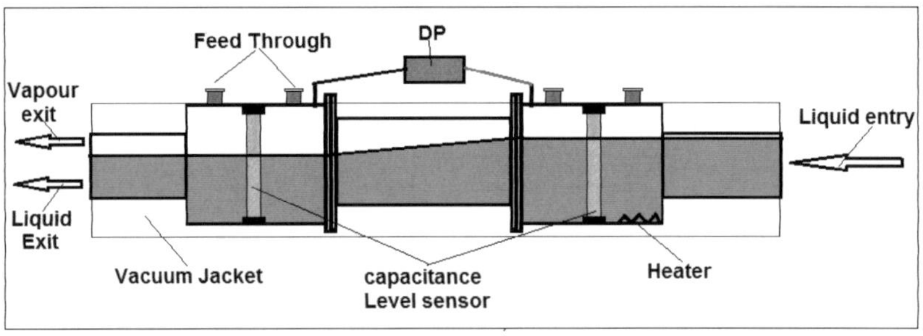

Figure 6: Typical Laminar flow stratified flow Meter Schematic

5. CFD Modeling

Due to the complex behaviour of two phase gas liquid or vapour liquid flow, they are mostly being studied experimentally. Large number of empirical correlations were developed for describing the flow but most of them are only valid in a limited range of operating conditions. The growing computer facilities has helped to sort this problem to some extend as it has provided much flexibility to construct and use large scale computational models to calculate these complex two phase flow types. Thus experimental research is not always needed [2]. CFD is a powerful tool to understand and analyze the characteristics of two phase flows. Modelling of two phase simulation involve three basic steps. The first step is the definition of the number of phases and consideration of the possible flow regime. Next step is the formulation of basic flow governing equations which represent the mathematical statement related to conservation laws of physics. The final step involves solving of these equations that describe the mass, momentum and energy conservation in the selected domain.

The CFD simulations were performed by using the commercial software ANSYS Fluent which is capable of doing all CFD process in a single interface. Various multiphase models are available in ANSYS of which VOF model is being opted for studying the two phase flow regime. VOF model is being considered as the direct method of predicting interface shape between the two immiscible phases. In two phase flow modelling only a single set of equation is used for the purpose of tracing the void fraction of individual phases in VOF model [1]. In the VOF method the sum of the volume fraction of gas and vapour phases are taken as unity. The evaluation is being done by solving the momentum and continuity equation.

5.1 Geometry and meshing

A three dimensional model of a horizontal rectangular channel with length, height and width as 305mm, 74mm and 1mm respectively has been created in ANSYS design modeler. The rectangular channel was chosen as the geometry so that the flow becomes similar to that of an open-channel. This means that the fluid flows in the channel, but the liquid surface is also exposed the vapour zone, similar to that of a canal or river. A fine quadrilateral mesh of the geometry is being done using ANSYS mesh modeler. Number of elements and nodes for meshing was 30960 and 25000 respectively.

5.2 Solution strategy and boundary conditions

The analysis is being carried out under no slip conditions. Since the two phase flow is time dependent an implicit unsteady study needs to be done. The effect of gravity on the flow characteristics were considered. The number of eulerian phases were given as two in the VOF model. The inlet condition for both the phases are being taken as velocity inlet and the outlet boundary condition was given as pressure outlet. The operating pressure is being taken as 1.01325 bar during the flow analysis. In this work air water two phase system is being considered for analysis with water as the primary phase and air as the secondary phase. The phase interaction needs to be specified for the VOF model. Effect of surface tension on the flow characteristics should be considered so for that one of the surface tension model need to be applied.

Continuum Surface Force (CSF) model needs to be applied for calculating the surface tension force. For both the phases velocity inlet condition was specified for the inlet surface. In this type of boundary condition both the velocity profile and volumetric fraction of the phases needs to be specified. The inlet surface is being face split for the above mentioned factor. Volume fraction of air at the upper half of the inlet surface was 1 and that at the lower half was 0. This means that air enters the domain through the upper half and water enters through the lower half. The outlet surface was modelled as pressure outlet with a constant pressure of 0 Pa. No slip condition was assumed for the channel wall. Solution method used for pressure velocity coupling was PISO algorithm and a second order upwind scheme is used for determination of flow characteristics. Modelling is done in transient state and details of time step can be found from the table shown below.

Table 1: Time step details used

No.	Title	Description
1	Time step method	Fixed
2	Time step size	0.001
3	Number of time steps	20000
4	Max iteration / time step	20

6. Results and discussion

Analysis was done with the velocity of water and air taken as 0.121 m/s and 6.56 m/s respectively. These velocities were selected as per the pressure drop correlation ensuring that both the liquid and gas phase are having laminar flow. The use of tall narrow rectangular channel and the velocity mentioned above has forced the flow into laminar two phase flow. In this laminar flow, the liquid and gas phases get separated into stratified flow as shown in figure 7. Since the mixture density is proportional to its phase composition, the distribution of air and water in the channel can be clearly seen.

In the volume fraction of air contour red colour refers to the presence of air while the dark blue colour refers to the presence of water. Meanwhile in the volume fraction of water contour red colour shows water and dark blue colour shows air. Such a flow leads to the development of a slope down the channel length due to the difference in viscosity between the liquid and gas and the friction with the narrow channels. The viscous force between the liquid and the wall was more than that between the gas and the wall and this leads to a negative slope of the liquid level as it travels. The stratified flow is observed as a liquid medium flowing at the bottom of the tube and the gas medium occupying the remaining part. Gravity plays a vital role in stratified flow. When superficial gas velocity is high and superficial liquid velocity was very low, stratified flow occurs.

a) Volume fraction air

b) Volume fraction water

Figure 7: Volume fraction contour of air and water for stratified laminar flow

Development of stratified flow during the flow time of 1 second is shown in figure 8 and the formation of negative slope can be clearly seen from the contour given below. Figure 9 shows the mass flow rate plot of air water mixture against the flow time. Effect of viscosity, friction and gravity leads to the separation of the two phases by a smooth air water interface leading to the evolution of stratified flow. It is found that the mass flow rate of air shows an increasing trend whereas the mass flow rate of water shows a decreasing trend. The total flow time taken for the simulation was 1 second. The mass flow rate of air water mixture also shows an increasing trend as the flow time increases. It is mainly because of the higher velocity of air when compared with velocity of water. The mass flow rate of the air water mixture was found to be 0.49 kg/s.

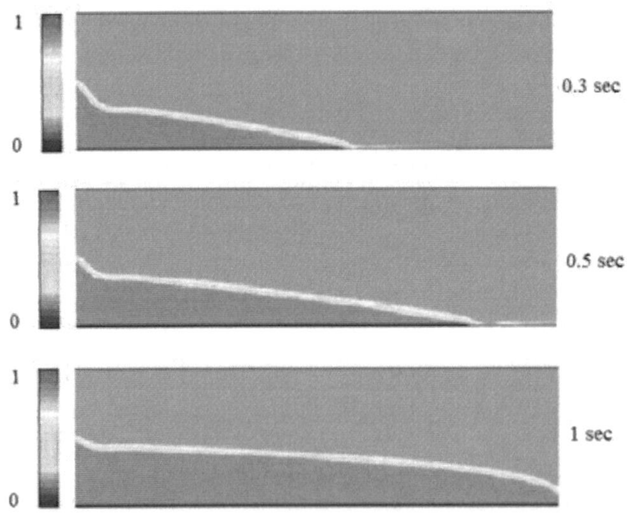

Figure 8: Contour of volume fraction obtained after a flow time of 0.3, 0.5 and 1 second

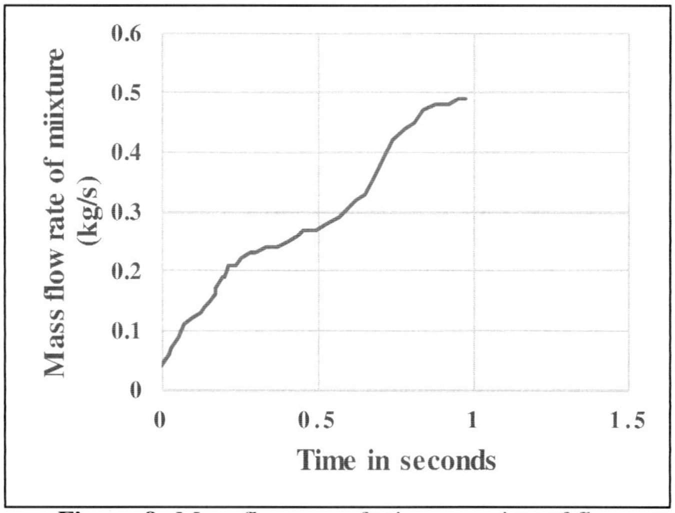

Figure 9: Mass flow rate of mixture vs time of flow

Volume fraction for water was plotted against time of flow as shown in figure 10. Total time of flow is being taken as 1 second and with increase in flow time the volume flow rate of water shows an increase. When the flow time reaches 1 second the volume fraction of water was found to be 0.35. The volume fraction of air has dropped from a value of 1 to 0.65 during the flow time whereas the volume fraction of water increases from 0 to 0.35 during the flow time. Figure 11 shows the pressure contour for the stratified two phase flow and from the plot we could find the pressure drop during the flow as 600 Pa. Highest pressure region was shown at the water inlet and the pressure at the air inlet was found to be around 340 Pa.

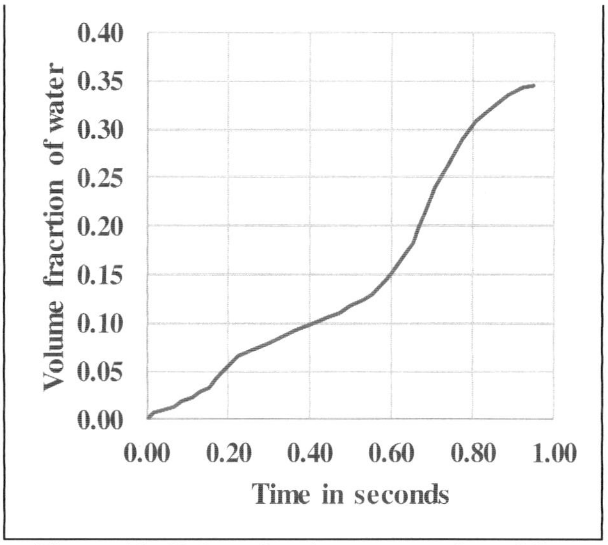

Figure 10: Volume fraction water vs time of flow

Figure 11: Pressure contour for stratified flow

7. Conclusion

The two phase stratified air water flow pattern is obtained when the inlet velocity of air and water was 6.56 m/s and 0.121 m/s respectively. Mass flow rate of the mixture shows an increasing trend with increase in flow time. Pressure contour was analysed and pressure drop during the two phase flow was found to be 600 Pa. Volume fraction of water after a flow time of 1 second was found to be 0.35. Simulation was only done with one set of velocity and for validating further simulation needs to be done with other set of velocities. The problem with the CSF model is the discontinuity in the domain, such as an interface and cannot be solved by a finer mesh. So proper meshing needs to be done for improving the accuracy of results. The geometry used for simulation is not provided with any development section and buffer zone. Development section will help in making the flow fully developed before entering the test section. Buffer zone will reduce unwanted fluctuation in the simulation result at the pressure outlet section.

Reference

[1] De Schepper, C K Sandra, G J Heynderickx and G B Marin, 2008 CFD modeling of all gas liquid and vapor–liquid flow regimes predicted by the Baker chart *Chemical Engineering Journal* **Vol 138(1-3)** 349–357.

[2] I Lun, R K Calay and A E Holdo, 1996 Modelling two-phase flows using CFD *Applied Energy* **Vol 53** 299–314.

[3] S Ghorai and K D P Nigam, 2006 CFD modeling of flow profiles and interfacial phenomena in two-phase flow in pipes *Chemical Engineering and Processing* **Vol 45** 55–65.

[4] J N Al-Sheikh, D E Saunders and R S Brodkey, 1970 Prediction of flow patterns in horizontal two-phase pipe flow *Canadian Journal of Chemical Engineering* **Vol 48** 21-29.

[5] J M Mandhane, G A Gregory and K Aziz, 1974 A flow pattern map for gas liquid flow in horizontal pipes *Int. Journal of Multiphase Flow* **Vol 1(4)** 537–553.

[6] H A Johnson and A H Abou-Sabe, 1952 Heat Transfer and Pressure drop in a Horizontal Pipe *Trans. ASME*, **Vol 74** 977–987.

[7] G E Alves, 1954 Co-current liquid-gas flow in a pipe-line contactor *Chemical Engineering Progress* **Vol 50** 449–456.

[8] R F Barron, 1999 *Cryogenic heat transfer* Louisiana Tech University Ruston Louisiana.

[9] M Guilizzoni, 2013 Flow pattern identification in gas liquid flows by means of phase density imaging *Int. Journal of Multiphase Flow* **Vol 51** 1–10.

[10] S B Kumar, D Moslemian, M P Dudukovic, 1995 A γ-ray tomographic scanner for imaging voidage distribution in two-phase flow systems *Flow Measurement and Instrumentation* **Vol 6 (1)** 61–73.

[11] T Ikeda, K Kotania,Y Maeda and H Kohno,1983 Preliminary study on application of X-ray CT scanner measurement of void fractions in steady state two-phase flows *Journal of Nuclear Science and Tech.* **Vol 20(1)** 1–12.

[12] A Jaworek, A Krupa and M Trela, 2004 Capacitance sensor for void fraction measurement in water/steam flows *Flow Measurement and Instrumentation* **Vol 15** 317–324.

[13] D Ruixi, Y Da, W Haihao, G Jing, L Ying, Z Tong and Z Lijun, 2013 Optical method for flow patterns discrimination, slug and plug detection in horizontal gas liquid pipe *Flow Measurement and Instrumentation* **Vol 32** 96–102.

[14] S Arunkumar, J Adhavan, M Venkatesan, S K Das and A R Balakrishnan, 2015 Characterization of gas liquid two phase flows using dielectric sensors *Flow Measurement and Instrumentation* **Vol 45** 274–279.

[15] S Arunkumar, J Adhavan, M Venkatesan, S K Das and A R Balakrishnan, 2016 Two phase flow regime identification using infrared sensor and volume of fluids method *Flow Measurement and Instrumentation* **Vol 51** 49–54.

Effect of Orifice Geometry on Heat Transfer Characteristics of Jet Array

Leena R[1], Arunraj B[2] and M. Jose Prakash[3]
T.K.M college of Engineering, Kollam, Kerala, India

E-mail: [1]leenakalidas@gmail.com, [2]arunrajbabu.95@gmail.com, [3]jpmmech@yahoo.co.in

Abstract. Jet impingement is one of the best methods for achieving high heat transfer coefficient in a single phase and has been a topic of active research for several decades, involving both experiments and computations. Most of the research on multiple jets have been carried out for an array of jets in the high Reynolds number regime. Experimental and numerical investigations were carried out to study the effect of inline three jet array impinging on a flat plate with varying Reynolds Number. Air was considered as the cooling medium. This study investigates the fluid flow and heat transfer characteristics of circular and square jet arrays impinging orthogonally on a flat-plate with inline arrangements. The number of jets is 3. The Reynolds number is varied from 3000 to 6000. The numerical methodology is validated using the experimental local Nusselt number distribution for inline three jet array with the same geometry and boundary conditions. In this work, circular and square of equal hydraulic diameters were selected for a comparative study. The results reveal that square jet array has better heat transfer characteristics compared to a circular array. The optimum value of jet to target plate distance is 6 times the jet diameter for both circular and square jet array. Also, the optimum value of pitch to diameter ratio is 2.

Keywords: Jet impingement; inline array; Nusselt number

1. Introduction

Thermal management has become very important in industries due to the advancement of technology. The primary objective of thermal management is to remove heat from the thermal devices to ensure that they operate properly and avoid triggering the temperature-activated failure mechanisms. This is generally accomplished by conducting the heat away from the thermal bodies to a gas or liquid coolant provides significantly high local heat transfer coefficient. In this method, a thin boundary layer is formed due to the collision of the fluid with the wall surface. But the major disadvantage of jet impingement is that the local heat flux can be highly non-uniform. Processes such as cooling of gas turbine components, tempering of glass, drying of paper and textile, cooling of high heat flux supercomputers, power devices and advanced military avionics require a promising technique like jet impingement. A single jet is usually employed to produce localized heating and cooling. In many applications a large area is required to be heated or cooled or enhancement of global heat transfer is needed. Thus, it is necessary to apply multiple jet system or array of jet system. Uniform temperature distribution can be obtained by array of jets. Since multiple jets cover a large area the key factors to be considered are the high average heat transfer coefficient and uniformity of heat transfer over the target surface. In a multi jet configuration, individual jet can be affected by essentially two types of interactions. The first is the jet-to-jet interaction between pairs of adjacent jets prior to the impingement. Secondly, there is interaction between the impinging jets and the flow formed by the spent air of the neighbouring jets. Therefore, studies on multiple jets have been mainly focused on identifying the optimum multiple conditions by considering the interaction among the jets.

Ozmen [1] found that the interaction between the twin jets decreases with the increase in nozzle-to-plate distance and jet-to jet spacing. Goodro et al. [2] conducted experimental studies on the thermal characteristics of circular jet array of 8d and 12d spacing. Local and spatially-averaged Nusselt number show a strong dependence on the jet Reynolds number for both situations, with negligible variations between Mach number = 0.1 and 0.2.

Chander et al. [3] conducted experimental studies on three interacting circular jets arranged in triangular configuration at low Reynolds number of 800 with inter-jet spacing (S/d) of 3, 4, 6 and 7.58. The interaction between the jets becomes stronger when the jet-to-jet spacing is 3d. It was found that the average heat flux is more uniform at a moderate separation distance of 5d and for a small inter jet spacing of 3d. Katti et al. [4] studied the influence of streamwise pitch on local heat transfer characteristics for in-line arrays of circular jets. They concluded that the stagnation Nusselt number decreases with increase in nozzle-to-plate distance and with the reduction in jet spacing due to increase of interaction between the jets and cross flow. Katti et al. [5] studied the effect of stream wise pitch on the local, stagnation and strip average heat transfer coefficients of multiple round jets.

Puneet Gulati et al. [6] studied the effect of jet-to plate spacing and Reynolds number on the local heat transfer distribution to normally impinging submerged circular air jet on a smooth and flat surface. San and Lai [7] studied the effect of jet to jet and jet to wall distances to optimize the heat transfer in stagnation point and obtained a correlation as a function of Reynolds number, s/d and H/d ratios. Heyerichs and Pollard [8] assessed the performance of turbulent models k–ω and several versions of the k– ε by considering both separating and impinging turbulent flows with heat transfer. Lance Fisher [9] conducted a numerical study to investigate the heat transfer to an axisymmetric circular impinging air jet using the k-ϵ turbulence model.

Srivalli et al. [10] analysed numerically laminar flow of multiple staggered array of square jets for Reynolds number ranging from 100 to 500. Leon et al. [11] has revealed that the thermal pattern in an in-line arrangement of circular jets. They reported that the interaction with the neighbouring jets affects the flow more strongly for the jet from a sharp-edged orifice than that from a contoured orifice. Yoshisaburo et al. [12] conducted experimental studies with square in-line array of circular jets for minimum cross flow condition. They found that the average Nusselt number shows its maximum value for a jet-to-jet spacing of 4d compared to 6d and 8d.

2.Experimental Setup

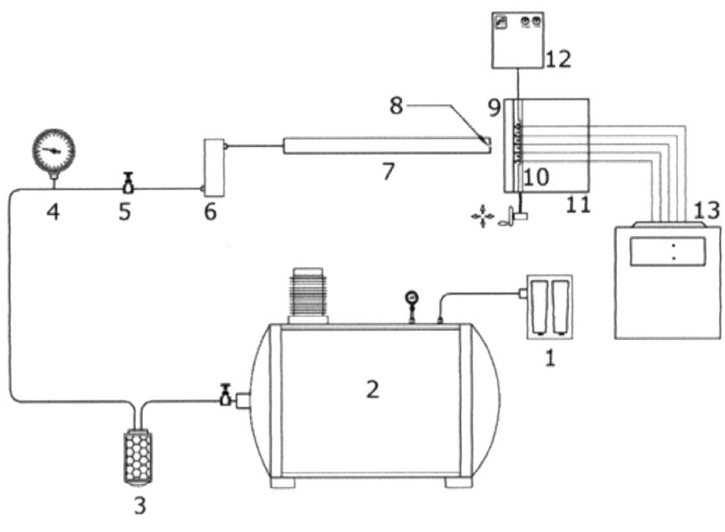

1.Dryer 2. Compressor with storage Tank 3. Air Filter 4. Pressure Regulator 5. Needle Valve 6. Flowmeter 7. Stainless steel pipe 8. Nozzle 9. Target Plate 10. Heater11. Insulation12. Power Supply 13. Data Acquisition System.

Figure .1 Schematic layout of jet impingement system

Figure 1 shows the schematic diagram of the experimental setup. Air is supplied by an air compressor through a calibrated flowmeter. Air filter and pressure regulator are installed upstream of the flow meter to filter the air and to maintain the downstream pressure at 2 bars. The thermal environment of the microprocessor is simulated using a copper plate of size 100 X 100 ×6 mm with a heater placed underneath. A thin heating coil having the same dimensions as that of the cross section of the target plate is used for ensuring uniform heating. The bottom of the heater is well insulated to prevent heat loss. The target plate is placed on a two-dimensional traverse system so that the relative position of the target plate with the nozzle plate can be varied conveniently. The power input to the heater is varied using a dimmer stat and the voltage and current are measured using voltmeter and ammeter respectively. T- type thermocouples were used to measure the temperature at various locations. The temperature data were acquired using a temperature scanner.

Power is supplied from DC power source through a voltage stabilizer, a variac and a current transformer. The voltage across the heater and the current are measured by digital meters whose ranges and the accuracies are of 0 to 20 ± 0.5% V and 0 to 400 ± 0.5% A, respectively. The velocity of the jet exit air close to the nozzle was measured with a calibrated anemometer. In the experimental and numerical studies, the size of the orifice is 3 mm. Three circular holes, each with a diameter of 3 mm are drilled on the nozzle

3.Numerical Setup

In this section, an overview on the numerical setup investigations is presented. The numerical grid density is taken into consideration

3.1. Boundary and Initial Conditions

The flow is assumed to be steady, incompressible, three dimensional and turbulent. Heat transfer across the thickness of the plate, buoyancy and radiation heat transfer effects are neglected. Properties of the fluid such as density, viscosity and thermal conductivity are assumed to be constant. The impinging plate was kept at constant heat flux of 2100W/m2. The three-dimensional Navier-stokes and energy equation with standard turbulent model are solved using Ansys Fluent software. The numerical simulation has been done using SST k-ω-turbulent model for jet impingement application. This model used to predict capture gradient effect and heat transfer distribution near the wall accurately with moderate computation cost. The schematic diagram 3-D computational domain and section view of three jet circular and square array is shown in Figure 2.

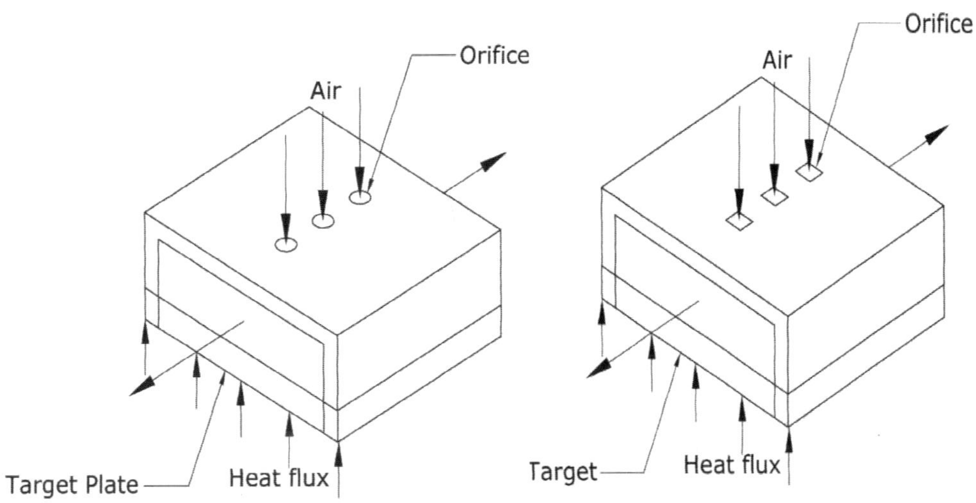

Figure 2. computational domain of circular and square jet array of hydraulic diameter 3mm

3.2. Numerical grids independence study

Figure 3 shows the grid independent study performed for the physical model to obtain the optimum number of elements. The number of elements is varied from 0.05 million to 0.35 million. Result from the calculation using 0.25 million cells agrees well with the case using 0.3 million cells. From the study between 0.25 million cells and 0.3 million cells there is a slight variation which is below the order 5%. So, the mesh with 0.3 million cells has been chosen as the suitable mesh for further investigations.

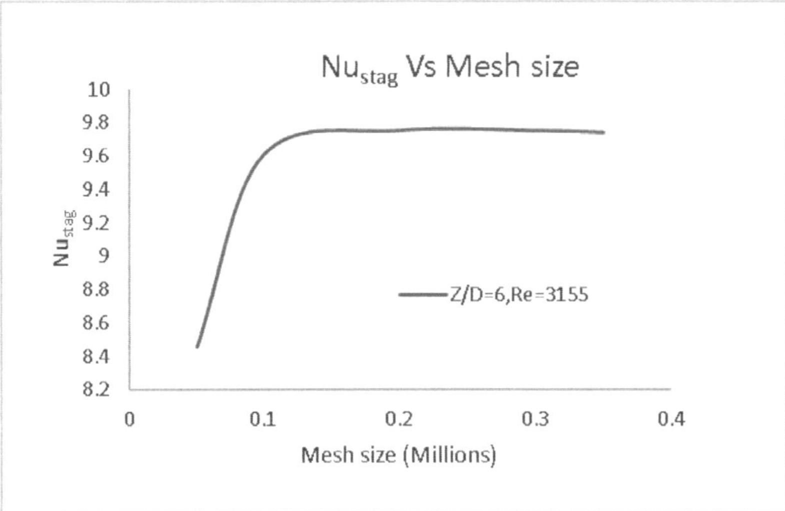

Figure 3 Grid independence study.

3.3 Numerical solver

The simulation was performed using Computational Fluid Dynamics (CFD) Ansys Fluent that solves equations of continuity, momentum and energy using the Reynolds-Averaged Navier-Stokes approach.
Fluid properties (Air):

- Density = 1.225 kg/m3
- Dynamic Viscosity = 1.983e-5 Pa-s
- Thermal Conductivity = 0.024 W/m-K

CFD model:

- Software: Ansys Fluent
- Steady analysis of single jet impingement on the flat plate
- Dimension of plate (100mmx100mmx6mm)
- Diameter of nozzle jet (3mm)
- Side of square jet (3mm)

The computational domain is modelled as a 3D steady system in the present investigation. Multiple circular jets and square jets are impinging on a flat surface. A finite volume based CFD package can be used for solving the governing continuity, momentum, energy and turbulence model equations. K-ω SST turbulence models are used for closure and power law scheme is used for discretization. A second order spatial discretization with SIMPLE algorithm can be used for pressure velocity coupling. The solution convergence is achieved when the sum of the normalized residuals was on the order of 10^{-4} for continuity and momentum. Convergence was achieved generally in 4,000 iterations with the use of under-relaxation parameters.

4. Results and discussion
This section provides the analysis of the flow field behavior of the array of circular and square impingement jets. Finally, the heat transfer characteristics in terms of the Nusselt number is presented.

4.1 Validation of Experimental results.
Figure 4 shows the variation of stagnation temperature at the top surface of the plate obtained from numerical studies. The result obtained from experimental studies is also shown for comparison. Here orifice diameter is 3 mm, jet to jet spacing is 6D and the Reynolds number is 3155. There is very good agreement between the values obtained from numerical and experimental studies and the maximum difference is only 2%. The results from present experiments and numerical analysis show that the heat transfer depend on distance from stagnation point in the stagnation region. The presence of the optimum nozzle-to-plate distance can be explained by the relationship between the fluid flow and heat transfer rates.

The heat transfer rate depends on the turbulence intensity of the jet nearer to the impingement surface. The turbulence intensity depends on the velocity of jet. At low values of the nozzle-to-plate spacing the jet axial velocity decreases as soon as it emerges from the nozzle and it increases as the distance increases. Thus, due to the increase of turbulence, the heat transfer rate increases with the increase of nozzle-to-plate distance up to the optimum value. At the optimum nozzle-to-plate distance, the mean velocity of the jet at the centre remains constant. This distance is almost equal to the length of the potential core. When the nozzle-to-plate spacing increases further, the fluid velocity decreases due to the spreading of jets due to the presence of surrounding air. But the presence of the confinement can reduce the spreading rate of the jet by restricting the flow of surrounding air.

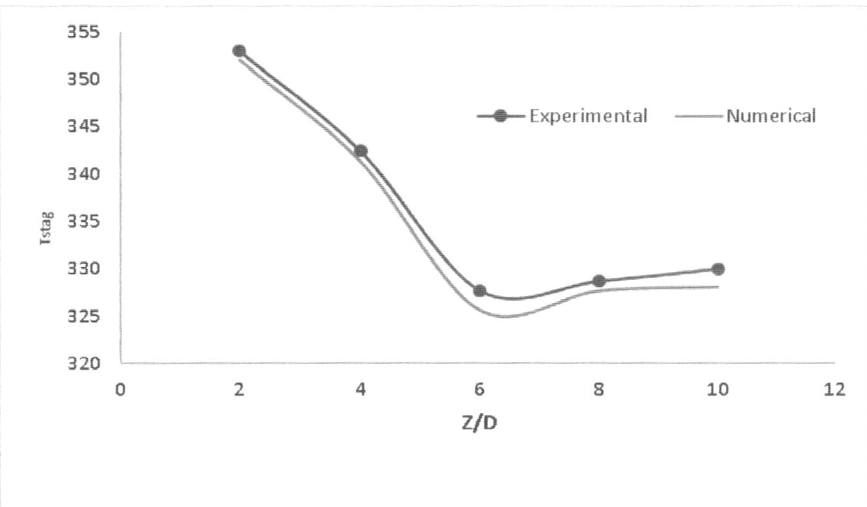

Figure 4. Variation of stagnation temperature with orifice-to-plate distance for Re=3155

4.2 Effect of stagnation Nusselt number with Reynolds number for Z/D=6
Figure 5 shows the effect of Reynolds number on stagnation Nusselt number. As the graph indicates, the stagnation Nusselt number increases with the Reynolds number for the range of values tested. The heat transfer rate of the je impingement depends on the turbulence of the impinging jet. The turbulence of the jet depends on the velocity of the jet. Thus, the stagnation Nusselt number increases with the jet Reynolds number.

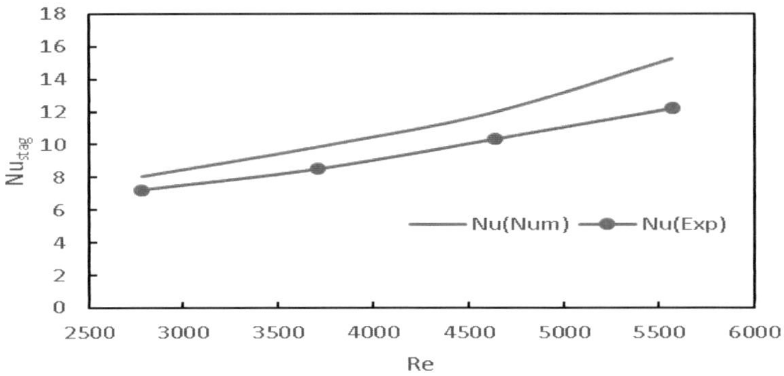

Figure .5. Variation of stagnation Nusselt number with Reynolds number for Z/D=6

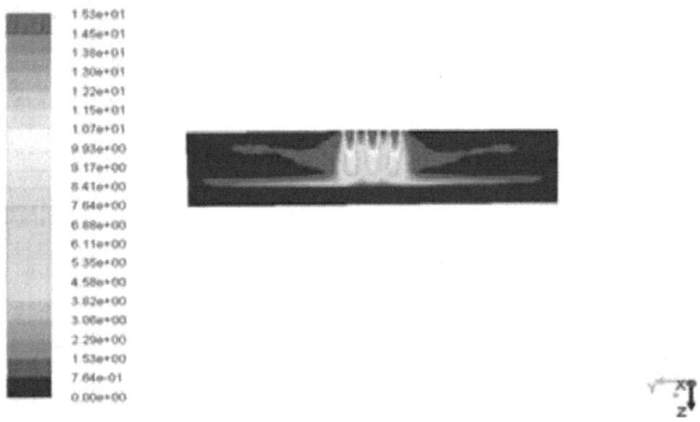

Figure 6 Velocity distributions of 3 inline jet arrays impinging on the surface for Z/D=6.

4.3 Effect of jet to jet spacing on stagnation Nusselt number

Figure 7 shows the variation of the local Nusselt number with the jet-to-jet spacing for the Reynolds number equals 3155 and for the optimum orifice-to-plate distance of Z/D equals 6. The graph indicates that the Nusselt number reaches maximum values when the jet-to-jet spacing is 2D. Thus, the stagnation Nusselt number increases with the increase in jet-to-jet spacing until spacing between the jets up to 2D. The Nusselt number decreases with further increase of the jet-to-jet spacing the cooling area becomes smaller at larger jet-to-jet spacing. Thus, the stagnation Nusselt number shows a negative slope. Two relative maxima of Nu exist for the jet array considered. First relative maximum occurs at S=4mm and second relative maximum at larger S/D ratio. For small S/D ratio an increase in its value results in reduction in the jet interference, so the Nu increases. Further increase at S=6mm due to the fountain effect. The impingement introduces an up-wash motion which is called fountain because of which a recirculation occurs between the jets.

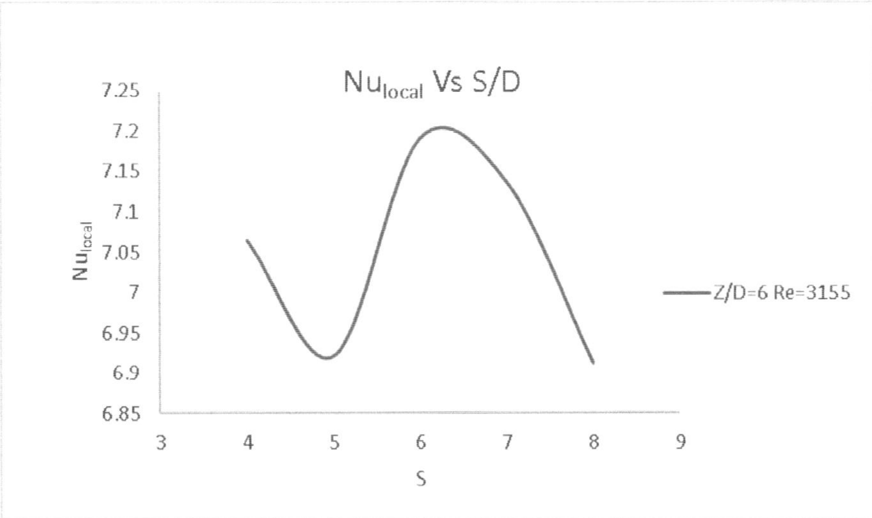

Figure 7 Variation of Nusselt Number with jet-to-jet spacing

4.4 Effect of Nusselt number with Reynolds Number for Circular and square orifice

Figure 8 shows the variation of Nusselt number with Reynolds number of different orifice shape (square and circular) at stagnation point. Figure shows the variation of Nusselt number with axial distance for various nozzle shape having equivalent diameter and constant Reynolds number. The square orifice has larger heat transfer coefficient has compared to circular orifice this is because of larger covered area of square et. Due to larger entrainment the square orifice attains a large mass flux at the same distance from nozzle. It is evident that the square orifice has 15% more heat transfer rate as compared to circular orifice.

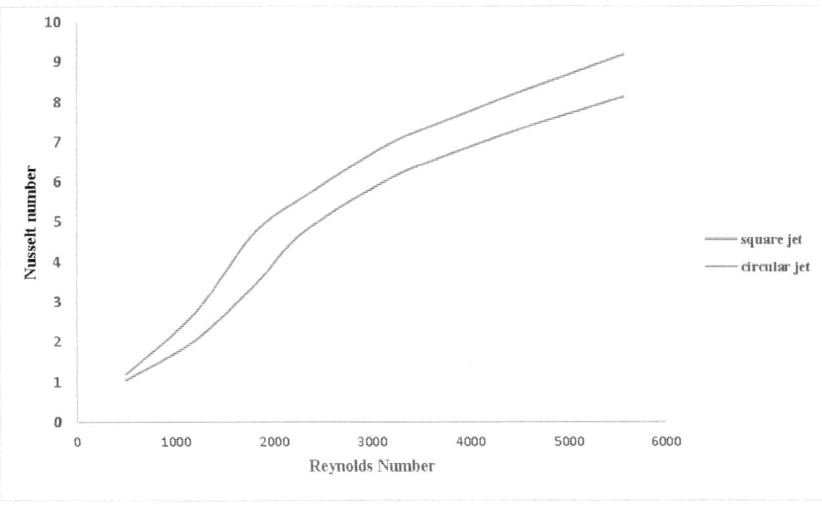

Figure 8 Variation of Nusselt number with Reynolds Number for Circular and square orifice.

5.Conclusions

Numerical and experimental investigations were carried out to study the effect of geometrical and flow parameters on the heat transfer characteristics of multiple jets. The present investigations provide a good understanding of the fluid flow and heat transfer characteristics of air impinging jets. The numerical results of the present study agree satisfactorily with experimental data. Experimental and numerical results reveal that the stagnation Nusselt number attains its maximum value when the nozzle-to-plate distance is six times the diameter of the orifice.

The heat transfer rate is a strong function of the jet Reynolds number, nozzle-to-plate distance, jet-to jet spacing. The experimental results for 3-inline arrangement showed excellent agreement with numerical results computed by the theoretical analyses. Maximum percentage error of approximately 2% occurs of temperature value at stagnation point. From the numerical study for circular and square jet array it is evident that the square orifice has 15% more heat transfer rate as compared to circular orifice.

6.Nomenclature

D Jet or nozzle diameter (mm)

Z distance between orifice and target plates (m)

T_{stag} Stagnation Temperature (K)

Re Reynolds number

S Jet to jet spacing (mm)

Nu_{local} Local Nusselt Number

7. References

[1] Y Ozmen 2011 Confined impinging twin air jets at high Reynolds numbers *Experimental thermal and fluid science* 35pp 355-363

[2] Matt Goodro, Jongmyung Park, Phil Ligrani, Mike Fox, Hee-Koo Moon 2008 Effects of hole spacing on spatially-resolved jet array impingement heat transfer *International Journal of Heat and Mass Transfer* 5pp 6243–6253.

[3] Subhash Chander, AnjanRay 2007 Heat transfer characteristics of three interacting methane/air flame jets impinging on a flat surface *International Journal of Heat and Mass Transfer* 50pp 640–653.

[4] VadirajKatti, S.V. Prabhu 2008 Influence of span wise pitch on local heat transfer distribution for in-line arrays of circular jets with spent air flow in two opposite directions *Experimental Thermal and Fluid* Science 33pp 84–95.

[5] VadirajKatti, S.V. Prabhu 2009 Influence of stream wise pitch on local heat transfer characteristics for in-line arrays of circular jets with cross flow of spent air in one direction *International Journal of Heat Mass Transfer* 45pp 1167-1184.

[6] Puneet Gulati, VadirajKatti, and S.V. Prabhu 2009 Influence on the shape of the nozzle on local heat transfer distribution between smooth flat surface and Impingement Air Jet *International Journal of Thermal Sciences* 48pp 602-617.

[7] San, J.-Y. and M.-D. Lai 2001 Optimum jet-to-jet spacing of heat transfer for staggered arrays of impinging air jets *International Journal of Heat and Mass Transfer* 44pp 3997-4007.

[8] K. Heyerichsand A. Pollard 1996 Heat transfer in separated and impinging turbulent flows *International Journal of Heat and Mass Transfer* 39pp 2385–2400.

[9] Lance Fisher 2001 A Numerical Investigation of Jet Impingement on a Heated Flat Plate *The Pennsylvania State University* ME pp 513-16.

[10] G. Srivalli, B. Raghavarao, M.R.CH. Sastry 2012 Fluid flow and heat transfer analysis of laminar multiple square jets impinging on a flat plate *International journal of engineering research and applications* 2pp 048-056.
[11] Leon F. G. Geers, K. Hanjalic, M. J. Tummers 2006 Wall imprint of turbulent structures and heat transfer in multiple impinging jet arrays *J. Fluid Mech* 546pp 255–284.
[12] Yoshisaburo Yamane, Yoshiyasu Ichikawa, Makoto Yamamoto and Shinji Honami 2012 Effect of Injection Parameters on Jet Array Impingement Heat Transfer *International Journal of Gas Turbine, Propulsion and Power Systems* 4pp 27-34.

Acknowledgments
The authors acknowledge the financial support by CERD research project.

Experimental investigation on the effect of geometry on cryogenic transfer line chilldown

Jesna Mohammed[1], Rizwan Rasheed[2], Faraz P Junaid[2] and K E Reby Roy[2]
[1, 2] Assistant Professor, Department of Mechanical Engineering, TKMCE, Kollam.

E-mail: jesnamohammed@tkmce.ac.in

Abstract. In order to analyze the effect of geometry on cryogenic chilldown, experiments were performed on straight, one twisted and twin twisted copper tubes using liquid nitrogen. Data presented in this paper is for the experiments conducted under terrestrial gravity conditions at supply pressures of 1.2 bar and 1.3 bar. Test sections considered were copper tubes (7.9375 mm OD, 0.8128 thick) of straight, one twisted and twin twisted geometry traversing a distance of 220mm under Polyurethane foam insulation. Pressure measurements at test section inlet and outlet was monitored using transducers. Temperature measurement at four equidistant sections along the length were tracked in real-time using a data acquisition system. Temperature and pressure profiles were obtained. The results of the experiment indicated that, for a given mass flux, the transition during chilldown occurs earlier in one twisted transfer lines compared to straight and twin twisted tubes. This indicates the earlier occurrence of film boiling for the one twisted channels. Thereby, changing the transfer line geometry from straight tubes to one twisted sections, the chilldown time can be reduced to a considerable extend further resulting in reduction of cryogen consumption during the initial stages of transfer process.

1. Introduction

Cryogenics, a scientific branch concerning with liquefaction of gases has outgrown conventional gas supply industry and has been instrumental to the extent that today we can find its applications over a wider horizon such as cryosprays, food preservation, space simulation chamber, cryonics, chemical and glass industries, superconducting magnets etc. Optimized transportation of cryogen from its storage to intended use point with minimum losses, still remain an area of concern, owing to the energy intensive nature of cryogen production. When the cryogen initially navigate through the transfer line which is in thermal stability with the ambient, it gets vaporized until the transfer line is brought to a steady state temperature near the saturation temperature of the fluid so that consequent flow would be in liquefied stage. This initial phase characterized as chilldown of cryogenic transfer line is accompanied with ravenous evaporation, pressure surges and temperature fluctuations. From chilldown studies it is inferred that an amount of 90% of the cryogen initially supplied to the transfer lines are consumed, [1] or rather wasted in the initial period of chilldown where, film boiling dominates. Any initiative in the way of reducing chilldown time in cryogenic transfer lines can result in energy conservation and thereby reducing the cost of overall performance of any cryogenic system. The first remarkable work in this field dates back to 1960s when Burke et al.[2] probed pressure chilldown of cryogenic transfer lines using liquid nitrogen as the cryogen and proposed the existence of single phase convective heat transfer and film boiling. In 1962, Bronson et al.[3] observed the existence of flow patterns in two phase flow using liquid nitrogen while conducting visual assessment of chilldown. His conclusions validated the fact that flow stratification led to the oscillations in peripheral temperature. Chilldown experiments employing liquid hydrogen through copper tubes were conducted by Chi and Vetere[4] in 1963. They associated transient wall temperature with the visual inspections to identify the regimes in flow transitions. Demonstration of flow regimes grouping into single phase convective boiling, film boiling and nucleate boiling were reported by Chi [5] in 1965. He inferred that the film boiling dominates the total chilldown time. Klimenko [6] investigated the influence of channel orientation and geometry on heat transfer with two-phase forced flow of nitrogen and explained the need of conducting experiments in the same laboratory for identifying the variations associated with factors governing chilldown. Yuan et al.[7][8] studied the effect of mass flow rate and gravity on cryogenic chilldown and associated the former results concerning visual inspections and peripheral temperature fluctuations. He inferred that the driving force behind chilling of the bottom wall of transfer lines was liquid film-wall interaction while, forced convection of superheated vapor caused cool down in upper wall. Through their chilldown experiments at low and high mass flow rates to find its impact on various flow patterns, Hu et al. [9] studied the variation of heat flux and chilldown time through a vertical pipe under upward and downward flow. They found that chill down time was longer for upward flow while downward flow had higher critical heat flux. The existence of an optimum upward line inclination that can minimize chilldown time was reported by Johnson et al.[10] through their work on horizontal and inclined pipes. Inverse heat transfer technique to the transient chilldown period was employed to derive heat flux and local heat transfer coefficient. Darr et al.[11] established that optimal design of cryogen transfer lines can be accomplished by identifying methods for minimizing cryogen consumption. Effects of mass flux, pressure, flow direction, equilibrium quality and inlet subcooling in relation to gravity was investigated, enabling them to arrive at a correlation for the heat transfer coefficient in straight transfer lines. Apart from these parameters, geometry of transfer lines and the accompanying effect of other relevant factors are to be investigated into. Hence, optimization of chilldown parameters has been an area of research interest for decades.

Vashisth et.al.[12] in their review paper on the applications of curved geometries in process industries , demonstrates that helical geometries exhibit similar and sometimes better performance at lower energy consumption and reduced maintenance requirements over conventional configurations. Mohammed et al. [13] identified the earlier occurrence of film boiling for the helical channels when compared to straight tubes. In spite of the aforementioned works, experimental studies focused on the effect of geometry variations on chilldown are minimal. Those available are limited to straight transfer lines. With the added advantage of secondary flows in curved geometry, the authors have investigated the effect of providing slight twist to straight transfer lines and henceforth, investigate the heat transfer characteristics during chilldown of one twisted and twin twisted transfer lines. Moreover, literature lacks information concerning data on direct comparison of heat transfer intensity in different geometries obtained in the same laboratory.

2. Experiment work

A systematic layout of the experimental setup can be shown in figure 1. The entire setup consists of liquid nitrogen storage and supply system, test sections and instrumentation with Data Acquisition system.

Cryogen for this experimental study was Liquid nitrogen. It was selected because of its non-corrosive, relatively inexpensive, chemically inert, easily available, non-flammable characteristics and it does not pose any major hazards. The liquid nitrogen was stored in CRYOCYL 120LP which is a low pressure Dewar vessel with 120 liters of capacity. It expels liquid nitrogen by self-pressurization. LN2 is forced through the test section by using the pressure exerted by the gaseous nitrogen generated. Flow rate of liquid nitrogen was controlled manually by controlling the Dewar pressure by employing a pressure regulator.

Copper tubes (specific heat capacity of 401 J/kgK and thermal conductivity of 0.14 W/mK at 300 K) were selected for the study in regard with the low flow rates from the Dewar. Test section under

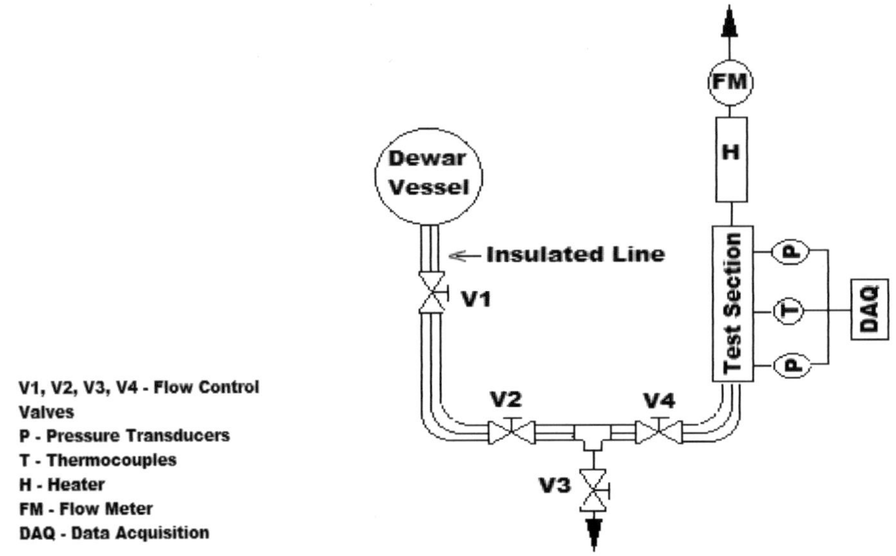

Figure 1. Layout of Experimental setup for chilldown studies.

consideration were 5/16" Copper tubes (7.9375 mm OD 0.8128 mm Thick) of for straight channel (220 mm) and copper tubes of one twisted (coil length 275 mm) and twin twisted (coil length 385 mm). Details of the test section used are shown in figure.2. Liquid nitrogen was supplied to the test section through 1/2" SS 304 grade pipes and brass fittings. A bypass line was provided before the test section to bypass the initial vapour produced during the initial cool down of the supply system. A transparent line provided at the bypass line ensures a steady state of liquid being supplied and only after that the line to the test section is opened. Heat in leak at the inlet section was minimized by employing a combination of asbestos rope, nitrile rubber and urethane modified poly-isocyanurate foam (thermal conductivity of 0.14 W/mK).

The wall temperatures of each geometry was measured by employing16 T-type thermocouples. Temperature measurements were taken from four equidistant sections till a distance of 220 mm (along the tube) from inlet. Each section was attached with four equidistant thermocouples along its circumference. Two calibrated piezo-electric pressure transducers were used to measure the inlet and outlet pressure of the test section. An insulation was provided with 140 mm OD with urethane modified poly-isocyanurate foam to reduce the heat in leak at the section. Keysight 34972A Data Acquisition / Data Logger Switch Unit was adopted for logging data. Mass flux was measured by a single phase mass flow measuring Rota meter at the exit line kept after a hot water bath maintained at 100^0 C.

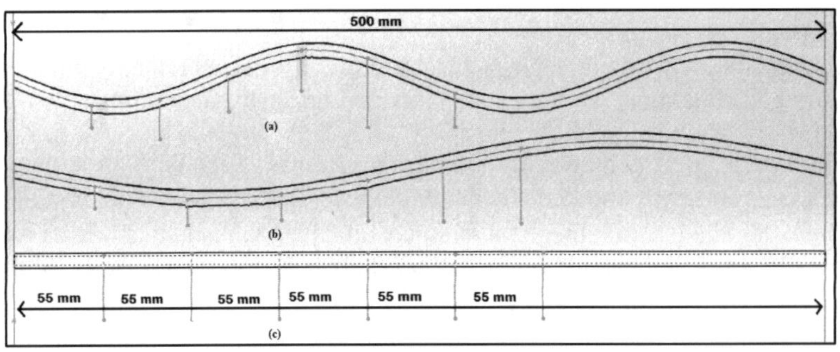

Figure 2. Test sections (a) Twin Twisted (b) One Twisted (c) Straight.

3. Experiment Procedure

Entire turning along with the test section was purged with gaseous nitrogen to avoid moisture condensation inside the experimentation setup. The inlet lines before the test section were allowed to chilldown before commencement of the experiment and the vapor generated were vented to the atmosphere via bypass line. The data acquisition program was initiated to record data as soon as activated. The required supply pressure was obtained by regulating the pressure regulator connected to the Dewar vessel. Then the flow towards bypass line is closed and the flow is introduced to the test section, by that time the DAQ should also started for recording various thermocouple readings. Liquid nitrogen was allowed to flow through the test section until all the thermocouples that connected on the wall of the test section read a steady value corresponding to the saturation temperature of liquid nitrogen. By that time we can conclude that chilldown process was completed. Wall temperature and Flow rate and supply pressure were recorded. The above steps were repeated with the different supply pressures and at various flow direction so that a wide parameter range may be investigated.

4. Results and Discussion

For single sampled experiment, Kline and McClintock [14] method is widely used for uncertainty measurement. Simple equation relating the measured data to the heat flux does not exist. For this reason, this section contains only the uncertainty of the instruments used for the experiment. 'T' type thermocouples have an uncertainty of ±0.50 C. for obtaining a fast response time the wire of the thermocouple used is of 0.25 mm and the response time corresponding to it is less than 0.2 seconds. Data Acquisition Systems have an accuracy of $\pm 1^0$C between -100^0C to 100^0C and $\pm 1.5^0$C between -200^0C to -100^0C. Mass flow rate have an accuracy of 0.5L/min. the uncertainty from third metal is avoided by using the same materials for connection and attention was provided to avoid large temperature gradients to avoid feed through errors

4.1 Time temperature profile of straight, one twisted and twin twisted transfer lines

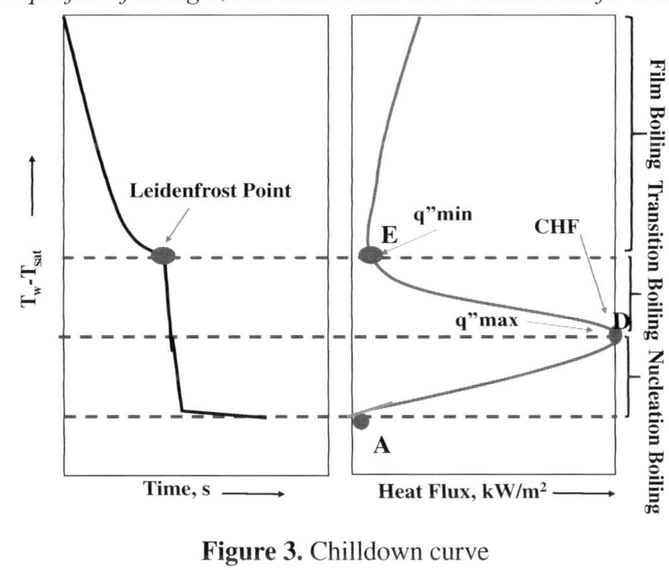

Figure 3. Chilldown curve

Experiments were conducted for straight, one twisted and twin twisted channels under two different supply pressures of 1.2 bar and 1.3 bar amounting to mass fluxes of 162 Kg/m^2s and 176 Kg/m^2s respectively. The results obtained are provided here. Wall temperature readings obtained were used for predicting the heat transfer and flow characteristics and was subjected to comparison. Jackson et al. [15] proposed three different phases during chilldown phenomena, namely film boiling regime, where the flow structure can be either stratified flow or inverted annular film flow. Transition boiling regime, where a sudden drop in temperature is observed as liquid droplets begin to wet the wall. Nucleate boiling regime, where flow can be either bubbly or slug flow. Figure 3 depicts a chilldown curve with the three regimes demarked. As observed from the works of researchers in this field, the present study also regards the initiation of transition regime as the basis for the comparison of chilldown parameters. This is because the temperature is found to be almost steady after the transition regime. Three regimes observed during previous studies of chilldown in straight tube was also observed in curved geometries. However, the observed chilldown transition temperature and time were different from the straight geometry.

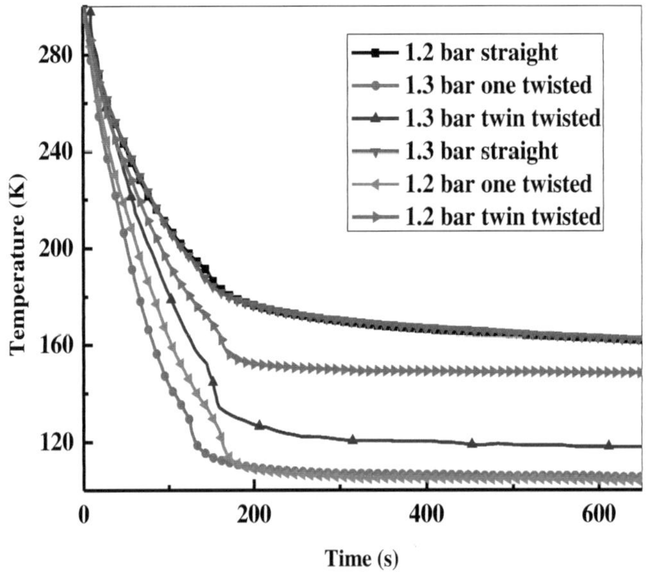

Figure 4. Temperature profile.

Variation of outer wall temperature of the section with respect to time is provided in figure 4. The different flow regimes can be identified by comparing with the chilldown curve given in figure 3. It is evident that even though, straight channels completed the film boiling and transition regimes at higher wall temperatures, time taken for onset of transition and nucleate boiling is lesser for one twisted tube whereas twin twisted lagged behind straight channel. This leads to the existence of optimized curved geometries that can perform better than straight transfer lines. Another observation from the results is that, as supply pressure increased, the chilldown time reduced. While comparing the twisted tubes separately, it can be seen that, under all flow conditions, one twisted tube performed better than twin twisted tubes in terms of chilldown time. In one twisted tubes the liquid phase comes in contact with the tube wall earlier than that in twin twisted tubes. This can be credited to the longer dwelling time of the liquid phase in one twisted tubes.

4.2 Heat Flux variation

In the present study outer wall temperature measurements were the only direct temperature measurements. In similar studies conducted by Reid Shaeffer et al. [1] and Hong Hu et al. [9] used Burggraf correlations [16]. Burgraff's method is applicable to unsteady heat conduction with a single boundary condition. Burggraf developed a method that allows the estimation of the inner wall surface temperature by the measured outer wall temperature history through the solution of the inverse heat conduction problem (IHCP).

The method uses lumped capacitance analysis and can produce accurate results with fewer terms. Inside wall temperature can be given by the correlation[16]

$$T_i = T_0 + \left(\frac{r_0^2}{4\alpha}\left(\left(\frac{r_i}{r_0}\right)^2 - 1 - 2\ln\frac{r_i}{r_0}\right)\right)\frac{dT_0}{dT} + \left(\frac{1}{64\alpha^2}\left(r_i^4 - 5r_0^4\right)\frac{r_0^2 r_i^2}{8\alpha^2}\ln\frac{r_i}{r_0} - \frac{r_0^4}{16\alpha^2}\ln\frac{r_i}{r_0}\frac{r_0^2 r_i^2}{16\alpha^2}\right)\frac{d^2T_0}{dt^2} + .. \quad (1)$$

Inner wall surface heat can be calculated with the equation[16]

$$q_i'' = \rho c\left(\frac{r_i^2 - r_0^2}{2r_i}\right)\frac{dT_0}{dt} + \left(\frac{(\rho c)^2}{k}\left(\frac{r_i^3}{16} - \frac{r_0^4}{16r_i} - \frac{r_0^2 r_i}{4}\ln\frac{r_i}{r_0}\right)\right)\frac{d^2T_0}{dt^2} + \frac{(\rho c)^3}{k^2}\left(\frac{r_i^5}{384} - \frac{3r_0^4 r_i}{128} + \frac{3r_0^2 r_i^3}{128} - \frac{r_0^6}{384r_i} - \frac{r_0^2 r_i^3}{32}\ln\frac{r_i}{r_0} - \frac{r_0^4 r_i}{32}\ln\frac{r_i}{r_0}\right)\frac{d^3T_0}{dt^3} + \quad (2)$$

Heat transfer coefficient can be found using the equation

$$h_i = \frac{q''}{(T_i - T_{sat})} \quad (3)$$

The average temperature was used for finding the inner wall surface heat flux. Heat flux is plotted as a function of wall superheat of straight tube at different mass flux. The curve is similar to boiling curve in pool boiling conditions. Critical heat flux and Leidenfrost point are clearly identified and time for reaching these points are also determined. Figure.5 represents the variation of heat flux with time for the test sections considered. Critical heat flux (CHF) and Leidenfrost point are usually recognized as key points in a boiling curve, dividing it into three distinct flow regimes. During a quenching process, heat flux diminishes to a minimum value at the Leidenfrost point and it increases drastically in transition boiling regime until the CHF point.

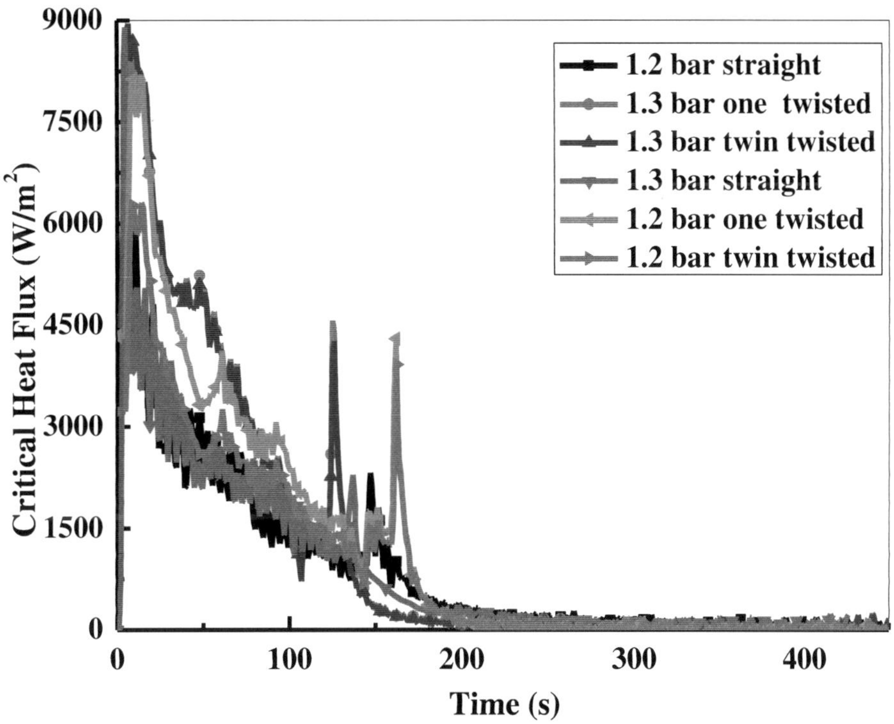

Figure 5. Variation of heat flux with time

In laminar flow, previous studies showed that curved geometries generally have superior heat transfer characteristics than straight channels. But this may not be true in turbulent flow to which our study belongs.

Table 1. Critical heat flux values.

Geometry	Critical Heat Flux (W/m^2)	
	1.2 bar Supply pressure	*1.3bar Supply pressure*
Straight tube	2272	2319.864
One twisted	4299.66	4433.437
Twin Twisted	3921.358	4261.197

The variation may be due to the combined effect of added turbulence due to the curvature and the effect of body forces present in the system. This is also coincident with the results got through this work which suggests that, CHF was consistently higher in twisted tubes but, its occurrence was delayed in twin twisted tubes when compared with straight channel. Table 1 provides the Critical Heat Flux values for the test sections considered. Gravitational effect must have played a significant role in chilldown of a horizontal straight line. In case of straight geometry, the gravity forces and pressure force are perpendicular to each other. In the present scenario the pressure force changes the direction depending on the alignment of the twisted tube. Pressure force was at a positive inclination with the horizontal during the first half turn and was in a negative inclination in the second half turn. With increase in mass flux, twin twisted tubes had an increase of 7.8% in its CHF while, the increment in one twisted was limited to 3%. This leads to an expectation that with further increase in mass fluxe, twin twisted tubes might over perform one twisted tubes.

4.3 *Heat Flux variation at a section*

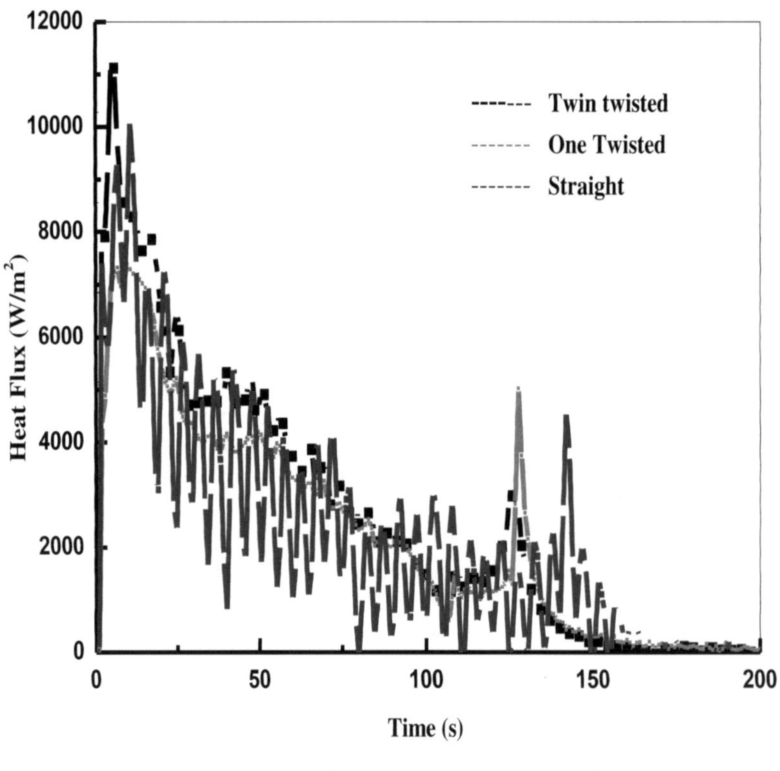

Figure 6. Heat Flux variation at a point 220mm from inlet.

The variation in heat flux at a distance 220mm from inlet is provided in figure 6. It can be seen that the twisted tubes complete their transition regime earlier than the straight section considered. However, the critical heat flux values for one twisted is nearly 15% higher than twin twisted, and has an increment of 5% when compared to straight tubes. It can be said that, by providing means to increase swirl resulting in secondary flow during the initial phases of cryogen transfer, time and energy saving can be envisaged. From the results it can also be inferred that there exists an optimum condition for transfer lines that can lead to faster chilldown with minimum time consumed for film boiling.

4.4 Variation of fluid pressure and heat flux with time

Figure. 7 (a), (b) and (c) shows the variation of fluid pressure and corresponding variations in heat flux along with time during the chilldown process for twin twisted, one twisted and straight tubes respectively. It can be seen that in all the three cases there is a pressure reduction during the initial phases of film boiling where the vapor phase dominates. As the vapor fraction decreases, the pressure difference reduces and shows a deviation from inlet conditions again during the onset of nucleate boiling regime. The fluid pressure levels back to the inlet conditions as the flow proceeds to single phase nucleate boiling. Thus, in twisted tubes the swirl flow created has of course a cost in terms of pressure loss. Pressure losses in the tubes are produced by the additional turbulence resulting from swirling. However, experiments show that the increased pressure loss is relatively small compared to the critical heat flux

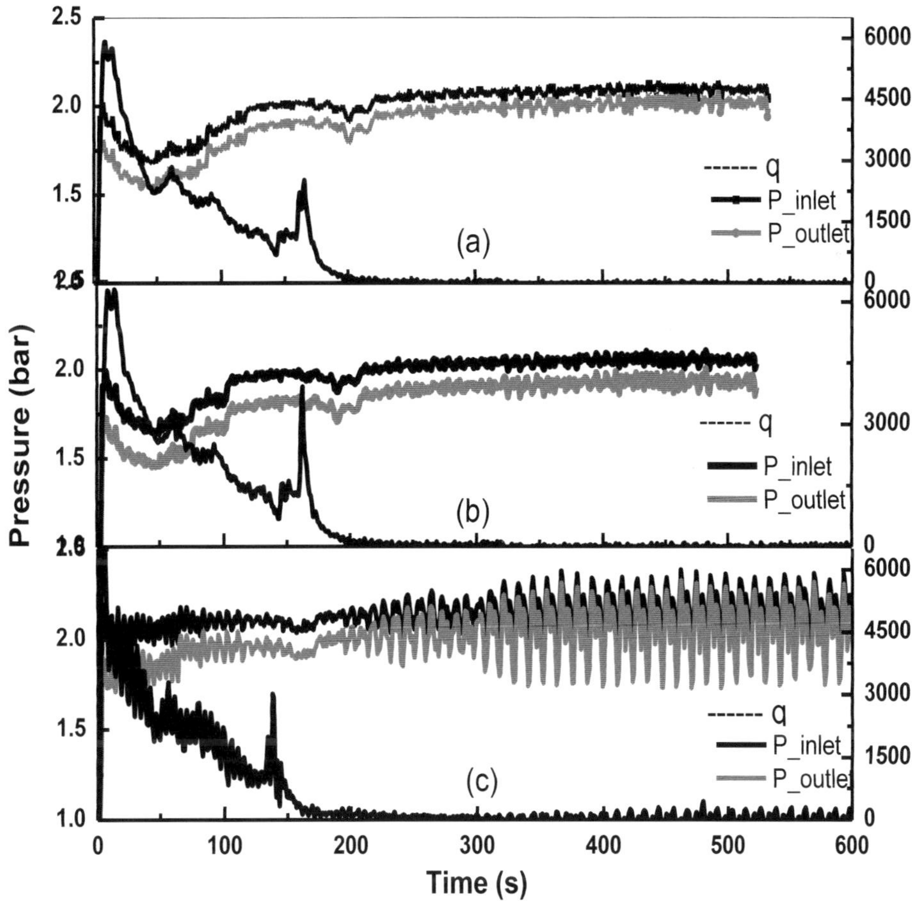

Figure 7. Variations in fluid pressure and heat flux along with time.

gain. From figure 7(c), it can be realized that the deviation of inlet and outlet fluid pressure follows the same pattern but the percentage deviation is not much evident during initial flow regimes and unlike in twisted tubes, the pressure fluctuations are severely varying through the onset of nucleate boiling and turns more rigorous as the flow takes on to single phase convection

Conclusion

This paper presents the results of experiments performed on straight, one twisted and twin twisted copper tubes using liquid nitrogen at supply pressures of 1.2 bar and 1.3 bar. For analyzing heat flow characteristics during chilldown process the transient temperatures for different mass flux conditions were measured. It was identified that the chilldown characteristics of copper transfer lines with a simple insulation like Polyurethane foam were similar to that of previously published studies. The chilldown characteristics of twisted tubes were found to be different from the straight tube due to the variation in direction vectors of pressure force with positions in them. There is an inverse relation existing between the mass flux and chill down time for each geometry. Straight channels completed the film boiling and transition regimes at higher wall temperatures, time taken for onset of transition and nucleate boiling is lesser for one twisted tube whereas twin twisted lagged behind straight channel. This leads to the existence of optimized curved geometries that can perform better than straight transfer lines. In straight tubes, deviation of inlet and outlet fluid pressure follows the same pattern but unlike in twisted tubes, the percentage deviation is not much evident during initial flow regimes and the pressure fluctuations are severely varying through the onset of nucleate boiling and turns more rigorous as the flow takes on to single phase convection. At an instance during the process, the critical heat flux values for one twisted is nearly 15% higher than twin twisted tube, and has an increment of 5% when compared to straight tubes. From the results it can also be inferred that there exists an optimum condition for transfer lines that can lead to faster chilldown with minimum time consumed for film boiling. Unlike the temperature difference between upper and lower portion of the straight pipe, there is uniform distribution of wall temperatures at each sections in twisted tubes. This may be due to additional turbulence caused by the curvature. Future works on identifying the avenues for reduction in cryogen consumption by varying flow parameters can be investigated into.

Acknowledgments

Authors would like to acknowledge Space Technology Laboratory in TKM College for engineering for setting up the experimental setup. They would also like to express gratitude for Technical Education Quality Improvement Programme Phase-II (TEQIP-II) promoted by National Project Implementation Unit, MHRD, Government of India for their support.

Nomenclature

d	Diameter, m
q"	Heat flux, W/m^2
r	Radius, mm
T	Temperature, K
t	Time, s
ΔT	Wall super heat, K
ρ	Density, kg/m^3
C	Specific heat, J/(kgK)
hi	Heat transfer coefficient, W/m^2K
α	Thermal Diffusivity

Abbreviations

CHF	Critical heat flux
HTC	Heat transfer coefficient
FB	Film boiling
NB	Nucleate boiling
TB	Transition boiling

Subscripts

i	inner surface of wall
o	outer surface of wall
sat	saturation conditions

References

[1] Shaeffer R, Hu H and Chung J N 2013 An experimental study on liquid nitrogen pipe chilldown and heat transfer with pulse flows *Int. J. Heat Mass Transf.* **67** 955–966.

[2] Burke J.C., Byrnes W.R., Post A.H R F . 1960 Pressurized Cooldown of Cryogenic Transfer Lines. *Adv. Cryog. Eng.* **4.**

[3] Bronson, J.C., Edeskuty, F.J., Fretwell, J.H., Hammel, E.F., Keller, W.E., Meier, K.L., Schuch, A.F and Willis W L 1962 Problems in Cool-Down of Cryogenic Systems *Adv. Cryog. Eng.* **7** 198–205.

[4] Chi J. W. H., and Vetere A.M., Chi J. W. H. and V A M 1963 Two-phase flow during transient boiling of hydrogen and determination of non-equilibrium vapour fractions *Adv. Cryog. Eng.* **9** 1656–62.

[5] Chi J W H 1965 Cooldown Temperatures and Cooldown Time during Mist Flow *Adv. Cryog. Eng.* **10** 330–40.

[6] Klimenko V V., Fyodorov M V. and Fomichyov Y A 1989 Channel orientation and geometry influence on heat transfer with two-phase forced flow of nitrogen *Cryogenics.* **29** 31–6.

[7] Yuan K, Ji Y and Chung J N 2007 Cryogenic chilldown process under low flow rates *Int. J. Heat Mass Transf.* **50** 4011–22.

[8] Yuan K, Ji Y and Chung J N 2009 Numerical modeling of cryogenic chilldown process in terrestrial gravity and microgravity *Int. J. Heat Fluid Flow* **30** 44–53.

[9] Hu H, Chung J N and Amber S H 2012 An experimental study on flow patterns and heat transfer characteristics during cryogenic chilldown in a vertical pipe *Cryogenics.* **52** 268–77.

[10] Johnson J and Shine S R 2015 Transient cryogenic chill down process in horizontal and inclined pipes *Cryogenics.* **71** 7–17.

[11] Darr S R, Hu H, Glikin N G, Hartwig J W, Majumdar A K, Leclair A C and Chung J N 2016 An experimental study on terrestrial cryogenic transfer line chilldown I. Effect of mass flux, equilibrium quality, and inlet subcooling *Int. J. Heat Mass Transf.* **103** 1225–42.

[12] Vashisth S, Kumar V and Nigam K D P 2008 A review on the potential applications of curved geometries in process industry *Ind. Eng. Chem. Res.* **47** 3291–337.

[13] Jesna Mohammed; Abdul Mohizin A M; Sachin Sasikumar; Sulav Kafle; Muhammed Shameer; K.E Reby Roy 2018 Experimental investigations on Helical and straight cryogenic transfer lines under different mass flux *Int. J. Sci. Eng. Res.* **9** 698–703.

[14] Kline, S. J. McClintock, F. A. 1953 Describing the uncertainties in single sample experiments *Mech. Eng.* 3–8.

[15] Jackson J, Liao J, Klausner J F and Mei R 2005 Transient Heat Transfer During Cryogenic Chilldown *Heat Transf. Vol. 2* 253–60.

[16] O. R. Burggraf 1964 An Exact Solution of the Inverse Problem in Heat Conduction Theory and Applications *ASME J. Heat Transf.* **86** 373–82.

Journal of Physics: Conference Series **1355** (2019) 012017 doi:10.1088/1742-6596/1355/1/012017

Study of Modified Fuel and Advanced Zeolite-Copper Oxide Nanocomposite Catalytic Converter in a Diesel Generator

Kurian George[1], B A Mohamed Azeem[1], Rijin Raju[1], Sujith Surendran[1], Neethu Divakaran[1], A Fazil[2]

[1] Department of Chemical Engineering, TKM College of Engineering, Kollam, Kerala, India
[2] Assistant Professor, Department of Chemical Engineering, TKM College of Engineering, Kollam, Kerala, India

E-mail: fazilchemical@gmail.com

Abstract. The issue of environmental pollution from diesel engines is a cause of major concern, even though diesel engines provide high thermal efficiency. This problem can be reduced to an extent through the use of modified fuel and catalytic converters. This study focus on improving the energy density of fuel by blending it with metal particles and hence increasing the thermal efficiency and ignition within the combustion chamber. Diesel was blended with nano and micron aluminium particles separately using ultrasonicator. Sorbitan monooleate was used as the surfactant. The catalytic converter was developed by making modification inside the silencer by placing stacks of mesh, which were coated with zeolite copper oxide nanocomposite. The nanocomposite matrix was prepared using copper nitrate solution and zeolite powder in an alcoholic medium, treated with a reducing agent. Engine performance test was carried out separately for both nano and micron aluminium blended diesel in a single cylinder four stroke, air cooled, direct injection Kirloskar engine. The experiment was carried out by varying the load at a constant speed of 1500 rpm to evaluate the performance characteristics. Emissions from the diesel generator fuelled by diesel and modified fuels were recorded by OPAX gas analyser both in the presence and absence of catalytic converter. The results indicated an increase in brake thermal efficiency by 8% and 72% for micro Al blended diesel and nano Al blended diesel respectively, in comparison with ordinary diesel. The emission test using catalytic converter gave 50%, 34% and 17% decrease in CO emission for ordinary diesel, micro-Al blended diesel and nano Al blended diesel respectively.

1. Introduction

Air pollution generated from mobile source is a cause of great concern. This environmental problem is due to the fact that, most of the engines employ combustion of fuels derived from crude oil as the source of energy. Burning of hydrocarbon ideally leads to the formation of water and CO_2 but due to imperfect combustion control and the high temperatures reached during combustion, exhaust contains significant amount of pollutants which need to be converted to harmless compounds. The concept of making alternative fuels is aimed at reducing the percentage emissions of such harmful gases and to increase the efficiency of engine. Diesel is a liquid fuel derived from crude oil. It has an average reported density of 840kg/m^3 and calorific value of 36.9 MJ/L. Diesel engines are self- ignition and do not require any spark to start. Diesel engines produce high torque at low engine speeds and are more reliable than petrol engines. The main concern of a diesel engine is its emission rate of greenhouse gases and hydrocarbon soot due to incomplete combustion. Diesel can be used as a base fuel for the preparation of alternative fuel. Aluminium is the most abundant metal in the earth's crust and is relatively safe to use. Aluminium is known for its reaction with oxygen in the presence of heat leading to an exothermic reaction. It also has high combustion energy. Due to these factors, aluminium of various size ranges (nano and micron) can be used as an additive in the base fuel to form the alternative fuel. Nano aluminium (n-Al) and micron aluminium (μ-Al) were separately dispersed in diesel in an ultrasonicator to form modified fuels[1]. Sorbitan monooleate (span 80) was used as surfactant[2]. One of the main issues in the combustion of μ-Al particles is its high ignition temperature. For a particle with a diameter greater than 100-micron metre, ignition is achieved only upon the melting of oxide shell at high temperatures[5]. The molten oxide shell forms a cap because of surface tension and exposes the aluminium core, thereby allowing ignition of the particle. The application of micron scale energetic metal particle additives in liquid fuel is a concept yet to be explored to its full potential.

Higher energy density of metal particles results in the increase of overall calorific value of the liquid fuel, eventually improving the performance of the engine by boosting power output. The study on evaporation and ignition probability hints to a very important role in determining two critical properties, viz; ignition delay and ignition temperature which probably characterizes the performance of a diesel engine, very much instrumental in controlling emissions.

The most effective treatment for reducing engine emissions is the use of a catalytic converter[13]. Thermal converters require a temperature of about 700°C to oxidise CO & HC to CO_2 and water. But if certain catalysts are present, the temperature needed to sustain these oxidation reactions is reduced to 250-300°C. CuO is a good candidate for replacement of the costly metal catalyst since it is cheap[19]. The effectiveness of the CuO can be heavily improved by making it to nanometer scale. The catalyst is not consumed in the reaction and so it functions indefinitely unless it gets degraded by heat, age or other factors. Catalytic converters consist of chambers mounted in the flow system through which exhaust gases pass through. Catalytic material incorporated in the meshes, which is stacked in the chamber, promotes the oxidation of emissions contained in the exhaust flow.

2. Experiment

2.1 Fuel Preparation

The Aluminium nanoparticles (from Nanoshell LLC) of 100nm average particle size were mixed with liquid fuel by first stirring them vigorously. Sorbitan monooleate was used as the surfactant to promote chemical stabilization of the suspension. It is a typical surfactant used to enhance the stability of metal particles in n-alkanes and it is also commonly used in water/fuel emulsion. Then an ultrasonic disrupter was used to disperse particles evenly and to avoid agglomeration[1]. Sonication was performed in a water bath to maintain a constant temperature for the mixture. The sonication generates a series of 4-s-long pulses 4s apart, was turned on for about 5 minutes. The figure 1 shows the blending of fuel in an ultrasonicator.

Modified diesel was prepared by blending diesel, μ-Al (from Mepco- dust grade du10) of average particle size 40 μm and Sorbitan monooleate in the weight ratio of 200:1:10. The use of 5 (weight) % surfactant will help to improve the settling time of μ-Al in diesel from 5 minutes [1] to 23 minutes, thus ensuring the stability of the blended diesel. The same procedure is repeated to prepare n-Al blended diesel in the weight ratio of 1000:5:2.

2.2 Zeolite Copper Oxide Nanocomposite Preparation
Zeolite powder was added to 250ml $CuNO_3$ solution and was constantly agitated for two hours at a temperature of 150-200°C, until the colour of the $CuNO_3$ solution (blue colour) fades. The supernatant liquid was removed and residue was dried. The resultant matrix was treated with reducing agent (hydrazine hydrate) in an alcoholic medium for 1 hour[18]. It was then vaccum filtered and heated at 250°C. Figure 2 shows the prepared zeolite copper oxide nanocomposites.

2.3 Diesel Generator Modification
For the prevention of coagulation caused by the modified diesel, the nozzle size of the diesel generator was modified to 0.3mm. Modification was made inside the silencer by placing stacks of 4 meshes (with negligible pressure drop) which were coated with zeolite copper oxide nanocomposite as shown in Figure *3.*

2.4 Engine Performance Test
A study of engine performance was conducted on a single-cylinder, four stroke, constant speed (1500 rpm) direct injection Kirloskar KBM-108 Diesel Four-stroke, CI, vertical, air cooled, single-cylinder Bore stroke 87.5 mm x110 mm Diesel engine. Compression ratio was 17.5:1. Capacity 661 cc and rated output of 7.5 kVA were at 1500 rpm diesel generator. The engine test was performed initially with pure diesel at fully throttled and no-load conditions and then micron/nano fuel was fed through a separate fuel feed line. During the engine test for μ-Al blended diesel, the stability of the suspension was ensured by conducting the test well before its settling time. All experiments were conducted by varying the loads at a constant speed of 1500 rpm to evaluate the performance characteristics such as specific fuel consumption (SFC), brake power (BP) and brake thermal efficiency (BTE). As the engine was used for electric generation purpose, electric loads were used for testing the engine. Engine was stepped up for each loading and the loading test was carried out. Each step up of 500W was carried out. The engine was started by cranking after taking the necessary precautions. Engine was allowed to run for 5-10 minutes to get heated up. Time was noted for the consumption of 10cc fuel at no load. The load test was carried out for different modified fuels.

Figure 1. Ultrasonication of blended fuel

Figure 2. Zeolite copper oxide nanocomposites

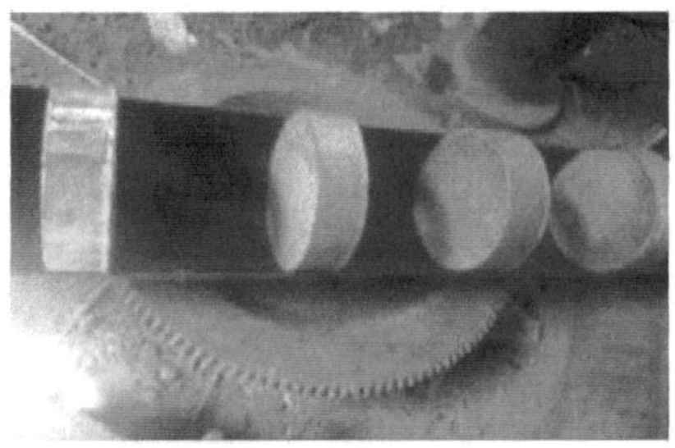

Figure 3. Modification of Exhaust

2.5 Engine Emission Test
The emission test was carried out for the exhaust gas from the single cylinder, four-stroke Kirloskar diesel generator as shown in Figure 4. The emissions from diesel generator fuelled by diesel and modified diesel (μ-Al & n-Al particles blended with diesel) samples was recorded by gas analyzer (OPAX 2000 II/DX) shown in Figure 5. The test was carried out under no load condition. The exhaust gas emission was recorded both in the presence and absence of advanced zeolite-copperoxide nanocomposite catalytic converter.

Figure 4. Emission test with catalytic converter

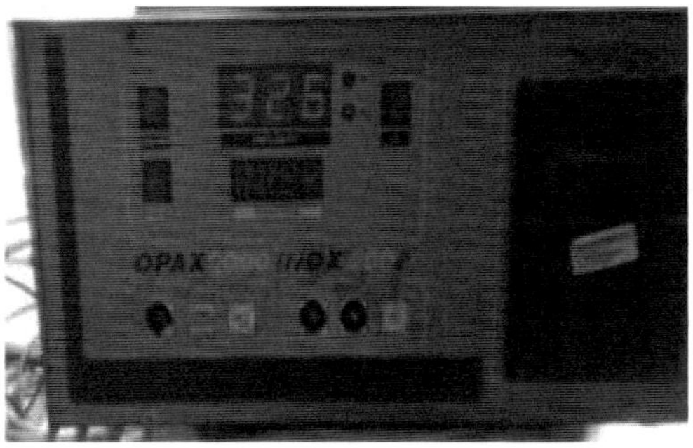

Figure 5. OPAX Gas Analyzer

2.6 Droplet test

The droplet test was carried out for both n-Al & μ-Al fuels. A Silica crucible was heated upto 270°C in a Bunsen burner, the temperature was measured using Arduino program. A drop of nano/micron fuel was introduced into silica crucible. A video camera (Canon power shot A 1800 ISO-1600) was setup to observe the micro explosion of the fuel. Figures 6 and figure 7 shows the images of droplet tests carried out with μ-Al and n-Al blended diesel, respectively.

The droplet test shows that, combustion time is inversely proportional to size of particle. Presence of thicker aluminium oxide coating is the main reason for this. Thickness depends on size of the particle and for lesser size more area is available for reaction[5].

1. Ignition t=0s
2. Disruption t=0.17s
3. Particle Ignition & burning t=1.4s
4. Micro explosion t=1.5s
5. Streak of burning particle t=1.6s
6. Extinction t=1.7s

Figure 6. Droplet test for diesel + μ-Al

1. Droplet formation t=0s
2. Preheating & ignition t=0.2s
3. Micro explosion t=1.3s
4. Steady flame t=1.6s
5. Flame extinction t=1.9s
6. Combustion residue t=2.3s

Figure 7. Droplet test for diesel + n-Al

3. Result and Discussion

The result of the engine performance test is shown in tables 1, 2 and 3. It is observed that the brake thermal efficiency changes on account of aluminium metal particle size. On comparing the load test values, the efficiency increases from the ordinary diesel values to diesel+ n-Al modified fuel as shown in Figure 8.

Table 1. Engine performance test for ordinary diesel

Sl: No	Load (kW)	Brake power (kW)	Time (Sec)	Total fuel consumption (kg/hr)	Specific fuel consumption (kg/kw.hr)	BTH (ŋ)
1	0	0	41.30	0.951	-	-
2	0.5	0.961	30.76	0.994	1.035	7.688
3	1	1.925	27.62	1.108	0.575	13.83
4	1.5	2.886	22.19	1.378	0.477	16.66
5	2	3.846	20.25	1.511	0.392	20.267
6	2.5	4.811	15.29	2.001	0.415	19.146

Table 2. Engine performance test for diesel + μ-Al

Sl: No	Load (kW)	Brake power (kW)	Time (Sec)	Total fuel consumption (kg/hr)	Specific fuel consumption (kg/kw.hr)	BTH η
1	0	0	39.43	0.776	-	-
2	0.5	0.960	32.11	0.952	0.992	8.026
3	1	1.925	28.78	1.063	0.552	14.42
4	1.5	2.886	23.40	1.307	0.453	17.58
5	2	3.846	21.96	1.393	0.362	21.98
6	2.5	4.811	17.01	1.799	0.374	21.29

Table 3. Engine performance test for diesel + n-Al

Sl: No	Load (kW)	Brake power (kW)	Time (Sec)	Total fuel consumption (kg/hr)	Specific fuel consumption (kg/kw.hr)	BTH η
1	0	0	42.6	0.709	-	-
2	0.5	0.959	41.5	0.728	0.759	10.484
3	1	1.923	40.4	0.748	0.389	20.459
4	1.5	2.882	38.9	0.777	0.269	29.527
5	2	3.842	34.5	0.876	0.228	34.903
6	2.5	4.805	26	1.163	0.242	32.902

From table 2 and table 3, it is clear that n-Al blended fuel has more brake thermal efficiency than its micron counterpart. The expected reason is that, larger Al_2O_3 coating of μ-Al compared to n-Al hinders the micro explosion of Al metal particle[5]. In case of ordinary diesel there is no micro explosion and the energy density increases with the addition of metal particles[5]. Therefore, efficiency of ordinary diesel is less, compared with the other two.

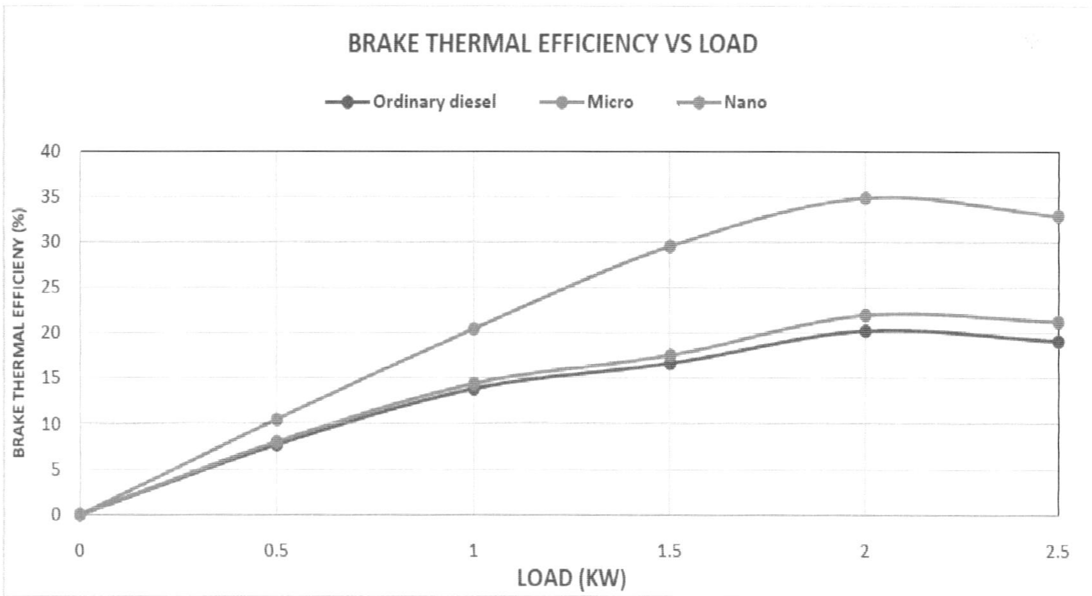

Figure 8. Break thermal efficiency vs load

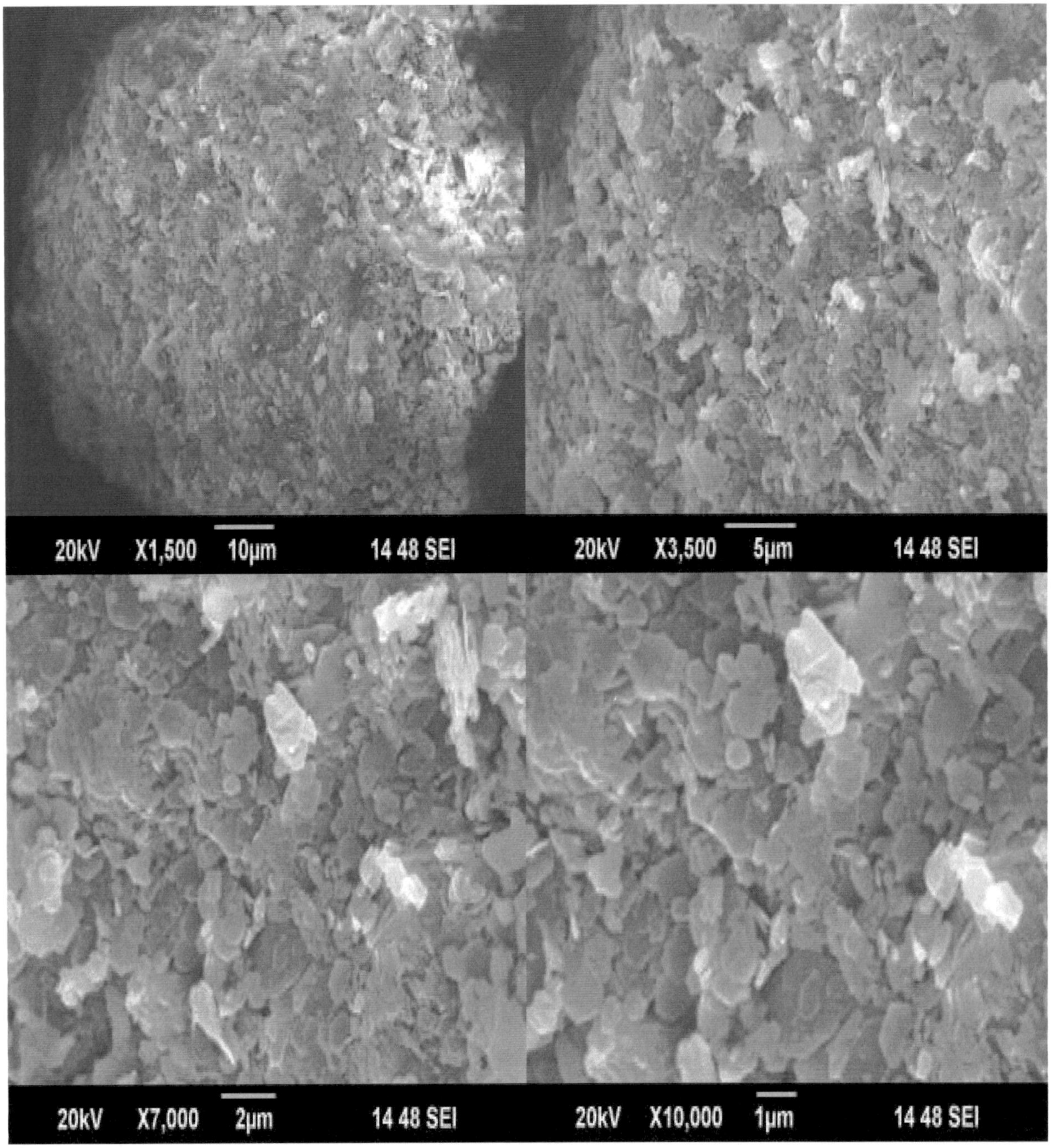

Figure 9. SEM images of zeolite-copper oxide nanocomposite

Figure.9 shows the SEM image of CuO-Zeolite mixture and it reveals the nature of the matrix. The CuO particles have a size range of 1000nm to 50nm. Here nano particles are incorporated into Zeolite sites. From XRD shown in Figure 10, it can be seen that the crystallite formations are in the range of 27 to 45 nm size.

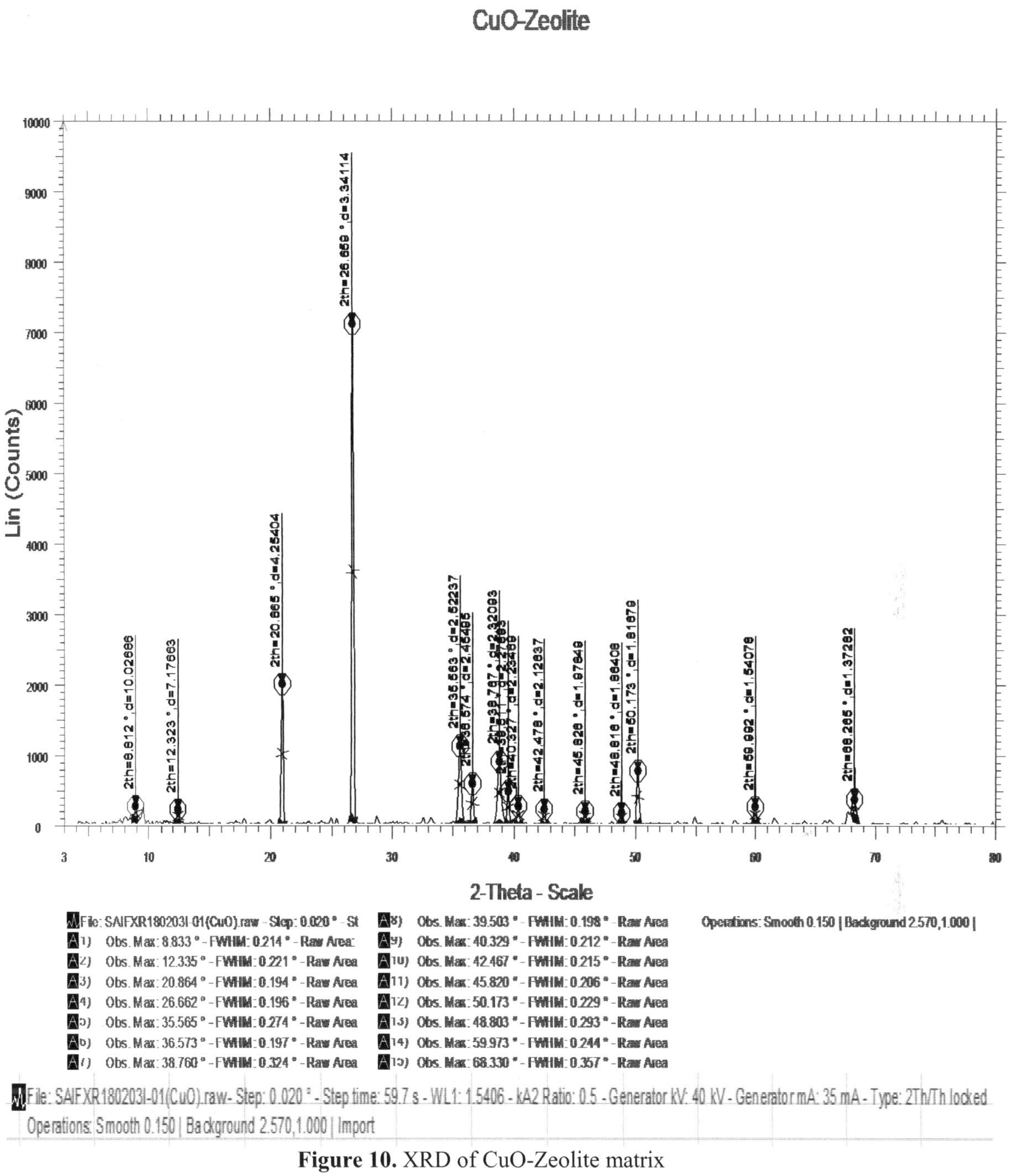

Figure 10. XRD of CuO-Zeolite matrix

Table 4 shows the pollution assessment of the studied system. There is a decrease in the pollutant level with the use of diesel containing metal Al (μ-Al, n-Al) due to the increased combustion rate and the performance of the catalytic converter. n-Al blended fuel has lowest CO emission, because of effective air mixing due to increased turbulence as a result of micro explosion than its micron counterpart[4].

Table 4. Pollutant assessment

	Without catalytic converter		With catalytic converter	
	% Smoke	% CO	% Smoke	% CO
Ordinary diesel	58.2	1.2	26.3	0.8
	47.1	1.2	24	0.7
	49.5	1.2	22.1	0.7
	42.2	1.1	20.2	0.7
	35.7	1	21.2	0.7
Average:	46.54	1.14	22.76	0.72
Diesel + μAl	36.6	0.9	26.6	0.8
	30.4	0.8	25.7	0.7
	36.3	0.9	22.6	0.7
	34.8	0.9	20.1	0.7
	33.9	0.8	18.5	0.6
Average:	34.4	0.86	22.7	0.7
Diesel + n-Al	26.07	0.8	24.5	0.7
	23.6	0.7	24	0.7
	27.3	0.7	26	0.6
	30.1	0.8	20	0.74
	29	0.8	18	0.6
Average:	27.214	0.76	22.5	0.668

4. Conclusion

The use of the nano-particles improved the functioning of diesel fuel. The droplet test gave more details about micro explosions occurred in engine cylinder. There is a large difference between the combustion behaviours. A sharp increase in efficiency can be seen, while using nano fuel as compared sswith other fuels due to its higher combustibility and the promoter action of nano aluminium and its oxides.

Aluminium particles have significant effect on the combustion of diesel. Its ability to decrease the delay in combustion is reported in literature. The experiment shows that, application of aluminium particles could increase the break thermal efficiency (BTE) by 8% and 72% for micron and nano aluminium fuels respectively.

Copper acts as a catalyst for the conversion of pollutants in exhaust but in a limited proportion. Experimental results show that, by using copper oxide zeolite based catalytic converter, CO reduces by 51% for ordinary diesel. Experiments show that catalytic converter always makes emissions at a range of 22.5-22.7 (volume)% smoke concentration and 0.65-0.73 (volume)% CO concentration.

Therefore, it can be concluded that, development of copper oxide zeolite based catalytic converter is feasible since it gives satisfactory results for given operating conditions and reduction of CO emissions. Thus, the copper oxide catalyst system can be the effective approach instead of expensive noble metal based catalytic converter.

5. References

[1] Yanan Gan and Li Qui 2011 *Combustion characteristics of fuel droplets with addition of Nano and micron-sized aluminum particles* 158 pp 354–368
[2] D Ganesh and G Gowrishankar 2011 *Effect of Nano-fuel additive on emission reduction in a Biodiesel fueled CI engine* 978 pp 3453–3459
[3] Matthew Jones Calvin H Li Abdollah Afjeh GP Peterson 2011 *Experimental study of combustion characteristics of nanoscale metal and metal oxide additives in biofuel (ethanol)* 6 pp 246–258
[4] Rakhi N Mehta Mousumi Chakraborty Parimal A Parikh 2014 *Nano fuel: Combustion, engine performance and emissions* 120 pp 91–97
[5] D S Sundaram V Yanga V E Zarkob 2015 *Review: Combustion of Nano Aluminum Particles* 51 pp 173–196
[6] H Soukht Saraee S Jafarmadar H Taghavifar S J Ashrafi 2015 *Reduction of emissions and fuel consumption in a compression ignition engine using nanoparticles* 245 pp 321–333
[7] R Friedman and A Macek 1962 *Ignition and Combustion of Aluminium Particles & Hot Ambient Gases* 6 pp 563–568
[8] S M Khopkar 2005 *Environmental pollution Analysis* (New age International (P) Ltd, Publishers, New Delhi) 5 pp 180–190
[9] V S Srivasthava 2010 *Metallic and organic nano-materials and their use in pollution control: A Review, Archives of Applied Science Research* 2(6) pp 82–92
[10] P V Walke and N V Deshpande 2011 *Cost effective catalytic converter for diesel engine after treatment* (International journal of engineering research and technology) vol 4 pp 9–20
[11] Bari A Rosso A Miniciardi M R Troiani 2001 *Analysis of heavy metals in atmospheric particulates Environmental monitoring and Assessment* 69 pp 205–220
[12] G A Mansoori G R Vakli Nezhaad A R Asrafi 2005 *int ` J. Pure & applied math Sci* 2 p 58
[13] Harmut Presting and Ulf Konig 2003 *Future nano technology developments for automotive applications Material science and engineering* 23 pp 737–741
[14] G A Mansoori 2001 *Nano thermodynamics & Phase Transition in Nano systems* (The 4th int'l Conf. Fluids & Thermal Energy Conversion) 7
[15] G A Mansoori 2003 *Phase Transitions in small systems* (Proceed. Nano Sci. Tech. Workshop Kashan Univ)
[16] David M Kennedy 2006 *The role and future of nano-technology in research, industry and education* (Conference on materials, Processes, Friction and Wear MATRIB'06 Vela luka)
[17] Subramanian and T Pitchumani K 2012 *Transfer Hydrogenation of Carbonyl Compounds and Carbon-Carbon Multiple Bonds by Zeolite Supported Cu Nanoparticles* (Catal. Sci. Technol.) 2 pp 296–300
[18] Brandenberger S Krocher O Tissler A Althoff R 2008 *The State of the Art in Selective Catalytic Reduction of NOx by Ammonia using Metal-Exchanged Zeo lite Catalysts* (Catal. Rev. Sci. Eng.) 50 pp 492–531
[19] Chung L C and Yeh C T 2008 *Synthesis of Highly Active CuO-CeO2 Nanocomposites for Preferential Oxidation of Carbon Monoxide at Low Temperatures* (Catal. Commun.) 9 pp 670–674.
[20] Jia A P Jiang S Y Lu J Q Luo M F 2010 *Study of Catalytic Activity at the CuO-CeO$_2$ Interface for CO Oxidation.* (J. Phys. Chem. C) 114 pp 21605–21610

6. Acknowledgment
We thank Centre for Engineering Research and Development (CERD) under KTU (Proceedings No. KTU/RESEARCH/ 2/2744/2017) for the financial support. We also thank Department of Chemical Engineering and Department of Mechanical Engineering, TKM College of Engineering, Kollam for providing the lab facilities.

A study of factors affecting supply chain performance

Joby George[1] and V. Madhusudanan Pillai[2]
[1]Department of Mechanical Engineering, Viswajyothi College of Engineering and Technology, Vazhakulam, Muvattupuzha, Ernakulam, Kerala-686 670, India
[2]Department of Mechanical Engineering, National Institute of Technology Calicut, Calicut, Kerala-673 601, India

E-mail[1]: jobycg2005@gmail.com
E-mail[2]: vmp@nitc.ac.in

Abstract. In the present global market, the competition is not between the companies but between the supply chains. The comparison of performance measures of supply chains helps to identify a good supply chain. The best performing supply chains may survive over a long period. The performance of a supply chain is affected by various internal and external factors. The objective of this paper is to identify and conduct a detailed study of the major factors affecting the supply chain performance. The factors are identified by reviewing various literature in the supply chain field. In this respect, a total of 54 literature are reviewed. The major factors identified as supply chain structure, inventory control policy, information sharing, customer demand, forecasting method, lead time and review period length. The optimum selection of parameters of these factors improves the supply chain performance.

1. Introduction

A Supply Chain (SC) can be defined as a network of members linked together by flow of materials, information, and funds with an objective to reduce overall system cost [1]. The supply structure may differ based on the number of intermediate facilities and position of the facilities within the supply chain. In general, the facilities in the supply chain are supplier, manufacturer, distributor, wholesaler, and retailer. Each facility in the supply chain processes order based on the available information and places order to the immediate upstream facility. From available stock, the upstream facility meets the demand of the downstream facility. The downstream facility represents the direction towards the end customer, and the upstream facility is the direction towards the end supplier. Finally, customers collect the product from the last facility of the supply chain. Figure 1 shows a representative supply chain. The goal of an SC is to maximize overall profitability, and it is calculated as the difference between the revenue generated from the customer and total cost incurred across all stages of the SC. The higher the SC profitability, the more successful is the SC. Nowadays the SC success is measured in terms of SC profitability and not in terms of the profits at an individual stage. SC profitability may decrease due to the lack of strategic fit between the competitive strategy and SC strategy. The competitive strategy defines the set of customer needs that it seeks to satisfy through its products and services. The SC strategy determines how the SC should perform with respect to efficiency and responsiveness. To achieve a strategic fit, a company must first understand the uncertainties and capabilities of the SC in terms of efficiency and responsiveness. So the right balance between responsiveness and efficiency leads to achieve strategic fit [3].

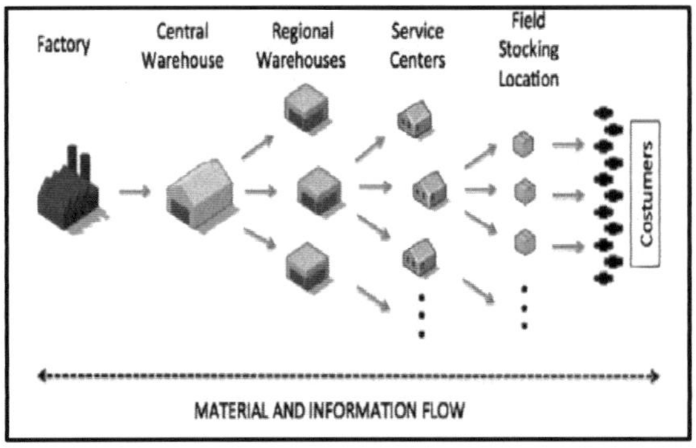

Figure 1. A representative supply chain [2]

In the present global market, the competition is not between the companies but between the SCs [4]. Hence, the best performing SCs may survive over a long period. The better performance of an SC and also the success of SC members can ensure by the high degree of coordination among members. When the members are not coordinated, it may lead to poor performance of the entire SC. Coordination among members can be achieved by information sharing mechanism [5], and some of the objects in the sharing may include supplier's capacity, customer demand, inventory policies and inventory levels of other members [6]. Some of the factors affecting SC performance are operating conditions or environmental factors and SC decision parameters or internal factors [7], [8]. The paper is organized as follows: Section 2 explains various supply chain performance measures. Section 3 describes major factors that affect the supply chain performance. Finally, in Section 4, we give a few concluding remarks.

2. Supply chain performance measures

“Performance measurement is defined as the process of quantifying the effectiveness and efficiency of an action" [9]. The comparison of performance measures helps to identify a good SC. The performance measures can be classified into qualitative and quantitative categories. Qualitative performance measures are those measures for which there is no direct numerical measurement (such as customer satisfaction, products quality, SC vulnerability, SC resilience) and quantitative measures can be described numerically (such as fill rates, costs, inventory levels, resource utilization) [10]–[12]. The various performance measures noticed in the literature to analyse SC's are: SC fill rate [3], [13], order variance and Bullwhip Effect (BWE) [14], [15], order variance ratio [13] and Bullwhip Slope (BwSl) [13], [16], inventory variance and zero replenishment phenomenon [17], risk of shortage [18] and Total Cost of the Supply Chain (TCSC) [13], [19].

- SC fill rate: It is an indication of the service level of a company for a product, and it is defined as a ratio of demand met to demand arose in the lost sale environment [3]. The average of fill rates of all the retailers is considered as SC fill rate or customer service level [7]. The fill rate of an SC depends on the inventory policy used [20], [21].
- Bullwhip effect: The increase in demand variability from the downstream stage to upstream stage in an SC is called bullwhip effect [22]. A high value for the bullwhip creates an adverse impact on the performance of an SC [23]. Previous researches confirmed that it could not be eliminated completely, but it can be reduced or controlled [5], [24]. One of the causes of its occurrence is the replenishment strategy or inventory policy used [25], [26]. BwSl is an entire SC measure showing the presence of the bullwhip effect. A high value of the slope means a fast

propagation of the bullwhip effect through the SC, while a low value means a smooth propagation [16].

- Risk of shortage: It can be found by counting the number of stockouts and dividing it by the total number of periods for which demand arises. The retailer risk of the shortage is considered as SC risk of shortage [18].
- Zero replenishment phenomenon: It is defined as the period in which a member does not place an order [27].
- Inventory variance: This metric quantifies the fluctuations in actual inventory in every period [17]. The increased inventory variance will in result higher holding cost and shortage costs.
- Total cost of the supply chain: This cost is related to inventory management and the costs of inventory holding, shortages, ordering, and transportation vary based on the inventory policies used. Hence, the TCSC is equal to the sum of the inventory holding, shortages, ordering, and transportation costs for whole members in all stages of an SC [13]. The order cost is associated with issuing a purchase order to an outside supplier or from internal production setup costs. The major cost ingredients are: making requisitions, analyzing vendors, writing purchase orders, receiving materials, inspecting materials, following up orders and doing the paperwork necessary to complete the transaction. The setup cost comprises the cost of changing over the production process to produce the ordered item. The cost incurred to transport items from one member to another is termed as transportation cost. The cost associated with carrying items is termed as inventory holding costs. Storage and other related costs are included in this cost. The major cost ingredients are: capital, storage, insurance, taxes, damage, theft and obsolescence. The cost associated with not having enough inventory to meet demand is shortage cost. The major cost ingredients are: penalties, lost sales, expediting, loss of goodwill, customer irritation and backorder.

3. Factors affecting supply chain performance

The supply chain performance is influenced by various factors. Some of the main factors are supply chain structure, inventory control policy, information sharing, customer demand, forecasting method, lead time and review period length. These factors are described below:

3.1. Supply chain structure

The number of facilities, the number of stages, and the structure of the material and information flow contribute to the complexity of the chain. The SC structures can be classified as: dyadic, serial, convergent, divergent, conjoined and network. The various SC structures are shown in figure 2.

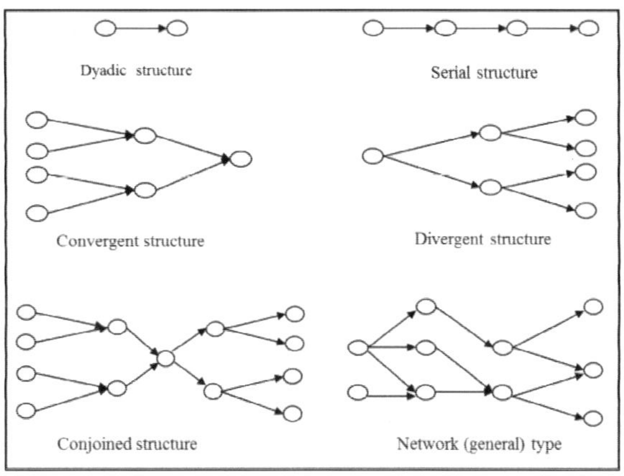

Figure 2. Various supply chain structures [28]

- Dyadic: The dyadic structure consists of two business entities (e.g. buyer-vendor) [29].
- Serial: Cascading of several dyadic structures form a serial SC structure, and generally, it consists of retailer, distributor, wholesaler and manufacturer entities or stages [29].
- Convergent: Convergent structures are assembly-type structures in which each node (or facility) in the chain has at most one successor, but may have any number of predecessors [30].
- Divergent: Each node in a divergent has at most one predecessor, but any number of successors [30]. A divergent structure can be observed in an SC in which one supplier distributes stock to its several downstream entities. Mineral processing organizations tend to have divergent structure.
- Conjoined: It is a combination of convergent and divergent SC structures and is observed in web-based retailing [30].
- Network: If an SC structure which is not falling in any of the above structures come under the category of network or general type [28]. It is a complex structure, observed in electronics manufacturing SCs.

As the supply chain structure complexity increases the bullwhip effect generated in the supply chain also increases [31]–[33]. Bottani and Montanari [32] observed that total inventory cost of the supply chain and bullwhip effect would increase significantly as the number of stages in the supply chain increases.

3.2. Inventory control policy

Inventory control means to ensure that the business has the right goods on hand to avoid stock-outs, to prevent shrinkage, and to provide proper accounting. There must be an economic balance between the costs incurred and the costs saved by holding the material in stock. There are two basic decisions that must be made for every item that is maintained in inventory. These decisions have to do with the timing of orders for the item and the size of orders for the item. So inventory control mechanism involves decisions regarding 'when' and 'how much' to order.

There are two independent-demand inventory systems viz., periodic review system and continuous review system. In the periodic review system, the inventory position is reviewed at regular intervals (at review periods), and an appropriate quantity is ordered. So, periodic inventory review involves counting and documenting inventory at specified times. For example, a retail store operating under a periodic review policy might count inventory at the end of each month. Periodic inventory review reduces the time a business owner or manager spends analyzing inventory counts, which allows more time for other aspects of running the business. However, it may not provide accurate inventory counts for businesses with high-volume sales. The owner or manager must make assumptions between inventory review periods regarding inventory counts. This can make it difficult to ascertain when reordering an item is necessary. It also can make accounting less accurate. However, in a continuous review system, reviews are often carried out continuously (after each transaction), and a fixed quantity of the item is ordered if the inventory position reaches the reorder level. Continuous inventory review, also known as perpetual review, involves a system that tracks each item and updates inventory counts each time an item is removed from inventory. Perpetual inventory review permits real-time updates of inventory counts, which can make it easier to know when to reorder items to replenish inventory. This method of inventory review also facilitates accurate accounting, since the inventory system can generate real-time costs of goods sold. The main disadvantage of this type of inventory review is the cost of implementation-bar code scanners, inventory software and computer systems are all necessary to maintain perpetual inventory review.

Inventory policies consider inventory position for placing orders (either for order decision or order size determination) are called inventory position-based policies and the policies that do not consider inventory position are called non-inventory position-based policies. Inventory position at a period is equal to on-hand plus on-order inventory minus backorders where on-hand means the quantity immediately available to meet demand, on-order means the quantity yet to arrive, and backorder means demand that could not be met in the previous periods [34].

3.3. Information sharing

Information sharing can be called as the biggest driver of performance in the SC [3]. Information connects various SC partners and allows them to coordinate activities. Information is crucial to the daily operations at each stage of the SC. An information system can enable a firm to get a high variety of customized products to customers rapidly and to understand the changing customer's tastes and preferences. Continuous sharing of new information with key individuals in a timely and qualitative manner help managers in better decision making, resulting in improved performance [35]. Information sharing in SCs offers many benefits such as reduced bullwhip effect, better coordination among various activities, taking better decisions and reduced uncertainties in SCs [5], [36]. Making information available and sharing it with the partners, an organization can speed up the flow of information within SC, improve productivity and effectiveness and decrease the response to the market time. Thus, information sharing brings competitive advantage for SC partners [35]. Information sharing reduces uncertainty in SC and thus reduces the need for safety stock. Cachon and Fisher [6] studied the impact of information sharing in a divergent SC and the result showed that the information sharing reduce the cost by 22% on average.

3.4. Customer demand

Customer demand pattern is one of the environmental factors affecting the performance of the SC. In most industrial contexts, demand is uncertain and hard to forecast. When customer demand is wildly fluctuating, the member in an SC sends a highly variable order pattern to the associated member in the upper stage, which may cause amplification of order variance (which is termed as bullwhip effect) through the SC. Customer demand volatility also results in high capacity and inventory costs on the manufacturer [37]. Different demand patterns such as constant, seasonal, seasonal with increasing trend, seasonal with decreasing trend are generated from a common demand generator for different values of the various parameters such as base, slope, season and noise influences the service level and total cost of SC [7], [36]. The safety stock required for the case of non-stationary demand is much greater than for stationary demand [38]. The increased visibility of inventory and consumer demand pattern for the players reduces SC inventory costs by enabling the players in developing improved and efficient operations plans at each level in the SC. The inventory related benefits are particularly sensitive to demand variability, the service level provided by the supplier, and the degree to which the order and production cycles are out of phase [39].

3.5. Forecasting method

Lee et al. [40] identified one of the main causes of the bullwhip effect in the supply chain is the use of demand forecasting. In a supply chain, the members need to forecast its future demand, and it is impossible to predict demand with certainty. This uncertainty will result in distorted order quantity and via order variance amplification [7]. The accuracy of forecast highly influences the supply chain performance measures such as inventory cost, backorder cost, lost sales cost, and customer's goodwill. An inaccurate forecast results in underutilization of the factory capacity [41]. Forecast method with appropriate ordering policy can alleviate the bullwhip effect up to 55% [42]. The most commonly used forecasting techniques include simple exponential smoothing and moving average. The simple exponential smoothing technique has an advantage of easy implementation in computer systems as it required fewer data storage [43]. However, the increase in order variance under simple exponential technique is greater than the increase in the order variance under moving average forecast technique [44]. Some researchers [8], [45] analysed the impact of information sharing and forecasting models under different levels of demand patterns and suppliers capacity tightness on a divergent SC having one supplier and four retailers. They found that all the factors significantly affect the SC performances, such as service level and total cost.

3.6. Lead time

The replenishment parameters such as lead time and review period can also affect the performance of SC [46]. The time gap between the receipts of the order to delivery of the product is referred as lead time, which is the sum of order lead time and delivery lead time. One of the causes of the bullwhip effect is the long lead time [47]. Under long lead time, the fluctuation in orders is more, which may results bullwhip effect [48]. The experimental study of Steckel et al. [49] found that lesser the lead time, the lesser the bullwhip effect. Long lead also results in either a shortage or accumulation of total inventory in the supply chain. Shorter lead time helps the efficient operation of the entire SC. Heydari et al. [50] analysed the impact of lead time variability in BWE for a serial four-stage SC where the members followed lead-time-sensitive ordering policy with zero safety stock and is operated under lost sales environment. Chaharsooghi and Heydari [51] developed a simulation model to analyse the effect of mean and variance of lead time in a serial four-stage SC and found that the effect of lead time variance on SC performance is much greater than the effect of the mean lead time. Khosroshahi et al. [52] noticed that the parameter of moving average method and lead time have a positive impact of bullwhip effect.

3.7. Review period length

Review period length refers to the time between the successive evaluations of inventory status to determine whether to reorder or not. The optimum selection of the review period depends on the nature of the demand of the product. Sezen [53] studied the changes in performance of a two-stage supply chain under various lengths of review period with lost sales environment and found that for products having high fluctuating demand shorter review periods are preferable to avoid long-lasting stock-outs. Also, for fast-moving consumer goods having a high carrying or shortage cost long review periods should be avoided. The study of Movahed and Zhang [54] for a serial three-stage SC under uncertain review period showed that a shorter review period is optimum to reduce the order variance; whereas a longer review period should be chosen to reduce the total expected cost. So a trade-off between the length of the review period and cost of the supply chain is essential.

4. Conclusion

The performance of the supply chain is affected by various factors. Some of the factors identified in the present study are the supply chain structure, inventory control policy, information sharing, customer demand, forecasting method, lead time, and review period length. All factors are supply chain decision parameters or internal factors except the fourth factor, which is an operating condition or environmental factor. The bullwhip effect in the supply chain increases as the supply chain structure complexity increases. Inventory control policy used in a supply chain plays a vital role in the performance of the supply chain. Information sharing is the biggest driver of performance in the supply chain. Continuous sharing of information helps managers in better decision making, which improves the performance of the supply chain. The highly fluctuated customer demand increases the bullwhip effect and inventory cost in the supply chain. The accuracy of forecast highly influences the supply chain performance. Shorter lead time helps the efficient operation of the entire supply chain. A longer review period should be chosen to mitigate the total inventory cost of the supply chain. The optimum selection of parameters of all the factors results in better performance of the supply chain.

5. References

[1] D. Simchi-Levi, P. Kaminsky, E. Simchi-Levi, and R. Shankar, *Designing and managing the supply chain: concepts, strategies and case studies*, Third edit. Tata McGraw-Hill , New Delhi, 2008.

[2] E. F. Israel, A. Albrecht, E. M. Frazzon, and B. Hellingrath, "Operational Supply Chain Planning Method for Integrating Spare Parts Supply Chains and Intelligent Maintenance Systems," *IFAC-PapersOnLine*, vol. 50, no. 1, pp. 12428–12433, 2017.

[3] S. Chopra, P. Meindl, and D. V. Kalra, *Supply chain management: strategy, planning, and operation*, Third edit. Pearson Education, 2010.

[4] M. Christopher, *Logistics and supply chain management*, Third edit. Prentice Hall, 2005.

[5] T. Moyaux, B. Chaib-Draa, and S. D'Amours, "Information sharing as a coordination mechanism for reducing the bullwhip effect in a supply chain," *IEEE Trans. Syst. Man Cybern. Part C Appl. Rev.*, vol. 37, no. 3, pp. 396–409, 2007.

[6] G. P. Cachon and M. Fisher, "Supply chain inventory management and the value of shared information," *Manage. Sci.*, vol. 46, no. 8, pp. 1032–1048, Aug. 2000.

[7] X. Zhao, J. Xie, and R. S. M. Lau, "Improving the supply chain performance : Use of forecasting models versus early order commitments," *Int. J. Prod. Res.*, vol. 39, no. 17, pp. 3923–3939, 2001.

[8] X. Zhao, J. Xie, and J. Zhang, "The impact of information sharing and ordering co-ordination on supply chain performance," *Supply Chain Manag. An Int. J.*, vol. 7, no. 1, pp. 24–40, 2002.

[9] F. T. S. Chan, "Performance measurement in a supply chain," *Int. J. Adv. Manuf. Technol.*, vol. 21, no. 7, pp. 534–548, May 2003.

[10] B. M. Beamon, "Measuring supply chain performance," *Int. J. Oper. Prod. Manag.*, vol. 19, no. 3, pp. 275–292, 1999.

[11] F. T. S. Chan and H. K. Chan, "Simulation modeling for comparative evaluation of supply chain management strategies," *Int. J. Adv. Manuf. Technol.*, vol. 25, no. 9–10, pp. 998–1006, 2005.

[12] F. Longo and G. Mirabelli, "An advanced supply chain management tool based on modeling and simulation," *Comput. Ind. Eng.*, vol. 54, no. 3, pp. 570–588, 2008.

[13] M. Pillai, P. C. Talari, and P. V. Elluri, "Performance analysis of some supply chain replenishment strategies," *Int. J. Logist. Res. Appl.*, vol. 17, no. 5, pp. 357–376, Dec. 2013.

[14] C. Zhang and C. Zhang, "Design and simulation of demand information sharing in a supply chain," *Simul. Model. Pract. Theory*, vol. 15, no. 1, pp. 32–46, 2007.

[15] E. Bayraktar, S. C. Lenny Koh, A. Gunasekaran, K. Sari, and E. Tatoglu, "The role of forecasting on bullwhip effect for E-SCM applications," *Int. J. Prod. Econ.*, vol. 113, no. 1, pp. 193–204, 2008.

[16] R. Dominguez, S. Cannella, and J. M. Framinan, "On bullwhip-limiting strategies in divergent supply chain networks," *Comput. Ind. Eng.*, vol. 73, no. 1, pp. 85–95, 2014.

[17] C. Salvatore, E. Ciancimino, and J. Ashayeri, "On the Significance of Demand and Inventory Smoothing Interventions in Supply Chain," 2010.

[18] B. Sezen and H. Kitapci, "Spreadsheet simulation for the supply chain inventory problem," *Prod. Plan. Control*, vol. 18, no. 1, pp. 9–15, Jan. 2007.

[19] T. C. Pamulety and V. M. Pillai, "Impact of information sharing in supply chain performance," *Technol. Syst. Manag.*, vol. 145, pp. 327–332, 2011.

[20] T. C. Pamulety, J. George, and V. M. Pillai, "An inventory position-based policy for better supply chain performance," *J. Ind. Prod. Eng.*, vol. 34, no. 3, pp. 180–198, 2017.

[21] T. C. Pamulety and V. M. Pillai, "Statistical process control-based inventory management in serial supply chains," in *International Conference on Advances in Supply Chain and Manufacturing Management*, 2011, p. 29.

[22] H. L. Lee, V. Padmanabhan, and S. Whang, "Information distortion in a supply chain: the bullwhip effect," *Manage. Sci.*, vol. 50, no. 12, pp. 1875–1886, 2004.

[23] V. M. Pillai, K. Lavanya, and T. C. Pamulety, "Bullwhip effect analysis using supply chain role play game and ranking of supply chains," *Int. J. Procure. Manag.*, vol. 7, no. 3, p. 299, 2014.

[24] F. Chen, Z. Drezner, J. K. Ryan, and D. Simchi-Levi, "Quantifying the bullwhip effect in a simple supply chain: the impact of forecasting, lead times, and information," *Manage. Sci.*, vol. 46, no. 3, pp. 436–443, Mar. 2000.

[25] P. Kelle and A. Milne, "The effect of (s, S) ordering policy on the supply chain," *Int. J. Prod. Econ.*, vol. 59, no. 1, pp. 113–122, 1999.

[26] S. M. Disney and D. R. Towill, "A methodology for benchmarking replenishment-induced bullwhip," *Supply Chain Manag. An Int. J.*, vol. 11, no. 2, pp. 160–168, 2006.

[27] S. Cannella and E. Ciancimino, "On the bullwhip avoidance phase: supply chain collaboration and order smoothing," *Int. J. Prod. Res.*, vol. 48, no. 22, pp. 6739–6776, Jan. 2010.

[28] T. C. Pamulety, "Analysis of information sharing and inventory policies on supply chain performance," PhD Thesis, National Institute of Technology Calicut, 2014.

[29] R. Croson and K. Donohue, "Impact of POS data sharing on supply chain management: an experimental study," *Prod. Oper. Manag.*, vol. 12, no. 1, pp. 1–11, 2003.

[30] B. M. Beamon and V. C. P. Chen, "Performance analysis of conjoined supply chains," *Int. J. Prod. Res.*, vol. 39, no. 14, pp. 3195–3218, 2001.

[31] S.-K. Paik and P. K. Bagchi, "Understanding the causes of the bullwhip effect in a supply chain," *Int. J. Retail Distrib. Manag.*, vol. 35, no. 4, pp. 308–324, 2007.

[32] E. Bottani and R. Montanari, "Supply chain design and cost analysis through simulation," *Int. J. Prod. Res.*, vol. 48, no. 10, pp. 2859–2886, May 2010.

[33] R. Dominguez, S. Cannella, and J. M. Framinan, "The impact of the supply chain structure on bullwhip effect," *Appl. Math. Model.*, vol. 39, no. 23–24, pp. 7309–7325, Mar. 2015.

[34] R. N. Boute and M. R. Lambrecht, "Exploring the bullwhip effect by means of spreadsheet simulation," *INFORMS Trans. Educ.*, vol. 10, no. 1, pp. 1–9, Sep. 2009.
[35] J. P. Hatala and J. G. Lutta, "Managing informantion sharing within an organizational setting: a social network perspective," *Perform. Improv. Q.*, vol. 21, no. 4, pp. 5–33, 2009.
[36] X. Zhao, J. Xie, and J. C. Wei, "The Impact of Forecast Errors on Early Order Commitment in a Supply Chain," *Decis. Sci.*, vol. 33, no. 2, pp. 251–280, Mar. 2002.
[37] R. N. Boute, S. M. Disney, M. R. Lambrecht, and B. Van Houdt, "An integrated production and inventory model to dampen upstream demand variability in the supply chain," *Eur. J. Oper. Res.*, vol. 178, no. 1, pp. 121–142, Apr. 2007.
[38] S. C. Graves, "A single-item inventory model for a nonstationary demand process," *Manuf. Serv. Oper. Manag.*, vol. 1, no. 1, pp. 50–61, 1999.
[39] K. E. Bourland, S. G. Powell, and David F Pyke, "Exploiting timely demand information to reduce inventories," *Eur. J. Oper. Res.*, vol. 92, pp. 239–253, 1996.
[40] H. L. Lee, V. Padmanabhan, and S. Whang, "The bullwhip effect in supply chains," *Sloan Manage. Rev.*, vol. 38, no. 3, pp. 93–102, 1997.
[41] J. S. Shang, S. Li, and P. Tadikamalla, "Operational design of a supply chain system using the Taguchi method, response surface methodology, simulation, and optimization," *Int. J. Prod. Res.*, vol. 42, no. 18, pp. 3823–3849, 2004.
[42] D. Wright and X. Yuan, "Mitigating the bullwhip effect by ordering policies and forecasting methods," *Int. J. Prod. Econ.*, vol. 113, no. 2, pp. 587–597, Jun. 2008.
[43] S. M. Disney and D. R. Towill, "The effect of vendor managed inventory (VMI) dynamics on the bullwhip effect in supply chains," *Int. J. Prod. Econ.*, vol. 85, no. 2, pp. 199–215, Aug. 2003.
[44] F. Chen, J. K. Ryan, and D. Simchi-Levi, "The impact of exponential smoothing forecasts on the bullwhip effect," *Nav. Res. Logist.*, vol. 47, no. 1, pp. 269–286, 2000.
[45] W. K. Chiang and Y. Feng, "The value of information sharing in the presence of supply uncertainty and demand volatility," *Int. J. Prod. Res.*, vol. 45, no. 6, pp. 1429–1447, 2007.
[46] T. Kelepouris, P. Miliotis, and K. Pramatari, "The impact of replenishment parameters and information sharing on the bullwhip effect: a computational study," *Comput. Oper. Res.*, vol. 35, no. 11, pp. 3657–3670, 2008.
[47] P. Berling and M. Farvid, "Lead-time investigation and estimation in divergent supply chains," *Int. J. Prod. Econ.*, 2014.
[48] X. W. X. Wang, Z. L. Z. Liu, C. Z. C. Zheng, and C. Q. C. Quan, "The impact of lead-time on bullwhip effect in supply chain," in *Proceedings of 2008 ISECS International Colloquium on Computing, Communication, Control, and Management*, 2008, pp. 93–97.
[49] J. H. Steckel, S. Gupta, and A. Banerji, "Supply chain decision making: will shorter cycle times and shared point-of-sale information necessarily help?," *Manage. Sci.*, vol. 50, no. 4, pp. 458–464, 2004.
[50] J. Heydari, R. B. Kazemzadeh, S. K. Chaharsooghi, R. Baradaran Kazemzadeh, and S. K. Chaharsooghi, "A study of lead time variation impact on supply chain performance," *Int. J. Adv. Manuf. Technol.*, vol. 40, no. 11–12, pp. 1206–1215, Feb. 2009.
[51] S. K. Chaharsooghi and J. Heydari, "LT variance or LT mean reduction in supply chain management: Which one has a higher impact on SC performance?," *Int. J. Prod. Econ.*, vol. 124, no. 2, pp. 475–481, Apr. 2010.
[52] H. Khosroshahi, S. M. M. Husseini, and M. R. Marjani, "The bullwhip effect in a 3-stage supply chain considering multiple retailers using a moving average method for demand forecasting," *Appl. Math. Model.*, vol. 40, no. 21, pp. 8934–8951, 2016.
[53] B. Sezen, "Changes in performance under various lengths of review periods in a periodic review inventory control system with lost sales: A simulation study," *Int. J. Phys. Distrib. Logist. Manag.*, vol. 36, no. 5, pp. 360–373, 2006.
[54] K. K. Movahed and Z.-H. Zhang, "Robust design of (s , S) inventory policy parameters in supply chains with demand and lead time uncertainties," *Int. J. Syst. Sci.*, vol. 0, no. 0, pp. 1–11, Nov. 2013.

Optimizing the exhaust emission from logistics and packing industries, by means of green logistics

P Benrajesh[1], A John Rajan[2]
[1] Department of Mechanical and Production Engineering,
Sathyabama University, Chennai, India
[2] Department of Manufacturing Engineering,
School of Mechanical Engineering (SMEC),
Vellore Institute of Technology, Vellore, India

E-mail: benrajesh@gmail.com

Abstract. The product packing, is an essential process for preserving the product but it also emits higher percentages of green house gases to the atmosphere, while decomposing it. Therefore, it leads to a new science named 'Green packing', which deals with the process of identifying Eco-compatible materials for packing. This article aims at detecting: the high performing packing factor for green house gas emission and also to develop an optimized values for packing enterprises to mitigate the emission. The results achieved shows that packing material selection has a huge influence on green house gas emission.

Keywords: Green house gas emission, Taguchi method, green logistics.

1. Introduction

Green logistics is defined as an initiative to control the [1] harm made to the environment by logistics by using cutting edge tools. It can be strategically executed by the tactical planning of merging the tri-poles namely: economic field, social field and environmental field of an industry. A survey made on 2009, reveals that china's manufacturing goods overwhelms the competitive nations, but it drives them to deplete and exploit the available natural resources. This leads the policy makers to introduce a jargon named 'Circular Economy,' which connects the business demand of each product with the shortage of resources, through an interface called Green logistics management [2] [3]. There after it is renamed as "environmental logistics" by the china's national standard organization [4]. Moreover, logistics companies from european union countries have established a standard interface between customers and manufacturers; to facilitate the dialogue in the domains of electronic wastes recycling, eco-friendly packing, vehicle clinic centers for reducing pollution [6].

Environmental regulations, are also an useful motivating tool for implementing green logistics management. Some of the Eco-policies in the developed nations are: discouraging the unprocessed waste disposal by heavy taxes, incentives are added along with every purchase of nature-friendly raw material, tax subsidies for every environment-friendly product [7]. The aim of this article, is to identify the high ranking parameter of an packed product, which can trigger the green house gas emission and also to determine the optimized settings of packing parameters, to mitigate the green house gas emission.

2. Literature Review:

In many developing countries, environmental pollution policies are introduced in multiple domains, since 1990. Moreover, it is observed that obsolete monitoring equipment in use, reduces the pace of green practices and therefore the governing committee exhorts the enterprises to use RFID's, electronic logistic information technology such as: bar code, magnetic card and so on[8] [9]. Along with that, few functional non-green parameters in the logistics domain are detected. Based on that countermeasures, are evolved to reduce: pollution emission and fresh resource intake as raw materials [10].

Based on the evidence, it is inferred that eco-friendly policy is a seed for any environmental initiatives. To nurture a green logistics field, from a sapling stage to a deep-rooted tree; few critical goals are to be pursued, and they are given below [11], [12];

- To develop a new green logistics theory, along with empirically proven policies.
- To revamp the operating mode of an existing practices in preferred enterprises.
- To develop a clean fuel.
- To optimize a distribution route.
- To utilize the transport efficiently.
- To reduce the delivery error.
- To promote the green storage and green packaging.
- To protect the urban environment.
- To organize the disorderly competition, exists between the primary logistics players.

3. Green House Gas Emission From Packing

Most of the non-green packing materials (such as white plastics) comprises of: polyvinyl chloride, a resource for green house gases (CO2, N20, CH4). The other alternative of non -green packaging are: mushroom packaging (MP), carbon positive packaging (CPP) and cardboard packaging.

Carbon positive packaging is a byproduct, while sugar manufacturing. It has a function ability to absorb carbon dioxide and emits oxygen to atmosphere. Cardboard packaging is made by chlorine chemical structure, and it is also an environmentally digestive product [13].

4. Methodology

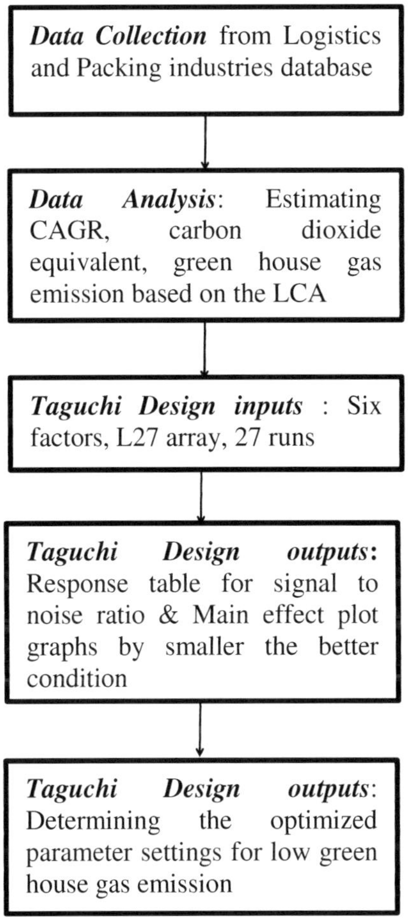

Figure-1: Methodology Of Problem

The above methodology chart (Fig-1), depicts the progression of the problem. The initial step is data collection from packing industries, followed by data analysis using Taguchi method. L27 array is selected for analysis with six parameters and two levels.

5. Green Packing

Green packing is defined as the process of providing essential layers to a product with bio-degradable simple layers to improve recyclable nature and quick disposal of materials. White plastic utility usage [14], have to be avoided due to non-degradable nature .

5.1 Principles of green packaging (4R+1D):

- Reduce (R1) : Design of green packaging should be thin and light weight.
- Reuse (R2): By basic and simple treatment, utilized containers can be further used for multiple times.
- Reclaim/Reusable (R3): Accumulation of waste materials for combustion should produce green energy rather than harmful gasses.

- Recycle (R4): A partial quantity of reusable raw materials have to extracted from the used materials, and re-inducted in the production, rather than using fresh raw materials on every time.
- Degradable (D): Recyclable packing waste should have an ability for easy degrading, after it is utilized for several times.[15]

5.2 Strategies for green packing:

5.2.1 Federal Level:

- Regulations for green packing have to be formulated, with an intent of environmental purification.
- Design and development of dedicated green storage cells for daily consuming liquids like water, beer et al; are required.
- Biodegradable empty cans could be collected back by the retail outlets, with an incentive of cash back offers and discount coupons.
- Promotion of new green packing material through online media, daily newspapers, and hoarding boards have to be done.

5.2.2 Firm Level:

- The symbols and signs have to be printed on the green packing box.
- Green packet design have to done based on an intent of disassembly and recycling.
- A green packing material preference is based on following parameters namely: customer's green demands; light weight; thin fluorine-free and high-performance materials.

5.2.3 Adverse impacts of Non green packing:

Solid waste:

* The total gross of solid waste generated annually are approximately ten million tons in the world; besides one-third of its waste are generated from packing. Some of the packing waste materials are paper,white plastic, metal, glass and other ingredients.

Bacteria and pest spreading:

* The crop pest and various bacteria packed in the less protected packing will be a potential for an emerging damage to the safety stock and humans.

5.2.4: Packing Cycle

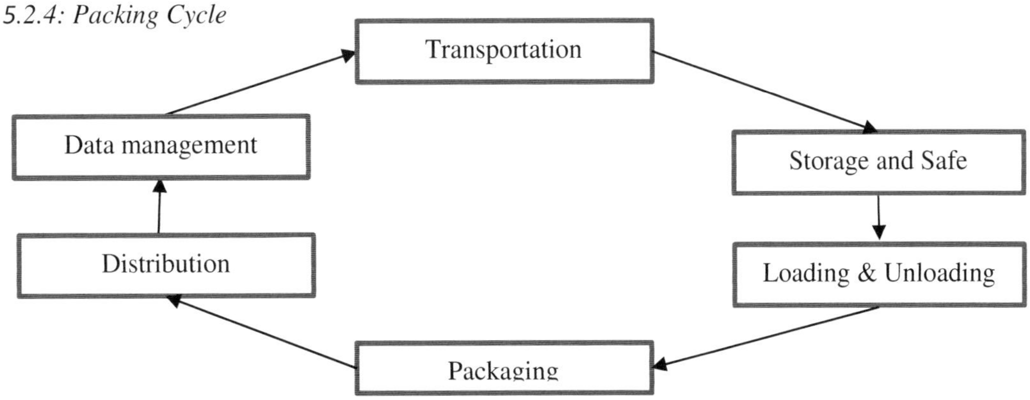

Figure-2: Cycle of Packing

The above flow chart depicts the packing cycle (Fig-2) involved in the logistics. In the early 19th century, logistics term is defined only for the activities namely: inbound and outbound transportation. But in the later half, the logistics scope is extended to multiple domains namely: Warehousing, automated loading and unloading, packing, distribution and data management. This article focused on packing domain of logistics.

5.2.6: Terminologies used in greenhouse gas emission calculation,

(i) Compound annual growth rate (CAGR)

Compound annual growth rate is a relative term, used to calculate the annual progression (revenue/sales) between two different time frames, of an firm. It is defined as the ratio of recent years revenue to the last year revenue.

$$[CAGR]^{2015}_{2003} = \left(\left\{ \frac{\text{(Revenue during 2015 year)}}{\text{(Revenue duing 2003 year)}} \right\}^{(1/12)} -1 \right) \mathbf{X}\ 100$$

(ii) Carbon dioxide equivalent (CO2e)

Carbon dioxide equivalent (CO2e) is defined as a product of green house gas acquired from the process and the global warming potential. (Note: Global warming potential is an constant term assigned for measuring the potential of green house gas. Eg. GWP for carbon dioxide is one ; GWP for methane is 25; GWP foras Nitrogen dioxide is 298.)

$$\mathbf{(CO_2e)^{gas}_1} = \{\{\text{Total green house gas (CO2/CH4/N2O) acquired from the process}\}\ \mathbf{X}\ \{\text{GWP (global warming potential for CO2/ CH4/N20)}\}$$

6. Data Analysis

6.1: Taguchi Method

The steps involved in Taguchi Method are

a) Data collected are arranged systematically, and then outcome variable and control parameters are calculated from the available data.

b) Taguchi design done, by identifying the optimum orthogonal array for the given inputs.

c) The levels of each factors are found out, based on the available data.

d) Experimental results are analyzed and based on the result, and then the optimized values for each factor was recognized..

e) The obtained results are then compared with the experimental values.

In Taguchi method, the L27 (6 X 3) orthogonal array is selected for analysis. The inputs given are: six factors; three levels, 27 iterations; whereas ‘Smaller the better’ condition is selected for the ranking method. Entire lot, of Taguchi calculation (Table-6) is done by Minitab software tool.

6.2: Taguchi Design:

The input parameters required for processing taguchi method are organized under three levels namely: Level-1, 2, 3. The parameters identified are equivalent to six and they are arranged in a tabular form.

TABLE-1: Selected Input Parameters at Three Levels

SNO	CONTROLLED FACTORS	LEVEL-1	LEVEL-2	LEVEL-3
1.	A : Revenue attained in packing industries for an year 2015 (Million Dollars)	52	171	287
2.	B: Packing units Sold (Billion)	127	1494	2861
3.	C: Compound annual growth rate (2015)	0.77	16	30.34
4.	D:Materials used for packing	1.5	2.5	3.5
5.	E: Quantity consumed in Kilo tons	5581	43666	81750
6.	F: Carbon dioxide equivalent of Packing materials	1	1.23E7	2.46E7

The levels of a factor (Table-1) is defined as multi range of (finite) values, which can trigger the response of an experiment (Fuel consumption).

Taguchi design (Table-2) will provide a optimized proposition for all six parameters, in-order to achieve the minimum outcome (i.e., smaller the better for green house gas emission).Table-3 is also utilized as a reference table for detecting the optimum values, furnished by six key parameters.

TABLE-2: L27 Design Array of the experiment

SNO	A	B	C	D	E	F
1	52	127	0.77	1.5	5581	1
2	52	127	0.77	1.5	43666	1.23E+07
3	52	127	0.77	1.5	81750	2.46E+07
4	52	1494	16	2.5	5581	1
5	52	1494	16	2.5	43666	1.23E+07
6	52	1494	16	2.5	81750	2.46E+07
7	52	2861	30.34	3.5	5581	1

SNO	A	B	C	D	E	F
8	52	2861	30.34	3.5	43666	1.23E+07
9	52	2861	30.34	3.5	81750	2.46E+07
10	171	127	16	3.5	5581	1.23E+07
11	171	127	16	3.5	43666	2.46E+07
12	171	127	16	3.5	81750	1
13	171	1494	30.34	1.5	5581	1.23E+07
14	171	1494	30.34	1.5	43666	2.46E+07
15	171	1494	30.34	1.5	81750	1
16	171	2861	0.77	2.5	5581	1.23E+07
17	171	2861	0.77	2.5	43666	2.46E+07
18	171	2861	0.77	2.5	81750	1
19	287	127	30.34	2.5	5581	2.46E+07
20	287	127	30.34	2.5	43666	1
21	287	127	30.34	2.5	81750	1.23E+07
22	287	1494	0.77	3.5	5581	2.46E+07
23	287	1494	0.77	3.5	43666	1
24	287	1494	0.77	3.5	81750	1.23E+07

TABLE-3: S/N Ratio Formulations

SNO	S/N RATIO CONFIGURATION	FORMULAE
1.	The Smaller-The better	$S/N = -10 \log (\sum Y^2/n)$
2.	The higher-The better	$S/N = -10 \log (\sum(1/Y^2)/n)$
3.	The more nominal-the better	$S/N = -10 \log (\sum Y^2/S^2)$

The formulae for calculating the Signal to Noise ratio values are given in the Table-3. This article follows the condition Smaller the better. Hence, the formulae ($S/N = -10 \log (\sum Y^2/n)$) is used in calculating those values.

TABLE-4: Response Table for Signal to Noise Ratios (Smaller is better)

Level	Revenue in packing industry for an year (2015) -{Million dollars} (A)	Packing Units sold (Billion) (B)	CAGR-2015 (C)	Packing Materials (D)	Quantity consumed in Kilotons (E)	CO_2e of Packing materials (F)
1	-89.57	-99.58	-104.66	-107.42	-89.57	-87.02
2	-97.1	-73.04	-83.43	-78.98	-97.1	-126.69
3.	-98.16	-112.22	-96.74	-98.44	-98.16	-71.12
Delta	8.59	39.18	21.23	28.44	8.59	55.57
Rank	5	2	4	3	6	1

The response values for each parameter is calculated by the formula: $S/N = -10 \log (\sum Y2/n)$ based on the condition : Smaller the better. Based on the highest scalar value of Delta, (Table-5) the ranking is done. Based on the delta ranking, parameter-F(CO2e), B (Units sold), D (Packing materials) occupies the top three spaces respectively.

6.3 Data Outcome

This section, have captured the outcome of Taguchi model and it is organised in Table-5. Two outcomes of Taguchi model are S/N ratio (Signal to Noise ratio) and Means (Average).

TABLE-5: Experimental values for the Green House gas emission

SNO	Green house gas Emission (Kilotons)	
	S/N ratio	Means
1	-68.018	2517
2	-135.807	6170930
3	-145.349	18512790
4	0	1
5	-120.542	1064408
6	0	1
7	-147.848	24683720
8	-120.542	1064408
9	-68.018	2517
10	-135.807	6170930
11	-145.349	18512790
12	0	1
13	-120.542	1064408
14	0	1
15	-147.848	24683720
16	-120.542	1064408
17	-68.018	2517
18	-135.807	6170930
19	-145.349	18512790
20	0	1
21	-120.542	1064408
22	0	1
23	-147.848	24683720
24	-120.542	1064408
25	-68.018	2517
26	-135.807	6170930
27	-145.349	18512790

The values of the above table (Table-5) are extracted from the graph (figure-3). The graph (figure-3) was processed with the response condition: smaller the better. The inference from the graph based on the optimum levels of parameters are: A3-B3-C2-D1-E3-F2. These optimized level values, are then translated to optimized experimental values; which are also captured in the 27th iteration of Table-2 & Table-6. The experimental value attained is the smallest value of all the 27 iterations. The optimum parameters achieved (based on the principle: Smaller the better) are: A3(287) - B3(2861) -

C2(16)- D1(1.5)- E3(81750) - F2(1.23E7) value. The predicted Optimum Signal to noise ratio and experimental value are extracted from the table-7 (26th Iteration).

TABLE-6: Optimum Settings For Low Emission Of Green House Gas

SNO	CONTROLLED PARAMETERS	Green House gas Emission
1.	A : Revenue attained in packing industry for an year (Million Dollars)	287
2.	B: Packing units Sold (Billion)	2861
3.	C: Compound annual growth rate (2015)	16
4.	D:Materials used for packing	1.5
5.	E: Quantity consumed in Kilo tons	81750
6.	F: Carbon dioxide equivalent of Packing materials	1.23 X 10^7
	Experimental Outcome	1.85 X 10^7 Kilotons
	Signal to Noise ratio	-145.349
	Means	1.85 X 10^7

7. Graphical Inferences

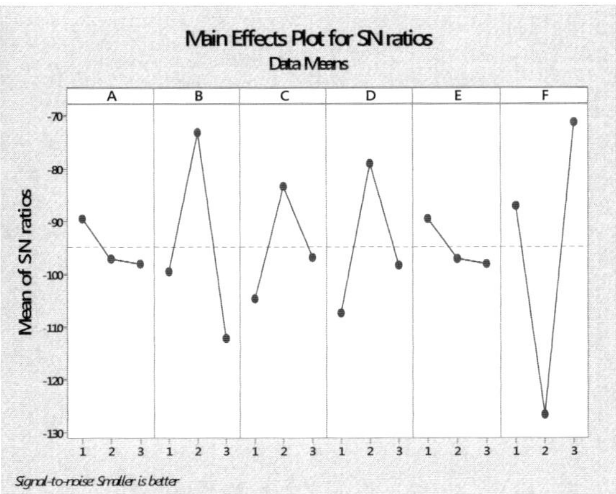

Figure-3: Main effect plot for SN ratios

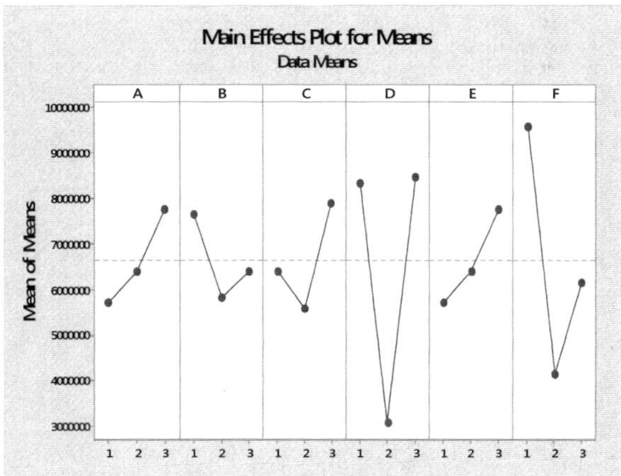

Figure-4: Main effect plot for means

Graphical figures (Fig-3 & 4) mentioned in this section, estimates the optimum settings; in-order to achieve the target of reducing the green house gas emission. Figure-3, portrays the optimized level values for SN ratios (Signal to Noise). Based on the condition of smaller the better, these level values (A3-B3-C2-D1-E3-F2) are considered as the optimum settings. Figure-4 portrays the optimized settings for the mean values. The optimized level setting of packing system for means are: A1-B2-C2-D2-E1-F2. The plot for figure-3 are considered for optimum value estimation, because of its alignment with objective function. (i.e. reducing Green house gas emission.)

8. RESULTS AND DISCUSSIONS

The list of packing parameters identified are: revenue attained in packing industry for an year; packing units sold; compound annual growth rate; materials used for packing; packing quantity consumed in kilotons; carbondioxide equivalent of materials. Based on delta values (Table-5), high ranking packing parameter are: carbondioxide equivalent of packing material; packing units sold per year and packing materials used for logistics. Packing material selection is a key packing factor as it has an dominion over carbon dioxide equivalent also. The packing materials sales trend on recent years shows that white plastics are showing a steep high whereas recyclable materials (namely carton box) are not in appreciable growth. It is necessary to conduct promotional events on green packaging for end users, during every quarter of the financial year. Logistic service providers can mitigate the green house gas emission, by adopting optimized value settings of packing parameters mentioned in the tabular column (Table-7). Based on the two results attained above (Table-5 & 7), the green house gas emission can be mitigated by two poles namely: customer's wish for green marketing and logistic service provider's packing material selection .

REFERENCES:

[1]. Tianyulan;2008 "A Comparitive study on green logistics"; *Logistics technology* **5**: 131-132.

[2]. Zhu.Q. etal; 2008 "Firm level correlates of emergent green supply chain management practices in Chinese context"; *Omega* **36**(4):577-91.

[3]. Xue Guangming;2010 "Construction of green logistics network for renewable resources"; *Shanghai economy* **9**:37-39.

[4]. Zhang huanmei,2007 "Develop environmental logistics to establish an economical society"; *Logistics management*; **8**: 24-25.

[5]. Zhang Guirong, Gao qing, Wei bo, Lidbhua; 2012 "Green logistics and sustainable development"; in the proceedings of "*2012-International Conferences on information management, innovation management & Industrial engineering*; IEEE, doi-978-1-4673-1931-7/12

[6]. Mayers.C.K.; 2007; "Strategic, financials & Design implications of extended producer responsibility in Europe: a producer case study"; *Journal of Industrial ecology*; **11(3)**:113-131.

[7]. Yu.J., Williams.E, Ju.M; 2010 "Analysis of materials and energy consumption of mobile phones in china"; *Energy policy* **38(8)**:4135-41.

[8]. Xu ping;2008; "Research on friendship between green logistics and environment"; *China storage and transport*; **12**: 80-81.

[9]. Modern logistics research centre of Tsinghua university graduate school;2010; "Green logistics development abroad and the gap to china"; *Science and technology* **(5)**:19.

[10]. Zhang Lisha, Qi Xiuli & Cheng Weijian; 2006; "Study on the establishment and development of modern green logistics in our country"; *Journal of Shandong University of science and technology, (Natural science)*, **3**: 78-80.

[11]. Qiu ailian, Yuan Feng & Wang hong jun;2009; "On development strategy of green logistics based on ecological civilization perspective"; *Journal of Shenynag University of technology (Social science edition)*; **10**:359-362.

[12]. Chen Honglin;2010; "The mechanism of the development of green logistics innovation" *Economist* **(8)**:23

[13]. Deepti Chhabra, S.K.Garg, Rajesh Singh (2017), "Analysing alternatives for green logistics in an Indian automotive organisation: A case study", *Journal of cleaner production* **167**: 962-969

[14]. Yan Qiumei; 2011 "Green logistics: Selection of logistics & environment"; *China Logistics and purchasing*, **1**: 48-49.

[15]. Gao Qing, Zhang guirong; "The Green packing management for the logistics enterprise"; Proceedings of "*2012-International conference on Information Management, Innovation Management & Industrial engineering*"; IEEE, **Doi:** 978-1-4673-1931-7/12

POGO Stability Analysis of a typical Launch vehicle for studying the effectiveness of POGO corrector in cryo stage

Raji R[1], Jancy Rose K[2], P R Murali[3] and R Neetha[4]
[1] Engineer, RAS, SDSD/SRLG/SDAG/STR & Vikram Sarabhai Space Centre
[2] Section Head, RAS, SDSD/SRLG/SDAG/STR & Vikram Sarabhai Space Centre
[3] Division Head, SDSD/SRLG/SDAG/STR & Vikram Sarabhai Space Centre
[4] Group Head, SRLG/SDAG/STR & Vikram Sarabhai Space Centre

E-mail: r_raji@vssc.gov.in

Abstract. POGO is the longitudinal instability of liquid propellant launch vehicles due to the interaction of structural dynamics with the propulsion system. This interaction may give rise to the initiation of vibrations of large magnitudes when the propulsion and structural frequencies are nearer. This can be avoided by the separation of structural and propulsion frequencies by means of a POGO corrector. The present paper describes the propulsion structural interaction studies for the cryo stage of a typical launch vehicle to avoid POGO oscillations. The propulsion systems are represented as discrete elements described by a set of second order equation and the structure is represented by the normal mode parameters. Assembly of propulsion elements along with structural equations will define the structure propulsion system. Stability of system is studied by evaluating the roots of the assembled equation. The analysis showed that there will not be any POGO oscillations during the flight.

1. Introduction

POGO refers to the dynamic instability problem faced by launch vehicles with liquid propellant engines. The instability arises due to the interaction of any structural mode with the thrust oscillation of the liquid propulsion system. The structural vibration causes vibrations in the propellant flow which will enhance the structural vibration through the oscillations in the thrust chamber. On the other hand the pressure oscillations can cause structural vibrations which will reinforce the pressure oscillations. The entire process can be expressed as a closed-loop interaction problem as shown in Figure 1.

When the global longitudinal modes and propulsion frequencies are closer, the structural vibrations and thrust oscillations can reinforce each other, causing a spontaneous increase in the accelerations in vehicle and propellant structures, instability and subsequent failure. The occurrence of POGO can be avoided by employing a POGO Corrector which will separate the propulsion and structural frequencies. The illustration of POGO phenomenon is explained in Figure 2.

Eventhough the POGO Corrector is employed, a launch vehicle has to be ensured as POGO free for a successful mission. For that POGO Stability analysis is to be carried out. The studies are carried out based on the methodology developed by Bohdan W Appenhein and Sheldon Rubin[1],[2].

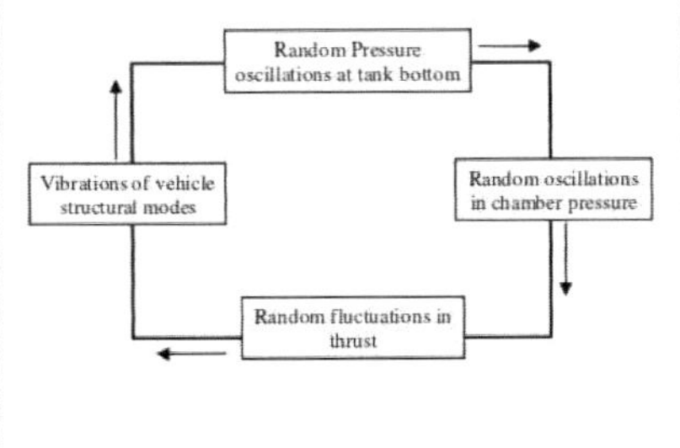

Figure 1. Block diagram showing the closed loop process of POGO.

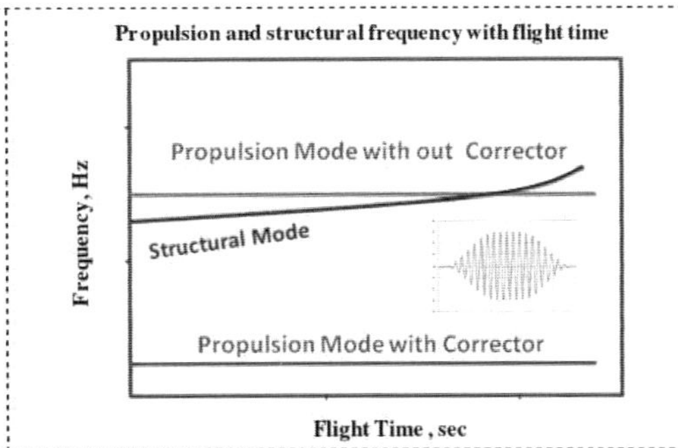

Figure 2. Block diagram showing the illustration of POGO phenomenon.

The present paper describes the propulsion structural interaction studies for the cryo stage of a typical vehicle. The stability is analysed during the entire flight regime using a time invariant, linear mathematical model of the structure-propulsion system of the launch vehicle.

2. Propulsion modelling

The propulsion system for POGO analysis is represented by the discrete element approach. The propulsion system is divided into different propulsion elements. These elements are associated with one or more nodes. The propulsion model for the POGO studies is obtained by joining elements through the nodes. The different types of elements are tank bottom, compressible duct, incompressible duct, POGO corrector, pump and chamber[1]. The schematic diagram of the propulsion model for the analysis is shown in Figure3.

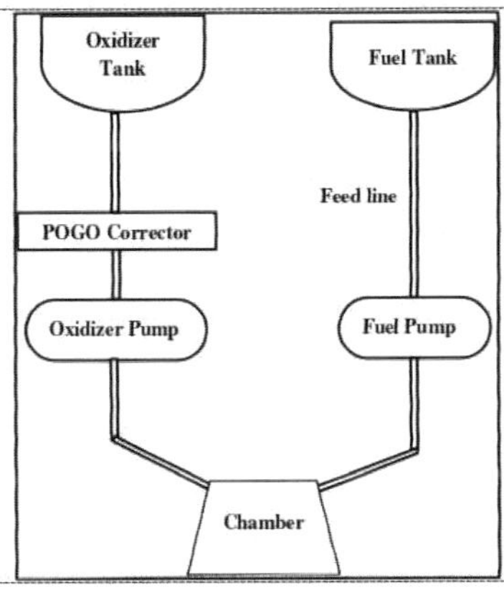

Figure 3. Sketch of the propulsion system model

Each element type is associated with a set of second order linear equation in terms of state variables perturbation static pressure 'p', fluid mass displacement 'm' at the nodes and generalised absolute displacements of the structural modes 'q'. These equations represent the input output relationship of pressure/ mass flow and flow reactions of each element. The coupling between the structure and propulsion system is through the forces generated by the propulsion system on the structure. The coefficients of the equations are in terms of the basic propulsion coefficients resistance 'R', inertance 'L' and Compliance 'C'. Each propulsion elements is modelled using these coefficients which are supplied by the propulsion agency.

The flow is described using the angles in the three directions θ_x, $\theta_y$and θ_z. The formulation for a typical propulsion element 'l' is shown in Figure 1. The flow is from the Node i to Node j. The element is represented by the propulsion coefficients, direction cosines of the nodes , head vectors of the element in the x , y and z directions. The element is associated with the propulsion degrees of freedom p_i, p_j, m_i and m_j as shown in the Figure 3.

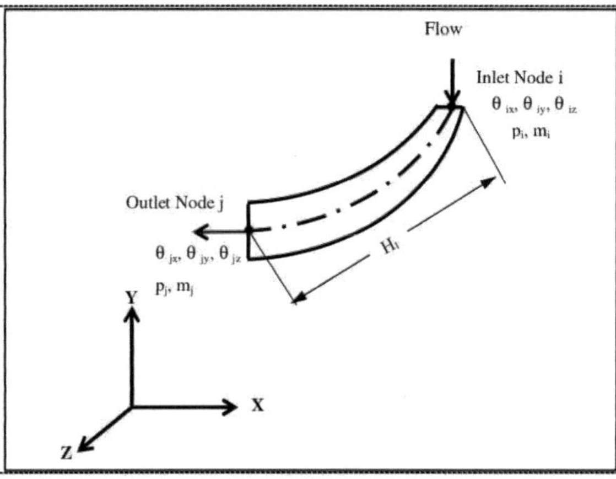

Figure 4. Typical element definition

The propulsion equation for a typical propulsion element can be written in the form

$$[A]\{\ddot{P}\} + [B]\{\dot{P}\} + [C]\{P\} = [S]\{q\} \tag{1}$$

Where A, B, C and S are the coefficient matrices of the propulsion system and structure, q is the generalised co-ordinate and $\{P\}$ is the vector of pressure and mass flow.

3. Structural modelling

The structure is represented in terms of normal mode parameters of the structure (frequency, mode shape and generalised mass).The frequencies and mode shapes from a separate analysis of 3D Finite Element model is used to model the structure in the interaction studies. The fuel and oxidiser tanks are modelled with the fluid inside. he detailed modelling of feed lines are included. The modal parameters such as frequencies and modal amplitudes at the appropriate locations are obtained through a free vibration analysis. Modes significant for POGO are selected based on criteria given in NASA SP-8055[3].

The structure can be represented using **N** normal modes. Displacement '**r**' in any direction at location '**i**'can be expressed in terms of normal co-ordinate $q_n(t)$ and modal amplitude ϕ_{ni} as

$$r_i = \sum_{n=1}^{N} \phi_{ni} q_n(t)\,,\ \text{where } i = 1,2 \ldots . N \tag{2}$$

. The structural equation of motion can be written as

$$[M]\{\ddot{q}(t)\} + [C]\{\dot{q}(t)\} + [K]\{q(t)\} = \{Q(t)\} \tag{3}$$

where M, C and K are the generalised mass, Modal damping and generalised stiffness matrices, q are the normal co-ordinates and Q is the generalised force. The forces are due to the contributions from the propulsion elements as a result of fluid flow, combustion etc.

4. Coupled Propulsion structural dynamics modelling

The propulsion dynamics equation (1) and structural dynamics equation (4) are assembled together to represent the combined propulsion structure dynamics. The aggregate of the two equations, yields the a set of equations equal to the number of state variables. The state variables are the pressure and mass flow at each node of propulsion elements and generalised co-ordinates of the modes considered. The coupled system equation can be expressed as

$$[GA]\{\ddot{x}(t)\} + [GB]\{\dot{x}(t)\} + GC\{x(t)\} = \{0\} \tag{4}$$

$$\{x(t)\} = \begin{Bmatrix} p \\ m \\ q \end{Bmatrix} \tag{5}$$

The coefficient matrices GA, GB and GC are square matrices of order 'm' which equal to the number to state variables. These set of m x m second order equations can be converted to a set a 2m x 2m first order equation by defining state vector v(t)

$$\{u(t)\} = \{\dot{x}(t)\}\text{and}\{v(t)\} = \begin{Bmatrix} x(t) \\ u(t) \end{Bmatrix} \tag{6}$$

By this transformation equation (5) can be transformed to a first order equation

$$[P]\{v(t)\} - [Q]\{\dot{v}(t)\} = \{0\} \tag{7}$$

The roots of the equation can be obtained by converting equation (8) in the Laplace domain as

$$[P]\{\tilde{v}(s)\} - s[Q]\{\tilde{v}(t)\} = \{0\} \tag{8}$$

Eigen value solutions of equation (9) determine the stability of the system. The roots of the equation are in general complex in nature. The complex eigen value s can be expressed in terms of natural frequency$\tilde{\omega}$ and fraction of critical damping $\tilde{\xi}$ as

$$s = -\tilde{\xi} \pm i\left(1 - \tilde{\xi}^2\right)^{1/2}\tilde{\omega} \tag{9}$$

where

$$\tilde{\omega} = |s|\text{and}\tilde{\xi} = \frac{-Re(s)}{|s|} \tag{10}$$

The oscillation of the system in a particular natural mode is of the form

$$\tilde{v}(t) = \tilde{v}(t)e^{-i\tilde{\xi}\tilde{\omega}t}\cos\left(1 - \tilde{\xi}^2\right)^{1/2}\tilde{\omega}t \tag{11}$$

Stability of the system is determined based on whether the real part of the roots is positive or negative. Instability occurs in a mode when the closed loop damping of the system,$\tilde{\xi} < 0$. The system has to satisfy 6dB gain margin for the nominal cases and 0dB gain margin for the worst case as per NASA guidelines[3].

.

5. Propulsion structure interaction studies

Propulsion modelling of cryo stage consists of oxidiser, fuel tanks, feed lines, pumps, oxidiser chamber pressure and thrust chamber is carried out and propulsion frequencies are estimated. From the model of the system propulsion frequencies are computed. The dynamics of each propulsion elements are modelled using the equations specified by Rubin[1]. The mode shapes corresponding to the propulsion frequencies are obtained. It is observed that with the introduction of POGO corrector reduced the propulsion frequency well below the first axial mode and the structural and propulsion

Journal of Physics: Conference Series **1355** (2019) 012020 doi:10.1088/1742-6596/1355/1/012020

frequencies are well separated. Typical mode shapes are shown in Figure 5 and Figure 6. The flight also showed the same frequencies in the pressure channels.

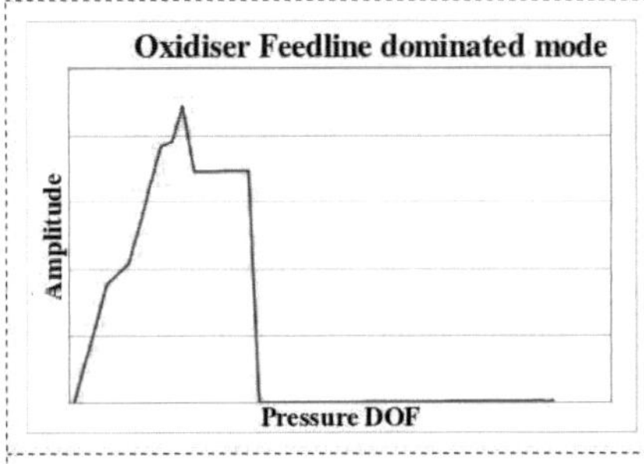

Figure 5. Mode shape corresponding to propulsion frequency1.

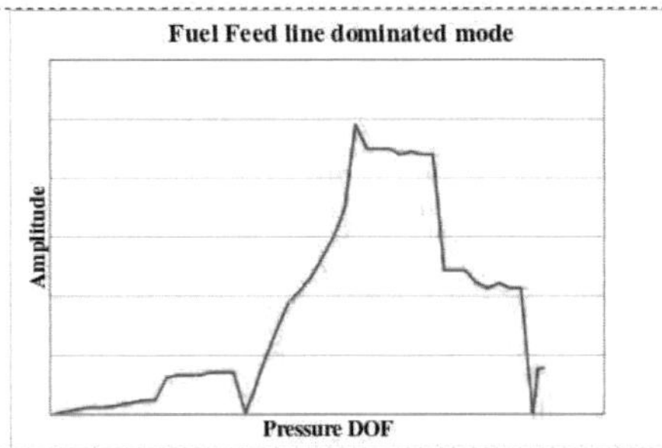

Figure 6. Mode shape corresponding to propulsion frequency 2.

Structural model is coupled to the propulsion model using the equation (4). The propulsion interaction studies are carried out through out the flight regime. Closed loop damping of the system is estimated for all structural and propulsion modes. POGO stability analysis showed that all the modes satisfy the stability margin. The closed loop damping at a particular time instant is showed in Figure 7.

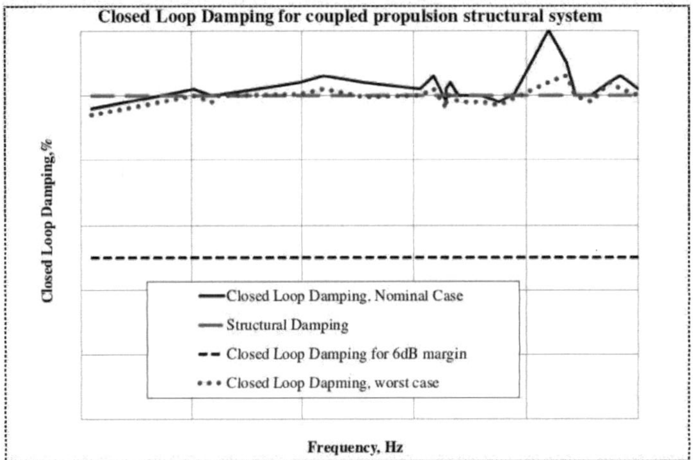

Figure 7. Closed loop damping of the coupled propulsion structural system at a particular flight time

Flight data analysis of the pressure channels and accelerometer showed that there is no signature of POGO in the flight. The estimated propulsion frequencies are in good match with the frequencies measured by the pressure channels.

6. Summary

The propulsion structure interaction of the cryo stage of the launch vehicle based on the methodology developed by S Rubin [1] is studied. The stability of the system is computed from the roots of the coupled propulsion structural equation. The introduction of POGO Corrector reduced the propulsion frequency of the oxidiser feedline mode well below the first axial mode. The analysis showed that the vehicle will be stable against POGO oscillations.

Acknowledgments
Authors acknowledge the guidance given by Shri:M.N.G.Elayathu, former GH, SDG and Shri Senthil Visagam for the basic module of the software. The authors also acknowledge Dr T Sundararajan , Shri AG Balamurali and Smt BM Jeyajanaki for providing the structural data and Shri Sajeev P, LPSC for the propulsion data input.

7. References

[1] Bohdan W. Oppenheim and Sheldom Rubin 1993 "Advanced POGO stability analysis for liquid rockets"AIAA-92-2454-CP/ Journal of spacecraft and Rockets Vol.30, No.3, pp. 2048-2058
[2] Sheldom Rubin 1966 "Longitudinal instability of liquid rocket due to propulsion feedback `POGO)" J.Spacecraft and rockets, Vol.3, No.8, PP.1188-1195
[3] Sheldom Rubin 1970 NASA Space Vehicle Design Criteria (Structures): "Prevention on Coupled Structural Propulsion Instability POGO",NASA SP-8055

Methodology for preparing a Cost-Value Stream Map

Rajesh Menon B[1], Shalij P R[2], Kiron K R[3], Sajith V S[4], Sreejith J[5]

[1]Associate Professor, Department of Mechanical Engineering, NSS College of Engineering Palakkad, Kerala 678008, India.
[2]Associate Professor, Department of Production Engineering, Government Engineering College, Thrissur 680009, India.
[3]Associate Professor, Department of Mechanical Engineering, NSS College of Engineering, Palakkad, Kerala 678008, India.
[4]Assistant Professor, Department of Science and Humanities, NSS College of Engineering, Palakkad, Kerala 678008, India.
[5]Research Scholar, Amrita School of Engineering, Amrita Vishwa Vidyapeetham, Coimbatore, Tamilnadu, India

E-mail: brmenonnssce@gmail.com

Abstract. This paper proposes the methodology for preparing the Cost Value Stream Map (CVSM) which possesses the advantage of measuring the cost involved and time consumed in the production process stream. A value stream map addresses the value addition process in terms of time domain in a production process. CVSM addresses both the time and cost information simultaneously in a production process. This CVSM method equips the study of the production process in detail and helps to figure out the wastages in the process directly in terms of the cost involved in the production. The cost line added alongside the time line in the modified VSM is a simple tool which can be utilised by lean experts in identifying the wastage in production process in terms of the cost involved. CVSM is prepared by collecting data from company in the prescribed format. All this data thus collected can be consolidated to prepare CVSM. A case study is conducted and a CVSM is prepared in a scaffolding pipe manufacturing unit in Kerala.

1. Introduction

Lean was first introduced by Taichii Ohno and Shigeo Shingo into a concept called Toyota Production System (TPS)[1]. Lean manufacturing is a concept used to reduce cost and hence enhancing the firm's productivity by eliminating non-value added items or wastes in the production line.'The machine that changed the world' is the book which describes the entire lean concept. Lean tools include 5S, just-in-time (JIT), cellular manufacturing and total productive maintenance (TPM)[2].

One of the major lean tools developed for reducing waste is value stream mapping (VSM) and it is developed by Rother and Shook in 1998[3]. Value stream mapping is a systematic approach of visual mapping and communication technique to eliminate the wastes in a company. Hence the competitiveness of the company will get improved because of enhanced productivity. Environmental value stream mapping (E-VSM) is the extended technique of conventional VSM and in this case, environmental factors are also taken into consideration[4]. The two major phases involved in value stream mapping are mentioned below [5].

- Value stream analysis: - Present value stream is plotted and studied.
- Design of value stream: - Non-value added wastes are minimized by implementing lean tools.

Value stream mapping is a paper pencil tool to identify non value added activities in the production process. It gives a visual representation of the material and information flow associated with the production stream. Value stream mapping technique utilizes current state map to assess the present condition of the manufacturing system and future state map for proposing the changes that are to be incorporated for addressing the issues found with the help of the first current state map.
The application of value stream mapping is widely spread into variety of areas, like paint[6], steel[7,8], aircraft maintenance industries[9]. Figure 1 show the procedural steps for implementing VSM in an industry.

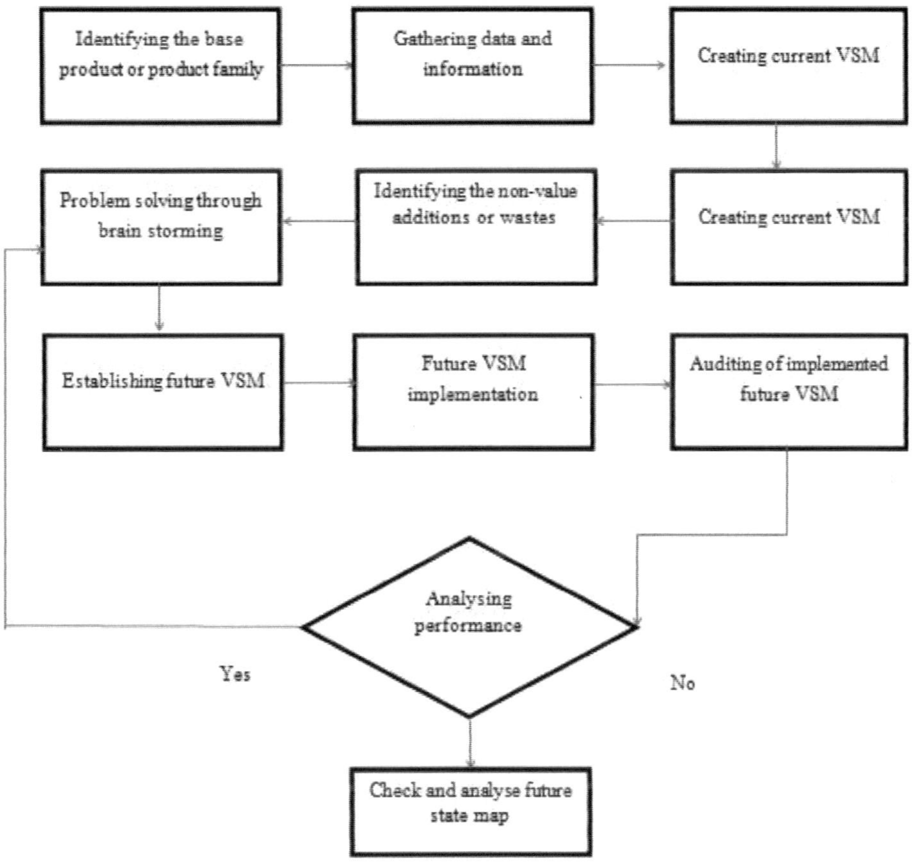

Figure 1. Steps for implementing VSM in an industry.

Studies were also carried out on the field of value stream mapping in small and medium enterprise which are manufacture-to-order[10]. A new product is planned, designed, created and marketed in the field of product development (PD). Value stream mapping is implemented in the product development field also to minimize the lead-time[10]. Single Minute Exchange of Die (SMED) can be used as a Kaizen tool to implement future VSM to reduce lead-time in small-medium enterprise[11,12]. Also, simulation models can be used to identify the bottlenecks in the production line and modified algorithms were used to reduce the lead time in a midsize glass fabrication company[13]. Value stream mapping can be combined with the decision- making tools like discrete event simulation to improve the productive performance[14]. One of the major drawbacks of VSM is the iteration process to implement future state. To overcome this difficulty, an immersive virtual reality approach is presented to visualize the image of real models[15]. Other major problems related to VSM is mentioned below[16].

i. Incomprehensibility of procedures.
ii. Collection of data in processes is difficult.
iii. Current state map becomes obsolete.
iv. VSM is compromised in the case of flexible production line.
v. It is difficult to implement when standardisation and stability of the process is poor.

In this paper, a new tool is developed which combines both value stream mapping and Cost Time Profile (CTP)[17]. Value stream mapping gives the information of the processes in units of time and with the help of this we can reduce or eliminate the mistakes and hence reduces lead time and value added time. In this paper, value stream mapping combines with activity-based costing to develop a new tool to estimate the process costs. By using this technique, it will be easy to identify the processes which possess highest cost and greater lead time and these processes can be called as critical processes. Hence, both cost and time can be visualised in a single figure and improvements can be done to reduce both cost and time of the critical process.

2. Development of CVSM

A case study has been conducted in a company which manufactures scaffolding pipe assembly in an MSME unit in Palakkad, Kerala, in one shift per day. The product consists of the production of two major parts namely the Exterior and Interior pipes assembled together using a nut-bolt and locking lever mechanism. Exterior pipe and interior pipe used for the construction of scaffolding pipe assembly is made through two parallel processes as shown in figure 2. Time is recorded on a normal working day with all breaks and all operators present for duty. Cost of operators, Resources required at each process, Machine Cost, Maintenance cost, Tooling Cost etc. are accounted annually and arrived at the cost rate of each of the above mentioned parameters per hour. Along with the standard value stream map which shows traditionally the time line is modified by showing the cost line which gives a direct measure of non-value-added cost associated with all processes.

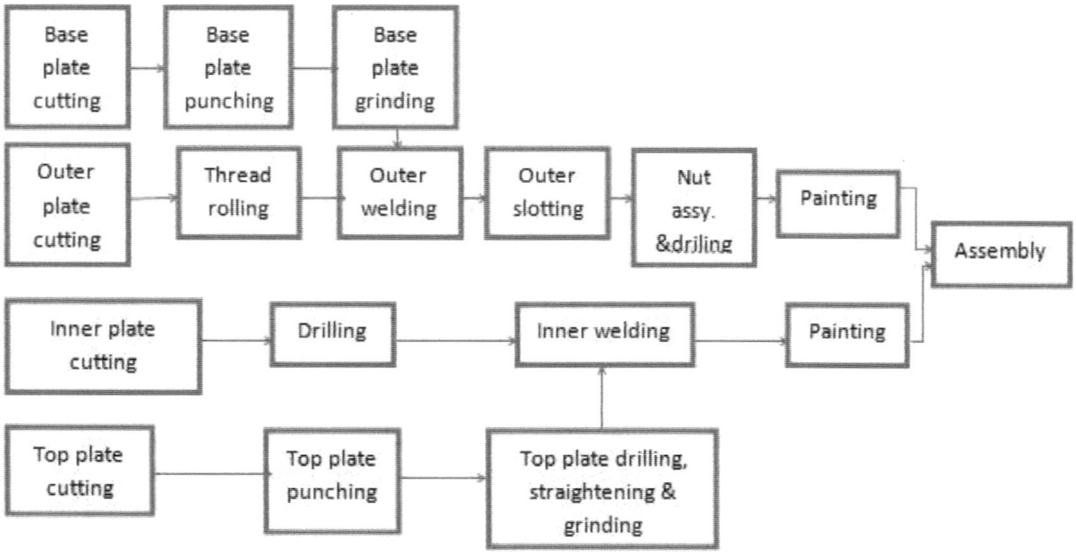

Figure 2. Layout of production processes

After creating the process layout, next step is to collect data for the above mentioned processes. From the historical data of shipments available in the company, the daily demand of the selected product was identified. After that, the net available time was calculated and finally with the help of above collected data, takt time was calculated and it is mentioned in the equation (1).

2.1 Primary data to be collected from the company

- Daily average demand
- Working hours per shift
- Break time
- Net available time = (Working hours – break time) minutes.

After collecting the primary data about the product demand, product volume per day and the net available time per day we measure each process by its cycle time. From each activity of the manufacturing process the time data such as Loading time / Changeover time, Cycle time and Waiting time is to be recorded. Number of semi-finished goods in between each work station is to be accounted for further analysis. Material flow and information flow is to be recorded in each machining centre. The time recorded as per the above list is refined by applying the weighted averaging procedure for more accuracy. Here the time taken for each event is recorded three times at random (most likely) and then it is approximated for more realistic measure by considering the most pessimistic and most optimistic time as reported by the operator of each machine as shown in table 1.

Table 1. Template for measuring time

Most likely (Average of three random values)	Most Optimistic	Most Pessimistic	Activity Duration (Seconds)

Next step to draw current VSM is the collection of data of each processes like cycle time, change over time, number of operators. Work in process (WIP) inventory data collection is the next critical step in the VSM. All the necessary data were collected from the company when there is a continuous flow of production. Also, the mode of transaction within and across the company was obtained from the top management level. Table 2 shows the collected information for VSM from the company.

Table 2: Collected data for current state VSM.

Activity	Process Description	Cycle time (mins)	Change over time (mins)	Waiting time (mins)	Number of Operators	Work In Process Inventory

In this work, a new term called 'cost investment' is added and this gives the waiting cost for the work in process (WIP) inventory which is a major non-value added cost. For the calculation of overall cost, both total cost and cost-investment is calculated. The summation of both above mentioned costs gives the overall manufacturing cost of the product. Again, the total cost is divided into material cost and activity cost. The summation of both operator-cost and resource cost implies the activity cost. More precisely, the resource cost is again divided into machine depreciation cost and machine maintenance cost. Hence, almost all value added and non-value added costs are taken care of by this cost analysis method. Both machine depreciation cost and machine maintenance costs are calculated using equations (2) & (3) respectively.

Takt time = (Net available time/ daily average demand) minutes/unit. (1)

$$\text{Machine depreciation cost} = \text{Price of the equipment}/(\text{Years of life} * \text{Time units it is used in a year}) \quad (2)$$

$$\text{Machine maintanence cost} = \text{Maintanence cost per year}/(\text{Time units it is used in a year}) \quad (3)$$

Resource cost = Machine depreciation cost + Machine maintenance cost (4)

Equation (5) gives the activity cost calculation and the total cost was calculated using equation (6).

Activity cost = Operator cost + Resource cost (5)
Total cost = Activity cost + Material cost (6)
Waiting cost is termed as cost investment in this research work and finally the direct cost calculation is shown in equation (7).
Direct cost = Total cost + Cost investment (7)
For the calculation of cost investment, the rate of return for the company was assumed as 15%. Table 3 shows the resource cost calculation.

Table 3. Resource cost calculation

Activity	Machine cost (Rs)	Machine depreciation (Rs/min)	Machine maintenance(Rs/min)	Resource cost rate

Table 4 depicts the whole cost calculated for the scaffolding pipe assembly. Here the cost rates are calculated by considering the total emoluments for each operator annually and then converting this data to calculate the cost rate per hour for maintaining such an operator. Then the cost rate of each operator depending upon his position and experience in that company is found out. Activity cost rate is calculated by combining the operator cost rate and resource cost rate.

Table 4. Value and non-value added costs.

Activity	Operators	Operator cost rate (Rs/min)	Resource cost rate(Rs/Min)	Activity cost rate(Rs/Min)	Activity cost(Rs)	Material cost(Rs)	Total cost(Rs)	IRR (Internal Rate of Return)(%)	Cost investment(Rs)	Direct cost(Rs)

From the above mentioned table, it is easy to understand and differentiate all the value added and non-value added costs. After the identification of critical process which has greater non-value added costs, control measures can be applied to reduce these costs. The figures 3 and 4 show typical CVSM for Exterior and Interior pipes. From this map the non-value-added processes can easily be identified directly because of its cost-domain representation.

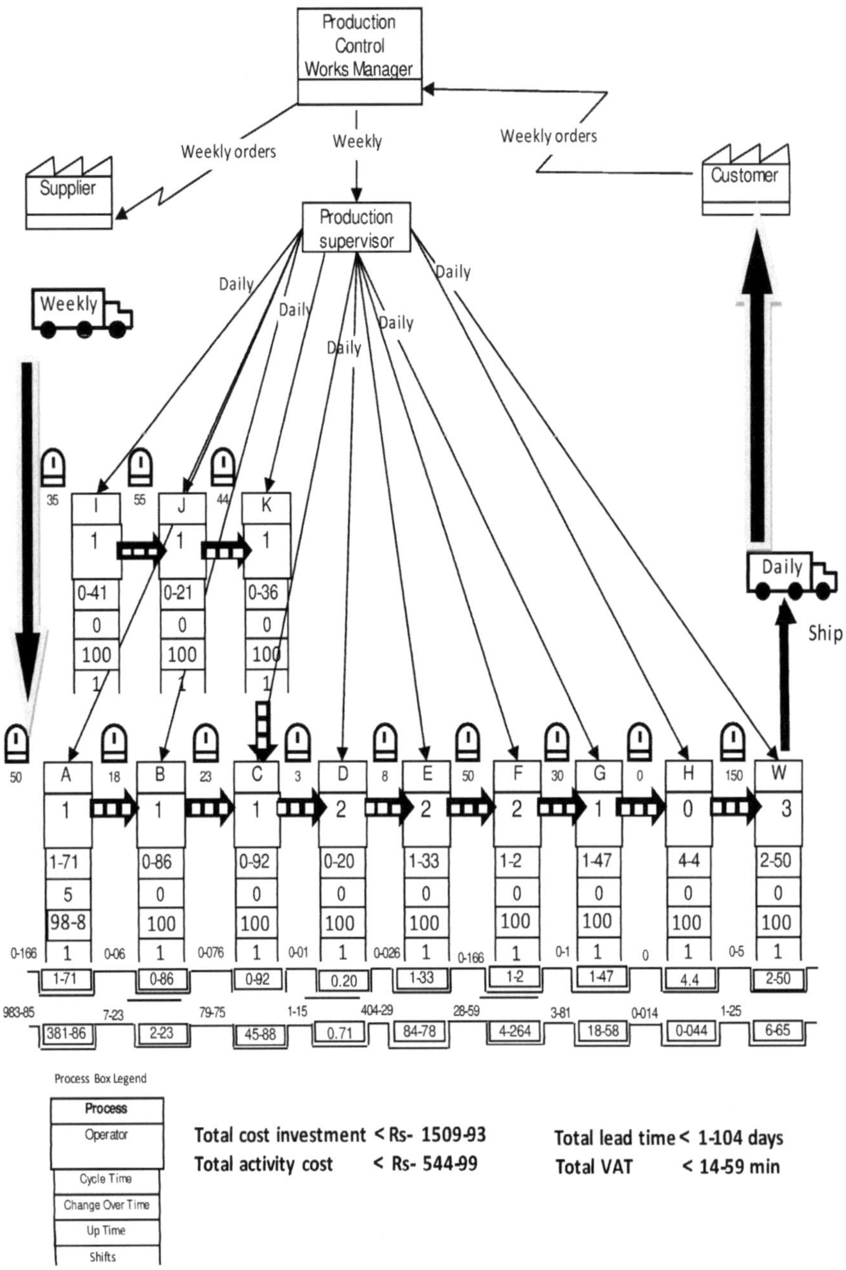

Figure 3. CVSM of exterior pipe.

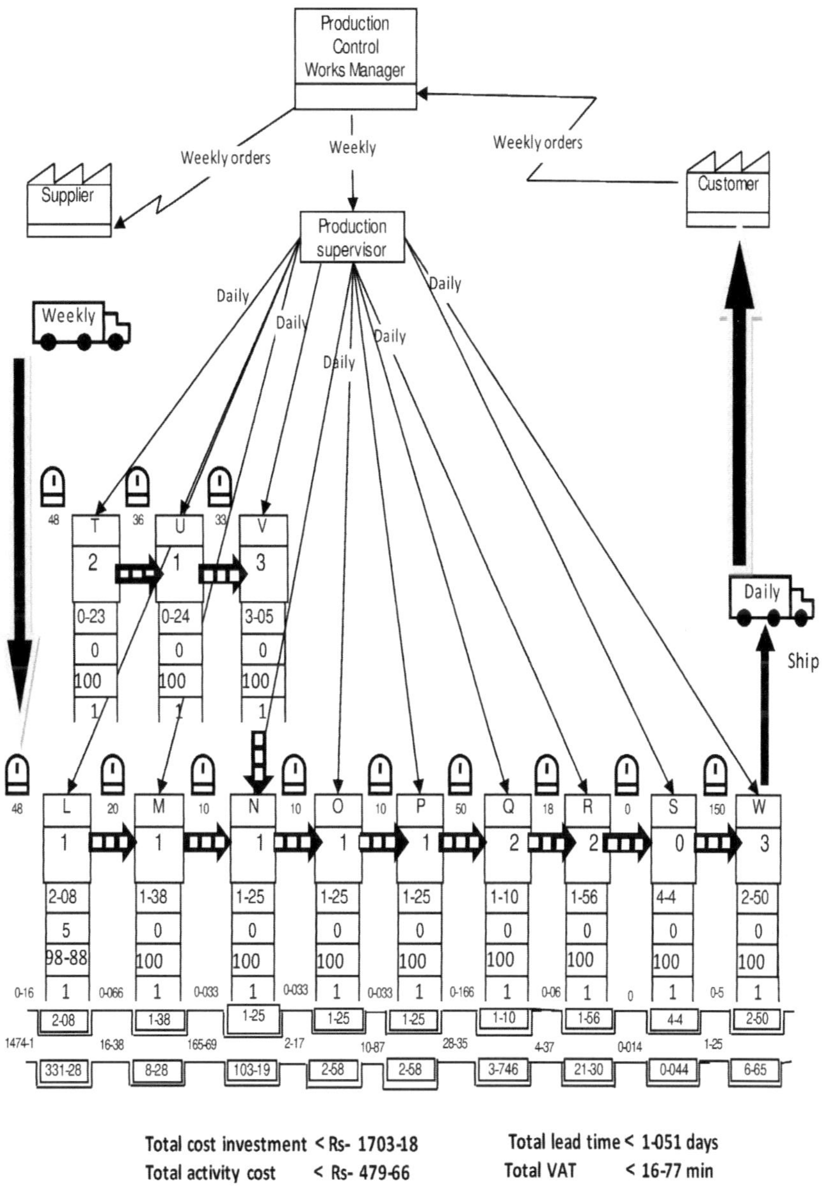

Figure 4. CVSM of interior pipe.

3. Conclusions

This work proposes a methodology which combines both value stream mapping and costing. The general tools required to collect the data from a production firm is designed and explained for the preparation of the conventional VSM and modifying it further to finally arrive at the CVSM. These general procedures can be conveniently employed by anyone who is familiar with lean manufacturing tools. A scaffolding pipe company was selected for the implementation of this new tool. The result obtained from the case study is shown in Fig 3 and Fig 4 which clearly depicts the cost-line along with the conventional time-line of VSM. The results in cost-line combines the material cost, activity cost and the waiting cost of production for the in process inventory as well as all other inventories waiting to get processed. The waiting cost account for the cost investment of all resources which are involved in the production process. Therefore the CVSM can be utilized as a very simple and effective assessment tool for identifying the non-value-added (NVA) cost involved in the process of production. From the cost-line it is easy to identify the most NVA process in the order of their financial liability and then to recommend a suitable lean tool as a remedial measure.

References

[1] Ohno T 1988 The Bible of Toyota Production system
[2] Abdulmalek F A and Rajgopal J 2007 *Int. J. Prod. Econ.* **107** 223.
[3] Rother M and Shook J 1998 *Learning to See :Value stream mapping to create value and eliminate muda* Brookline,MA: Lean enterprise institute
[4] Huang Y and Tomizuka M 2007 *Procedia CIRP* **61** 446.
[5] Anon The Bible of Toyota Production - Ohno's Manuscript.pdf
[6] Rohani J M and Zahraee S M 2015 *Procedia Manuf* **2** 6.
[7] F. Adullah, "Lean Manufacturing tools and techniques in the process industry with a focus on steel", 2003.
[8] Mishra K C and Khokhar E V 2017 *Int. J. Mech. Eng. Inf. Technol* **05** 616.
[9] Luciana L and Lestari Y D 2015 *J. Bus. Manag* **4** 1119.
[10] Tyagi S, Choudhary A, Cai X and Yang K 2015 *Int. J. Prod. Econ* **160** 202.
[11] Azizi A and Thulasi M 2015 *Procedia Manuf* **2** 153.
[12] Das B, Venkatadri U and Pandey P 2014 *Int. J. Adv. Manuf. Technol* **71** 307.
[13] Helleno A L, Pimentel C A, Ferro R, Santos P F, Oliveira M C and Simon A T 2015 *Int. J. Adv. Manuf. Technol* **80** 1059.
[14] A L Helleno, Pimentel C A, T S, R F, M C O and A T S 2015 Integrating value stream mapping and discrete events simulation as decision making tools in operation management
[15] Tyagi S and Vadrevu S 2015 *Int. J. Adv. Manuf. Technol* **81** 1259.
[16] Forno A J D, Pereira F A, Forcellini F A and Kipper L M 2014 *Int. J. Adv. Manuf. Technol* **72**, pp.779.
[17] Rivera L and Chen F F *2007 Robot. Comput. Integr. Manuf* **23** 684.

FINITE ELEMENT ANALYSIS OF L3-L4 VERTEBRAE AT VARIOUS BODY POSTURES UNDER OSTEOPOROTIC CONDITION

ResmiS.L[1],Ananthu Mohan[2],Firoz N[3], Hashim V[4] and Dileep P.N[5].
[1],[3]Assistant Professor, Department of Mechanical Engineering, TKMCE, Kollam.
[2]Under graduate student,Department of Mechanical Engineering, TKMCE, Kollam.
[4]Associate Professor, Department of Mechanical Engineering, TKMCE, Kollam.
[3]Professor, Department of Mechanical Engineering, TKMCE, Kollam.

E-mail: resmisaji@gmail.com

Abstract. The present study gives detailed quantitative information on the biomechanical behavior of osteoporotic vertebra which can be used in the evaluation of osteoporotic bone fracture. A three –dimensional finite element model has been developed and applied for the analysis of the lumbar L3-L4 disc -body unit under various body postures like normal standing, extension,weight lifting, stoop lift and squat lift .The stress and deformation of disc and vertebrae are compared with the corresponding values at 10% and 20% osteoporotic condition. It is observed that for weight lifting, stoop lifting, and squat lift under 20% osteoporotic condition the maximum value of stress obtained is close to compressive strength of the bone. Also noted that fewer than 20% osteoporotic condition the deformation of the disc is increased to 20.08% under weight lifting postures. Therefore ,we come to the conclusion that people at old age should avoid risky postures which result in low back pain or lumbar fracture.

Keywords:*Osteoporotic vertebra, CT based FE model*

1. Introduction

Age related, osteoporotic fractures of spine are more common in women and have associated health care costs of lakhs of rupees. Probabilistic analysis can predict the variability of failure when the stress concentration exceeds the yield stress. Finite element (FE) methods have become an important tool to evaluate mechanical stresses and strains in bone and have been widely used to investigate the mechanical behavior of bone tissue.Asa research tool, finite element analysis can be used to analyze the mechanisms of age-related fractures. The research potential of CT-based finite element analysis of bone structures has been recognized by many investigators [1, 2], but few have used this technique to predict the absolute failure loads of bones. Finite Element studies concluded that the highest compressive loads act on L3-L4 lumbar spine segment [3].

The present study focuses on the effect of osteoporotic behavior on the strength of human lumbar vertebra. The strength of human bone can be evaluated using FE analysis. The CT image in DICOM format is decoded to develop 3D finite element model of human bone.The material property of lumbar vertebrae and intervertebral discs is assumed to be isotropic.The various body postures considered for this study are extension, weight lifting,stoop lift and squat lift. The finite element model is developed using ABAQUSsoftware.Initially, the finite element model of a normal human vertebra-disc assembly subjected to various body postures like weight lifting, extension, stoop lift and squat lift was analyzed and its strength is evaluatedfrom the stress and displacement values of disc obtained using ABAQUS.The Osteoporotic behavior is simulated using a finite element model of vertebra-disc assembly with gradual reduction in bone density.

Table1. Body postures and the corresponding loads on l3-l4 vertebrae

Body posture	*Compressive Force on L3-L4 (N)*	*Shear Force (N)*	*Moment on L3-L4 (Name)*
Extension (20^0)	500		7.5
Weight Lifting(7.4kg)	518	233	1
Stoop Lifting (11.5kg)	2819	436	
Squat Lift (11.5 kg)	2932	237	

2. Geometric modeling

In computational biomechanics exclusively for orthopedic applications, CT quantitative images are more suitable for bone modeling since hard tissue (bone) has a high contrast relative to soft tissue. The DICOM image of femur bone of a healthy female patient with a body weight of 70kg is decoded to develop a computer Aided 3D model for the FE Analysis. The software simpleware (scan IP) was used to process CT image [4,5].

In the next step, finite element model is developed using ABAQUS by importing the geometric model in SAT format. Then, the geometric model is meshed into nodes and elements. The element used in the present study is the 8-noded brick element available in ABAQUS.

2.1. Loading and Boundary Conditions

For a man with 70 kg weight, the force acting on the L3-L4 vertebrae is 450N. In which 70% of the force is acting on the vertebrae column and remaining 30% on the pedicle [3]. A friction coefficient of 0.1 is provided in between all surfaces. The lower surface of the lower vertebrae is made to be fixed. The density of the bone is 1990 kg/m³ and the young's modulus of the bone is 12000MPa and with a Poisson's ratio of 0.3. The density of intervertebral disc is 1050 kg/m³ with a young's modulus and Poisson's ratio of 100MPa and 0.3 respectively. [6]The validated model has been extended for various body postures under osteoporotic condition .The table.1 shows the five body postures and the corresponding loads on l3-l4 vertebrae.[7,8]

3. Results and discussions

In the present work, strength of human lumbar vertebra subjected to axial compression at various body postures is evaluated using CT based finite element method. In this study, two types of vertebrae i.e., a healthy vertebra and an osteoporotic vertebra, are modeled to study the effect of osteoporotic behavior on the strength of human lumbar vertebra. In order to predict stress and strain distributions within vertebra and disc for different phases of osteoporosis, a gradual reduction of bone density was used to model the bone.

3.1 Disc deformation during extension

On conducting Static analysis on L3-L4 segment, it is found that during extension the disc deformation in the posterior direction is found to be 1.865×10^{-2} mm and at that of 20% osteoporotic condition is found to be 1.982×10^{-2} mm that is, deformation increased by 6.2%.But this deformation is far less to produce any pain.The maximum stress at 20% osteoporotic condition is to found to be 79.44 MPa which is below the compressive strength of the vertebrae.

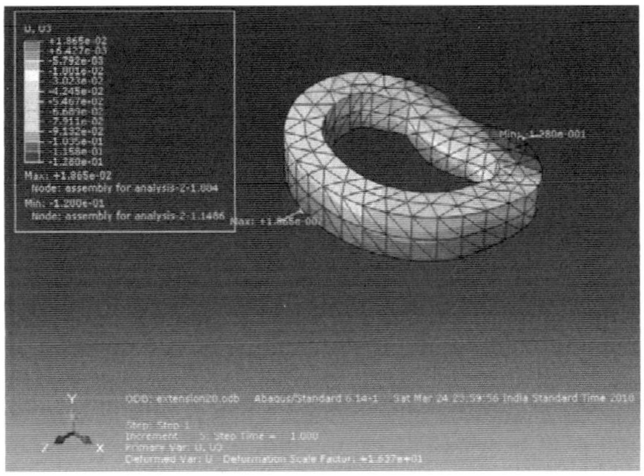

Figure.1. Deformation on annular disc during standing at healthy condition

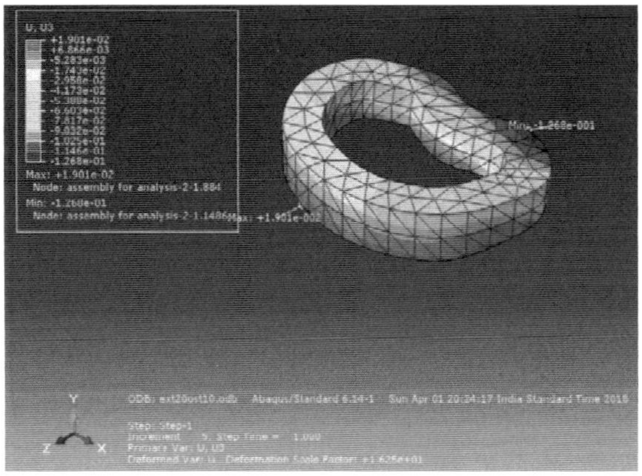

Figure.2. Deformation on annular disc during standing at 10% osteoporotic condition

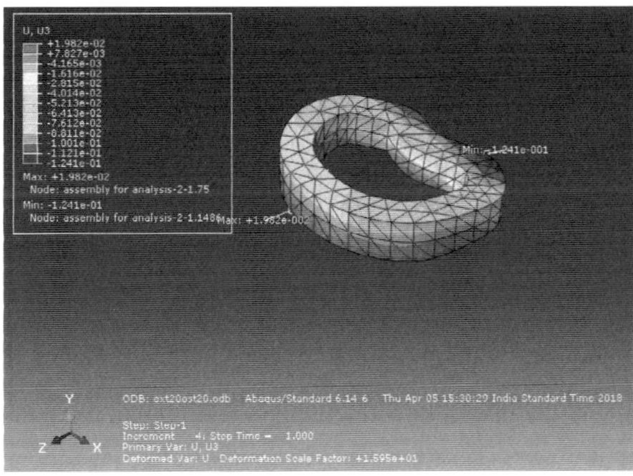

Figure.3. Deformation on annular disc during standing at 20% osteoporotic condition

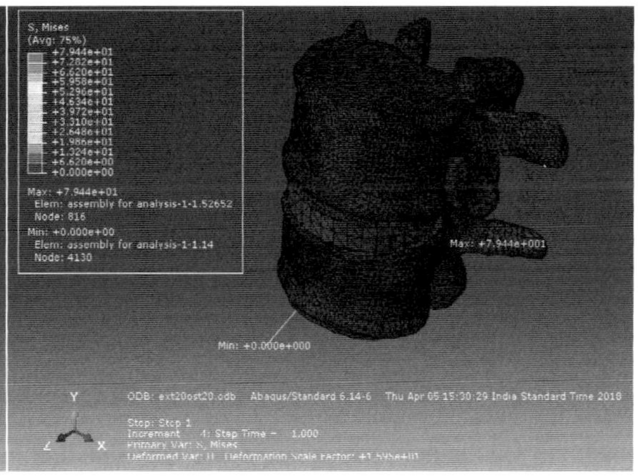

Figure.4. Von mises stress distribution lumbar vertebrae during standing at 20% osteoporotic condition

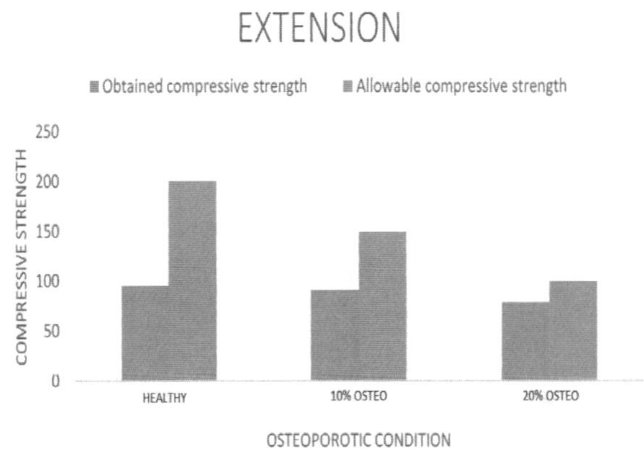

Figure.5. Variation of compressive strength with percentage osteoporotic condition (Extension)

3.2. Disc deformation during weight lifting

During weight lifting the deformation in the anterior direction of intervertebral disc is found to be 0.1693mm and that of in 20 % osteoporotic condition it found to be 0.2033mm. But for 20% osteoporotic condition deformation increased by 20.08%.The maximum stress acting on the vertebrae is found to be 96.98 N in 20% osteoporotic condition. Since the maximum stress value is very close to the allowable strength of the vertebrae. Hence chance of vertebral fracture is high.[9]

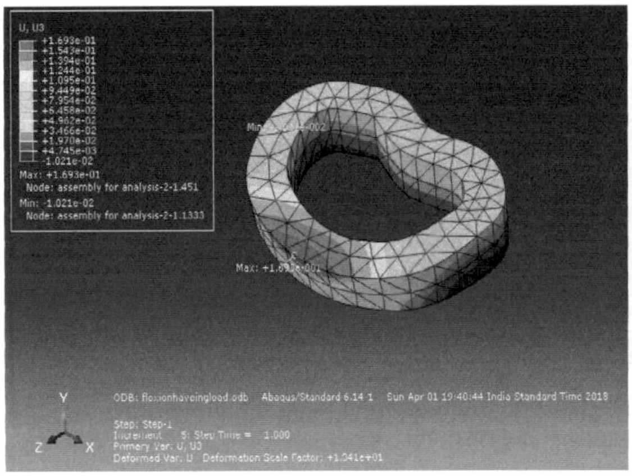

Figure.6. Deformation on annular disc during weight lifting at healthy condition

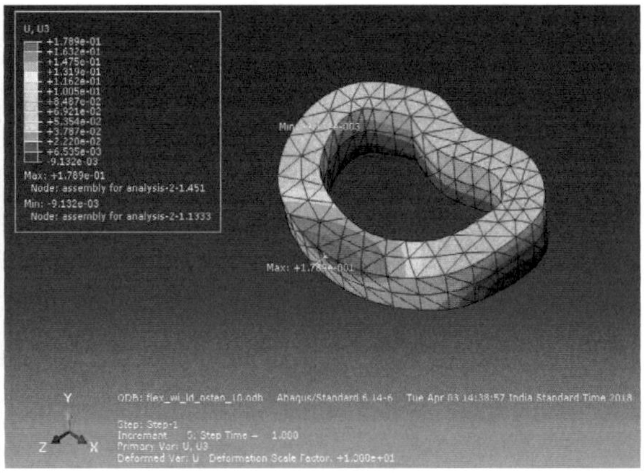

Figure.7. Deformation on annular disc during weight lifting at 10% osteoporotic condition

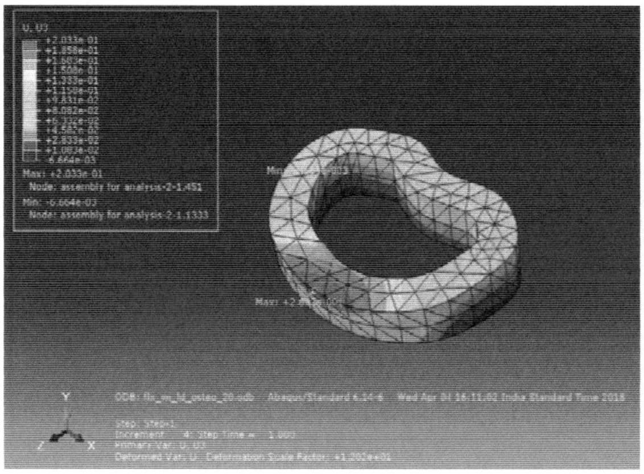

Figure.8. Deformation on annular disc during weight lifting at 20% osteoporotic condition.

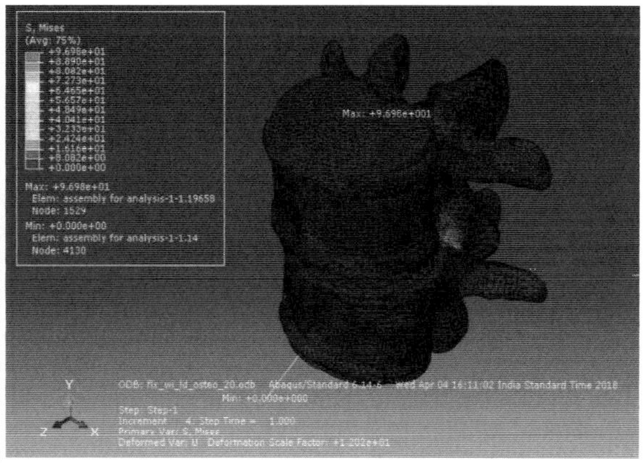

Figure.9. Von mises stress distribution lumbar vertebrae during weight lifting at 20% osteoporotic condition.

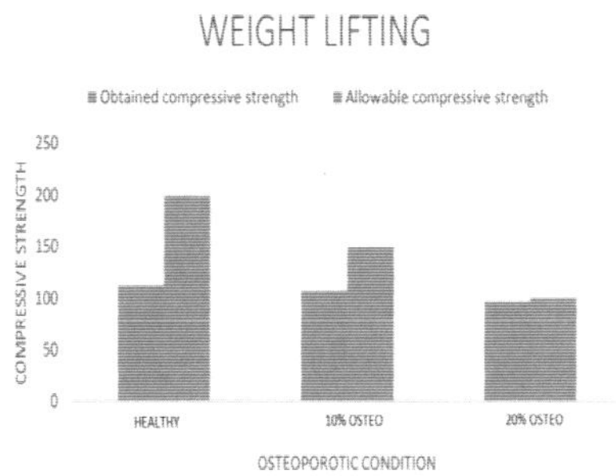

Figure.10. Variation of compressive strength with percentage osteoporotic condition (Weight lifting)

3.3. Disc deformation on stoop lift

While lifting 11.5kg load using stoop lift posture the deformation of disc obtained as 2.905×10^{-1}mm. Under same loading condition for 10% osteoporotic bone properties the percentage increase in deformation of disc is 4.8%. And for 20% osteoporotic bone property conditions percentage increase in deformation of disc is 17.2%. For 20% osteoporotic condition maximum allowable stress on the vertebrae is 100MPa, but as per the results obtained the stress value is 138MPa which is greater than 100MPa. Since there are chances of fracture in vertebrae, people of old age should avoid this posture during weight lifting. Disc deformation increases with osteoporotic conditions, hence the chance of occurrence of pain is high.

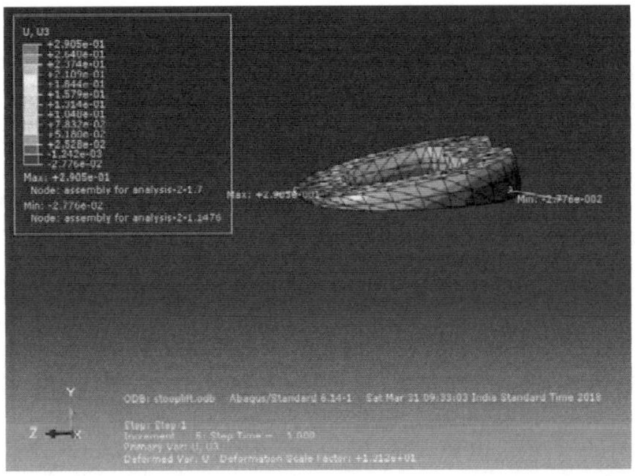

Figure.11. Deformation on annular disc during stoop lift at healthy condition

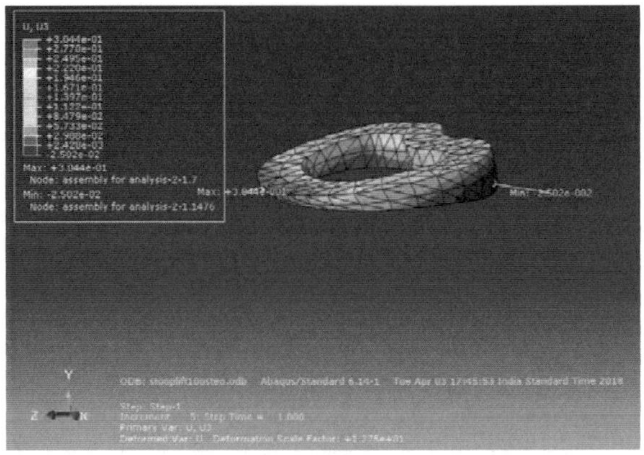

Figure.12. Deformation on annular disc during stoop lift at 10% osteoporotic condition

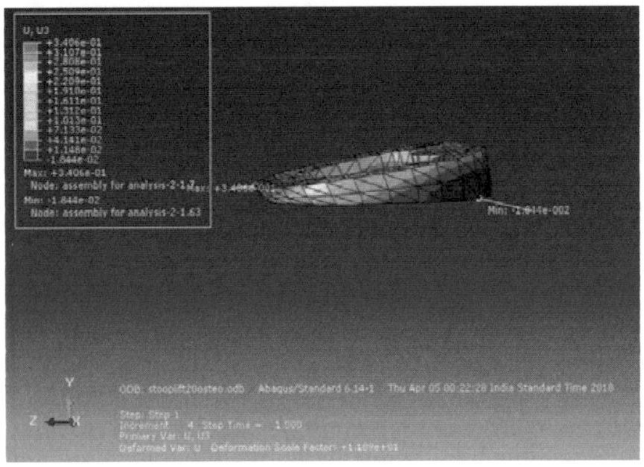

Figure.13. Deformation on annular disc during stoop lift at 20% osteoporotic condition

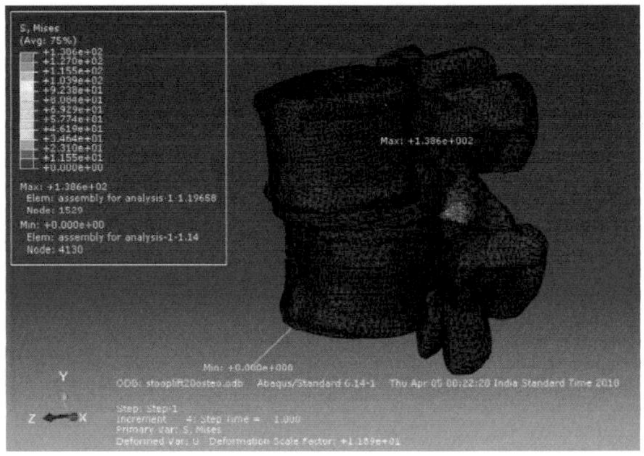

Figure.14. Von mises stress distribution lumbar vertebrae during stoop lift at 20% osteoporotic condition.

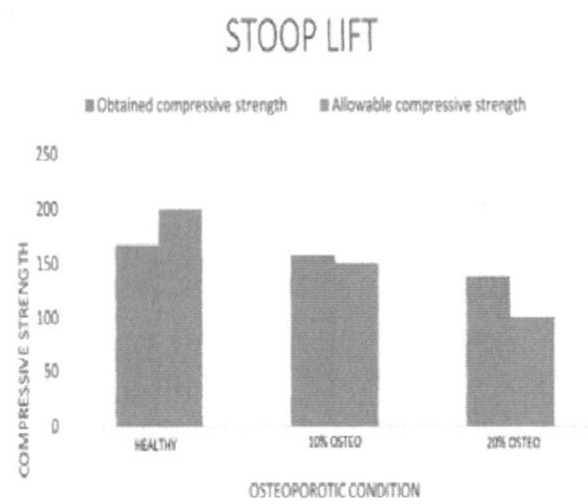

Figure.15. Variation of compressive strength with percentage osteoporotic condition (Stoop lift)

3.4. Disc Deformation in Squat lift

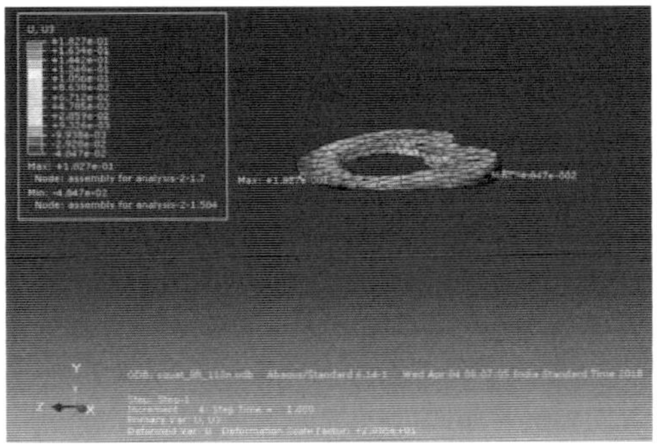

Figure.16. Deformation on annular disc during squat lift at healthy condition

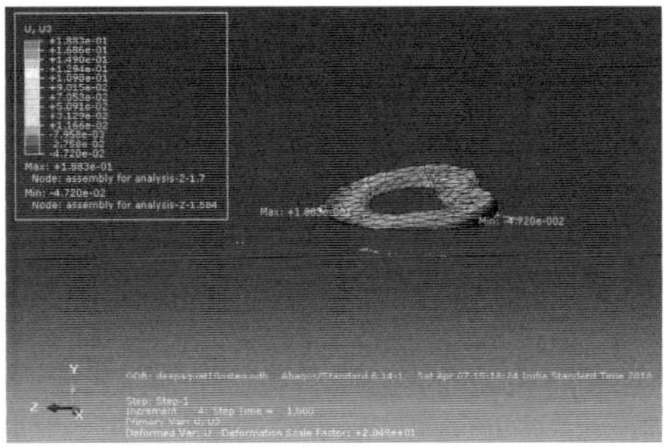

Figure.17. Deformation on annular disc during squat lift at 10% osteoporotic condition

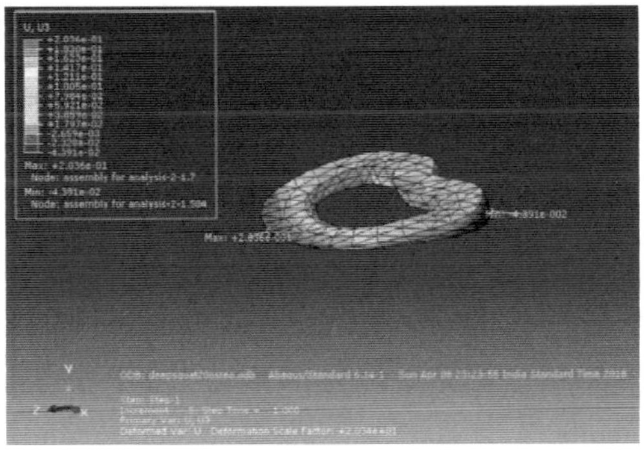

Figure.18. Deformation on annular disc during squat lift at 20% osteoporotic condition

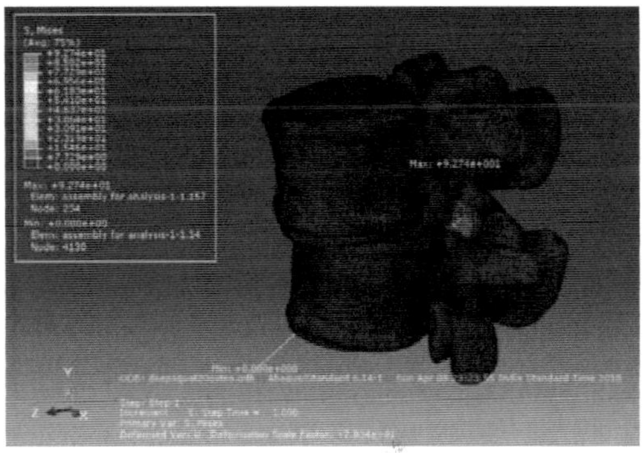

Figure.19. Von mises stress distribution lumbar vertebrae during squat lift at 20% osteoporotic condition

Figure.20. Variation of compressive strength with percentage osteoporotic condition (Squat lift)

While lifting 11.5kg load using squat lift posture the deformation of disc obtained as 1.827×10^{-1}mm.Under same loading condition for 10% osteoporotic bone properties the percentage increase in deformation of disc is 3%. And for 20% osteoporotic bone property conditions percentage increase in deformation of disc is 11.4%. For 20% osteoporotic condition maximum allowable stress on the vertebrae is 100MPa, but as per the results obtained the stress value is 92MPa which is closer to the allowable values. Deformation of disc increases with the increase in osteoporotic conditions and thereby increases the chance of compression of nerves which results in pain.

4. Conclusions

Lower back pain is common clinical compliant in these present days. From the analysis it is found that, as the osteoporotic condition get worsen the maximum allowable stress acting on the vertebrae get decreased and the strain on the vertebrae get increased. From the present study it is observed that as bone mineral density decrease the deformation of the disc increases which result in compression of nerves .Hence low back pain result in old age people.

- For a 20% osteoporotic bone maximum value of compressive strength is 100MPa[9], the analysis results show that the maximum value of stress is 96MPa for weight lifting, 76MPa during extension, 138MPa for stoop lift and 92MPa for squat lift.
- Also the deformation of disc under 20% osteoporotic conditions is increased to 20.08% under weight lifting posture.

From these studies it is clear that due to osteoporosis, compressive strength of vertebrae get reduced with age. There are chances of bone failure with aging. So old age people should avoid risky postures like stoop and squat lift. The present study gives detailed quantitative information on the biomechanical behavior of the vertebra and these results are considered to be a very useful analysis tool in the risk evaluation of fracture including osteoporotic bones. The resulted information will be helpful for medical practitioners to suggest proper prevention and precaution to the patients. The integration of technologies such as computed tomography, computer aided modeling, FEA is important in medical field to reduce the cost and risk to patients and strengthening the decision making capacity of medical practitioner.

References

[1] Georg Osterhoff, Elise F. Morgan, Sandra J. Shefelbine, Lamya Karim, Laoise M. McNamara, Peter Augat (**2016**), “Bone mechanical properties and changes with osteoporosis”, *Injury, Int. J. Care Injured* 47S2, S11–S20

[2] Faulkner KG, Cann CE, Hasegawa BH, “Effect of bone distribution on vertebral strength: assessment with a patient-specific nonlinear finite element analysis”,*Radiology*, Vol. 179, pp. 669-674, 1991

[3] Jong Ki Shin, Tae SikGoh, Myung-Sung Kim, Keunyoung Kim, Myung Jun Shin, Seung Min Son, Hee Jin Lee, Jung Sub Lee, Chi-Seung Lee (**2017**), “Stress and strain analyses of single and segmental lumbar spines based on an accurate finite element model for vertebrae”, *Biomedical Research 2017; Special Issue*, S602-S609

[4] M.Vineeth, S.L.Resmi, M.Shamnadhand P.N.Dileep(2013),“A study on the osteoporotic behaviour of human vertebra using FEM”,In Proceeding of *Second Internatinal Conference On Science and Innovative Engineering*.Chennai

[5] Justin Thomas, K Pranav Krishna, Resmi S L, Dileep P N (2013), “A Study on the mechanical behaviour of lumbar vertebrae of aSkater using FEM” ,In Proceeding of *Third International Conference on Materials for the Future*,GEC Thrissur.

[6] Sofia Pedroso de Faria(**2015**), "Biomechanical Analysis of the Human Lumbar Spine - An Experimental and Computational Approach", *LAETA*, IDMEC, IST, Lisbon, Portugal, 1-10

[7]Shaobai Wang, Won Man Park, Yoon Hyuk Kim, Thomas Cha, Kirkham Wood, and Guoan Li (**2014**), "In vivo Loads in the Lumbar L3-4 Disc during a Weight Lifting Extension", *journal of cilinical biomechanics*,29(2), 155-160

[8] Jaap H. van Dieen, Marco J.M. Hoozemans, Huub M. Toussaint (**1999**), "Stoop or squat: a review of biomechanical studies on lifting technique", *Journal of Clinical Biomechanics*, 14, 685-696

[9] Dennis R Carter, Wilson C Hayes (**1976**), "Bone compressive strength: The influence of density and strain rate", *Journal of science*, 194, 1174-1176

Simulation and Analysis of Solar Powered Induction Motor Drive with V/f Control for Electric Vehicle Applications

Sheik Mohammed[1], Muhammedali Shafeeque K[2], Muhammed Shanir P P[3] and Shihabudheen K V[4]

Assistant Professor, Department of Electrical & Electronics Engineering, TKM College of Engineering, Kollam, Kerala

E-mail:sheikmsulthan@gmail.com[1], kshafeeq007@gmail.com[2], shanirpp@gmail.com[3]

Abstract. Induction Motor (IM) drive with V/f control for electric vehicle application is presented in this paper. The DC-AC converter of induction motor drive is supplied by a Solar PV system. The complete system model consists of solar PV array, DC/DC converter and three phase Inverter with V/f control is developed in MALTAB/Simulink. A 4 kW Induction motor is selected as drive. The simulation model developed in MATLAB is operated under different conditions, the results are obtained. The results are presented and discussed.

1. Introduction

Photovoltaic (PV) based power systems had significant developments during the last few decades. There is a noticeable progress solar PV based power generation around the world due to its advantages such as low maintenance, easy installation, zero pollution. Total capacity of first solar PV power station "*Arco Solar*" installed in California in 1982 is 1 MWp[1] . At present, "Tengger Desert Solar Park" in China is the world's largest PV station with the installed capacity of 1547 MW[2]. The solar power plant with a capacity 648MW at Kamudhi, Tamilnadu is largest solar PV installation of its kind in India. The total worldwide PV capacity has increased by 95 GW, with a 34% growth in 2017 and the cumulative installed capacity of Solar PV exceeded 401 GW by the end of the year 2017. This is sufficient to supply 2.1 percent of the world's total electricity consumption [3].

Electric vehicles have a century long history back from 19th century. Introduction of Internal Combustion (IC) engines suppressed the electric vehicle industry. But, in 21st century, the technological developments and the environmental concerns due to the usage of fossil fuels have turned the focus towards EVs and Renewable energy [4,5].

Automobile Industry has been migrating into electric powered vehicle and this revolution has received positive impact worldwide. The legends in automobile industry like Tesla, Mercedes Benz, Toyota and Chevrolet have already launched their variants into the market. Tesla's EV is driven by Induction Motor. One of the major limitations in the developing countries which restrict the promotion of electric vehicle is the charging stations. Charging Stations are the important requirements for the electric vehicle industry since, the vehicle need to be charged in a regular interval.

Solar Powered charging stations are emerging along with EV industry [6,7]. India's first solar powered charging station has recently installed in Mumbai. The charging station is the associated project of Magenta Power and Exicom Power Solution. A solar powered charging station at Netherlands is equipped with 20 charging points. The charging points are connected to a solar PV system consists of 200-panel solar array and a 400kW/800kWh Tesla Power pack battery system.

In this paper, solar powered induction motor drive for electric vehicle application is attempted. The Induction motor drive system is directly fed by the power generated using solar PV system. The paper is organized as follows; Section 2 describes the structure of the complete system. Functionality of each component in the system is also explained. The simulation model and the results are presented and discussed in Section 3. The presented work is concluded in section 4.

2. Structure of Induction Motor Drive System

Figure 1 shows the block diagram of the proposed solar powered electric drive system for electric vehicle. The main components of the system are PV array, DC-DC Converter with MPPT Controller, 3 phase Induction Motor drive.

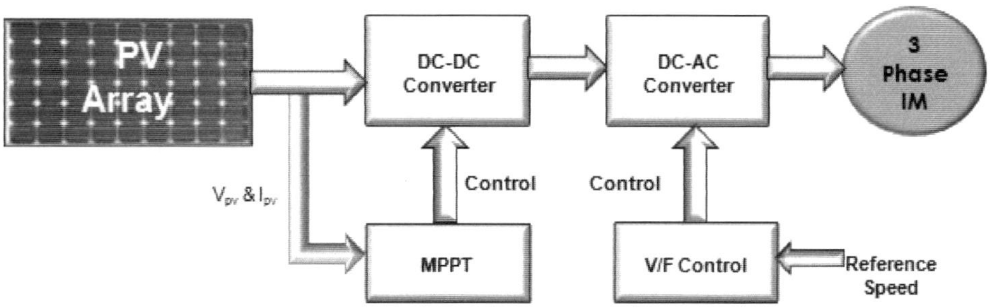

Figure 1. Structure of Induction Motor Drive System

Photovoltaic (PV) modules convert sun light into DC electricity. Solar PV has non-linear characteristics and the output power of PV mainly depends on the weather conditions such as irradiation and temperature of light falling on the cells [8-9]. DC-DC converters are used in PV systems to regulate the voltage generated by the PV modules [10-11]. DC-DC boost converters are used in grid connected applications to step up the module voltage. The MPPT algorithms are used to control the duty cycle of the converter or power conditioning circuit connected in between the PV and load to extract the maximum power from the source and to feed it into the load. The P&O and InC algorithm are widely used algorithms among the MPPT algorithms. Some of the other MPPT algorithms are soft computing MPPT techniques such as Fuzzy, ANFIS based MPPT algorithms and hybrid MPPT algorithms are also proposed [12-16]. Soft computing technique based control provide efficient computation with minimum effort [19], [20]. The P&O algorithm is simple and easy to implement and hence, P&O MPPT algorithm is used in this work. The MPPT controller continuously tracks the output voltage and output current of the PV module, the change in output power (ΔP) is calculated from them. The duty cycle of the converter switch is adjusted according to the change in output power after each perturbation. The MPPT controller continuously adjusts the duty cycle of the converter until the Maximum power is attained.

Induction motors are widely used in Electric vehicle applications and in many other domestic and industrial applications due to its low cost, less maintenance and robustness. To achieve desired performance IM should work under rated stator flux. This can be achieved by maintaining the voltage to frequency ratio constant [17]. To control the voltage and frequency independently, Inverters are used in AC drive applications with V/f control strategy [18]. Hence, Induction Motor with V/f controlled inverter is adopted for the system development.

2.1 Operation of the System

The output of PV array is supplied to the load through DC-DC Converter. The DC-DC converter is controlled by P&O MPPT controller based on the output of PV module. The duty cycle of DC-DC converter switches are adjusted by P&O MPPT controller. The DC-DC boost converter output is fed into the load.

In this case, the load is an Inverter which drives the induction motor. The DC input is converted into 3 phase AC and fed to the induction motor. Inverter switches are controlled by V/f method to maintain the stator flux constant. The reference speed is set by the user and the V/f controller generates the control signal based on the reference value. The PWM controller provides switching pulses to the inverter switches. The speed of IM is controlled under different load torques.

3. Simulation and Analysis

The simulation model developed in MATLAB/Simulink is presented in figure 2. A 4 kW induction motor is selected as drive. Hence, Solar PV system for 4kW is designed. A solar PV string consists 16 solar PV panels of 250 W (ASW-250P) is built. The electrical specification of solar PV module is given in Table 1. The DC-DC converter is designed to produce DC output of 816V at 4kW. The design procedure presented in [8] is followed to design the converter. The switching frequency f_s of DC-DC Boost converter is 10 kHz. The components values are calculated as follows L=8.1 mH, C=10μF.

Table 1. Electrical Specification of Solar PV Module

Parameters	Unit	Value
Power at STC	W	250
Power at PTC	W	227.8
Vmp: Voltage at Max Power	V	35.2
Imp: Current at Max Power	A	7.1
Voc: Open Circuit Voltage	V	43.22
Isc: Short Circuit Current	A	7.76
Nominal Operating Cell Temp	°C	43.5
Open Circuit Voltage Temp Coefficient	% / °C	-0.303
Short Circuit Current Temp Coefficient	% / °C	0.035

PV panels has non-linear characteristics and the power generated by PV varies depends on the change in weather conditions [9]. Figure 3 and Figure 4 shows the V-I and V-P characteristics of 4kW PV string under different irradiation level respectively. The power output of PV string is 4000W at 25 °C and 1000 W/m^2.

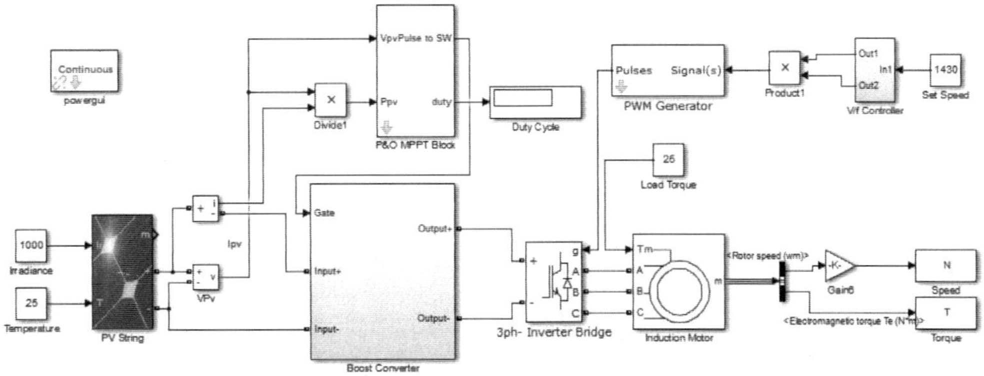

Figure 2. Simulink model of proposed system

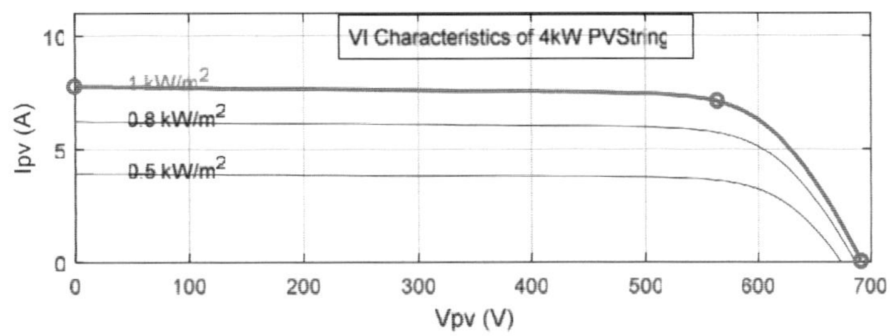

Figure 3. V-I characteristics of 4 kW PV String

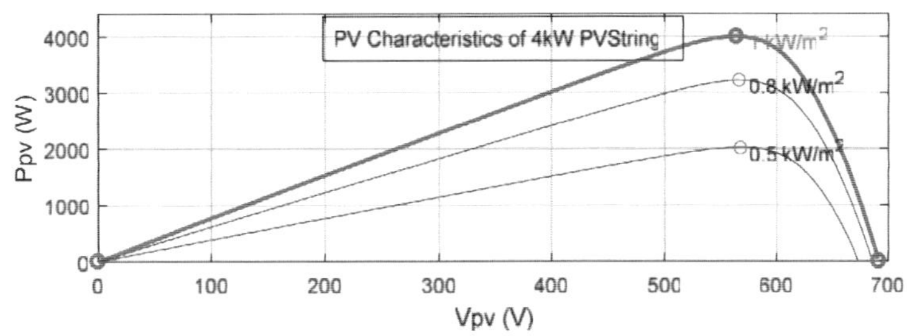

Figure 4. V-P characteristics of 4 kW PV String

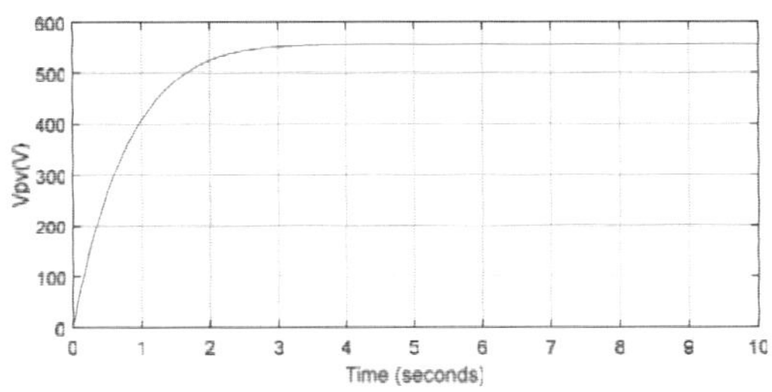

Figure 5.1 Output Voltage of 4kW PV String

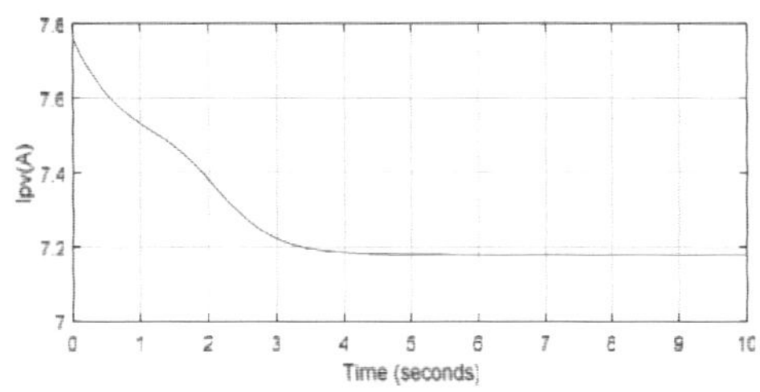

Figure 5.2 Output Current of 4 kW PV String

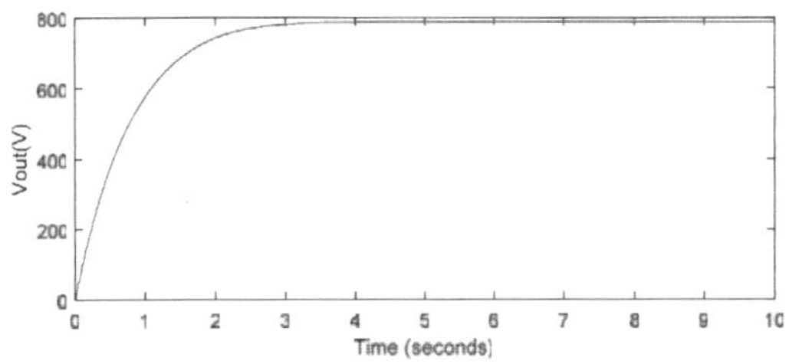

Figure 6.1 DC-DC Boost Converter Output Voltage

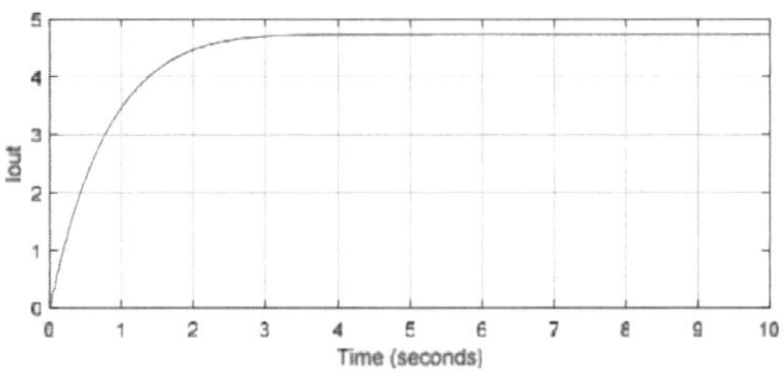

Figure 6.2 DC-DC Boost Converter Output Current

Figure 5.1 and Figure 5.2 shows the output voltage and output current of PV string respectively under standard test conditions of PV module. The PV output is fed as input to the DC-DC Boost Converter. The maximum power output voltage (V_{mp}) of the string is 563.2V and the current at maximum power (I_{mp}) is 7.1A. The DC-DC boost converter is designed for the output voltage of 804 V and the current output of the converter is 4.9 A. The output voltage and current waveforms shown in figure 6.1 and 6.2 respectively confirms that the converter outputs has very close match with the designed values.

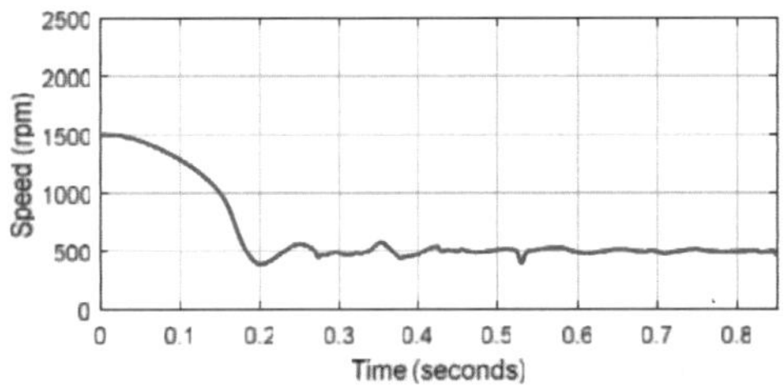

Figure 7.1 Speed of Induction Motor at starting condition

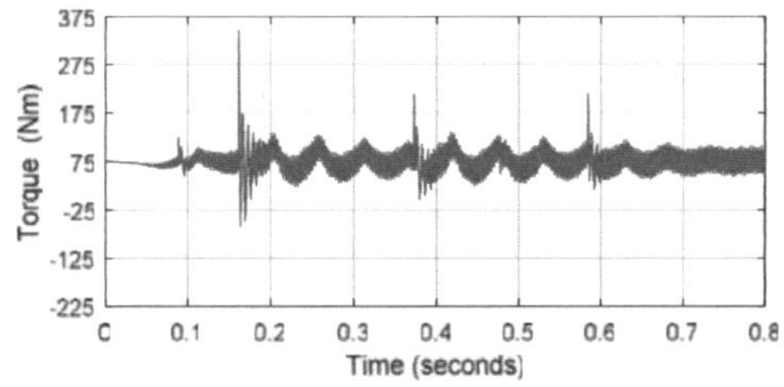

Figure 7.2 Torque of Induction Motor at starting condition

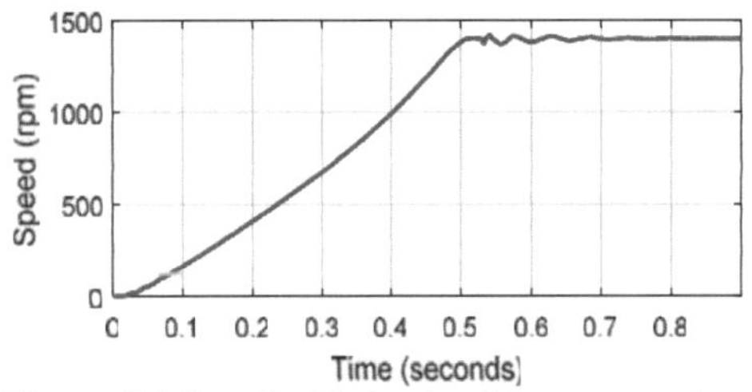

Figure 8.1 Speed of Induction Motor at running condition

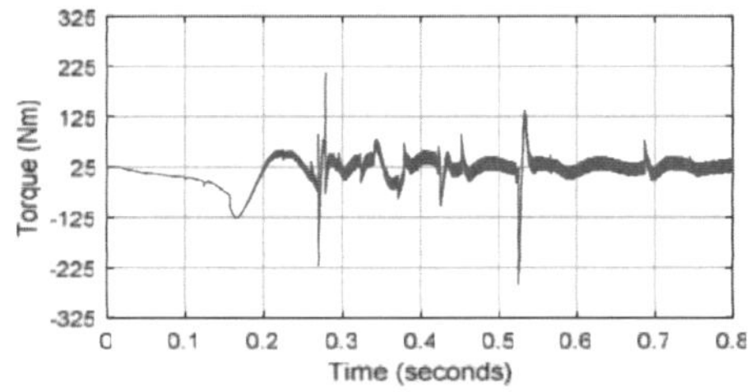

Figure 8.2 Torque of Induction Motor at running condition

Electric vehicle requires high torque at low speed during starting condition and the torque gradually decreases with increase in speed. Figure 7.1 shows the speed on induction motor drive under starting condition and the stating torque of the motor drive is depicted in figure 7.2. From the figures, it can be noticed that the speed of the drops at starting condition and the torque is high. Similarly, under running condition, speed increases until it reaches the rated speed (figure 8.1) and the torque is lower than that of the starting torque (figure 8.2).

4. Conclusion

A V/f controlled induction motor drive for electric vehicle application is presented in this paper. A 4kW induction motor drive with Solar PV system is designed and modelled in MATLAB/Simulink. The MATLAB simulation model is simulated under different conditions and the speed, torque of the drive motor is obtained in addition to the output of PV string and DC-DC Boost converter. The results are presented and discussed. The V/f controlled IM drive maintains the speed based on the reference set by the user and the torque of the motor is adjusted accordingly.

5. References

[1] Sheik M S, Devaraj D and Imthias A 2016 Modeling, Simulation and Analysis of Photovoltaic Modules under Partially Shaded Conditions. *Indian Journal of Science and Technology, 9*(16).

[2] Eric W 2017 IEA: Global Installed PV Capacity Leaps to 303 Gigawatts, greentechmedia

[3] Snapshot of Global Photovoltaic Markets 2017 (PDF). report. International Energy Agency. 19 April 2017. Retrieved 11 July 2017.

[4] Asif F, Christopher S and Michael P 1996 Air Pollution from Motor Vehicles: Standards and Technologies for Controlling Emissions. *World Bank Publications*. 227, ISBN 978-0-8213-3444-7.

[5] Guarnieri M 2012 Looking back to electric cars Proc. HISTELCON 2012 – 3rd Region-8 IEEE HISTory of Electro - Technology Conference: The Origins of Electrotechnologies.

[6] Ye B, Jiang J, Miao L, Yang P, Li J and Shen B 2015 Feasibility Study of a Solar-Powered Electric Vehicle Charging Station Model. *Energies*, *8*, 13265-13283.

[7] Saadullah K, Aqueel A, Furkan A, Mahdi S, Mohammad S and Siddiq K 2018 A Comprehensive Review on Solar Powered Electric Vehicle Charging System, *Smart Science*, 6:1, 54-79

[8] Sheik M 2011 Modeling and Simulation of Photovoltaic module using MATLAB/Simulink *International Journal of Chemical and Environmental Engineering*, 2011

[9] Nishad T and Shafeeque K 2016 A novel single stage buck boost inverter for photovoltaic applications *International Conference on Electrical, Electronics, and Optimization Techniques (ICEEOT)*, pp. 3067-3071.

[10] Sheik M and Devaraj D 2014 Simulation and Analysis of Stand-alone Photovoltaic System with Boost Converter using MATLAB/Simulink *International Conference on Circuits, Power and Computing Technologies (ICCPCT-2014)*, March 2014.

[11] Shafeeque K and Subadhra P 2013 A Novel Single-Phase Single-Stage Inverter for Solar Applications *2013 Third International Conference on Advances in Computing and Communications*, Cochin, pp. 343-346.

[12] Sheik M and Devaraj D 2015 Interleaved boost converter with Perturb and Observe Maximum Power Point Tracking Algorithm for Photovoltaic System *Proceedings of International Conference on Substantial Environmental Engineering and Renewable Energy (SEERE-15),* Abu Dhabi (UAE), pp. 22-30

[13] Sheik M and Devaraj D 2015 Simulation of Incremental Conductance MPPT based Two phase Interleaved Boost Converter using MATLAB/Simulink *IEEE International conference on Electrical Computer and Communication Technologies*, Coimbatore, India, pp 1-6, March 2015.

[14] Sheik M and Devaraj D 2015 Fuzzy based Maximum Power Point Tracking Controller for Stand-alone Photovoltaic system using MATLAB/Simulink *International Journal of Applied Engineering Research*, ISSN 0973-4562 10,55 , 3101-3106

[15] Sheik M, Devaraj Dand Imthias A 2016 Maximum power point tracking system for standalone solar PV power system using Adaptive Neuro-Fuzzy Inference System, *Biennial International Conference on Power and Energy Systems: Towards Sustainable Energy (PESTSE),* Bengaluru, India, pp. 1–4

[16] Sheik M and Devaraj D and Imthias A 2016 A novel hybrid Maximum Power Point Tracking Technique using Perturb & Observe algorithm and Learning Automata for solar PV System *Energy* 112, 1096–1106.

[17] Sreejeth M and Singh M and Kumar P 2014 Efficient operation of IM drive through selection of optimal V/f ratio *6th IEEE Power India International Conference (PIICON)*, Delhi, 2014, pp. 1-5.

[18] Muhammedali S and Subadhra P 2013 A novel single stage DC-AC buck boost inverter for solar power extraction *Annual International Conference on Emerging Research Areas and 2013 International Conference on Microelectronics, Communications and Renewable Energy,* Kanjirapally, 2013, pp. 1-6.

[19] Shihabudheen K and Pillai G 2018 Recent advances in neuro-fuzzy system: A survey, *Knowledge-Based Systems*, 152, 136-162

[20] Shihabudheen, K and Krishnama R, and Pillai G 2018 Control for grid-connected DFIG-based wind energy system using adaptive neuro-fuzzy technique *International Transactions on Electrical Energy Systems* 28, 5, e2526.

Managing Iterations in Product Development Process Using Dependency Structure Matrix

Shajahan C A[1], Firoz N[2], Mohammed Sadhikh, [3]V R Pramod N[4] and Ashim[5]

[1]Professor, Department of Mechanical Engineering, TKM College of Engineering, Kollam
[2]Assistant Professor, Department of Mechanical Engineering, TKM College of Engineering, Kollam.

[3]Professor, Department of Mechanical Engineering, TKM College of Engineering, Kollam
[4]Professor, Department of Mechanical Engineering, NSS College of Engineering, Palakkad
[5]Maintenance Engineer, SADAFCO,Saudi Arabia

E-mail: cashaji@yahoo.com

Abstract. The market is highly challenging with customers demanding a variety of products at competitive prices. To make a product competitive, it needs to be introduced quickly without compromising on its performance or quality. Thus, reduction of product development cycle time has become essential for companies to thrive and sustain in the market. Product Development (PD) is the set of activities beginning with the perception of a market opportunity and ending in the production, sale and delivery of a product. Companies are forced to develop a well-coordinated development plan to organize their processes and resources to develop competitive products. Today PD confronts complex development challenges. The difficulties in developing complex engineering products do not only arise from their technical complexity but also from the managerial complexity necessary to coordinate the interactions between the different engineering disciplines, which impose additional challenges on the development process. Traditional Project Management techniques are limited in application. Researchers and design managers are constantly searching for the effective and efficient methods for managing PD process. To address the limitations of traditional project management tools, Design Structure Matrix (DSM) has been identified as a sensible tool to represent PD process. In this work, the capability of DSM is demonstrated with a real life case example .The necessity of reducing iterations in PD process using PSM 32 software is also demonstrated

Keywords: *Product Development (PD), Design Structure Matrix (DSM), PERT, CPM.*

1. Introduction

Product Development process is the sequence of steps or activities, which an enterprise employs to conceive, design and commercialize a product [1]. Many of these steps and activities are intellectual and organizational rather than physical. Some organizations define and follow a precise and detailed development process, while others may not even be able to describe their processes. There are various categories of new products such as new to the world products, new category entries, addition to product lines, product improvements and repositioning [2] Product development includes various types of decisions made by engineers and managers. Some decisions are related to design while others are development decisions [3].

Design Structure Matrix is a generic matrix-based framework for information flow analysis introduced by Donald Steward in 1981. DSM consists of an N-Square diagram showing the interaction of each element with every other element in the model. By reading across a row, one can observe these interactions through the cell contents corresponding to each cross-referenced column. The use of DSMs in both research and industrial practice increased greatly in 1990s.

A sample task based DSM is shown in Figure 1. Tasks appear identically labeled in rows and columns of the matrix are arranged top-down according to their sequence of execution.

Tasks → ↓	Task 1	Task 2	Task 3	Task 4	Task 5	Task 6	Task 7	Task 8	Task 9
Task1									
Task 2									
Task3		X							
Task 4	X		X						
Task 5		X		X					
Task 6							X		
Task 7					X	X		X	
Task 8			X				X		
Task 9							X		

Figure.1. Sample task based DSM

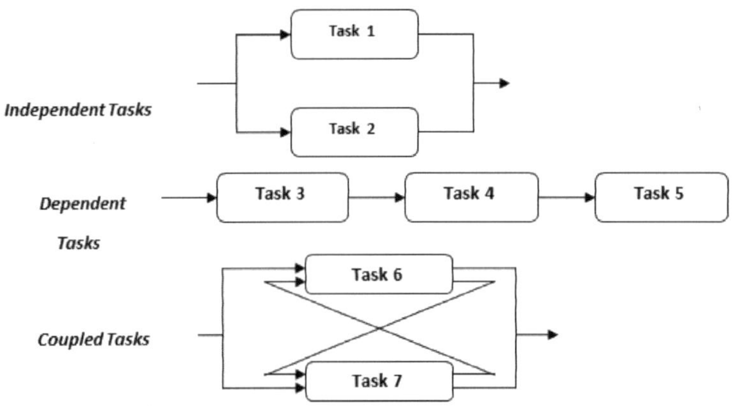

Figure.2. Sample task based DSM

Three types of task interactions can be observed from the matrix, namely, independent, dependent and coupled[4]. In Figure 3, Tasks 1 and 2 are "independent" since no information exchange is required between them. These tasks can be executed simultaneously (parallel). Tasks 3–5 are engaged in a sequential information transfer and are considered "dependent". These tasks would typically be performed in series (sequential). Tasks 6 and 7, however, are mutually dependent on information. These are *"interdependent" or "coupled"* task blocks and are often requiring multiple iterations to complete. Marked cells above the diagonal represent iterations in the process. This occurs when an activity is dependent on information from a task scheduled for a later execution. Such scenarios often lead to rework and are undesirable. A number of algorithms have been developed to minimize such instances of iteration by re-arranging the sequence of tasks in the process. Partitioning and tearing are two main processes that can manipulate the matrix and transform the matrix into lower triangular form. It is the process of rearranging the order of activities by moving an entire row and column on either side in such a way that the resulting matrix has marks either below the diagonal or close to the diagonal. Through the manipulations, the task execution sequence will be reordered, the number of iterations will be reduced, and fewer tasks will be involved in the iteration cycle and these will result in a faster development process.

2. DSM Applications in PD

Engineering design often involves a very complex set of relationships among a large number of coupled problems. It is this complex coupling that leads to iteration among the various engineering tasks in a large project. Iterations are fundamental to the engineering design process. PERT chart often created by managers to depict PD describes the process as a progression of series and parallel activities; however coupled and iterative PD processes are in fact quite common (Kline,1985). DSM is useful in identifying where iteration is necessary. Understanding design iteration is therefore fundamental to accelerating and improving product development practices. There are two ways to accelerate an iterative development process:

1. To execute faster iterations
2. To conduct less iteration.

Both approaches are important to consider. Faster iterations are achieved through several means, such as the use of engineering models or information technology. Less iteration may be experienced when the coupled development activities can anticipate each other's results, or when extraneous activities are removed from the iterative portion of the process. Models of design iteration can provide valuable insight into the iteration process. As iteration is common in most PD projects, especially in the product design stage, it will cause rework and require extra communication and negotiation that result in a prolonged development process. There are many approaches for handling these design iterations [5] [6] [7]. Iteration can be managed in many ways such as improving the sequence of design activities to stream line information flows which form the basis for task dependencies [8]developing new engineering automation tools to perform iterations faster [9] adding resources at bottleneck activities [10].One way of managing rework effectively is to avoid interdependent tasks. There are many possible ways to avoid the coupling of interdependent tasks by splitting a task into two or more tasks, combining two or more tasks into a single task, redefining the project tasks and their relationship itself, or sometimes breaking the relationship between the tasks. In order to speed up this iterative design process, the DSM methodology suggests the manipulation of matrix elements such that iterative behavior could be removed from the matrix, or at least minimized. Partitioning and tearing are two main processes that can manipulate the matrix and transform it into lower triangular form. Through the manipulations, the tasks execution sequence will be re-ordered, the number of iterations will be reduced and fewer tasks will be involved in the iteration cycle and these will result in a faster development process.

Several researchers have developed models of design iteration. Ha and Porteus [11] present a model of coordination between two coupled development activities. They address the frequency of design reviews at which times information is transferred between the activities. The sequential iteration model allows computation of the total lead-time for a group of tasks where each task has a probability of creating rework for the other tasks[12]. Each of these modeling efforts is an important contribution in this new and emerging area of management science because each model is able to explore a different facet of design iteration. Unfortunately many PD managers fail to plan for iterations; consequently PD projects incur schedule and cost overruns [13]. Browning identified cycle time reduction challenges and highlighted the capabilities of DSM based methods to manage projects for reduced schedule risk and shorter cycle time[14]. Maheswari et al. applied DSM representation for activity sequencing in concurrent engineering projects [15]. Using this representation, information dependency attributes can be organized and analyzed in a structured manner to identify activity groups and sequence for concurrent execution. The procedure is applied to the design phase of an induced draft cooling tower project. Gohlich et al [16] proposed an augmented DSM sequencing to support product development planning.

3. Case Example

For the illustration of the advantage of the DSM, a PD process which consists of 27 activities is modeled. Process modeled using DSM in PSM32 software is shown in figure 3.In activity dependency matrix, Activity 1 and 6 is marked as 0 which means Activity 6 produces an output that is an input to Activity 1.All such relationships are documented in the activity dependency matrix. The matrix is subjected to partitioning to rearrange the tasks and eliminate as many feedback loops as possible. Final partitioned matrix is shown in Figure 4. The sequence 1-2-3-4-5-6-7-8-9-10-11-12-13-14-15-16-17-18-19-20-21-22-23-24-25-26-27 is rearranged and new sequence is obtained as 1-2-6-4-5-7-3-8-9-10-11-12-13-14-15-16-17-18-19-25-20-21-22-23-24-26-27.

Table.1.List of activities in product development process

Sl No:	Activity	Sl No:	Activity
1	Idea Generation SWOT Analysis	15	Break Even Analysis
2	Idea Generation Brainstorming	16	Beta Production Run
3	Idea Screening Customer Analysis	17	Internal Testing
4	Idea Screening Market Analysis	18	Focus Group Interviews
5	Idea Screening Competitive Analysis	19	Determine Necessary Adjustments
6	Idea Screening Technical Feasibility	20	Initial Production run for test Market
7	Idea Screening Finalize Idea Selection	21	Market Testing
8	Concept Development	22	New program Initiation
9	Feature Specification	23	Quality Management System
10	Rapid Prototyping	24	Operational Planning
11	Production Optimization Analysis	25	Distributor Negotiation
12	Concept Testing	26	Market & Advertising Campaign
13	Pricing Analysis	27	Product Launch
14	Sales Forecast		

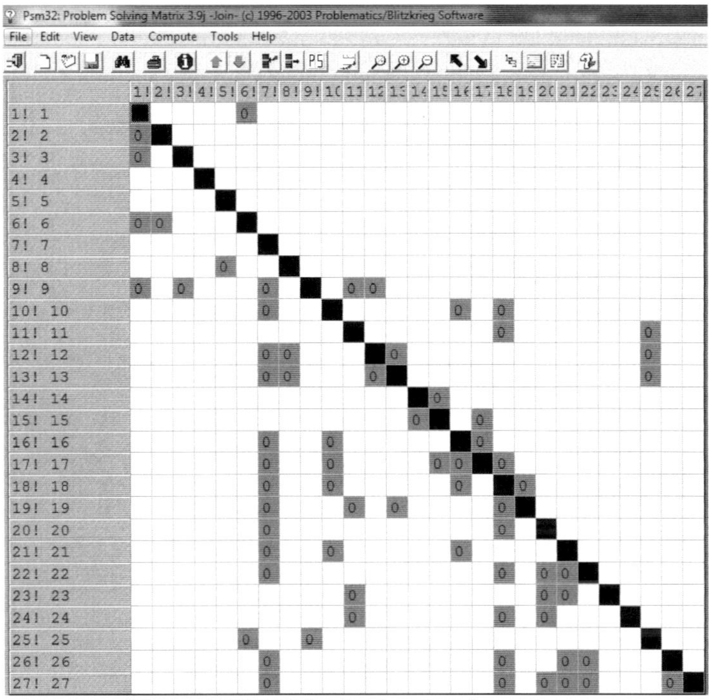

Figure.3.DSM Matrix showing dependency

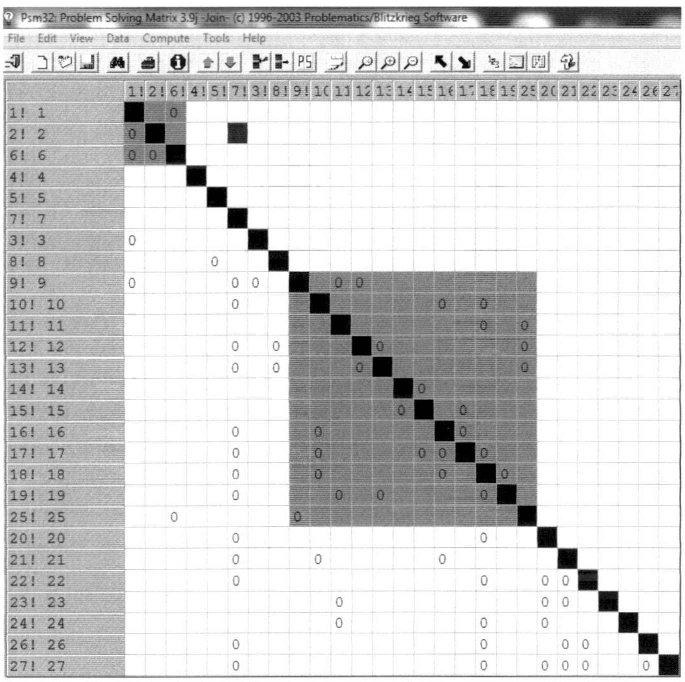

Figure.4. DSM after partitioning using PSM 32

4. Conclusion

Managing and scheduling the complex product development projects is a difficult exercise that requires much skill and experience. The traditional project management tools are weak in the areas, such as sequencing, controlling and managing the product development projects. In this work, Design Structure Matrix is used to tackle the problems encountered in scheduling the complex and interrelated product development projects. The re-sequencing of tasks is done using PSM32 software.

References

[1] Ulrich, K.T. and S.D. Eppinger Product Design and Development. Boston Irwin McGraw-Hill, 2000.

[2] Crawford,M. and A.D.Bendetto New Products Management McGraw-Hill Irwin, 2003

[3] Jeffery, W. H. and Linda, C. S., (2002) Viewing Product development as a decision production system. Proceedings of DETC 2002, ASME 2002 Design Engineering Technical Conferences and computers and information in engineering conference, Montreal, Canada, Sept 29-oct 2,1-10

[4] Eppinger, S.D., D.E.Whitney, R.P. Smith, D.A. Gebala (1994) A model based method for organizing tasks in product development Research in Engineering Design, 6, 1-13.

[5] Adler,P.S., A.Manbelbaum, V.Nguyen, and E.Schwerer (1995) From Project to Process Management: An Empirically-based frame work for analyzing product development time. Management Science, 41(3), 458-462

[6] Bell, D.G., S.M.Kannappan, and D.L.Taylor (1992) Product Development Process Dynamics in Proceedings of ASME Design Theory and Methodology Conference, DE- VI. 2, 257-266.

[7] Christian, A. D., and W. P. Seering (1995) A Model of Information Exchange in the Design Process, in Proceedings of ASME Design Engineering Technical Conferences, DE- 83, 323-328.

[8] Yassine, A. A., and D. Braha (2003) Complex concurrent engineering and the design structure matrix method, Concurrent Engineering Research Applications, 11 (3), 165-176.

[9] Yassine, A.A., D.E. Whitney, J. Lavine and T. Zambito (2000) 'Do it Right first Time approach to DSM restructuring', Proceedings of DETC, ASME International Design Engineering Technical Conference, New York.

[10] Yassine, A. A., and D. Braha (2003) Complex concurrent engineering and the design structure matrix method, Concurrent Engineering Research Applications, 11 (3), 165-176.

[11] Ha, A.Y., and E.L. Porteus (1995) Optimal timing of reviews in concurrent Design for manufacturability Management Science, 41, 1431–1440.

[12] Smith, R.P., and S.D. Eppinger (1997a) A predictive model of sequential iteration in engineering design Management Science, 43, 1104–1120.

[13] Bashir, H. and V.Thomas (1999) Metrics for Design Projects: A Review Design studies, 20 (3), 263-277.

[14] Browning, T.R. (2001) Applying the design structure matrix to system decomposition and integration problems: A review and new directions IEEE Transactions on Engineering management, 48(3), 292-306.

[15] Maheswari, J.U., Varghese, K., and Sridharan, T (2006) Application of Dependency structure Matrix for Activity Sequencing in Concurrent Engineering Projects Journal of Construction engineering and Management, 132, 482-490

[16] Gohlich D, Hildebrand and Schellert D, D (2018) " Augmented DSM sequencing to support product development planning" , International Design Conference 2018., 1139-1148.

Effect of Strut thickness on the Inflation Behaviour of Coronary Artery Stent using FEM

Hashim V.
Associate Professor
TKM College of Engineering Kollam, Kerala
hashimtkmce@gmail.com

Resmi S.L.
Assistant Professor
TKM College of Engineering Kollam, Kerala
hashimtkmce@gmail.com

Dr. Dileep P.N.
Professor TKM College of Engineering Kollam, Kerala
hashimtkmce@gmail.com

Abstract - Stent implantation is one of the most accepted method in treating coronary artery diseases due to their simplicity and faster patient recovery. Coronary stents are highly patient specific and the implantation success greatly relies on the design and clinical procedure. Studies confirm that, performance characteristics like radial strength, maximum expansion, stress concentration, recoil etc. are highly governed by the physical design of the stent. This paper focuses on finite element analysis using ABAQUS package to investigate the effect of varying strut thickness on the expansion behavior of a commercially available coronary stent. Envision Scientific Coronary Stent was adopted for this study. Total of five stent designs, one with actual size, two models with a strut thickness lesser than actual thickness and two with higher values were considered. The expansion behavior of the models under consideration was compared to the model with original dimensions. It may be noted that the thickness is one of important design parameter that affects the expansion and stress directly. With 15% reduction in thickness, the expansion increases by 10%. While, for an increase of 15% in thickness resulted in decrease in expansion by 10%. It may be noted that the percentage of increment in stress is linear and inversely proportional. Whereas, the increment in expansion is non linear and is minimal for low thickness. The present approach can be used for selecting the design parameters of stent to arrive in an optimum design.
Keywords - Coronary stent, FEM, Stress concentration, Strut design.

I. INTRODUCTION

Coronary artery disease is considered as one of the leading cause that claims lives of millions every year worldwide. Development of atherosclerotic plaques (stenosis) within the coronary arteries leads to heart blockage or heart attack. Stent implantation is the most popular non-surgical method and got wide acceptance, because of its simplicity and high success rate. Coronary stents acts as scaffold to the artery and open the stenosed artery to restore blood flow. The implantation of the stent in a human artery is a complicated process. Often, the type of stent selected or the methods involved in stent implantation determines the success rates. [1]. Since, physical test methods are very difficult, Finite element analysis provides a relatively cost-effective and extremely beneficial research tool to optimize the mechanical properties of stents. Numerical modeling of stents has been widely used to investigate the influence of stent design on the stent and artery. Many researchers have used the finite element method to evaluate stent characteristics such as recoil, foreshortening and dog boning effects. S.N. David Chua, et al. and others have investigated the influence of the expansion mechanism on the artery wall on stenosed vessels [2]. N Eshghi et.al. analyzed Palmatz stent with artery and balloon and found that the maximum stress is at the four corners of the cell and at the open end [3] . C Lally et.al. made a computational study to investigate the mechanical behavior of stent and interaction between stent and artery. He has compared the performance of different design and developed a novel stent design [4]. Julian Bedoya et.al. made a study on the effect of stent parameters on normal artery wall mechanism. Strut spacing, the radius of curvature at the crown section and axial amplitude are the stent parameters considered for the study[5]. Arjun R. et al. investigated the behavior of the stent by varying the inflating pressure and realized that the degree of expansion is mainly related with the design and the inflating pressure [6]. In the present study, finite element analysis is carried out to assess the effect of strut thickness on the stress and expansion developed in the coronary stent. The geometry of the commercially available Envision Scientific stent is selected for this study. Finite element analysis software ABAQUS is used to analyze the stress concentration in different strut thickness by keeping the inflating pressure constant

II. FE MODEL OF THE STENT

Most of the stents are of the complicated design and commercially available stents rarely disclose the information regarding their geometrical features, the generation of a solid model of a stent is possible by cutting an actual stent along its longitudinal axis and measuring important geometrical properties such as strut width, thickness, and length using a microscope. Using these measurements, a model of the planar stent geometry can be created accurately defined within an FE package. In this analysis solid models of a stent, balloon and artery were created using SOLID WORKS software. Envision 3.5x28 stent is a ring type stent. There are total 21 rings and rings are connected with straight bridge or connector. The strut is having a straight portion and a crown portion. Straight portion is not uniform width, the width is maximum at the centre and it becomes minimum near the crown. The geometry of the stent is measured by using Leica microscope available at Sree Chitra Institute of Medical Sciences, Thiruvananthapuram. The actual geometry of the stent strut is shown in Fig.2 and the dimensions are shown in table 1.

Journal of Physics: Conference Series **1355** (2019) 012025 doi:10.1088/1742-6596/1355/1/012025

Table : 1

Stent	Length (mm)	Outer diameter (mm)	Strut thickness (mm)	Strut width (mm)	Crown radius (mm)	No. of strut rings
Envision 3.5x28	26	1.09 (crimped)	0.064	0.115 /0.068	0.057	21

The Model A has been created with an actual size of the strut with 8 rings (total length is 10mm) and thickness 0.064mm. As there are two connectors in each ring and are diametrically opposite, the stent is cut axially from the middle and only half of the stent is considered by taking the advantage of symmetry for the simulation. The model is shown in Fig. 1 below. Four other models (Model B, C, D, E) were also created by increasing the thickness 15% and 30% of the actual thickness and reducing the thickness 15% and 30% of the actual thickness (strut thickness 0.044, 0.054, 0.074 and 0.084 respectively).

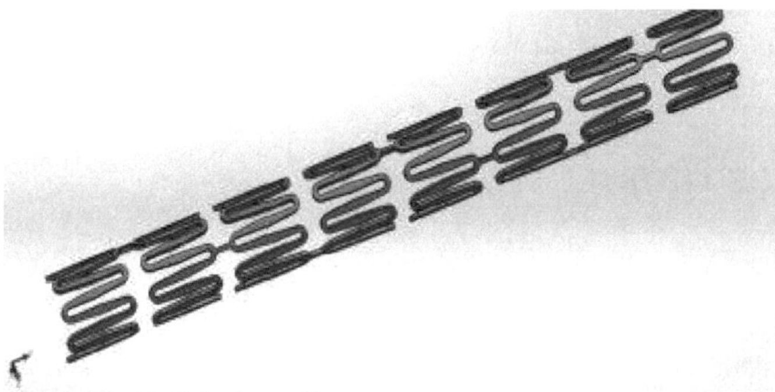

Fig. 1 FE model of stent

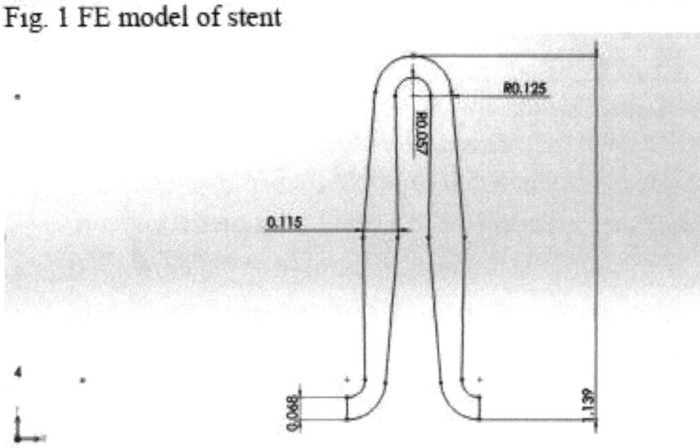

Fig. 2 geometry of the strut

III. METHODOLOGY

In the initial stage, a stent model A is created in Solid Works software. Due to the symmetry of the stent by utilizing the correct boundary conditions, half part was used to simulate the expansion process. An elastoplastic material was assumed as the stent material. Material properties taken were chosen to approximately represent the behavior of Cobalt Chromium L605 and is given in table 2. The plastic properties like plastic strain and corresponding stress were entered by taking values from the stress-strain graph assigned to the section of the model. The step time was given as 1.0ms. Symmetric boundary conditions were imposed on the nodes of the stent in the planes of symmetry where all the nodes perpendicular to y-axis were not allowed to move in the y-direction and all the nodes perpendicular to x-axis were not allowed to move in the x-direction. Both ends of the stent were free to move along its axis so that the expansion and shortening behavior of the stent would be observed.

Polyurethane rubber material was used to represent the balloon. The balloon as a medium to expand the stent was modelled as 11mm length, the outer diameter of the balloon as 0.962 mm and the thickness as 0.03mm. Angioplasty balloons are usually manufactured from materials such as polyethylene, polyamide, and their behavior is typically modelled using linear elastic material models based on data provided by the manufacturer. The material property of the balloon is given in the Table: 2. The boundary condition given for balloon is also same as stent except for the two ends. It was assumed that the ends of the balloon are fully tethered and hence only the expansion, in the radial direction was permitted. Surface to surface algorithm approach was selected in assembly of stent and balloon to cope with the nonlinear contact problem between the two component surfaces.

Table : 2

Part	Density (kg/m^3)	Young's modulus (GPa)	Poisson's ratio
Stent (L605)	$9.1e^{-6}$	243	0.30
Balloon (polyurethane)	$1.07e^{-6}$	0.09	0.49

IV. LOADING CONDITIONS

Table : 3

Part	Element geometry	No. of elements	No. of nodes
stent	Tetrahedral	28712	57159
Balloon	tetrahedral elements with hybrid formation	15052	29473

V. RESULTS AND DISCUSSIONS

During the inflation of the stent, the expansion up to 300KPa is quite neglible as the stent is in crimped state. The expansion from 1.09mm to 3mm was very fast above 300KPa. Because uncrimping happens around 320KPa. The further increment of pressure from 400 to 800KPa the increment in displacement is slow and steady. The stent deforms plastically and yielding takes place at this pressure. Due to strain hardening, again the expansion becomes slow above 700KPa. During expansion, the stress gradually increases. However, stress in some areas of the stent is typically high. These areas are found to be the crown region of the stent. This is because the struts are pulled apart from each other to enable radial expansion and there are only little changes in the cross-section area of the strut. But if the applied pressure exceeds 1000KPa the yielding starts and the weakest part of the stent collapses.

The maximum Von Mises stress obtained at an inflating pressure of 0.8MPa is 1117 MPa and the maximum deformation is 1.127mm. This is compared with the experimental results and the results are validated. The figures 3 and 4 gives the result of expansion analysis of the actual stent with the thickness 0.064mm (model A). Then same analysis was repeated to the other models also by changing the thickness (+15%,+30%,-15%,-30%) . The results of the models B, C, D, and E are given in table 4.

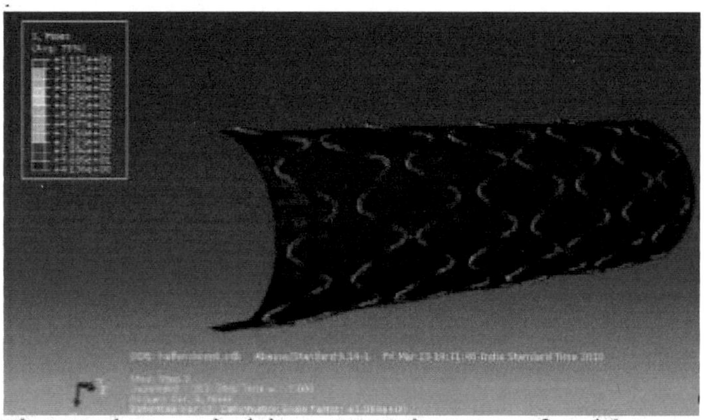

Fig.3 Maximum and minimum Von Mises stress of model A at 0.8MPa

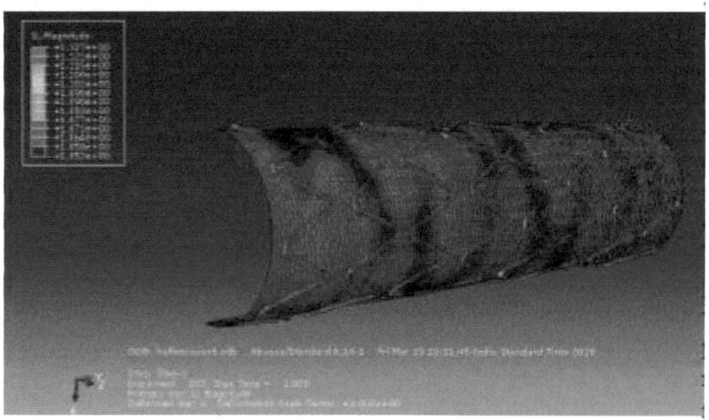

Fig.4 Maximum and minimum expansion of Model A at 0.8MPa.

Table: 4

Stent model	Strut thickness (mm)	Pressure load (MPa)	Stress (MPa)	Displacement (mm)
B (-30%)	0.044	0.8	1198	1.314
C (-15%)	0.054	0.8	1156	1.248
A (actual)	0.064	0.8	1117	1.127
D (+15%)	0.074	0.8	1069	1.005
E (+30%)	0.084	0.8	1024	0.789

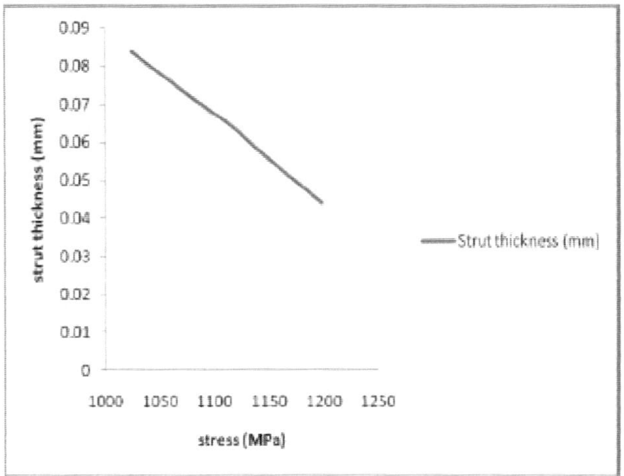

Fig. 5 Thickness Vs stress

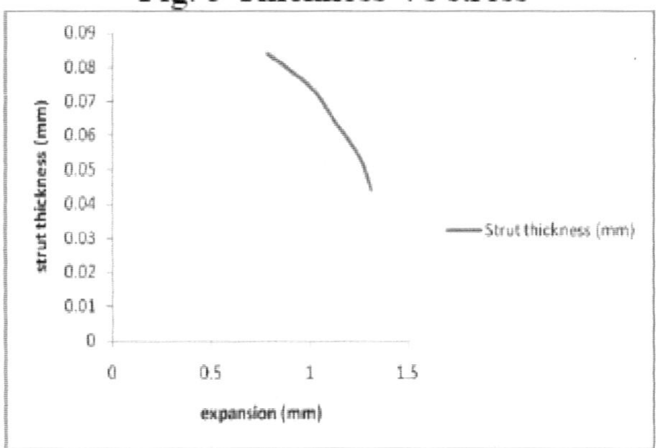

Fig.6 Thickness Vs Deformation

Graphs are plotted against Strut thickness Vs Stress and Strut thickness Vs Deformation as shown in Figs.5 and 6 respectevely. It shows that for a given design the stress induced due to the application of a pressure load depends on the strut thickness and it decreases with increase in thickness of strut. The stent B has maximum stress as its thickness is least among the five models. The yielding starts earlier in the strut with small thickness. Whereas the expansion or displacement of the stent decreases with increase in the strut thickness because yielding or plastic deformation is delayed when the strut thickness is increased. The stent B has maximum expansion of 1.314mm at a pressure of 0.80MPa. Also, the graph shows that the variation of the stress with thickness is linear and inversely propotional whereas the thickness Vs expansion is nonlinear. Percentage of increase in the expansion is minimal in small thickness compared to larger thickness.

VI. CONCLUSION

The mechanical behaviour of a balloon stent assembly with respect to the changes in strut thickness was analyzed in commercially available FEA software, ABAQUS. According to the results , the maximum stress and expansion are obtained at the crown region of the stent. This is due to the pulling effect during expansion. Moreover it was identified that, minimum thickness is better because with low thicknesss the stress is low and the expansion is high. A linear trend was observed in stress increment and is quite predictable but increment in expansion is nonlinear and is minimal for low thickness. It is concluded that the strut thickness which gives maximum expansion and sufficient strength is to be considered as optimum thickness of the strut. More over the results gives a clear indication that, the inflating pressure to be selected based on the strut thickness and the expansion required. The results can be used for further understanding of stent design and optimisation.

References

1. Francesco Migliavacca,Lorenza Petrini,Maurizio Colombo,Ferdinando Auricchio,Riccardo Pietrabissa "*Mechanical behavior of coronary stents investigated through the finite element method*".Journal of Biomechanics 35(2002): 803-811

2. David Chua SN, Mac Donald BJ, Hashmi MSJ. *Finite-element simulation of stent expansion.* Journal of Materials Processing Technology 2002; 120: 335-40.

3. N.Eshghi, M.H Hotjjati, M Imani ,A.M Goudarzi. "*Finite Element Analysis of Mechanical Behaviours of Coronary Stent*". Physics Engineering 10 (2011) :3056-3061

4. C.Lally, F. Dolan, P.J.Prendergast. "*Cardiovascular stent design and vessel stresses: a finite element analysis*. J.of Biomechanics 38(2005) 1574-1581

5. Julian Bedoya, Clark A Meyer " Effect of stent design parameters on nominal artery wall mechanics" J. of Biomechanical Engineering, oct 2006, vol 128/757

6. Arjun R., Hashim V.,Dileep P.N., " *Effect of inflating pressure on the expansion behaviour of coronary stent and balloon: A finite element analysis*", International Conference on Aerospace and Mechanical Engineering, 14-16 December 2015.

Risk Analysis in Implementation of Solar Energy Projects in Kerala

Mohamed Shahid P A[1], Firoz N[2], Mohammed Sadhikh[3], Mohith Dadu[4].

[1]Assistant Professor, Department of Mechanical Engineering, TKM College of Engineering, Kollam.

[2]Assistant Professor, Department of Mechanical Engineering, TKM College of Engineering, Kollam.

[3]Professor, Department of Mechanical Engineering, TKM College of Engineering, Kollam
[4]Assistant Professor, Department of Mechanical Engineering, TKM College of Engineering, Kollam.
E-mail: shahidvaidyar@gmail.com

Abstract. India's growth rate is slow in the deployment of solar power projects when compared to other developing and developed countries. The Ministry of New and Renewable Energy (MNRE) have often criticized Kerala for their inability to meet target for solar power projects installation. The aim of this study is to explore the state of solar projects in Kerala, with an objective to create a knowledge base of multifarious risks and control procedures. This study enables to identify the main risk in implementing solar energy projects and develop a guideline which could serve to reduce the risks associated with investments in solar projects. In this study, survey was conducted amongst the different players of solar power project installation and operation to identify the various risk factors and sub factors. These factors were categorised and further quantified using the decision making tool, Analytic Hierarchy Process (AHP). Political risk, technical risk, financial risk and environmental risk are the prominent risk factors involved in solar projects installation. The impact of the sub factors are identified using AHP. Payback period and non-approval of on grid small scale projects from KSEB are the critical factors. The inattention in reduction of risks at organizational level, and project level, can substantially affect the solar projects at a catastrophic level.Overall, the need for future research is established in this study for investigating the capability of Project Management Offices in risk assessment and management.

Keywords: Solar energy projects; Risk Analysis; AHP;

1. Introduction

The 6th largest energy consumer, India, consumes 3.4% of global energy consumption which accounts for more than 17% of population. The Public Sector Undertakings (PSUs) principally controls the energy policy of India, which is constituted by the Government of India's Ministry of Power, Ministry of Coal, and Ministry of New Renewable Energy. Renewable energy is defined as the continually replenished energy that comes from resources such as sunlight, wind, rain, tides, waves and geothermal heat. The energy tapped from renewable resources amounts to 16% of net global consumption, with a split up of 10% energy from traditional biomass and 3.4% from hydroelectricity. The rest 3% is supplied by swiftly growing sectors namely small hydro, modern biomass, wind, solar, geothermal, and biofuels. Solar Power, the remarkable potential energy resource with zero emission, clean and renewable energy can be connected using a variety of devices. The solar energy systems are easily accessible for industrial and domestic use with an additional advantage of minimum maintenance as a part of recent developments. The rebates and incentive plans on government tax can make solar energy financially feasible. The missing targets in completion of execution has led to criticism in the area of consulting, contracting, policy framing related to solar projects and the growth rate of India is found to be sluggish when compared to other countries. In the State of Kerala also, solar energy generation rate is very low, for the target of 9% in annual solar energy production the achievable is less than 2% till now. This prime reason is due to the risks associated with thc implementation of solar energy projects .The aim of our project is to analyses the risks in implementing solar energy projects. Risks are identified and analysed using survey and analytic hierarchy process.

2. Solar Technology Used in Large, Medium & Small Scale Projects

Sunshine is free mode of energy and never gets consumed. The solar radiation that hits on the earth surface in an hour has more energy potential than the energy consumption of people in a year. The solar cell can convert the solar radiation in to electric potential which is having zero harmful emission. It doesn't even make any noise. A solar panel is a group of solar cells that work together. The energy is carried through space as electromagnetic radiation. Light and radio waves are types of electromagnetic radiation. Electromagnetic radiation travels like waves in water. Like water waves, it is a series of ups and downs. Solar power is the conservation of sunlight into electricity, either directly using photovoltaic (PV), or indirectly using concentrated solar power (CSP). Concentrated solar power systems use lenses or mirror and tracking systems to focus a large area of sunlight into a small beam. The concentrated heat is then used as a heat source for conventional power plant. Photovoltaic (PV) is a method of generating electrical power by converting solar radiation into direct current electricity using semiconductors that exhibit the photovoltaic effect.

3. Literature Review

The current heavy dependence of human population on fossil fuels is considered as escalated factor in climate change and the piling difficulties in overcoming the renewable energy targets. The use of renewable energy has been considered as a solution to the formerly mentioned issues [1]. The poor risk-mitigating strategies has led to the prime reason for the deficiencies. The successful mitigation of the project-associated risks is a key factor to complete the projects on time and is well within the control of Indian contractors. An approach proposed by Rodney [2] is to model, simulate and to evaluate project risks in term of quality, delay and cost; which aims to reproduce the behaviour of the project, evaluate its performance and anticipate its probable drifts while valuing the following specifications applicable to the entire process of risk management with dynamic and multi-views which consider all aspects of the project and represent different levels in detail. Sudi Apak studied the evolving mechanisms in financial risk management that will cater the needs of the renewable energy sector in both the EU and Turkey [3]. The study analysed their renewable energy investments with reference to managerial and theoretical implications and concluded that the encouragement in increased number of renewable energy projects is driven by improvement of technology-specific insurance cover public and private financial institutions. Yurii Rolik determined that McKinsey expert rating supports to calculate the attractiveness of the area of activity and the competitiveness of the project in the form of weighted averages for subsequent positioning [4].

Goh H.H. And S.W. Lee clearly identifies the relationship states for the project development in wind power for Malaysia [5]. The wind speed prediction and site survey can calculate the annual wind power generated which is essential for the further analysis. Similar methodology could be adopted for the calculating the power generated from solar in reference to the geographical conditions. Renewable energy is not only to reduce the carbon dioxide to the atmosphere but also considered to be as more environmental friendly compared with other criteria's [6]. In the book "Project Management, Planning and Control" by Albert Lester designates the project management plan (PMP) should include an outline of the risk management plans, which defines the scope and applicable areas of risk management, particularly the investigation of risk types [7]. The government revised the target for renewables from 20 GW to 175 GW, including 100 GW of grid-connected solar projects, by 2022 [8].

In the Annual report by (MNRE 2014-15) clarifies that, over the years, renewable energy sector has emerged as a significant player in the grid connected power generation capacity in India. It has been recognized that renewable energy has to play a much profounder role in attaining energy security in the years forward and be a vital part of the energy planning progression.

4. Methodology

Risks associated with solar projects are identified by means of literature review and important ones are selected based on the survey at 10 institutes in Kerala. These risks are prioritised using AHP as the decision tool. Finally a framework is developed illustrating the relationship between each risk. The detailed steps are shown in Figure.1

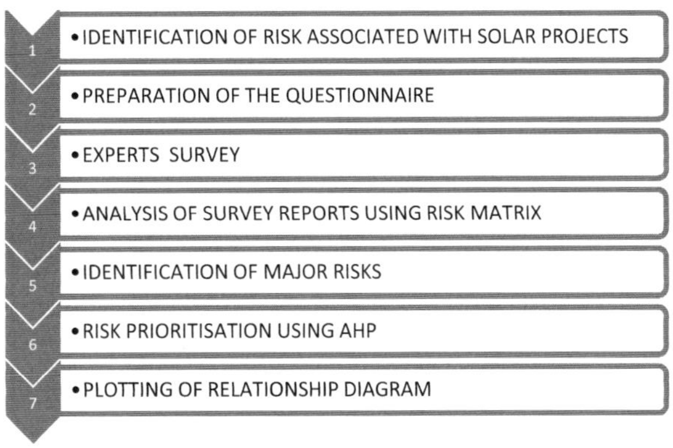

Figure.1. Methodology framework

4.1 Risks Associated with Solar projects

A research on solar projects based contractors, their capacity and practices in order to manage risks would be required to frame the state of the current situation. The main risks related with solar projects can be classified as political, financial, social and technical which is further classified into the following.

4.1.1 Political Risks

The delays in clearance and approval from government organizations is of prime importance as the risk of constraints in getting regulatory sanctions and achieving financial closure is relatively high. The failure to achieve financial closure inflates the required funds to execute a project if a project does not obtain the mandatory approvals. There are several political risks associated with solar projects in India which is illustrated in Table 1.

Table.1. Classifications of Environmental Risks

Classification of Risks	Risks/Barriers Description	Reference
Policy Risk	• High costs and unfavorable power pricing rules • Environmental externalities • Institutional and regularity barriers • High rate for consumers • Frequent changes in government policy	Beck&Martinot2004 Riedy 2008
Licensing, commissioning and approvals risks	• Payment guarantee mechanisms are not reliable. • Risks due to contracted parties not recognizing the contracts binding nature, and the judicial system is plagued by lengthy, intractable delays and "banked up" cases • Problems in obtaining clearances and certifications • Delay in receiving payments	Kurth 2007 Finley-Jones 2004
Contract	• Risks based on parties to a contract not recognizing their binding nature and the judicial system being plagued by substantial intractable delays "bank up" cases • Payment guarantee mechanisms are not reliable	Kurth 2007

4.1.1 Environmental Risk

There are several environmental risks related with solar projects in India which is demonstrated in Table 2.

Table.2. Classifications of Environmental Risks

Classification of Risks	Risks/Barriers Description	Reference
Environment	• Forthcoming environmental legislation that may affect the project • Biodiversity protection issues • Workers Health and Safety issues	Herbert 2010 , Stanton 1996 , Pasqualetti 2002, Gipe 1995

4.1.3 Financial and Ethical Risks

The list of financial risks associated with solar projects in India which is shown in Table 3.

Table.3. Classification of Financial and Equity Risks

Classification of Risks	Risks/Barriers Description	Reference
Financial and equity risk	• Unsteadiness of taxation environment in India with frequent changes to significant guideline in renewable energy sector of India. • Difficult to fund working capital because a) banks are hesitant to lend to rural entrepreneurs b) banks are reluctant to approve consent raw materials as collateral • Unpredictable rates in borrowing, lending and returns. • Skewed financial incentives • Inadequate funding in long –term basis • Loss in capital is high	Bansal, Bhatti & Kothari 2002

4.1.4 Technical Risks

The numerous technical risks associated with solar projects in India is listed in Table 4

Table.4. Classifications of Technical Risks

Classification of Risks	Risks/Barriers Description	Reference
Technology	• Technology tie-up/Joint venture, expected lead time issues if any	Lippert,Kishor& Dhingra 2006
Operation & Maintenance	• Issues such as machinery breakdown and downtime • Payback guarantee period is high	Reddy 2002
Procurement	• Indigenization percentage in reference with project cost and its impact on project cost • Overall hike in Indigenization percentage	ESMAP 2010
Human resource risk	• Difficulty in finding suitable skilled and unskilled employees to work in this field	Kurth 2007
Land feasibility	• Selection of appropriate site • Land acquisition issues	Reddy 2002
Transparency risk	• Prevalent corruption in administrative, bureaucratic and business circles.	Kurth

4.1.5 Identification of risks associated with Solar Energy Projects

From the above mentioned risks, about 200 questions are formulated for identifying the risks. For the easiness of survey from the above mentioned, about 60 questions are sorted for the survey. Survey are conducted in three stages that are

1. Before Installation 2.During Installation 3.After Installation

4.2 Questionnaire preparation

Questionnaire is prepared on the basis such that for each question ratings from 1-5 is given and for each question there have PRIORITY & IMPACT columns. PRIORITY - priority of each question towards the solar projects IMPACT - impact of each question towards the solar projects .5-5 rating means High priority & High Impact

4.3 Experts Survey

The following are the experts & selected organizations where the questionnaire based survey is carried out. ANERT, KELTRON, CIAL, PRANA GRIDS, 5. KC KOPAR ENERGY SOLUTIONS, KSEB (Kollam), DOMESTIC – 1(Kallada), DOMESTIC – 2 (Keralapuram) ,Technical Educational Institute.

4.4 Analysis of survey reports using risk matrix

The survey results are then analysed using matrix to find out the main risks specified by all the experts and then all the survey reports are merged together for Political/Policy risk, Financial Risk, Technical Risk Environmental Risk. The Weightage is given for each question and then from this collective result, the Weightage above 40 are taken as the most important Risk for the technical risk there are 19 questions. For each questions the survey results from each experts are different, in the table for each question relevant scores given by the experts are noted and the total Weightage is calculated.

4.5 Identification of Hierarchy Risk

From the data the most important technical risk were identified and the questions were chosen which has weightage greater than or equal to 40, then it is graphically represented by as shown in Figure 2.

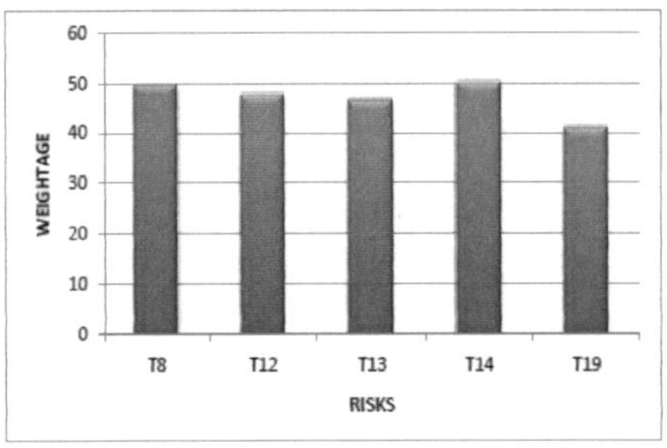

Figure: 2 Graphical representations of hierarchy risk

The most important risks from this method were identified from these 60 risks. The Aim is to prioritize the risks and compare the risks and need to plots a comparison chart, for AHP technique is being used.

4.6 A.H.P (Analytic hierarchy process)

Analytic Hierarchy Process, AHP belongs to the multi-criteria decision making methods (MCDM). In AHP, inputs like price, weight, or area, or even subjective opinions such as feelings, preferences, or satisfaction, can be translated into measurable numeric relations.

The basic steps in the formulation of solution for decision problem with the aid of AHP are not tedious: 1. Define the decision goals 2. Hierarchy structure formulation for the decision problem regarding the categories and criteria those figure into the decision 3. Comparison of pair criteria in each category 4.

Calculate the priorities and a consistency index in reference with logical comparisons and consistency 5. Evaluation of alternatives according to the identified priorities.

5. Results and Discussion

In this section survey results are analysed to determine the important risk factors in each category

5.1 Survey Results

The different political risks are evaluated and their weightage is shown in figure.3 and major risk factor identified and listed in Table 5

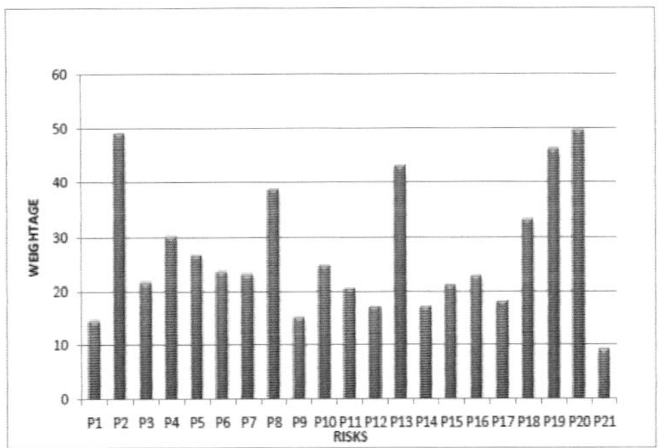

Figure: 3 Risk Matrix - Graphical Representation for Political Risk

Table.5. Major Political Risk

P2	Lack of Public Awareness about solar energy projects & government policies
P13	Payback period is larger in case of projects below 200kw
P19	Solar Scam badly affected the projects
P20	KSEB Do not approve on grid projects due to their technical issue

The different technical risks are evaluated and major risk factor identified and listed in Table 6.

Table.6. Major Technical Risks

T8	Availability of skilled technicians is very less in executing solar energy projects
T12	Feasibility of land for solar energy projects is very less
T13	Effect of weather changes (temperature, rain etc)is badly affect in solar energy projects
T14	Periodic maintenance schedule& cost for solar energy projects after installation (off grid) is very high
T19	Availability of technicians for rectifying complaints after installation is very less

The different financial risks are evaluated and major risk factor identified and listed in Table 7.

Table.7. Major Financial Risks

F1	The financial status of solar energy projects compared to other renewable energy projects is loss
F8	The return rate of investment compared to other renewable energy projects is less
F9	Capital investment of solar energy projects compared to other renewable energy projects is very high
F15	The initiatives taken by the agency for financial awareness to potential customers in implementing projects is less

The different environmental risks are evaluated and major risk factor identified and listed in Table 8.

Table.8. Major Environmental Risks

E5	Chance of health hazards to the workers/technicians is high
E7	Weather condition in Kerala to implement solar projects is negatively affected
E9	Availability of land for solar projects in Kerala is very less
E11	Recycling of solar modules is not possible

5.2 AHP Results

Using Analytic Hierarchial Process method weightage for each important risk were determined. The input of AHP,pairwise comparisons is performed by interviewing the experts from industry. Results are illustrated in Table.9, Table 10.Table 11 and Table 12.

Table.9.Political risk resulting weights for the criteria

Category		Priority	Rank
1	Restriction of KSEB towards on grid projects	36.40%	1
2	Lack of Awareness	36.40%	1
3	Solar scam negatively affects	18.20%	3
4	Larger payback period	9.10%	4

Table.10. Technical risk resulting weights for the criteria

Category		Priority	Rank
1	Off grid maintenance is high	31.20%	1
2	Less availability of skilled technician	27.30%	2
3	Land availability is very less	19.90%	3
4	Weather changes negatively affects	16.10%	4
5	Less availability of technician after installation	5.40%	5

Table.11. Financial risk resulting weights for the criteria

Category		Priority	Rank
1	Capital investment is high	59.10%	1
2	Financial awareness from agencies is less	25.40%	2
3	Payback period is higher	9.70%	3
4	Financial status is bad compared to other REP	5.70%	4

Table.12. Environmental risk resulting weights for the criteria

Category		Priority	Rank
1	Availability of Land is very less	40.00%	1
2	Recycling of module is not possible	20.00%	2
3	Chances for health hazards is more	20.00%	2
4	Weather changed negatively affects	20.00%	2

5.3 Major Findings

From the analysis 14 risks are sorted as the important risk in implementing solar energy projects they are

i. Lack of Public Awareness about solar energy projects & government policies
ii. Payback period is larger in case of projects of capacity below 200kw
iii. Solar Scam has badly affected the projects
iv. KSEB do not approve on grid projects due to additional technical liabilities
v. Availability of skilled technicians is very less in executing solar energy projects
vi. Periodic maintenance schedule & cost for solar energy projects after installation (off grid) is very high
vii. Availability of technicians for rectifying complaints after installation is very less
viii. The financial status of solar energy projects compared to other renewable energy projects is less
ix. The return rate of investment compared to other renewable energy projects is less
x. Capital investment of solar energy projects compared to other renewable energy projects is very high
xi. The initiatives taken by the agencies for the financial awareness to potential customers in implementing projects is less
xii. Chance of health hazards to the workers/technicians is high
xiii. Weather condition in Kerala to implement solar projects is negatively affected
xiv. Availability of land for solar projects in Kerala is very less

All the above 14 risks are equally important. For identifying the most important risk, again it is sorted by avoiding repeated risks with. Then the final risks analysed is as follows

i. Lack of public awareness about solar energy projects & government policies
ii. KSEB do not approve on grid projects due to their technical issue
iii. Availability of skilled technicians is very less in executing solar energy projects
iv. Periodic maintenance schedule & cost for solar energy projects after installation (off grid) is very high
v. Capital investment of solar energy projects compared to other renewable energy projects is very high
vi. The return rate of investment compared to other renewable energy projects is less
vii. Availability of land for solar projects in Kerala is very less
viii. Recycling of solar modules is not possible
ix. Effect of weather changes (temperature, rain etc) badly affect in solar energy projects
x. Chance of health hazards to the workers/technicians is high
xi. Solar Scam have badly affected the projects
xii. The initiatives taken by the agency for financial awareness to potential customers in implementing projects is less

5. Conclusion

Risk analysis in implementing solar energy projects in Kerala using AHP technique is conducted and following are the proposed solution for the reduction of the identified major risks.

1. Promote awareness about the policies and new schemes through the Newspapers, Social media, Televisions, etc.
2. KSEB needs to change their policy towards on grid connection from all type consumers irrespective of the capacity.
3. Government should promotes proper loan facilities, subsidies, incentives & tax reductions
4. Government should take initiative to conduct the training courses to develop skilled technicians, and also promote awareness programs for future generation
5. Promote more "On Grid connections" than "Off Grid connection"
6. Reduce the risk of "Availability of land for solar projects in Kerala is very less" by promoting floating panel projects , canal top projects , roof top projects , car port projects etc
7. Reduce the risk of "Chance of health hazards to the workers/technicians is high" by implementing safety rules , safety programs & by providing safety measures
8. Reduce the risk of " Recycling of solar modules is not possible" by providing R&D work on Recycling of modules to develop new technologies.

References

[1]. Siddarthgaurav, Nicolas Chileshe, Tony MA (**2011**), *Project risk analysis of solar energy project delays in India*, 49thAuSES Annual Conference

[2]. Rodney E, Ducq, Breysse D, Ledoux Y (**2015**), *An integrated management approach of the project and project risk,* IFAC- Paperonline-48-3(2015)535-540

[3]. Sudi Apak, ErhanAtey, Gungor Tuncer a (**2011**) *Financial Risk management in renewable energy sector: Comparative analysis between European Union and Turkey*, Procedia social and behavioral sciences 24(2011)935-945

[4]. Yurii Rolik (**2017**), *Risk management in implementing wind energy project*, Precedia Engineering 178(**2017**)278-288

[5]. Goh H H, Lee S W, Chua Q S et al (**2014**) *Renewable and sustainable energy review 38*(**2014**)929-930

[6]. Ahamad S, Tahar S M(**2012**), Selection of renewable energy source for sustainable development of electricity generation system using analytic hierarchy process; A case of Malaysia. Renewable energy 63, 548-566

[7]. Albert Lester, *Project Management, Planning and Control (Sixth Edition)* (**2014**) ISBN: 978-0-08-098324-0, 71-81

[8]. Sanjay Datta**(2017)**, Times of India, *Solar energy helps boost power generation to 10gW*, Mar 7.

[9]. PWC, "*Green tax incentives for a sustainable Malaysia*", N G T Council ed. Malaysia **(2010)**

Design Process Sequencing in Product Development Process Using Design Structure Matrix and Particle Swarm Optimisation Techniques

Shajahan C A[1],Subramanyan Embranthiri E[2],Kumanan S[3],Firoz N[4].
[1]Professor, Department of Mechanical Engineering, T K M College of Engineering, Kollam
[2]Consulting Engineer, TCE,Bangalore
[3]Professor, Production Engineering Department,NIT Tiruchirapalli
[4]Assistant Professor, Department of Mechanical Engineering, T K M College of Engineering, Kollam

E-mail: cashaji@yahoo.com

Abstract. Product development (PD) is a complex business process of transforming new ideas into marketable new products, processes or services. PD process is characterized by information flow and iteration. Traditional project management techniques are limited in representing interdependent activities. Design Structure Matrix (DSM) is a powerful tool for identifying and managing information exchange and iterations. In this work the process is modeled with DSM and the design activities are optimally sequenced using a Particle Swarm Optimization (PSO) technique. The methodology is exemplified with a case example

Keywords: *Product Development (PD), Design Structure Matrix (DSM), Product development iterative time (PDIT), Particle Swarm Optimization (PSO).*

1. Introduction

A profitable and successful product is the decisive winning factor in fast changing and competitive market. The intense competition forces firms to develop better products at an increasingly rapid pace. Product development is a process that guides a new product from ideation through product launch. It is an interdisciplinary and complex activity that requires inputs from numerous knowledge domains [1]. Engineering products become increasingly complex and the markets of the future are expected to become even more competitive than today[2].As a result, today's large product development projects may involve a complex set of tasks that require the participation of many engineers, managers, technicians and other professionals over a period of several years. The rate of advance of any task in a product development project may strongly affect the progress of a number of other tasks in the same project. The real challenge for a product development project under such circumstances is to overcome the tremendous complexity involved in scheduling and executing a large number of interconnected and dynamic tasks[3]. Besides technical complexity, the managerial complexity that is necessary to manage the interactions between

various engineering disciplines allocated in different geographical locations, imposes additional challenges on the product development process.
The dependency relationship among the activities in a product development project are of three types - independent, dependent, and interdependent [4]. Conventional tools like Program Evaluation and Review Technique (PERT) and Critical Path Method (CPM) are limited in application to represent interdependency. Researches have investigated DSM as a tool to identify and manage information exchange between activities [5] [6]. This work aims at bringing out an optimal activity execution sequence which will reduce the effect of iteration to a minimum level using DSM and PSO techniques.

1.1. Design Structure Matrix

Design structure matrix is a generic matrix based framework for information flow analysis and was introduced by Donald Steward. The basic representation of activity DSM is a square matrix containing a list of activities in the rows and columns in the same order in the matrix form. A DSM showing information dependency is shown in Figure 1(a) and 1(b). The order of activities in the rows or columns indicates the sequence of execution. The relationship between activities is represented with a '1' or 'X' mark in the off diagonal cells. The activities have to be read along the column as "gives information to" and along the rows as "needs information from". Marked cells above the diagonal represent iteration in the process.

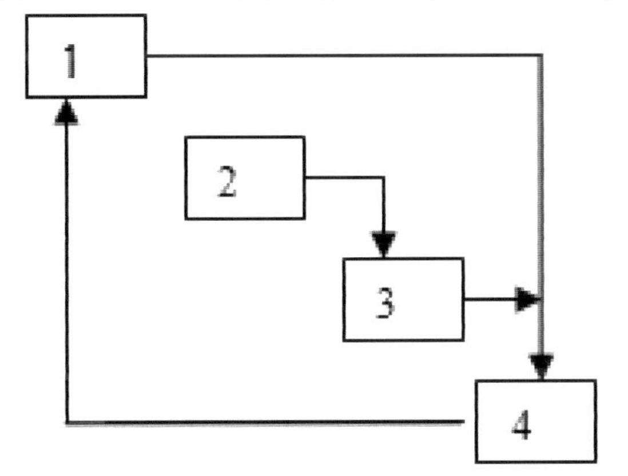

Figure 1.a. Information dependency

	1	2	3	4
1				1
2				
3		1		
4	1		1	

Figure 1.b. DSM showing Information dependency

Figure 1 (b) shows a typical DSM showing the information dependencies. This occurs when an activity is dependent on information from a task scheduled for a later execution. Such scenario often leads to rework and are undesirable. If any mark lies above diagonal, it implies that an assumption has to be made to execute the corresponding sequence.
A primary goal in basic DSM analysis is to minimize the number of feed backs and their scope by re-sequencing the activities to get the DSM into a lower-triangular form as possible. Partitioning and tearing are two main processes that can manipulate the matrix and transform the matrix into lower triangular form to the extent possible. It is the process of rearranging the order of activities by moving an entire row and column on either side in such a way that the resulting matrix has marks either below the diagonal or close to the diagonal. Through the manipulations, the task execution sequence will be reordered, the number of

iterations will be reduced, and fewer tasks will be involved in the iteration cycle and these will result in a faster development process.
There are different types of DSM like parameter based, component based, task based, numerical DSM etc. A numerical DSM can hold multi attributes like strength of dependence of task on specific data, sensitivity, evolution, variability accuracy of information etc. In this paper, a numerical DSM is used in which coupling marks are replaced with numbers (iteration factor).The strength of each coupling are quantified in the form of an iteration factor-representing the number of iterations required for convergence. Table 1 shows the available 7 coupling levels and their associated iteration factor default value.

Table 1. Iteration factor values

Coupling strength	Iteration factor
Extremely weak	2
Very weak	3
Weak	4
Nominal	5
Strong	6
Very strong	7
Extremely strong	8

It is very useful in planning the activity sequences and also in identifying and managing information exchange. A major advantage of the matrix representation over other lies in its compactness. Moreover matrices are easy to manipulate and store in computer. The utility of a matrix representation lie in its ability to illuminate complex relationships between process elements in a compact, visual and analytically advantageous format (www.problematics.com)[7].

1.2. Particle Swarm Optimization Algorithm

PSO is a population based optimization tool developed by Eberhart and Kennedy[8] and has been applied successfully in many areas like function optimization, artificial neural network, and fuzzy control system. Compared with other methods, the particle swarm concept was based on the premise of social behavior. A PSO algorithm mimics the behavior of flying birds and their means of information exchange to solve optimization problems. PSO has the advantage of being faster and cheaper. It can get the same quality results in significantly fewer fitness evaluations and constraint evaluations. The algorithm is intuitive and does not need specific domain knowledge to solve the problem. There is no transformation or any other manipulation needed to handle the constraints.

PSO shares many similarities with evolutionary computation techniques such as Genetic Algorithms (GA). The system is initialized with a population of random feasible solutions and searches for optima by updating generations. In PSO, the potential solutions, called particles, fly through the problem space by following the current optimum particles. Each particle keeps track of its coordinates in the problem space which are associated with the best solution (fitness) it has achieved so far. This value is called Pbest. When a particle takes into account of all the population it has already searched through, the best value is a global best and is called Gbst. At each time step, the velocity of (accelerating) each particle is changed towards its Pbest location. Acceleration is weighted by a random term, with separate random numbers being generated for acceleration toward Pbest location. The velocity of each string can be calculated by using the equation 'a' and the velocity can be updated to the present population by using equation 'b'.

$$Vid = Vid + C1 * rand() * (Pid - Xid) + C2 * Rand() * (Pgd - Xid) \quad \text{(a)}$$

$$Xid = Xid + Vid \quad \text{(b)}$$

Where, Vid is the velocity for the particle i , represents the distance to be travelled by the particle from its current position.

Xid represents the particle position, Pid and is called the Pbest, the local best solution represents i th particle's best previous position and Pgd is the Gbest, the global best solution.

C1 and C2 are two positive constants called accelerating constants, rand () and Rand () are two random functions in the range [0,1]. Equation (a) is used to calculate the particle's new velocity according to its previous velocity and the distances of its current position from its own best experience (position) and the group's best experience. Then the particle flies toward a new position according to equation (b). The performance of each particle is measured according to a pre-defined fitness function, which is related to the problem to be solved.

2. Proposed PSO Algorithm

The solution procedure for proposed PSO algorithm for Design process sequencing is developed in Turbo C. The flowchart of the process is given in Fig.2

2.1 Encoding scheme and initial swarm

Finding a suitable configuration between the problem solution and a PSO particle is a key factor in implementing the algorithm. This work encoded a feasible sequence of activity as a PSO particle. For a 'n' activity PD project, the PSO particle will be an array of 'n' segments, each segment containing different activity numbers in such a way that no activity could assume the same order. Or in other words activity repletion in the sequence is not permitted. The other hard constraints such as logical precedence relationship, fixation of first and last activities (if they are known) etc can be tailored according to the nature of the problem while generating feasible solutions. Hence in this problem, the initial swarm will be a set of N feasible sequences where N is the size of the swarm.

2.2 Fitness function

The circuits or feedback cycles require a finite number of iterations for its convergence and the associated time is called coupling iterative time (CIT). Product Development Iterative Time (PDIT) is nothing but the sum of all CITs. When the activities are changing its order of execution, the PDIT also changes accordingly. PDIT is the fitness function considered in this work. A heuristic is used to calculate the PDIT once the DSM is formed according to a particular sequence.

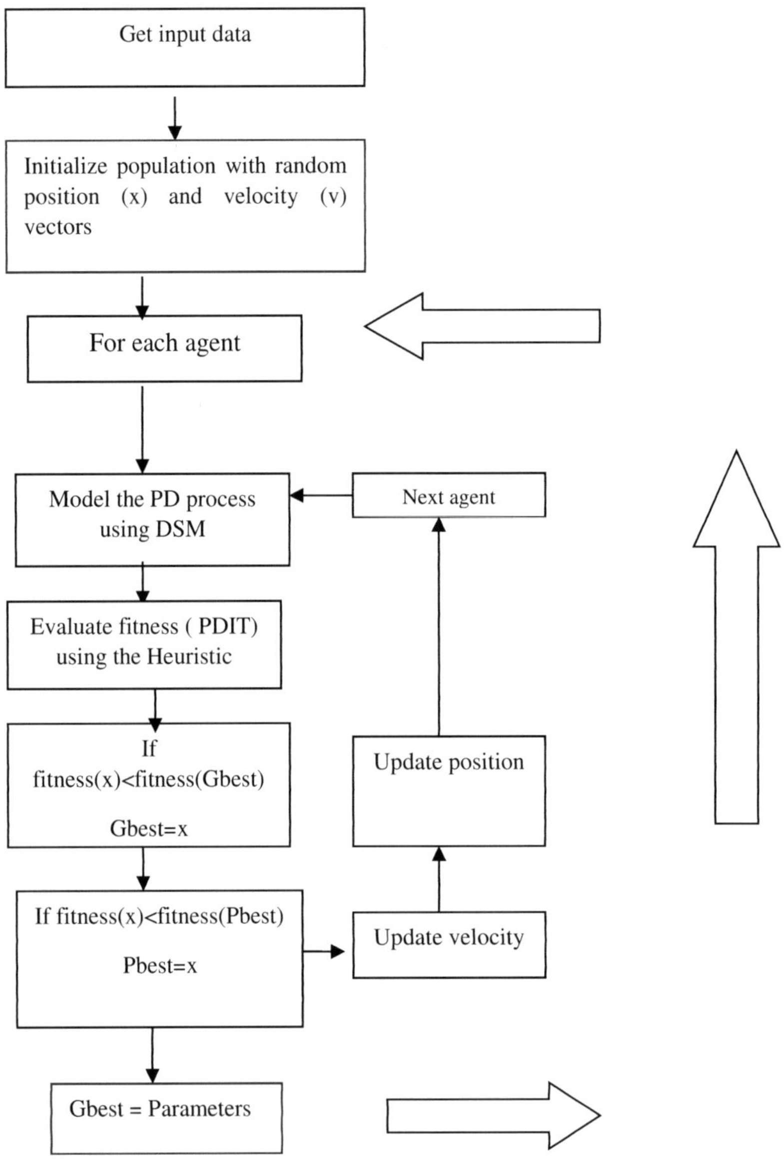

Figure 2. Proposed PSO algorithm for design process sequencing

2.3 Heuristic used for calculating PDIT

PDIT=0;{ Initial project iterative time}

$A=\{(X,T_x):1 \leq X \leq N$

$C=\{(Coup(i,j), IF_{i,j})\}$

Repeat

Select_Coupling(coup(i,j));

$CT_{i,j} = 0$;{ set initial coupling time }

$CIT_{i,j}=0$;{set initial coupling iterative time}

k=1.{set counter initial value}

Repeat

Select activity(k);

If $X_i \leq X_k \leq X_j$

Then $CT_{i,j} = CT_{i,j}+T_k$

k=k+1;

until k>N;{ all acivities in A are considered}

$CIT_{i,j}= CT_{i,j} \text{ x } IF_{i,j}$

$PDIT=PDIT+CIT_{i,j}$

Until all couplings in C are considered;

End

A	Set of sctivities in DSM and their associated times.
N	Number of activities
C	Set of feedback couplings and their associated iteration factors
Coup(I,j)	Coupliong from activity i to j
$IF_{i,j}$	Iteration factor of coup(i,j)
Select_coupling	A function that selects a coupling from set C
Select_activity	A function that selects an activity from set A
Tk	Time associated with activity k.

Journal of Physics: Conference Series **1355** (2019) 012027 doi:10.1088/1742-6596/1355/1/012027

		1	2	3	4	5
T =40	1			4		
T =10	2					6
T =30	3					
T =20	4					
T =50	5					

Figure 3.a . Initial Solution (PDIT=980)

		1	3	2	4	5
T = 40	1		4			
T = 30	3					
T =10	2					6
T =20	4					
T = 50	5					

Figure 3.b. Final Solution (PDIT=760)

A hypothetical PD process consists of 5 activities and 2 couplings are considered for the illustration as shown in Fig 3.a. and 3.b. Let the initial activity sequence is 1-2-3-4-5 and the corresponding activity times are 40,10,30,20 and 50 respectively. It consists two couplings or loops namely C1,3 and C2,5 respectively. The iteration factors considered for couplings C1,3 and C2,5 are IF1,3 = 4 and IF2,5 = 6 respectively.

CIT for the first coupling C1,3 =(40+10+30)*4 = 320. Similarly CIT for C2,5 = 660

Hence the associated PDIT = 980

By resequencing the activities as 1-3-2-4-5, the associated PDIT is calculated as (40+30)*4 + (10+20+50)*6 =760

3. Implementation of the proposed PSO algorithm

The implementation of the algorithm involves the following steps.

Step1: Input the number of activities, activity duration, iteration factors, constraints and the information dependencies.

Step 2: Generate the initial swarm which consists of N number of feasible solutions or particles. Each particle represents a feasible sequence which confine to the constraints. In the problem considered, first and last activities are fixed as hard constraints. Other hard constraints can be tailored according to the nature of the problem.

Step 3: For each feasible sequence (particle) corresponding DSM is modeled using the input data and the fitness of each particle (corresponding PDIT) is calculated. The Pbest and Gbest are to the identified. The sequence which is having minimum PDIT among the swarm of particles is selected as the Pbest. The Gbest corresponds to the sequence with minimum PDIT so far searched through.

Step 4: The velocity and position of each particle in the swarm is updated via Equation (a) and (b).

Step 5: Repeat step3 and 4 till the termination criteria met.

3.1 Setting of Parameters for the proposed PSO algorithm

The acceleration constants C1 and C2 in Equation. (a) adjust the amount of tension in the system. High values cause rapid movement towards the targeted regions. According to the experiences of other researchers, the values of C1 and C2 are taken as 2 for all the examples figured in this work. Through the application of Equations (a) and (b) the particle may overshoot the problem space. That is absolute value of Vid and Xid may be great. So in this work the maximum value of velocity Vid, denoted by Vmax , is set to 'n' (number of activities). Vid is a value in the range [-n,n]. The maximum value of Xid, denoted by Xmax is set to 'n'. Since Xid represents an activity number, it must be a positive integer. Thus Xid is an integer value in the range [1, n]. If the computational value of Xid is happened to be a real number, we round off it to the nearest integer. In this way we convert a continuous optimization problem to a discrete one.

The following are the values taken for different parameters of PSO algorithm.

Initial swarm size = 40; Cognition learning rate C1= 2; Social learning rate C2= 2 and Termination criteria is 20 iterations or no improvement in fitness function for 3 consecutive iterations, whichever occurs first.

4. Case Example

For the validation of the developed algorithm, a bench mark problem is taken from literature [9]. The project was taken from a larger conceptual design project. It consists of 22 activities and 39 couplings. The process is modeled using DSM and is shown in Figure 4. The iteration factor values are taken as shown in Table 1. By applying the proposed PSO algorithm, the activities are sequenced by keeping the logical constraint (that first and last activities are fixed.). The optimized sequence is obtained as 1-18-17-7-11-4-16-21-12-13-19-9-15-20-3-2-10-5-8-14-6-22. The DSM representation is shown in Figure 5 and the associated PDIT is 9560 units.

	Task1	Task 2	Task3	Task4	Task5	Task6	Task7	Task8	Task9	Task10	Task11	Task12	Task13	Task14	Task15	Task16	Task17	Task18	Task19	Task20	Task21	Task22
Task1												5										
Task 2			8				4															
Task3									8													
Task4							2															
Task5										7												
Task6																		6				
Task7	8																					
Task8									6													
Task9						2	6			2	8				4					7		
Task10		8							6													
Task11																	8					
Task12													3									
Task13														6								
Task14					3			7			2				8					4		
Task15																6						
Task16																	7				8	
Task17																6		8				
Task18				4													7		6			
Task19																		8		4		
Task20							4												8			
Task21							8															
Task22												5										

Figure 4. DSM using PSO algorithm

	Task1	Task18	Task17	Task7	Task11	Task4	Task16	Task21	Task12	Task13	Task19	Task9	Task15	Task20	Task3	Task2	Task10	Task5	Task8	Task14	Task6	Task22
Task1									5													
TAsk18			7			4					6											
Task17		8					6															
Task7	8																					
Task11			8																			
Task4				2																		
Task16			7					8														
Task21				8																		
Task12										3												
Tas13																			6			
Task19		8												2								
Task9				6	8								4	7			2				2	
Task15							6															
Task20				4							8											
Task3												8										
Task2				4											8							
Task10												6				8						
Task5																	7					
Task8												6										
Task14					2								8	4				3	7			
Task6		6																				
Task22									5													

Figure 5. DSM after partitioning using PSM 32

5. Conclusion

Iteration is inherent in PD process and has significant influence over development time and cost. Carrying out the activities in the optimum sequence will lead to considerable savings in terms of development time and cost since it minimizes iteration. This research work proposed a method which uses a DSM to model the design process and a PSO algorithm to obtain an optimized activity sequence. The method is capable of tolerating hard logical constraints in the activity sequence. The developed method is validated with a case example.

References

[1] Maoi, A. D.,Roberto, Vand Corso M.(1994), "Multi project management framework for new product development" , European journal of operational research, Volume 78, , pp178-191

[2] Gohlich D, Hildebrand and Schellert D, D (2018) " Augmented DSM sequencing to support product development planning" , International Design Conference 2018. Pp 1139-1148.

[3] Chun-Hsien Chun, Shih Fu Ling, Wei chen,(2003), " Project Scheduling for Collaborative Product Development Using DSM", International Journal of Project Management, Vol. 21 , pp 291-299.

[4] Yassine A, and Braha D (2003), "Complex concurrent engineering and the design structure matrix method", Concurrent. Engineering Management 11(3), pp165-176.

[5] Steward DV(1981), "The Design Structure system : a method for managing the design complex system", IEEE Transactions on Engineering management,Vol .28(3), pp 71-74

[6] Eppinger S,D, Whiteny , D E , Yassine A A and Romer T (2001) "The MIT design structure matrix –DSM homepage-" <http//web.mit.edu/dsm/>.

[7] Problematics Online ,http://www.problematics.com

[8] Kennedy, J., and Eberhart, R. C.(1995) , Particle Swarm Optimization, Proc. IEEE International Conference on Neural Networks (Perth, Australia), IEEE Service Center, Piscataway, NJ, IV: 1942-1948.

[9] Rogers James L (1997), "Reducing design cycle time and cost through process resequencing" International conference on engineering design(ICED), Aug 19-21.

Journal of Physics: Conference Series **1355** (2019) 012028 doi:10.1088/1742-6596/1355/1/012028

Experimental study on the influence of hydrophilicity on bacterial adhesion in bioimplants

V P Shiju[1], N V Abhijith[2] and U Sudeep[3]

[1, 2, 3] Department of Mechanical Engineering, N.S.S College of Engineering, Palakkad 678008, India

E-mail: usudeep@gmail.com

Abstract: Titanium and its alloys are widely used as dental and orthopaedic implants because of their desirable mechanical properties and biocompatibility. However, implant failures due to bacterial infections occur in many cases unexpectedly. Formation of a biofilm on the implant surfaces facilitates growth of colony of bacteria which in turn reduces the effectiveness of antibacterial drugs. Hydrophilicity of implant surfaces diminishes the chances of biofilm formation. Micro surface features are created on the Ti6Al4V implant surface using a pulsed Nd-YAG laser and its influence on water wettability (and thus on hydrophilicity) has been studied and compared with a polished surface. Strains of *Staphylococcus aureus* have been inoculated on the test specimens and the bacterial adherence have been compared. The results show that micro-textured surfaces of Ti6Al4V have improved water wettability and thus are more hydrophilic than polished surface. This has reduced the formation of biofilm and bacterial growth.

1. Introduction

Ti6Al4V (Grade 5 Titanium alloy) is widely used for bioimplants due to its excellent properties such as low toxicity, corrosion resistance, high specific strength, MRI compatibility and high biocompatibility [1].In many cases implantation fails due to reasons such as poor osseointegration, lack of adoption of proper procedures in placement of implants and infections developed after placing the implants in human body. Patients suffer due to implant associated issues which increases chances of co morbid conditions. Huge financial burden is also placed on patients and society due to more frequent implant failures. Bacterial infections are identified as a major cause of implant failure and usually require replacement of infected implants or treatment with prolonged use of antibiotics [2]. There are further issues such as bacterial drug resistance which is emerging.

Bacterial infections associated with implants are mainly attributed to the formation of a biofilm on the surface of implants. Biofilm consists of bacterial colonies, teichoic acids and extra cellular polymeric substances, which give immunity against antibiotics [3]. *Staphylococcus aureus (S.aureus)* is identified as the most common pathogen responsible to majority of the implant related infections. *S.aureus* is typically hydrophobic in nature due to the presence of highly negative charged teichoic and lipoteichoic acid as its cell wall constituents [4]. Hydrophobic cells prefer attachment to the hydrophobic surface [5]. So a hydrophilic surface can inhibit the adhesion and maturation of biofilm on the surface.

Recent research has indicated that physical and chemical surface modification techniques can improve wettability and bio-tribological performance of implant surfaces [6]. The micro/nano textures on the implant surfaces can enhance the wettability and bio-tribological properties [6].Hydrophilic properties of micro/nano textures depend on the surface energy, surface roughness and the topography of the surface [7]. Several surface modifications techniques are available in order to improve the surface properties such as surface coating, texturing patterns with variable geometries, creating micro/nano structures, combination of texturing and coating etc. [8]. Surface coating techniques has a disadvantage that the coating may disintegrate in the presence of biological fluids which will diminish their effectiveness. Among various surface topography modification techniques, laser surface texturing has emerged as one of the most popular methods for creating a super hydrophilic surface [9]. Periodic structures have been fabricated using laser surface texturing which will enhance the hydrophilicity and the surface free energy of the material. Laser texturing is found to have significantly reduced the adhesion of bacteria and reduced the formation of biofilm [10]. The possibility of using ultra short pulsed laser for surface texturing of grade 2 Titanium alloy surfaces and the subsequent formation of biofilm have also been investigated recently [11].

In this study, the effects of surface texturing on altering the hydrophilicity of Ti surface and its influence on enhancing the antibacterial properties of Ti based implant surface have been investigated. Micro dimples (having diameter and depth measuring few microns) are created on the implant surface using laser texturing and the hydrophilicity of surfaces is evaluated in comparison with a polished surface. *S.aureus* cultures have been introduced on the prepared specimen surfaces and the bacterial growth has been studied.

2. Experimental details

2.1 Selection of materials and surface texturing

Commercially procured Ti6Al4V (Grade 5) specimens have been used in the experimental study. The samples are made in the form of tiny blocks of flat surface area of 10mm x 10 mm and 3mm thickness using wire cut EDM method (figure 1). Surfaces were ground sequentially with a series of SiC papers from 120 to 2000 grits using a polishing machine. Three surfaces each of Ti6Al4V were further polished with lapping paste and were kept as polished surfaces for comparing the antibacterial performance with textured surface. The laser surface texturing was performed using a femto-second pulsed Nd-YAG laser system having a power of 200W. The wave length of the laser was 1064nm. The samples were irradiated with the laser beam using pre-selected processing parameters of laser power 40W, scanning speed of 25mm/s and a stand-off distance of 10mm.

Figure 1. Titanium Grade5 specimens (a) Polished (b) Textured

The laser-textured area on the specimen samples were 10 x 10 mm which is fully covered by the dimples having a depth of 2.5μm. The laser textured samples were ultrasonically cleaned in ethanol bath for 10 min, further rinsed in distilled water and air dried.

2.2 Evaluation of Surface Topography

The surface roughness and 3D topography of the polished and laser-textured samples were assessed using a laser confocal microscope (Olympus LEXT OLS4000). Arithmetic mean roughness (R_a) and maximum height of the profile (R_z) were measured and compared with that of the polished specimen.

Dimples of depth of 2.5μm and diameter of 10μm each cover the entire length of the specimen surface.

2.3 Wettability Analysis

The sessile drop method was used to measure the water contact angle of the polished and laser-textured samples using a video-based goniometer (First Ten Angstroms, FTA 200). The image capture and analysis were performed using the FTA 32 video software. Deionized water was used as the testing liquid, and the volume of each sessile drop was controlled at 5μl using a microlitre syringe. Droplet images were captured in the direction perpendicular to the laser track orientation at fixed time intervals, counting since the start of the droplet deposition to the cessation of droplet spreading for at least 60s. Average contact angle of eight measurements were taken at different locations for each sample at room temperature.

2.4 Bacteria strain

Specimens are inoculated with *S.aureus* which is cultured in Muller Hinton Broth (MHB) overnight (24 hrs) at 37 °C on a gyrotatory incubator with shaking at 100 rpm. After incubation, sterile MHB was used to adjust the overnight culture to an optical density of 0.3 at 550 nm and diluted 1 in 50 with fresh sterile MHB. This provided a bacterial medium of approximately 1 x 10^6 Colony Forming Units (CFU)/ml. 1 ml of culture was applied to each sample at an inoculums not exceeding 2.4 x 10^6 CFU/ml, as verified by viable count. The samples were incubated for 24 hrs at 37 °C on a gyrotatory incubator with shaking at 100 rpm. Three samples of each type of specimen surfaces (polished and textured) were tested to ensure the consistency of the results. The antibacterial performance is compared by measuring the area covered by dead (red colour) and live (green colour) cells.

3. Results and Discussions

3.1 Influence of surface topography

The Surface topography images of polished and micro-dimpled specimen are shown in figure.2The influence of surface roughness on microbial retention has been investigated in literature [12] Bacterial adhesion to surface depends on several factors, namely the topography, wettability and surface free energy, but the surface topography seems to be the most influential factor [13]. The R_a values for the polished and laser-textured samples extracted from the 3D roughness profiles are indicated in Table.1.

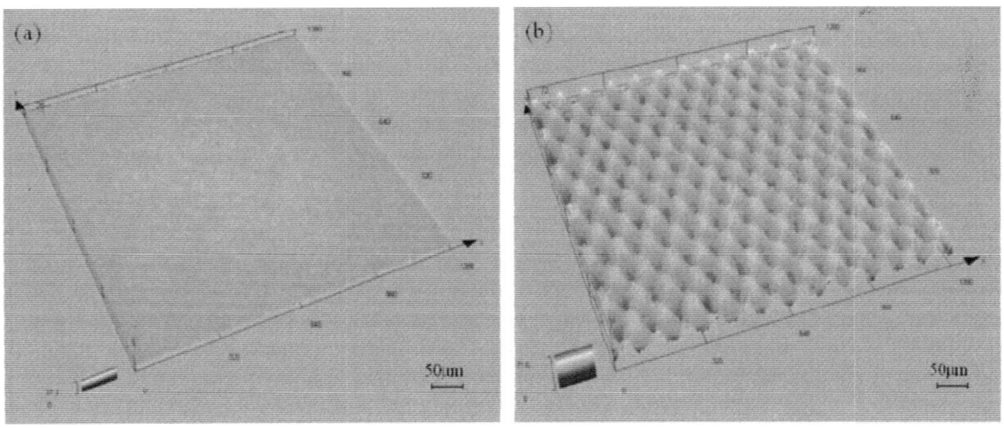

Figure 2. Surface topography of Ti6Al4V (a) Polished (b) Textured

The R_a values for the polished samples are in the range of 0.041 to 0.45μm.The R_a values increase significantly after laser treatment, lying between 3.6 and 3.7μm. Bacteria commonly adhere to low roughness polished surfaces, with R_a in the nanometre range, but also adhere to surfaces that present topographic features larger than the bacteria size [14]. It is found that surface roughness has a major role on cell adhesion and proliferation. Micro pattern supports the cell orientation, wound healing and repairing of cell defects [15].

Table 1 Surface roughness and contact angle values of specimens

No.	Specimen	Measured Values	
		R_a (μm)	Contact angle(^{0})
1	Ti6Al4V - Polished	0.451	121.2
2	Ti6Al4V - Textured	3.715	45.6

3.2 Influence of water wettability

The smaller value of contact angle of liquid drop is a direct measure of the hydrophilicity of the surface [16]. The values of contact angles measured from the polished and textured specimens are shown in figure 3.

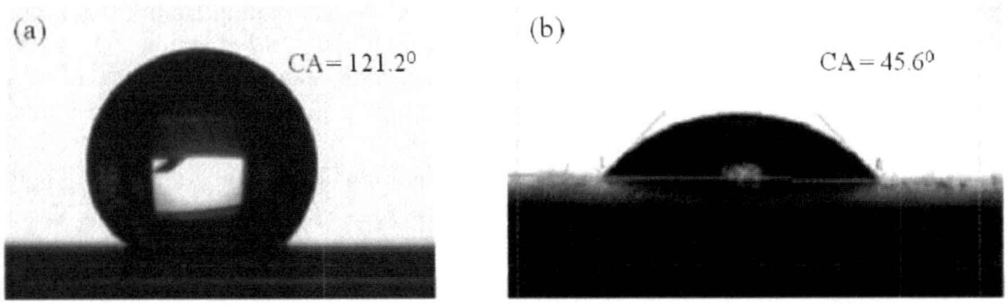

Figure 3. Water contact angle on Ti6Al4V specimen (a) Polished (b) Textured

Bacteria prefer to adhere on rougher surfaces but the surface is protected by unfavourable environmental disturbances such as shear force [17]. Laser textured specimen offers greater surface roughness than polished. From the Table.1 it is clear that the contact angle is significantly decreased after laser texturing and the surface become more hydrophilic. The hydrophobic nature of *S.aureus* cells is due to the presence of highly negatively charged and hydrophobic teichoic and lipoteichoic acid [18].

3.3 Influence on Bacterial adherence

The growth of Bactria on the specimen surfaces have been evaluated with the help of confocal florescence microscopy. The images obtained are as shown in figure 4. Bacterial luminescence on the polished (figure 4a) and laser textured (figure 4b) samples are shown for comparison. Green fluorescence indicates viable bacteria with intact cell membranes whilst red fluorescence indicates dead bacteria with damaged membranes. Hydrophobic interactions considered to be an important factor for initial attachment of bacteria.

The results show that bacterial adherence is greatly reduced in laser-textured specimens (figure 4b) than the polished specimens. Textured surfaces are found to be unfavourable for bacterial colonization. This is assumed to be due to the hydrophilic nature of the textured surface and increased surface roughness (R_a) value. This is evident from the wettability studies carried out and reported in section 3.2.Surface peaks act as a spacer between the bacterial cell and surface, decreasing the interaction energy of the bacteria. Ti6Al4V shows better antibacterial behaviour due to its inherent chemical composition [14].In Ti6Al4V, surface treatments convert the amorphous Titanium oxide to crystalline layer which reduces the chances of bacterial adhesion [9]. Bacteria produce Extracellular Polymeric Substances considered as the major cell to cell connecting substance, which are liable for the biofilm maturation and colonization of dead cells on the textured surface [4].

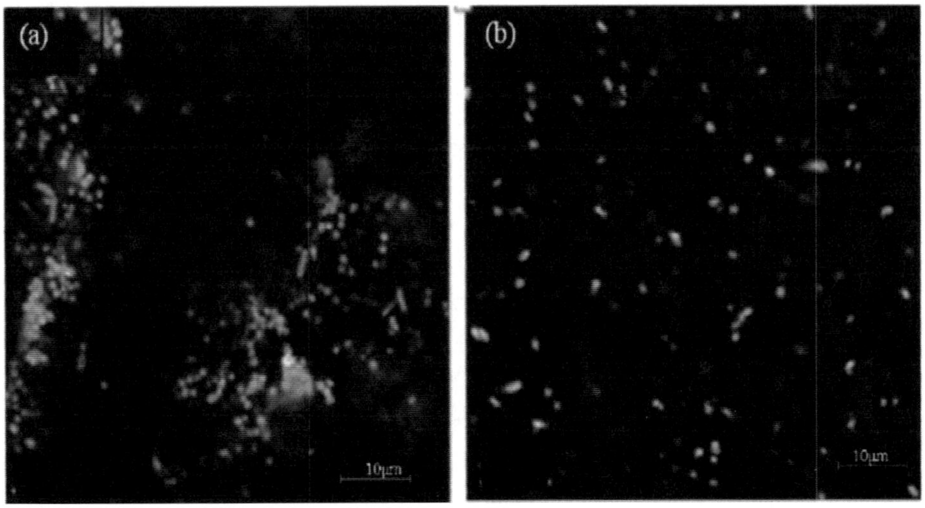

Figure 4. Microscopic images of bacterial luminescence in Ti6Al4V (a) Polished (b) Textured

Bacteria surface interaction through electrostatic, gravitational, Van der Waals forces, hydrophobic interaction of bacteria to matching biomaterial. The wavy nature of the laser textured surface reduces these interactions, through reductions in electrostatic and gravitational forces. *S.aureus* is typically hydrophobic in nature so they prefer hydrophobic surface for colonization [6].Here the surface is hydrophilic in nature, which is liable for accumulation of dead cells on the textured surface. Biofilm maturation is found to be inhibited by the presence of micro features.

4. Conclusions

The present work is focused on the study of antibacterial behavior of Ti6Al4V biomaterial surfaces under the influence of micro textures. The antibacterial properties of the polished and laser-textured samples are compared and analyzed. A comparative study of the performance of antibacterial performance of Ti6Al4V has been made by analyzing the bacterial adherence and visual identification of live/dead using the bacterial luminescence microscopic images. The surface roughness of Ti6Al4V increased significantly after laser texturing. Increased surface roughness decreases the area of contact between individual bacteria and textured surface. The water contact angle of textured specimen was greatly reduced by laser texturing and the surface became more hydrophilic. These effects significantly reduced the adhesion of *S.aureus* cells on laser-textured surfaces. Laser surface texturing is thus found to be a promising method for endowing dental and orthopaedic Titanium implants with antibacterial properties, reducing the risk of implant-associated infections

5. References

[1] C. Leyens and M. Peters 2003 *Titanium and Titanium Alloys Fundamentals and Applications*, (New York: Wiley–Interscience)

[2] A.G.Gristina1987 Biomaterial-centered infection: microbial adhesion versus tissue integration *Science* **237** pp.1588-95

[3] Carla Renata Arciola, Davide Campoccia, *et al.* 2012 Biofilm formation in Staphylococcus implants infections. A review of molecular mechanisms and implications for biofilm-resistant materials *Biomaterials* **33**, pp.5967-82

[4] Llinos G Harris, R .Geoff Richards 2006 Staphylococci and implant surfaces: a review *Injury* **37** pp.S3-S14

[5] Anna Krasowska1, Karel Sigler. 2014 How microorganisms use hydrophobicity and what does this mean for human needs *Cellular and infection microbiology* **4** pp.1-4

[6] Xuanyong Liu, Paul K. Chu, Chuanxian Ding. 2004 Surface modification of titanium titanium alloys and related materials for biomedical applications *Materials Science and Engineering* **47** pp.49-121

[7] Kumar G, Prabhu KN 2007 Review of non-reactive and reactive wetting of liquids on surfaces. *Adv.Colloid Interf. Sci.* **133** pp.61–89.

[8] D. Xiong, Y. Yang, Y. Deng, 2013 Bio-tribological properties of UHMWPE against surface modified titanium alloy, *Surface & Coating Technology* **228** S442-45.

[9] Chi-Wai Chana,Louise Carsonb, *et al.* 2017 Enhancing the antibacterial performance of orthopaedic implant materials by fibre laser surface engineering *Applied Surface Science* **404** pp. 67–81.

[10] A. Jemat, M. J. Ghazali, M. Razali and Y. Otsuka 2015 Surface Modifications and Their Effects on Titanium Dental Implants *Biomed Research International* **1** pp.1-8

[11] Donata Kuczyńska, Piotr Kwaśniak, *et al.* 2016 Laser surface treatment and the resultant hierarchical topography of Ti grade 2 for biomedical application *Applied Surface Science* **390**, pp.560-569.

[12] Alexandre Cunha and Anne-Marie-Elie, Laurent Plawinski, *et al.* 2015 Femtosecond laser surface texturing of titanium as a method to reduce the adhesion of Staphylococcus aureus and biofilm formation *Applied Surface Science* **360**, pp.485-493

[13] K.Subramani and R.T.Mathew 2012 Titanium Surface Modification Techniques for Dental Implants-From Micro scale to Nanoscale *Emerging Nanotechnologies in Dentistry* **2** pp.85-102

[14] X. Frank Walboomers and John A. Jansen 2001Cell and tissue behavior on micro-grooved surface *Odontology* **89** pp.2–11

[15] M.Järn, S.Areva and V.Pore, *et al.* 2006Topography and surface energy depended calcium phosphate formation on sol-gel derived \TiO_2 coatings *Langmuir* **22** pp.8209–8213

[16] L.C. Hsu, J. Fang, D.A. Borca-Tasciu, *et al.* 2013: 'Effect of micro-and nanoscale topography on the adhesion of bacterial cells to solid surfaces *Appl. Environ. Microbial* **79** pp. 2703–2712

[17] JaimeEsteban, RamónPérez-Tanoira, *et al.*: Microbiology for Surgical Infections (Academic press, Cambridge, 2014) pp.299-306

[18] M.Gross, S.E. Cramton, F. Götz, A. Peschel, 2001 Key role of teichoic acid net charge in Staphylococcus aureus colonization of artificial surfaces *Infect. Immun.* **69** pp.3423–3426

Acknowledgment
Authors gratefully acknowledge the financial support by APJ Abdul Kalam Technological University Kerala, India under the Research Seed Money (RSM) scheme.

Effect of Heat Input on Micro-Hardness and Shear Strength of Inconel 625 Hardfacing onto AISI 347 Steel pipes by GMAW Process

V.Rajkumar[1], T.V. Arjunan[2], Adithya Vasan[3] and Rajesh Kannan A[4]
[1, 2, 3] Department of Mechanical Engineering, CIET, Coimbatore, TN, India.
[4] Department of Mechanical Engineering, NIT, Tiruchirappalli, TN, India.

E-mail: **rajkmech42@gmail.com**

Abstract. Inconel 625 is one of the most accepted hardfacing material having good corrosion resistance properties at high temperature applications. Hardfacing is the deposition of a hard surface layer to the base metal by means of welding. The benefit of hardfacing is resistance to environmental stresses mainly corrosion at elevated temperatures. Thermal input plays a critical role for manufacturing good hardfaced components with respect to weld bead geometry, cooling rate, etc. In the present study, an L9 orthogonal array of input parameters were carried out with different heat input combinations using 98% Argon + 2% CO2 as shielding gas. Single layer hardfacing was done. The bead-on-pipe welding trials and macroscopic analysis were carried out to find the optimal welding parameters that yielded minimum dilution. The hardfaced layer thickness varies from 1.88mm to 2.58mm on substrate following optimum welding parameters. Visual inspection results reveal that most of welds of higher heat input conditions possess good quality. It is visually observed that width of weld increases with increase in welding heat input. It is noticed that in the range of 200-250 mm/min welding speed and 150A of welding current, good reinforcement and weld bead width at optimum penetration is observed in the experiments. Shear strength test and micro-hardness test were performed over test samples prepared from hardfaced part. Results revealed that both micro-hardness and shear strength of the hardfaced layer were more than that of base metal and values of both increased with increase in heat input. The Inconel 625 hardfaced layer exhibited 32 % higher hardness than the substrate material AISI 347.

1. Introduction

Hardfacing has become a major breakthrough for issues associated to applications for wear-resistant components, as it can increase the service life of engineered components. Hardfacing makes a new layer on the top surface of substrate with different compositions in which the deposited layer is harder than that of base material. Compared to other surfacing techniques, hardfacing has some distinctive advantages, such as enhancement of hardness and wear resistance of job surfaces, providing good bonding and favourable microstructure of hardfaced part. Hardfacing by welding has generally been used to improve the wear resistance in the agricultural, sugar, excavating and other industries to increase the hardness and wear properties of mechanical components [1]. Hardfacing is a method commonly used to treat surfaces subject to heavy wear and corrosion and for the manufacture of new components, their repair and their useful life for the most varied industries [2]. Various welding methods, such as metal inert gas welding (MIG), submerged arc welding (SAW), oxyacetylene gas welding (OAW) and shielded metal arc welding (SMAW) are widely used for depositing hard layers. The main difference between these methods are efficiency of the welding process, higher rate of deposition, dilution percentage on the substrate or base material and the filler material cost for laying hard deposits. The GMAW process is generally used to overhaul surfaces by hard surfacing as a result of its easy application, ability to weld at high densities of current and its potential to deposit hard layers of multiple filler material, specifically in repairing damaged parts of greater importance to the industries [3]. The demand of a joining process depends mostly on the operational considerations of the thermal distribution in the hard deposit which is beneficial for estimating the changes in metallurgical aspect and their effect on the mechanical integrity [4].

The selection of welding filler material and a welding process are first carried out in making the deposits by hardfacing [5]. Gopa et al. [6] investigated the percentage of dilution in the substrate by the hard deposit which alters the microstructure and their effect on hardness wear resistance of the hardfaced deposit. The microstructures and wear properties of the deposited layers of nickel-based hardfacings provide better abrasion wear resistance [7]. The studies on the single layer clad beads of Inconel 625 by SMAW produced low dilution and at lower current of 80A and wear resistance was enhanced under identical conditions of wear test [8]. The experimentations by Kotecki and Ogborn [9] shows that the most significant factor for wear resistance was the microstructure. The studies by John J. Coronado et al [1] states that the higher level of dilution by joining process reduced the wear characteristics of single layer hardfaced deposits and an increase in abrasion resistance was seen with multiple hardfacing layers applied one above other. Kesavan et al. [10] observed the metallurgical changes and wear performance of Colmonoy 5 hardfaced coating at high temperatures and noticed a decrease in wear loss and COF-coefficient of friction with increase in test temperatures. The investigations by Jin-Kim et al. [11] on nickel based (Deloro 50) deposits by hardfacing revealed higher wear loss at room temperature and suggested nickel based hard deposits for high temperature nuclear power plant applications with oxidative wear. In the present investigation, Inconel 625 filler wire is hardfaced on AISI 347 by GMAW process applying as shielding gas. Process parameters such as welding current and welding speed were selected in 98% Argon + 2% CO2 three levels, keeping welding voltage constant so that nine different heat input had been produced and nine weld bead samples were produced. Analyzing the weld bead geometry of the samples, 3 best samples with reference to bead geometry were selected and the corresponding process parameters were identified. Macro hardness testing and shear testing of hardfaced layer were carried out to check the effect of thermal input on mechanical behaviour of hardfaced layer.

2. Materials and Systems

The chemical composition of base metal and filler wire, i.e. AISI 347 austenitic stainless steel and ERNiCrMo-3 are shown in Table 1, and Table 2 respectively. ERNiCrMo-3 filler wire diameter is 1.2 mm. OTC Daihen FD-B6 model robotic MIG/MAG welding machine using DC source, with rotary positioner is shown in Fig. 1.

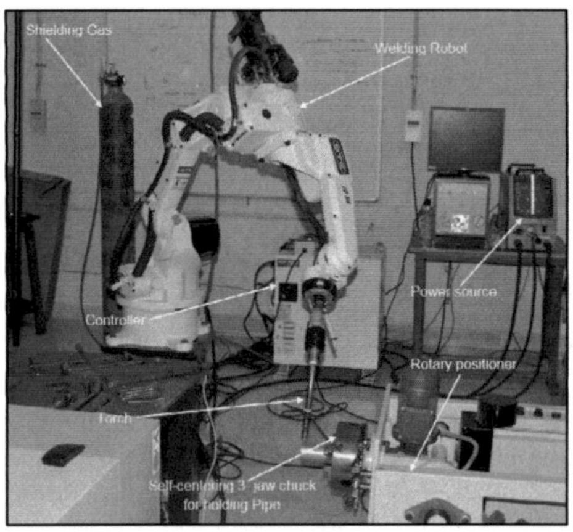

Fig. 1. Robotic welding setup for hardfacing

Table 1. Chemical composition of Base material (AISI 347 pipe)

Name of the element	C	Mn	Si	P	S	Cr	Mo	Ni	N	Fe
Composition in %	0.03	2.00	0.75	0.45	0.03	18.00	3.00	14.00	0.10	Balance

Table 2. Chemical composition of filler material (ERNiCrMo-3)

Name of the element	Cr	Co	Si	Mn	Fe	Mo	Nb	Other	Ni
Composition in %	21.00	0.90	0.40	0.40	5.00	9.00	3.50	0.90	Balance

3. 3. Experimental Procedure

3.1. Bead on plate experiments

Inconel 625 weld beads were hardfaced on AISI 347 base metal by GMAW process using 98% Argon + 2% CO2 as shielding gas. Welding Current and welding speed were taken in three levels such as 130 A, 140A, 150A and 200mm/min, 225mm/min, 250mm/min respectively so that nine number of samples were formed.

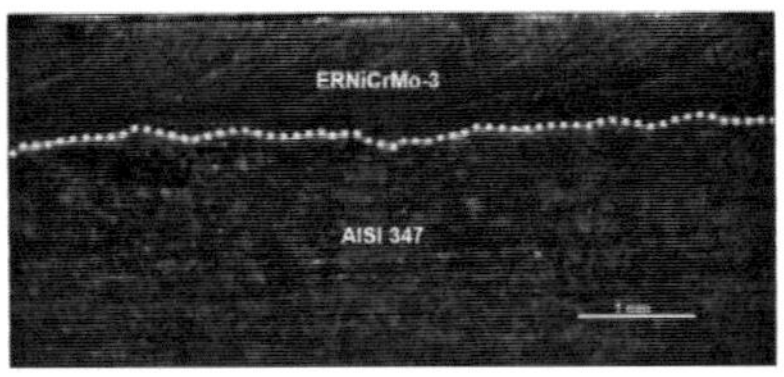

Fig. 2. Macrograph of hardfaced sample in longitudinal direction

Table 3. Heat input and input parameters for bead-on-plate trials

Sample No.	Welding Current, I (A)	Voltage, V (Volts)	Welding speed, S (mm/min)	Heat input (kJ/mm)
1	130	14.9	200	0.46
2	130	14.9	225	0.41
3	130	14.9	250	0.37
4	140	15.7	200	0.53
5	140	15.7	225	0.47
6	140	15.7	250	0.42
7	150	16.5	200	0.59
8	150	16.5	225	0.53
9	150	16.5	250	0.48

Welding voltage was kept constant during the experiment. The whole experiments were replicated twice. Equation (1) is used to find heat input, Q.

$$Q = \frac{V*I*60}{S*1000} * \eta \quad (1)$$

Where Q = Heat Input (kJ/mm), V = Voltage (V), I = Current (A), S = Welding Speed (mm/min) and η = Efficiency (In this experiment, it is taken as 0.8).

The heat input and other process parameters are shown in Table 3. The samples were cut in transverse direction by wire EDM using 0.18mm molybdenum wire, ground and etched with 2% nital solution and checked under Streuss weld expert. Weld bead geometry components like top reinforcement, bead width and depth of penetration were measured with the help weld expert microscope.

3.2. Hardfacing experiments
Single layer Inconel 625 hard deposits were performed on AISI 347 by GMAW process using 98% Argon + 2% CO2 as shielding gas. Three sets of heat inputs along with process parameters such as welding current and torch travel speed were selected on the basis of better results of weld bead geometry formation in previous experiments.

3.3. Micro-Vicker's hardness test
Micro-Vicker's hardness tests for the samples made were carried out by Matsuzawa MMT-X Vickers hardness tester by calculating the depth of penetration of the indenter into the specimen under fixed test conditions.

3.4. Shear test of hardfaced part
Shear strength of hardfaced sample is measured by a shear testing device. Shear strength is calculated by following formulae.

Shear strength = Shear load/ Area (2)

4. Results and Discussion

4.1. Results of bead-on-plate trials
Visual inspection results reveal that most of welds of higher heat input conditions possess good quality. It is visually observed that width of weld increases with welding heat input as expected. In all cases continuity in weld deposition is found. Spatters are varied from high range to low range with different heat input conditions. High spatter at lower current indicates the possibility of globular mode of metal transfer.

4.2. Weld bead characteristics of bead-on-plate trials
Table 4 shows the results of various weld bead geometry of bead-on-plate trials using 98% Argon + 2% CO2 as shielding gas. The observations from bead-on-plate trials clearly specifies that at higher input condition weld bead width, reinforcement and depth of penetration increases with increasing heat input. It is noticed that within the range of 200-250 mm/min welding speed and 150A of welding current, good reinforcement and weld bead width at optimum penetration is observed in both sets of experiments. Three heat input conditions (0.59 kJ/mm, 0.53 kJ/mm, and 0.48 kJ/mm) are selected for hardfacing based on visual inspection, weld bead geometry and their effects by thermal input from bead on plate trials.

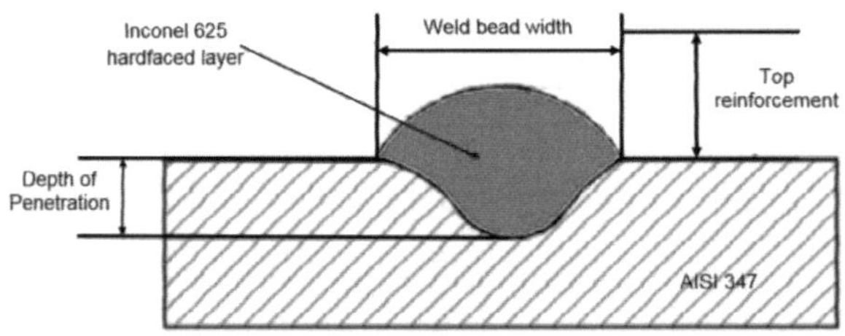

Fig. 3. Weld bead characteristics measured from macrographs

4.3. Micro-Vicker's hardness test

The three zones, on which the values were measured are the base metal, the deposited weld zone and the heat effected zone. Average hardness value of base material AISI 347 is 188 HV. Table 5 shows hardness values of three zones. From the result it is observed that hardness value is decreasing from hardfaced zone (237-255) to base metal (185-192). Hardness of the interface zone is 210-233. Fig.3 shows graphical representation of hardness at different zone.

Table 4. Measured values from macrographs of bead-on-plate trials

Sample No.	Welding Current, I (A)	Voltage, V (Volts)	Welding speed, S (mm/min)	Heat input (kJ/mm)	Top reinforcement (mm)	Weld bead width (mm)	Depth of penetration (mm)
1	130	14.9	200	0.46	2.21	3.51	1.92
2	130	14.9	225	0.41	2.04	3.59	1.97
3	130	14.9	250	0.37	1.88	3.64	2.06
4	140	15.7	200	0.53	2.43	4.09	2.05
5	140	15.7	225	0.47	2.29	4.15	2.12
6	140	15.7	250	0.42	2.17	4.28	2.19
7	150	16.5	200	0.59	2.76	4.57	2.18
8	150	16.5	225	0.53	2.58	4.72	2.35
9	150	16.5	250	0.48	2.49	4.98	2.59

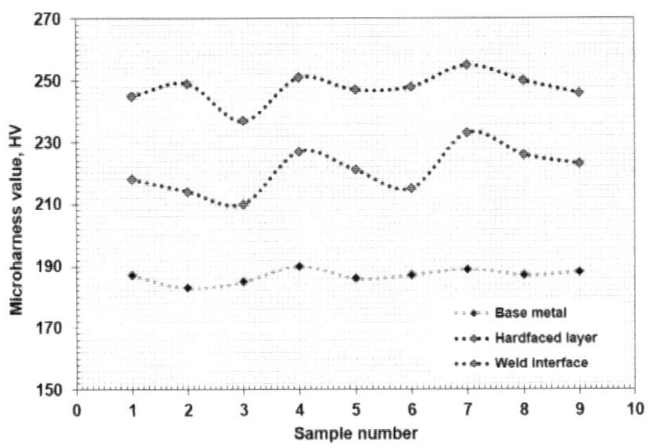

Fig. 4. Hardness at different zones of hardfaced part.

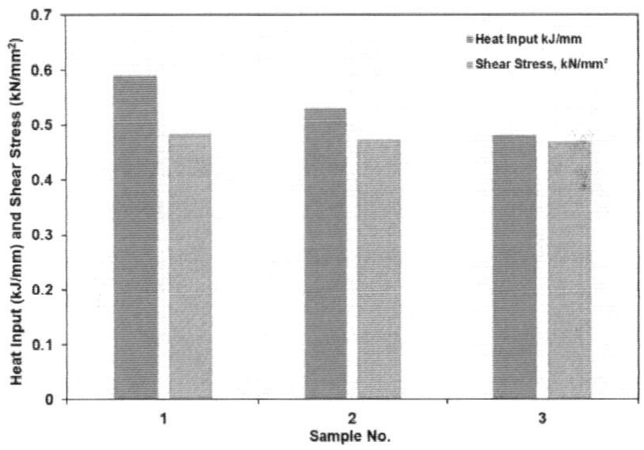

Fig. 5. Bar chart for heat input vs Shear stress of hardfaced samples

Table 5. Hardness values of bead on plate experimentations

Sample No.	Heat input (kJ/mm)	Hardness value on base metal (HV)	Hardness value on weld bead zone (HV)	Hardness value on the interface zone (HV)
1	0.46	189	245	218
2	0.41	192	249	214
3	0.37	188	237	210
4	0.53	189	251	227
5	0.47	187	247	221
6	0.42	190	248	215
7	0.59	187	255	233
8	0.53	185	250	226
9	0.48	189	246	223

4.4. Shear test

Table 6 shows the results of shear stress test on hardfaced samples. Shear strength of hardfaced samples are greater than base metal as standard ultimate shear strength of AISI 347 austenitic stainless steel at annealed condition is given by 430-460 MPa.

Table 6. Shear strength values of hardfaced layer

Sample No.	Heat Input (kJ/mm)	Shear area (mm2)	Shear force (kN)	Shear stress (kN/mm2)
1	0.59	45.43	21.94	0.483
2	0.53	38.06	17.96	0.472
3	0.48	33.92	15.87	0.468

Fig. 5 shows graphical representation of shear strength against heat input, which shows shear strength increases with increase in heat input within experimental domain. From the chart it is obtained that shear strength increases with increase of heat input. From the observations, it is seen that in all the specimens, shear occurs at the heat affected zone or at the base metal. In no case, shear occurs at weld bead. Inconel 625 filler wire contains alloying elements like Cr, Ni, Mo, Si, Mn, which are responsible to increase hardness and shear strength. The austenitic phase present in AISI 347 converts into retained austenite on cooling which has more mechanical strength than base metal.

5. Conclusion

From the results obtained from various tests performed it is observed from visual inspection that welds with higher input are found to be of good quality and high depth of penetration than lower heat input. The hardness of Inconel 625 hardfaced portion is found to be significantly higher than AISI 347 steel base pipe, as it has the presence of hardening elements like chromium, nickel, molybdenum, etc. Results of shear test clearly indicate that shear strength increases with increasing heat input. Shear strength of all the hardfaced samples is better than base metal. Comparing both sets of shear stress test, maximum shear stress is obtained in the present investigation at 0.59 kJ/mm heat input with 150A welding current, 16.5V weld voltage and 200 mm/min weld torch travel speed.

References

[1]. J. J. Coronado, H. F. Caicedo, A. L. Gómez, The Effect of Welding Process on Abrasive Wear Resistance for Hardfacing Deposits, Tribology International, 42 (2009), 745-749.

[2]. Agustín Gualco, Hernán G. Svoboda, Estela S. Surian and Luis A. de Vedia, Effect of welding procedure on wear behaviour of a modified martensitic tool steel hardfacing deposit, Materials & Design, 31 (2010), 4165–4173.

[3]. B Gulenc, N Kahraman, Wear Behaviour of Bulldozer Rollers Welded using a Submerged Arc Welding Process. Materials and Design, 24 (2003) 537–542.

[4]. D. V. Kiran, J. Cheon, N. Arif, H. Chung, S. J. Na, Three-dimensional Finite Element Modelling of Pulsed AC Gas Metal Arc Welding Process. International Journal of Advanced Manufacturing Technology, 86 (2016) 1453–1474.

[5]. R.J. Dawson, S. Shewchk, J.E. Pritchard. Selection and Use of Hardfacing Alloys. Weld. Journal, 61 (11) (1982), 15–23.

[6]. Gopa Chakraborty, N. Kumar, C.R. Das, S.K. Albert, A.K. Bhaduri, S.Dash, and A. K. Tyagi, Study on microstructure and wear properties of different nickel base hardfacing alloys deposited on austenitic stainless steel, Surface & Coatings Technology, 244 (2014), 180 - 188.

[7]. M. Morsy and E. El-Kashif, The effect of microstructure on high-Stress abrasion resistance of Fe-Cr-C hardfacing deposits, Welding in the World, 58 (2014), 491-497.

[8]. S. S. Singh and A.S. Shahi, Metallurgical, wear and fatigue performance of Inconel 625 weld claddings, Journal of Materials Processing Technology, 233 (2016), 1- 8.

[9]. D. J. Kotecki and J. S. Ogborn, Abrasion Resistance of Iron-Based Hardfacing Alloys. Welding Journal, 74(8) (1995), 269-288.

[10]. D. Kesavan and M. Kamaraj, The microstructure and high temperature wear performance of a nickel base hardfaced coating, Surface & Coatings Technology, 204 (2010), 4034-4043.

[11]. Seon Jin-Kim, Jun Ki-Kim, Effects of temperature and contact stress on the sliding wear of Ni-base Deloro 50 hardfacing alloy, Journal of Nuclear Materials, 288 (2001) 163-169.

Journal of Physics: Conference Series 1355 (2019) 012030 doi:10.1088/1742-6596/1355/1/012030

Establishing a relationship between machining parameters and output responses for cutting stainless steel ICSS 1218 by Wire EDM

Dr Rajeev N[1], Jishnu J[2]
[1] Associate Professor, Department of Mechanical Engineering, NSSCE Palakkad
[2] Mtech Student, Department of Mechanical Engineering, NSSCE Palakkad

E-mail: rajeevmaliyil@gmail.com

Abstract. Stainless steel ICSS 1218 have large potential in the aerospace sector due to their inherent characteristic and advantageous properties like strength, ductility, weldability and corrosion resistance. The Wire Electric Discharge Machining (Wire EDM) is a thermal machining process capable of accurately machining parts with high hardness or complex shapes. One of the statistical softwares Minitab 17 was used for conducting the analysis and develop a mathematical model of the relation connecting the input parameters pulse on time, flushing discharge, and wire feed rate and output responses surface roughness and material removal rate. This model is found reasonably accurate. Experimentation was conducted in a series of tests called *runs*, in which changes are made in the input variables in order to identify the reasons for changes in the output response.

1. Introduction

Machining is an important manufacturing process to obtain various shapes of the components. Over the decades conventional machining has sufficed the requirement of industries. But new exotic work materials as well as innovative geometric design of products and components are putting pressure on capabilities of conventional machining processes to manufacture the components with desired tolerances economically. This has lead to the development of non-conventional manufacturing processes in the industry as efficient and economic alternatives to conventional ones. Wire Electrical Discharge Machining (WEDM) is one of such nonconventional manufacturing processes. As the metal removal is based on complex thermal and chemical phenomena, in order to understand its effects on the machined components investigation has been made in this research. Wire-electrical discharge machining (WEDM) is a non-traditional machining process in which a pulsed voltage difference between a wire electrode and a conductive work piece initiates sparks which erode work piece material. Removing material in such away that it often advantageous when the work piece material would be difficult to machine with traditional machine tools due to high strength, hardness, toughness, etc.

1.1 Wire Electric Discharge Machining (WEDM)

It is probably the most exciting and diversified machining process developed for this industry in the recent years. Common materials such as steel, aluminum, copper, and graphite, to exotic space-age alloys including hastelloy, waspalloy, inconel, titanium, carbide, polycrystalline diamond compacts, conductive ceramics etc. are popularly processed by WEDM. "For the first time, we have indigenously developed and processed the aerospace grade stainless steel in Salem. The steel slabs were provided by ISRO. The end result was development of import-substitute exotic Russian grade (ICSS-1218) slabs that are used in the construction of the rocket launch vehicles. It will be shipped to ISRO's Liquid Propulsion System Centre (LPSC) soon," claimed by Salem Steel technical team. Seeking the help of ISRO scientists, a SAIL team processed electro-slag re-melted (ESR) and forged slabs. The final product conforms to ICSS-1218 that is equivalent to SS 321 grade stainless steel slabs. "We have been regularly processing stainless steel for various satellite launch vehicles, including the Polar Satellite Launch Vehicles that carried the Chandrayaan and Mangalyaan missions into the space," as per ISRO press release.

2 . Literature review

A through literature review was done to amass the knowledge about the recent advances in the area of WEDM and the following observations were made. Thomas R Newton [1] conducted an experiment to determine the main EDM parameters which contribute to recast layer formation in Inconel 718. It was found that average recast layer thickness increased primarily with energy per spark, peak discharge current and current pulse duration. Over the range of parameters tested , the recast layer was observed to be between 5 and 9 µm in average thickness, although highly variable in nature.

Sasan Khalaj Amineh [2] explored the feasibility of removing the recast layer formed on aluminium alloy cylindrical specimens machined by wire electric discharge machining by using magnetic abrasive finishing (MAF). This investigation demonstrated that MAF process can improve the quality of WEDM machined surfaces effectively by removing the recast layer. This work studied the effect of some parameters , ie, linear speed, working gap, abrasive particle size, and finishing time on surface roughness and recast layer thickness using full factorial analysis. Three level full factorial technique was used as design of experiments for studying the selected factors.

Bhosale Sachin [3] conducted an experiment to determine the main EDM parameters which contribute to recast layer thickness and micro hardness of MDN 250 alloy. Experimentation has been done using Taguchi's L9 orthogonal array. Each experiment was conducted under different combinations of pulse on time, pulse off time, and wire feed. The optimum machining parameter combination was obtained by using the analysis of signal-to-noise (S/N) ratio, analysis of means and analysis of variance (ANOVA) in Minitab 15 software.

B Naga Raju [4] made an attempt to study the effect of WEDM parameters like pulse on time, pulse off time, and peak current on surface roughness and material removal rate in aluminium metal matrix composites. Experimentation was conducted in a series of runs, in which changes are made in the input variables in order to identify the reasons for changes in the output response using Response Surface Methodology.

Rajesh Kumar Bhuyan [5] explored the effect of process parameters such as pulse on time, peak current, and flushing pressure on material removal rate, tool wear rate, and surface roughness during EDM of Al-SiC12% MMC. The experiment was followed by Central composite design (CCD) method under different combination of process parameters. Response surface methodology was used to develop the mathematical model and to correlate the process parameters with the response. Further, analysis of variance technique was employed to check the significance of the model and confirmation test is conducted to compare the experimental data with the predicted data to identify the effectiveness of the proposed method.

It was observed from the literature review that researchers have focussed on different combinations of input parameters and output responses for their studies, It was observed that there is a need to focus on a more sets of input parameters and explore their effect on multiple output responses, Therefore, the present paper investigates the effect of input parameters like pulse on time, flushing discharge, and wire feed rate on the output responses, surface roughness and material removal rate.

3. Experimental work

The details of the Work piece selected for the study is as follows. Stainless Steel ICSS 1218 of size 10mm x 10mm x 20mm. 27 similar pieces were prepared for the experiments, since three parameters with three levels were considered. A total 27 combinations of parameters were selected (3^3). The fig 1 shows the work piece. Choices of the factors that need to be varied in the experiment should be made. Control parameters are parameters which are directly affecting the response variables. From the available literatures, it was observed that the main parameters that affects Wire EDM are pulse on time, pulse off time, wire feed rate, and peak current. Flushing discharge also found to have significant impact on surface roughness.

Fig 1 27 specimens with size 10 mm x 10 mm x20 mm

Therefore, the test parameters selected for this work are pulse on time, flushing discharge, and wire feed rate. The parameter level selected as three. Proper selection of the cutting parameters was carried out based on the data available from the literatures and operator's opinion. The experiments were conducted at pulse on time 108, 110, 112 µs, flushing discharge 8, 10, 12 L/min, and wire feed rate 4, 6, 8 m/min respectively, as shown in Table 1.

Table 1 Level of Input parameters

Parameters	Symbol	Levels		
Pulse on time (µs)	T_{on}	108	110	112
Flushing discharge (L/min)	F_d	8	10	12
Wire feed rate (m/min)	W_f	4	6	8

4. Equipments used

4.1 CNC Wire EDM

Computer numerical control (CNC) machines are widely used in manufacturing industry. The CNC machine comprises of the computer in which the program is fed for cutting of the metal of the job as per the requirements. The main function of CNC machines is to remove some of the metal so as to give it proper shape such as round, rectangular, etc. ELECTRONIC ULTRACUT S1 CNC WIRE EDM was used for machining in this experiment. Figure 2 shows the ELECTRONICA ULTRACUT S1 CNC machine.

Surface roughness and material removal rate are the performance measures taken for this work. Inspection and assessment of surface roughness of machined work pieces was carried out by Taylor/Hobson Precision form Surtronic 3+. Material removal rate can be calculated by taking the weight of the work piece before and after machining and dividing the difference of weight by density of the material and machining time. The figure 3 shows the surface roughness tester.

Fig 2 Electronica Ultracut S1 CNC Wire EDM

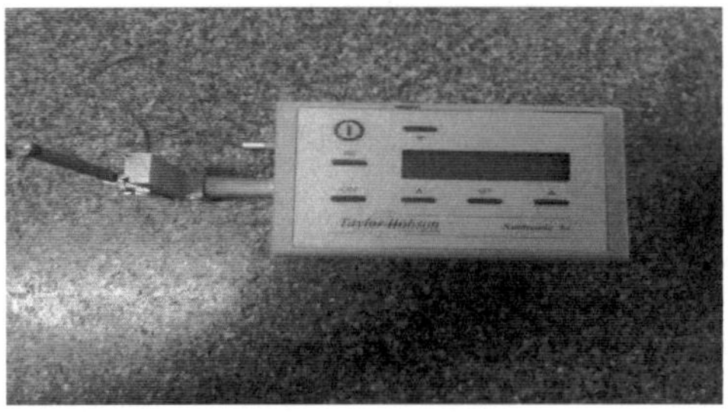

Fig 3 Taylor/ Hobson precision surface roughness tester

4.2 Material Removal Rate Calculation

Material removal rate is defined as the amount of material removed per unit time. Material removal rate is calculated as the ratio of amount of material removed to the time taken for machining.

$$MRR = (W_b - W_a)/\rho \times t \quad [7]$$

where

W_b – weight of the specimen before machining
W_a – weight of the specimen after machining
ρ – density of the material (7.92 g/cm^3)
t – machining time

4.3 Data Collection

Stainless Steel ICSS 1218 specimen were prepared first with the help of conventional milling machine. Twenty seven number of samples of same material and same dimensions were made. After that the weight of each samples have been measured accurately with the help of a weighing machine. Then using different levels of the process parameters specimens were machined in CNC Wire EDM accordingly. After machining, the surface roughness have been measured with the help of Taylor/Hobson Precision tester. The results were tabulated as shown in Table 2.

Table 2 Experimental Observation

Exp no	Pulse on time (µs)	Flushing discharge(L/min)	Wire feed rate (m/min)	R_a (microns)	MRR (mm^3 /min)
1	110	8	4	3.07	7.72
2	110	8	6	3.12	7.46
3	110	8	8	3.04	7.18
4	110	10	4	2.82	7.21
5	110	10	6	2.92	7.36
6	110	10	8	2.76	6.62
7	110	12	4	2.66	6.91
8	110	12	6	2.9	7.89
9	110	12	8	3.02	7.48
10	108	8	4	2.98	5.62

11	108	8	6	2.3	5.23
12	108	8	8	2.56	5.11
13	108	10	4	2.62	5.82
14	108	10	6	2.52	5.989
15	108	10	8	2.78	5.22
16	108	12	4	2.7	5.68
17	108	12	6	2.68	5.86
18	108	12	8	2.4	6.05
19	112	8	4	2.7	11.72
20	112	8	6	2.76	7.91
21	112	8	8	2.78	7.69
22	112	10	4	2.92	8.69
23	112	10	6	2.78	7.66
24	112	10	8	2.84	8.02
25	112	12	4	3.1	8.24
26	112	12	6	3.08	8.14
27	112	12	8	2.74	7.63

5 . Data Analysis

5.1 Multiple Regression Model

The relationship between a set of independent variables x and the response variable y is determined by a mathematical model called regression model. In general, a multiple-regression model with q independent variable takes the form of

For second degree model

$$y = \beta_0 + \sum_{i=1}^{k} \beta_i x_i + \sum_{i<j}\sum \beta_{ij} x_i x_j + \sum_{i=1}^{k} \beta_{ii} x_i^2 + \epsilon \quad [6]$$

Where y is the corresponding response

Mathematical relationship between the input parameters and surface roughness

For second degree

$R_a = -551 + 10.21 \times T_{on} - 1.89 \times F_d - 0.55 \times W_f - 0.0469 \times T_{on}^2 + 0.0093 \times F_d^2 + 0.0051 \times W_f^2 + 0.0154 \times T_{on} \times F_d + 0.0042 \times T_{on} \times W_f + 0.0015 \times F_d \times W_f$. The data analysis was done by Minitab 17 and the results were tabulated as shown in Table 3.

Table 3 Fitted value compared with experimental value

Exp no	Pulse on time (μs)	Flushing discharge(L/min)	Wire feed rate (m/min)	R_a(microns)	$R_a$predicted	% error
1	110	8	4	3.07	2.9862037	2.7295
2	110	8	6	3.12	2.9237037	6.2915
3	110	8	8	3.04	2.9023148	4.5291
4	110	10	4	2.82	2.9414814	-4.3078
5	110	10	6	2.92	2.8848148	1.2049
6	110	10	8	2.76	2.8692592	-3.9586
7	110	12	4	2.66	2.9712037	-11.699
8	110	12	6	2.9	2.920370	-0.7024
9	110	12	8	3.02	2.9106481	3.6209
10	108	8	4	2.98	2.7567592	7.4913
11	108	8	6	2.3	2.6775925	-16.417
12	108	8	8	2.56	2.6395370	-3.1069
13	108	10	4	2.62	2.6503703	-1.1591
14	108	10	6	2.52	2.5770370	-2.2633
15	108	10	8	2.78	2.5448148	8.4598
16	108	12	4	2.7	2.6184259	3.0212
17	108	12	6	2.68	2.5509259	4.8161
18	108	12	8	2.4	2.5245370	-5.1890
19	112	8	4	2.7	2.8400925	-5.1886
20	112	8	6	2.76	2.7942592	-1.2412
21	112	8	8	2.78	2.7895370	-0.3430
22	112	10	4	2.92	2.8570370	2.1562
23	112	10	6	2.78	2.8170370	-1.3322
24	112	10	8	2.84	2.8181481	0.7694
25	112	12	4	3.1	2.9484259	4.8894

26	112	12	6	3.08	2.9142592	5.3811
27	112	12	8	2.74	2.9212037	-6.6132

The following plots were generated by the software and they provide more insights in to the behavior of selected parameters.

5.2 Normal probability plot for Ra

The normal probability plot as shown below represents clear pattern indicating that all the factors and their interaction given in are affecting the Ra

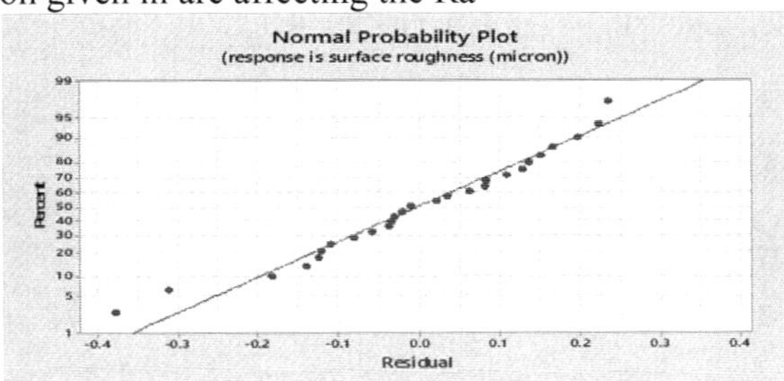

Fig 4 Normal Probability Plot

5.3 Versus Fits for Ra

The below graph shows the standardized residual Vs fitted value for surface roughness.

This indicate that the maximum variation which shows the high correlation that exists between fitted values and observed values.

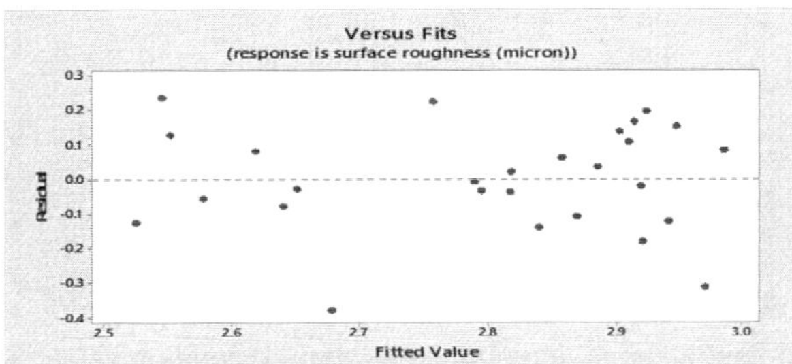

Fig 5 Versus Fits

The results were further analysed with ANOVA and F test was performed to check the adequacy of the model. The result obtained is shown in the Table 4, which shows that most influencing parameter that affect surface roughness is pulse on time. A P-value less than 0.05 indicate that the performance of the parameter is statistically significant (Kothari,2009).

Table 4 Results analysed with ANOVA and F test

Source	Sum of squares	DF	Mean square	F value	P value
Model	055451	9	0.061612	1.72	1.60
T_{on}	0.25902	1	0.259200	7.24	0.015
F_d	0.00005	1	0.000050	0.00	0.971
W_f	0.02347	1	0.023472	0.66	0.429
T_{on}^2	0.21156	1	0.2111563	5.91	0.026
F_d^2	0.00831	1	0.008313	0.23	0.636

W_f^2	0.00254	1	0.002535	0.07	0.793
$T_{on} \times F_d$	0.04563	1	0.045633	1.27	0.275
$T_{on} \times W_f$	0.00333	1	0.003333	0.09	0.764
$F_d \times W_f$	0.00041	1	0.000408	0.01	0.916

For second degree

$$MRR = -1135 + 18.7 \times T_{on} + 9.39 \times F_d + 8.89 \times W_f - 0.0747 \times T_{on}^2 + 0.0603 \times F_d^2 + 0.0224 \times W_f^2 - 0.1029 \times T_{on} \times F_d - 0.0952 \times T_{on} \times W_f + 0.1127 \times F_d \times W_f$$

The results were tabulated and value were obtained as shown in Table 5.

Table 5 Fitted value compared with experimental value

Exp no	Pulse on time (μs)	Flushing discharge(L/min)	Wire feed rate (m/min)	MRR(microns)	MRRpredicted	% error
1	110	8	4	7.72	8.340537037	-8.038
2	110	8	6	7.46	7.432925926	0.3629
3	110	8	8	7.18	6.704425926	6.6235
4	110	10	4	7.21	7.550703704	-4.725
5	110	10	6	7.36	7.093925926	3.6151
6	110	10	8	6.62	6.816259259	-2.964
7	110	12	4	6.91	7.243314815	-4.823
8	110	12	6	7.89	7.23737037	8.2716
9	110	12	8	7.48	7.410537037	0.9286
10	108	8	4	5.62	5.853481481	-4.154
11	108	8	6	5.23	5.326703704	-1.849
12	108	8	8	5.11	4.979037037	2.5628
13	108	10	4	5.82	5.475314815	5.9224
14	108	10	6	5.989	5.39937037	9.8452
15	108	10	8	5.22	5.502537037	-5.412
16	108	12	4	5.68	5.579592593	1.7677
17	108	12	6	5.86	5.954481481	-1.612
18	108	12	8	6.05	6.508481481	-7.578
19	112	8	4	11.72	10.2297037	12.715

20	112	8	6	7.91	8.941259259	-13.03
21	112	8	8	7.69	7.831925926	-1.845
22	112	10	4	8.69	9.028203704	-3.891
23	112	10	6	7.66	8.190592593	-6.926
24	112	10	8	8.02	7.532092593	6.0836
25	112	12	4	8.24	8.309148148	-0.839
26	112	12	6	8.14	7.92237037	2.6735
27	112	12	8	7.63	7.714703704	-1.110

5.4 Normal probability plot for MRR

The normal probability plot as shown below represents clear pattern indicating that all the factors and their interaction given in are affecting the MRR

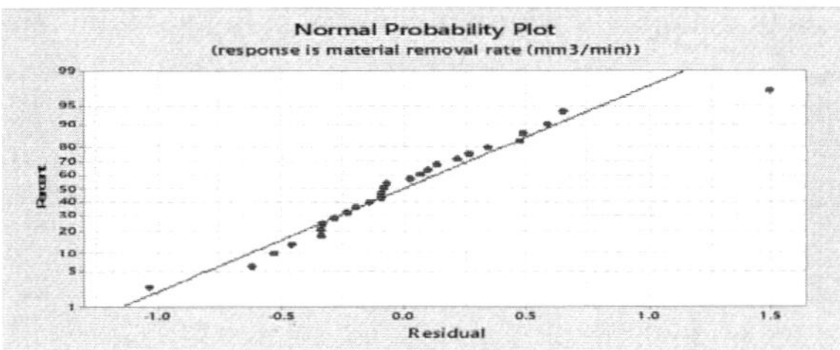

Fig 6 Normal Probability Plot

5.5 Versus Fits for MRR

The below graph shows the standardized residual Vs fitted value for MRR.

This graph also indicate the maximum variation which shows the high correlation that exists between fitted values and observed values.

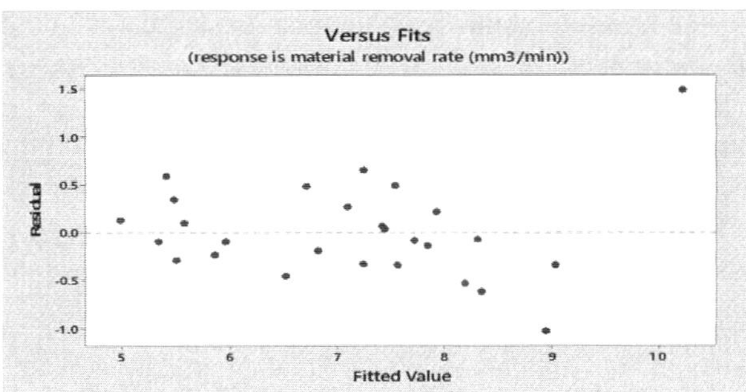

Fig 7 Versus Fits

The results were further analysed with ANOVA and F test was performed to check the adequacy of the model. The result obtained is shown in the Table 6, which shows that most influencing parameter that affect MRR is pulse on time, wire feed rate and interactions of pulse on time, flushing discharge and wire feed rate. A P-value less than 0.05 indicate that the performance of the parameter is statistically significant (Kothari,2009).

Table 6 Parameter that affect MRR

Source	Sum of squares	DF	Mean square	F value	P value
Model	44.8051	9	4.9783	13.58	0.000
T_{on}	35.0591	1	35.0591	95.61	0.000
F_d	0.1721	1	0.1721	0.47	0.503
W_f	2.4273	1	2.4273	6.62	0.020
T_{on}^2	0.5362	1	0.5362	1.46	0.243
F_d^2	0.3491	1	0.3491	0.95	0.343
W_f^2	0.0481	1	0.0481	0.13	0.722
T_{on} x F_d	2.0336	1	2.0336	5.55	0.031
T_{on} x W_f	1.7404	1	1.7404	4.73	0.044
F_d x W_f	2.4390	1	2.4390	6.65	0.020

6 . Results and Conclusions

Experiments were done successfully with considered input parameters and was evaluated using appropriate statistical techniques. Mathematical equation have been developed showing the relation between input parameters and responses with the help of Regression analysis. The normal probability plot and Versus Fits graphs were plotted successfully according to the research objectives. The association between input parameters and output responses are observed in literature. These associations were further consolidated using the regression models. Thus, the most important input parameters which influence MRR and Surface roughness during the WEDM process of ICSS 1218 steel were evaluated.

7.References

[1] Thomas R Newton, Shreyas N Melkote and Thomas R Watkins "Investigation of the effect of process parameters on the formation and characteristics of recast layer in wire EDM of Inconel 718", Journal of Material science and Engineering Elsevier (2009).

[2] Sasan KhalajAmineh, Alireza Fadaei Teharani and Aminollah Mohammadi "Improving Surface Quality in WEDM specimen by removing recast layer using MAF method",International Journal of Advanced Manufacturing Technology (2012).

[3] Bhosale Sachin, Sheldge Shrinivas, Bhagwat Vishal"Study and Analysis of Recat layer formation during WEDM process", International Journal of Innovative Research in science Engineering and Technology , (2015).

[4] B Naga Ragu, M Raja Roy, S Rajesh, K Ramji "Optimization of machining parameters for cutting AMMC's on Wire cut electric discharge machining using RSM" International Journal of Engineering Trends and Technology vol 23 (2015).

[5] Rajesh Kumar Bhuyan, B C Routara, Arun kumar Parida and A K Sahoo"Parametric Optimization of Al-SiC12% MMC machining by EDM" International Design and Research Conference (2014).

[6] Kothari R, "Research Methodology" New age publications (2009).

[7] Dr Rajeev N and Aswathy V G, "Effect of Machining Parameters on Surface roughness, Material removal rate and Roundness error during the wet turning of Ti-6Al-4V alloy"(2015)

Helmholtz resonator based metamaterials for sound manipulation

K Mahesh[1], R S Mini[2]
[1]M-Tech Student, Department of Mechanical Engineering, College of Engineering Trivandrum, Thiruvananthapuram-16, Kerala, India
[2] Assistant Professor, Department of Mechanical Engineering, College of Engineering Trivandrum, Thiruvananthapuram-16, Kerala, India

E-mail: maheshnilambur2012@gmail.com
E-mail: miniranjithrs@gmail.com

Abstract: Metamaterials are tailored artificial structures that received wide attention now a days due to their extraordinary properties like negative density, negative bulk modulus, negative permeability, negative refractive index etc.. Shape, geometry, size, orientation and arrangement of the basic units in metamaterials give them their smart properties capable of manipulating electromagnetic and acoustic waves by blocking, absorbing, enhancing, or bending waves to achieve benefits that go beyond what is possible with conventional materials. Acoustic metamaterial with Helmholtz resonators (HRs) is considered in this study. The behaviour of acoustic waves in the presence of array of Helmholtz resonators in series and parallel arrangement is studied numerically using finite element method. Series arrangement of Helmholtz resonators resulted in the extraordinary transmission of acoustic waves of certain frequencies and parallel arrangement resulted in the focusing of sound.

1. Introduction

Acoustic-wave propagation is controlled by mass density and bulk modulus of a material. In conventional materials both of these properties exhibits positive value and cannot be easily changed because the material properties are correlated with the chemical composition and bonding structures of the constituent atoms. However, if materials are made up of meta-atoms of sub wavelength scale that can enhance acoustic–matter interaction and hence it is possible for wave properties to obtain values that are not observed in nature. These are known as acoustic metamaterials and they can manipulate and control acoustic waves in ways that are not possible by conventional materials. Metamaterials with zero, or even negative refractive index offer new possibilities for acoustic imaging, lensing and controlling of acoustic waves at sub wavelength scales.

Generally metamaterials are useful for acoustic applications due to their extraordinary ability to manipulate sound and in particular they are useful for ultrasonic imaging. The conventional acoustic imaging using ultrasound exhibits lower resolution. This is due to the quickly diminishing evanescent waves which carry the sub wavelength attributes of materials. To succeed over this scenario, an acoustic super lens was proposed by Pendry et al. [1], which offered high resolution imaging of an object. This super lens is based on focusing the acoustic wave and retrieving the evanescent wave data through a flat negative-index slab. The negative bulk modulus system was experimentally demonstrated by Fang et al. [2]. They conducted an underwater ultrasonic experiment to study the transmission behavior of

Helmholtz resonators in series. Koju et al. [3] numerically demonstrated that a rigid barrier embedded with a Helmholtz Resonator can result in the extraordinary transmission of sound and it solely depends on the structure of the resonator. When acoustic waves pass through holes of sub wavelength scale, extraordinary acoustic transmission (EAT) was observed by them at resonant frequencies. They also created an acoustic lens using an array of Helmholtz resonators. Later the EAT through a solid barrier with an embedded Helmholtz resonator is experimentally demonstrated by Crow et al. [4]. From experiments they measured the amplitude and phase of transmitted wave. Maznev et al. [5] developed a metamaterial made up of soda cans that can focus sound at resonance frequency. They observed high transmission at the Helmholtz resonance frequency of the metamaterial.

Acoustic superlens was designed by Yang et al. [6] using Helmholtz-resonator-based metamaterials that can focus sound over a broad band of frequency with high resolution. They also conducted experiments which produce high resolution images of a double slits over a frequency range of 570 to 650 Hz. Wang et al. [7] developed an acoustic lens using tapered labyrinthine unit cells with sub wavelength thickness, planar profile, and broad operation bandwidth.

Later Amireddy et al. [8] experimentally demonstrated deep sub wavelength ultrasonic imaging of defects in metallic samples with a resolution of $\lambda/25$ using holey-structured metamaterial lenses. In the same manner, Laurety et al. [9] developed an acoustic lens by using additive 3D printing technique. A gradient acoustic metasurface is designed by Lan et al. [10] to manipulate acoustic wave front freely. High resolution imaging was achieved by them by using metasurface made up of series of unit cells.

2. Methodology

Sound transmission characteristics of array of HR embedded acoustic metamaterial is studied using finite element software, COMSOL multiphysics. Enhanced transmission of sound was observed by Koju et al. when a Helmholtz resonator was embedded within a rigid barrier. Extra ordinary transmission occurs at the resonance frequency and the transmitted waves goes through a phase change between 0 and π. In this study, influence of arrangement of double neck HR (necks on both sides of the cavity) arrays, in series and parallel modes within a rigid barrier is considered for the analysis. The HR system consists of two necks of length, l, and cross-sectional area A. The two necks are protruding from a closed cavity of volume, V, as shown in figure 1(a). It is similar to a simple spring- mass system because the air in the neck acts as a mass and the air in the cavity act as a spring. The resonant frequency of the double necked Helmholtz resonator system is given by,

$$f_r = \frac{v}{2\pi}\sqrt{\frac{2A}{L'V}} \tag{1}$$

where v is the speed of sound in air and L' is the corrected neck length. The corrected neck length, L', is longer than the actual neck length, l, and it varies with the radius of neck opening, r. $L'$is given by,

$$L' = l + 1.5r. \tag{2}$$

For validation, two configurations of HRs used by Koju et al [4] was simulated using COMSOL multiphysics and the results are compared. Each of them consists of a double necked HR embedded in a reflecting barrier. The resonant frequency, determined for each case using equation (1) is same. Each configuration of HR differed by the arrangement of necks. In the first case, the two necks are in line where as in the second case the necks are at offset position (figure 1(b)). The amplitude and phase of the transmitted waves is extracted using COMSOL multiphysics. For simulations a 2.6 m long waveguide was modeled with the HR positioned at the center of the waveguide. The geometric parameters used for creating the model is given in Table 1. The wave guide along with the resonator is shown in figure 1(c).

Table1. Geometrical parameters of model

Model Parameters	
Cavity length	0.02 m
Cavity diameter	0.1 m
Neck length	0.05 m
Neck radius	0.0079 m
Waveguide length	2.6 m
Waveguide radius	0.05m

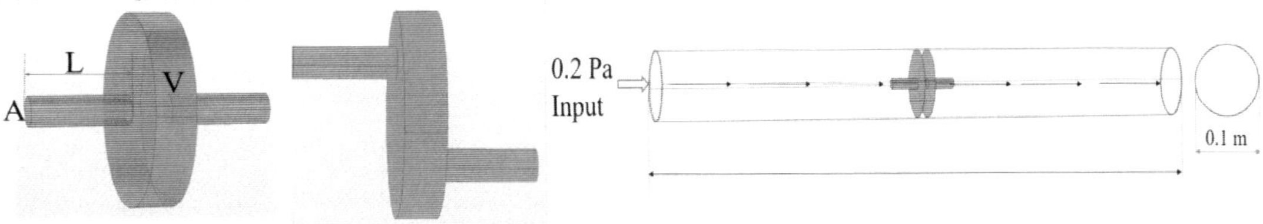

(a) (b) (c)

Figure 1- Schematic view of a) Helmholtz resonator with straight arrangement of necks. b) Helmholtz resonator with offset arrangement of necks c) waveguide with Helmholtz resonator embedded

In the simulations, the walls of HR and waveguide are selected as sound hard boundary walls. The simulation is carried out in a discrete frequency range of 200 Hz to 500 Hz in steps of 10 Hz. For all the simulations the highest frequency of wave chosen is well below the cut-off frequency, f_c

$$f_c = \frac{0.92v}{\pi d} \tag{3}$$

where d is the diameter of waveguide.

A plane acoustic wave of 0.2 Pa amplitude is given as the input at the left end of the waveguide, and the right end of the waveguide is set as a non-reflecting boundary (matched boundary condition), from which output is taken. The amplitude and phase of the output signal is determined from the complex transmission function by dividing the complex output signal obtained from the waveguide in the presence and absence of Helmholtz resonator. For the finite element simulation free tetrahedral element is used as the mesh element with an element size of 0.015m.

At each frequency, the output from an empty wave guide is taken as a reference signal ($S_{ref}(f)$). Then at same set of frequencies the output signal from the waveguide with HR is taken ($S_{HR}(f)$). T (f) is the complex transmission function obtained from the ratio,

$$\mathrm{T(f)} = \frac{S_{HR}(f)}{S_{ref}(f)} \tag{4}$$

The magnitude of transmission function represents the transmission amplitude at each frequency whereas the phase of transmission function gives the relative phase delay. For both configurations of Helmholtz resonator (offset and straight arrangement of necks) the graph between transmission and frequency is plotted in figure 2. The results obtained for Koju et al [4] is also plotted in the figure. As expected peak frequency is obtained at 350 Hz and it is very much similar to the theoretical value (349 Hz). From this it is clear that the EAT of metamaterial is solely based on structural parameters (cavity length, cavity radius, neck radius, neck length) of HR embedded barrier and resonance frequency. It does not depend up on the orientation of necks and line of sight of acoustic waves. By this validation we confirmed that our numerical model is correct. So we expanded our study to different arrangement of Helmholtz resonators.

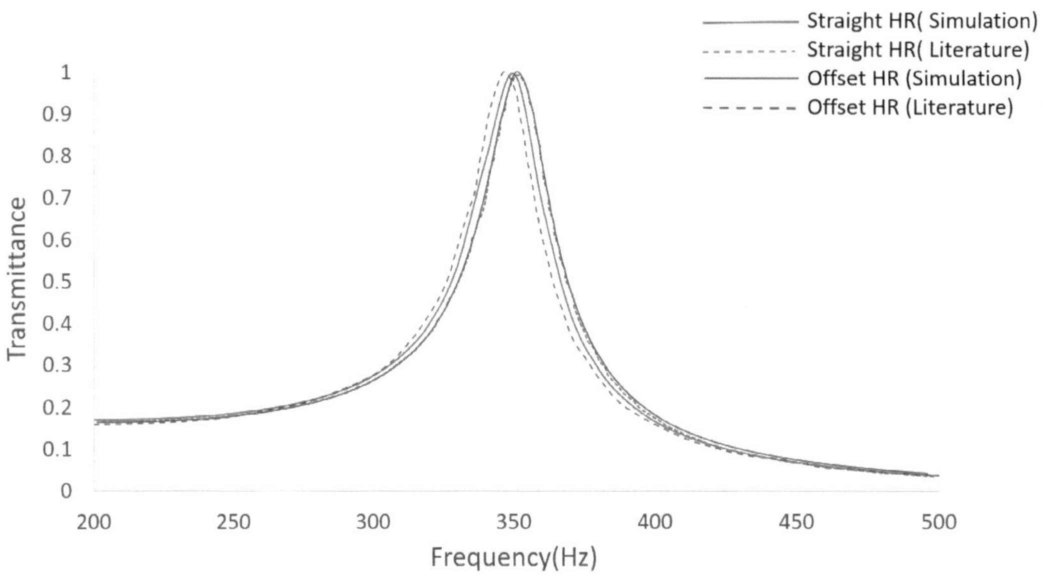

Figure 2- Comparison between simulation and literature results of normal HR arrangement and offset HR arrangement.

2.1. Horizontal array of Helmholtz resonators with normal arrangement and offset arrangement of necks

In order to study the effect Helmholtz resonator array on the sound transmission characteristics, seven HRs are arranged in series for both normal (figure 3 (a)) and offset placement of necks (figure 3 (b)). Similar to the earlier study a plane acoustic wave of 0.2 Pa amplitude is given as the input at the left end, where the at the right end is set as a matched boundary, from which output is taken. The transmittance through wave guide is measured using the equation (4). The simulation is carried out in a discrete frequency range of 200 Hz to 500 Hz in steps of 10 Hz.

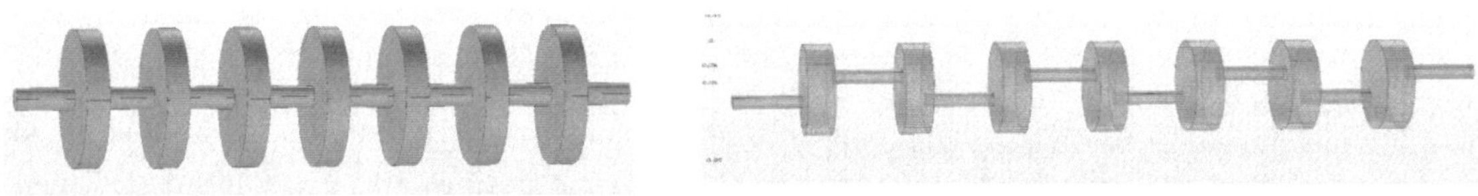

(a) (b)

Figure 3- Helmholtz resonators for a) straight arrangement of necks b) offset arrangement of necks

2.2. Vertical arrangement of Helmholtz resonators

The influence of vertical arrangement of Helmholtz resonators is studied by placing seven Helmholtz resonators in parallel mode in a rectangular waveguide. The cross sectional dimensions of the waveguide is 0.088 m x 0.62 m. All seven HR has equal cavity volume of $1.57\times 10^{-4}m^3$. One neck is protruding from each side of the resonator barrier with a cross sectional area of $2\times 10^{-4}m^2$. The neck is cylindrical in shape with a radius 0.0079 m. The neck length of HRs reduces from the center HR to the outer HRs with lengths 5.96 cm, 5.64 cm, 5.28 cm and 5.00 cm respectively. The acoustic waves passes through the central HR (having the highest neck length), is phase delayed compared to the acoustic waves passes through the outer HRs results in focusing. The schematic diagram of HRs in the rectangular waveguide,in vertical arrangement, is shown in figure 4.

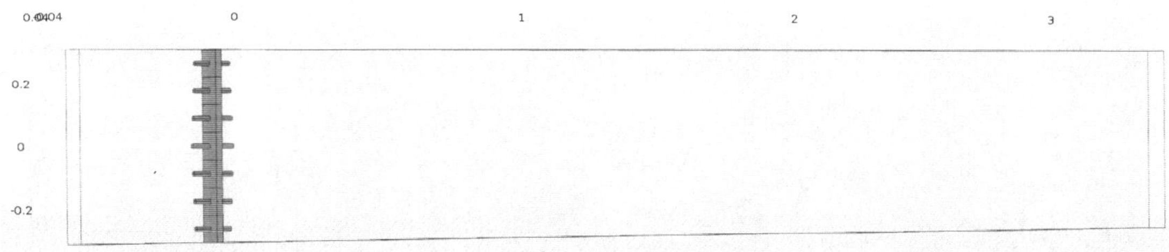

Figure 4- Schematic view of acoustic lens arrangement (HRs arranged vertically in the rectangular wave guide).

3. Result and Discussion

In the case of horizontal array of Helmholtz resonators extraordinary transmission is observed for both configuration at four different frequencies. For normal arrangement of necks EAT occurs at 268 Hz, 344Hz, 406 Hz and 448 Hz. For each of these frequencies corresponding transmittance values are 0.75, 0.88, 0.92, and 0.91 respectively. The increase and decrease behaviour of transmittance is mainly due to the phase shift occurring before and after the Helmholtz resonance frequency. Diffractive interference is the reason behind the continuous phase shift of acoustic waves. For offset arrangement of necks, EAT occurs at 254 Hz, 333Hz, 398 Hz and 438 Hz. For each these frequencies corresponding transmission values are 0.59, 0.94, 0.55, and 0.78 respectively. Similar to the normal arrangement it also exhibits continuous phase shift and it leads to the sudden decline in transmittance within the extraordinary transmission frequencies.

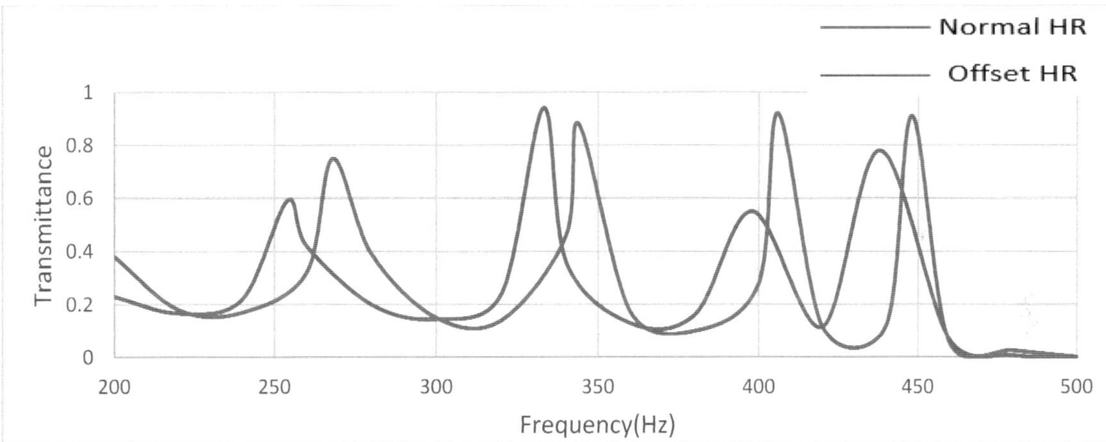

Figure 5 – Transmittance versus frequency plot for acoustic metamaterial with normal HR array and offset HR array

An acoustic lens is created and simulated its frequency response for a frequency range of 200-500 Hz. At 338 Hz the lens showed extraordinary transmitting and focusing capabilities. The pressure intensity plot obtained from the simulation is shown in figure 6. Plane wave is given as input at the left end of the waveguide and the focusing happens at a distance of 1.33 m in front of the lens.

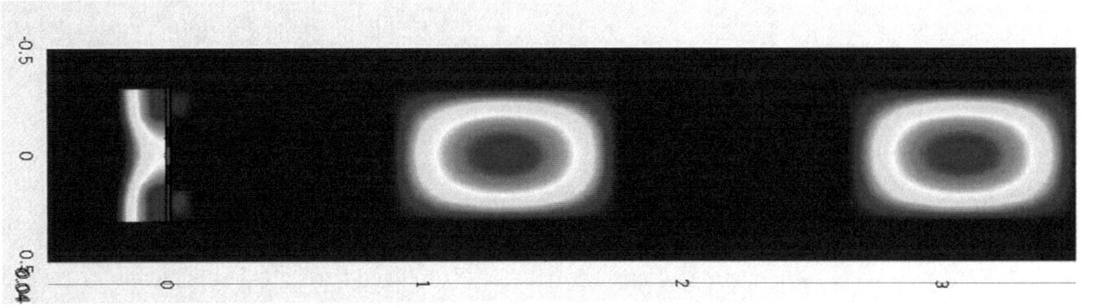

Figure 6- Pressure intensity plot of acoustic lens setup

4. Conclusion

Straight as well as offset arrangement of necks resulted in the transmission of acoustic waves over a broad range of frequencies. By increasing number of Helmholtz resonators in the horizontal arrangement we can increase the number of transmitted frequencies with extraordinary transmission. The vertical arrangement showed more practical significance. Since this kind of configuration enabled the focusing of sound to a particular point. But this focusing characteristics is observed over a narrow band of frequency. An important practical application of this type of arrangement is in the field of defect detection, which require the persistence of focusing over a broader range.

5. References

[1] J.B. Pendry, A.J. Holden, D.J. Robbins, W.J. Stewart, Magnetism from conductors and enhanced nonlinear phenomena. *IEEE Trans. Microw. Theory Tech.* **47**, 2075–2084 (1999)

[2] Nicholas Fang, Dongjuan Xi, Jianyi Xu, Muralidhar Ambati, Werayut Srituravanich, Cheng Sun and Xiang Zhang; Ultrasonic Metamaterials with Negative Modulus, *Nature Materials* **5**, 452–456, 2006

[3] Vijay Koju, Ebony Rowe, and William M. Robertson; Extraordinary acoustic transmission mediated by Helmholtz resonators, *AIP Advances*, **4**, 077132, 2014

[4] Brian C. Crow, Jordan M. Cullen, William W. McKenzie, Vijay Koju, and William M. Robertson; Experimental realization of extraordinary acoustic transmission using Helmholtz resonator , *AIP Advances* **,5**, 027114, 2015

[5] A A Maznev, Gen Gu, Shu- yuan Sun, Jun Xu, Yong Shen, Nicholas Fang and Shu-yi Zhang ; Extra ordinary focusing of sound above a soda can array without time reversal ,*New J.Phys.* **17**, 042001, 2015

[6] Xishan Yang, Jing Yin, Gaokun Yu, Linhui Peng, and Ning Wang; Acoustic superlens using Helmholtz-resonator-based metamaterial, *Applied Physics Letters* ,**107**, 193505 (2015)

[7] Wenqi Wang, Yangbo Xie, Adam Konneker, Bogdan-Ioan Popa, and Steven A. Cumme; Design and demonstration of broadband thin planar diffractive acoustic lenses; *Applied Physics Letters*, **105**, 101904 (2014)

[8] Kiran Kumar Amireddy, Krishnan Balasubramaniam & Prabhu Rajagopal; Deep subwavelength ultrasonic imaging using optimized holey structured metamaterials;*Nature Scientific Reports* 08036-4 (2017)

[9] S. Laureti, D. A. Hutchins, L. A. J. Davis, S. J. Leigh, and M. Ricci; High-resolution acoustic imaging at low frequencies using 3D-printed metamaterials; *AIP Advances* ,**6**, 121701 (2016)

[10] Jun Lan, Yifeng Li1, Yue Xu1 & Xiaozhou Liu; Manipulation of acoustic wavefront by gradient metasurface based on Helmholtz Resonators; *Nature Scientific Reports* 10781-5 (2017)

Influence of nano (h –BN) cutting fluid on machinability of Inconel 625

Pariniti Singh[1], Chinmaya P. Padhy[2]
[1] Research Scholar, Dept of Mechanical Engineering, School of Technology, GITAM University, Hyderabad, India
[2] Associate Professor, Dept of Mechanical Engineering, School of Technology, GITAM University, Hyderabad, India

E-mail: pariniti_singh@rediffmail.com

Abstract. Inconel 625 has got wide applications across industries (aerospace, petrochemical etc.) in various machining operations. During machining, large amount of heat is generated due to the friction, lowering the tool life and adding to various wear mechanism. This is sorted by conventional cooling/lubrication, but, with its harmful effect on environment and human health, it's imperative to pursue other cleaner, greener though efficient alternatives. In this regard, MQL assisted with h-BN nano fluid is an effective option. Hexagonal Boron Nitride (h-BN) having unique structure and composition displays advanced mechanical properties which enhances its versatility in applications. To investigate effectiveness of nano (h –BN) cutting fluid on machinability, first h-BN nanofluid was prepared and characterized for its mechanical properties for nanoparticle concentrations (0.05,0.1 % Vol.). Further, the machining performance and sustainability associated with turning of Inconel-625 under different lubricating conditions viz. dry, wet, MQL with conventional cutting fluid and MQL with (h-BN) nano cutting fluid (NCF) are investigated for machining characteristics. Results showed that MQL integrated with h-BN nano cutting fluid is an effective and sustainable option.

1. Introduction

Ni-Cr alloys such as Inconel 718 or 625 are mostly used in aerospace, petrochemical, nuclear, biomedical and shipping industries. Machining of Inconel 718 has been discussed by various researchers but there are very few literatures available for effective machining of Inconel 625. Like Inconel 718, Inconel 625 is a difficult to machine material. There were mainly two problems faced in machining of super alloys (difficult to machine materials); a). possibility of less tool life due to the work hardening and abrasion properties, b.) high surface roughness and deformation caused due to high machining temperature. For this reason, application of suitable cutting fluid during machining is a wise decision to reduce friction which helps in cooling the workpiece, and washes away the chips. However, number of drawbacks like its high cost, disposal problem, environmental and occupational health hazards are associated with the use of cutting fluids. To address these problems researchers have sought alternate methods of machining, e.g. near dry machining (NDM), combining cutting fluid with additives, minimum quantity lubrication (MQL) machining etc. MQL consists of misting limited quantity of cutting fluid at flow rate of 50 to 500 ml/hour in compressed air flow straight at the cutting zone [1]. Precise application of MQL technique has been seen in machining of hard materials, super alloys etc. in aerospace, automotive, medical and nuclear industries [2-3]. However, it's not as effective as to wet lubrication as excessive heat generation is involved during machining of hard metals. Hence it requires further improvement to meet demanding solution in high speed machining and hard machining. In this regard an attempt has taken to investigate machinability of Inconel 625 by using nano boron nitride (additives) cutting fluids under MQL condition to improve cooling and lubricating property. Decisive properties of boron nitride namely chemical inertness, non-hazardous and capability to lubricate at wide range of temperature and high thermal conductivity makes it a perfect option as an additive in MQL enabling enhancement of various machining parameters.

1.1. Background work on MQL with nano cutting fluid in machining

Past studies have indicated that nano MQL provide better heat transfer [4] and machining performance [5], surface roughness, tool wear and cutting temperature [6]. It is also seen that on machining of Inconel 718 with MQL improved cutting performance and lower surface roughness in comparison to dry and wet environments [7].

2. Experimental Methodology

2.1. Formation of nanofluids

Synthesis of BN nano cutting fluid is prepared with the help of dry BN powder (APS ~70-80 nm) obtained from nanoshel India. An attempt has been made to prepare the samples for BN nano-solution with 0.05 % and 0.1% vol. concentration. For this, the required quantity (measured by using eq.1) of nano particles were dispersed in deionized water and then magnetic stirred for 30 mins at 1200 rpm to obtain respective homogeneous solutions. These samples are further probe sonicated at amplitude of 40% with sonication pulse-on-off time of 50 seconds and 10 seconds respectively. The synthesized uniformly dispersed nano solutions are then added to conventional cutting fluid (Servo-S) in required quantity (95% nano fluid and 5% concentrate conventional cutting fluid) to obtain nano cutting fluid.

Boron Nitride (BN) volume percentage is calculated as,

$$\frac{V_{BN}}{V_{Total}} * 100\% = \frac{W_{BN}/\rho_{BN}}{W_{BN}/\rho_{BN} + V_{Water}} \qquad (1)$$

V_{BN} - Volume of BN nanofluid, V_{Total} - Total volume of nanofluids, W_{BN} - Weight of BN nanofluids,

ρ_{BN} - Density of BN nanofluids, V_{Water} - Volume of Deionized water, Calculated density of BN 2.27 gm/cm^3

2.2. Thermo- Physical properties of BN nano cutting fluid

It is important to know the thermal and fluid properties (thermal conductivity (k), viscosity (μ), surface tension(σ)) of nano cutting fluid, which largely effects the machining of hard metals. Therefore, properties were characterized for (0.05,0.1 % Vol.) h-BN nano particles using liquid thermal property analyzer equipped with thermocouple and digital temperature indicators, Saybolt Viscometer, dynamic contact angle and tensiometer instrument (DCAT) respectively.

The increase in thermal conductivity of nano cutting fluid in comparison to base cutting fluid is observed as the dispersed nanoparticles increase the surface area and thus better heat transfer is possible. For the considered h-BN nano cutting fluid the thermal conductivity is found to increase with the increase in Vol% of nanoparticles. The viscosity of the h-BN nanofluid was tested within the temperature range of 25°C to 70°C.The experiment showed that viscosity decreases with increase in temperature and increases with increase in concentration of BN nano particles (higher viscosity was achieved with 0.1% sample of BN nano cutting fluid) The surface tension of the samples was tested by dynamic contact angle and tensiometer instrument (DCAT). Last column in Table 1 displays % improvement in various thermo physical parameters of h-BN nano-cutting fluid in comparison with conventional cutting fluid. On an average 36% improvement was seen in kinematic viscosity, ~ 37% improvement in thermal conductivity and close to 2% improvement in surface tension.

Table 1. Thermo physical properties of cutting fluid and h-BN Nano cutting fluid of varying-BN concentration

Thermo-Physical Properties	Temperature (°C)	Conventional cutting Fluid	0.05 % by vol. of Nano cutting fluid	0.1 % by vol. of Nano cutting fluid	%Avg. enhancement in thermo physical properties (0.1% vol of nano fluid vs conventional(approx..)
Kinematic Viscosity (Stokes)	25	0.0945	0.0979	0.1006	36.3%
	40	0.0729	0.089	0.0894	
	60	0.0581	0.074	0.0769	
	70	0.0364	0.0514	0.0669	
Thermal Conductivity (W/m ℃)		1.818	2.064	2.4859	36.7%
Surface Tension(mN/m)		71.84± 0.030	71.309 ± 0.030	70.436 ± 0.020	2%

2.3. Experimental Condition and Procedure
To investigate the influence of BN nano solution on turning of Inconel 625 under various operating conditions (e.g. dry, wet, MQL conventional and MQL nano cutting fluid). Coated insert (Korloy insert: CCMT09T308-HMP) Model: PC9030 is used for machining of Inconel 625. Optimization of machining parameters is done by multiple responses based on taguchi orthogonal array integrated with grey relational analysis to maximize material removal rate and minimize surface roughness.MQL technique was performed at pressure of 4 bar and flow rate of 50ml/hr during machining. The nozzle was positioned to inject cutting fluids along the formation of chip.

3. Results and discussion

3.1. Optimization of machining parameters
Optimization of machining parameters (Cutting velocity, Depth of cut, feed rate) in turning of Inconel 625 with multiple responses based on orthogonal array with grey relational analysis to maximize material removal rate and minimize surface roughness is discussed. The levels of the process parameters are selected in order to cover a sufficient wide range of possible cutting conditions. Table 2. shows different parameters considered for machining. Based on taguchi concept a L9 array table 3. was prepared for experimentation. The combined Grey and taguchi methods analysis gave optimal parameters with multiple characteristics. Material removal rate and surface roughness were significant characteristics considered.

Table 2. Selected Parameters in turning process

Factors	Units	Levels		
		1	2	3
Cutting speed(V)	Rpm	500	770	1200
Feed Rate(f)	mm/ rev	0.05	0.1	0.2
Depth of Cut(d)	mm	0.7	1.0	2.0

Table 3. Orthogonal L9 Table for Response

Trial No	Cutting Speed	Feed Rate	Depth of Cut	Material Removal rate	Surface Roughness
1	500	.05	0.7	.126	1.63
2	500	0.1	1.0	.252	1.25

3	500	0.2	1.2	.270	1.133
4	770	.05	1.0	.180	1.01
5	770	0.1	1.2	.312	.663
6	770	0.2	0.7	.612	.998
7	1200	.05	1.2	.360	1.255
8	1200	0.1	0.7	.318	.865
9	1200	0.2	1.0	1.14	.834

The average grey relation grade calculated for Taguchi response. Parameters at which maximum material removal rate and minimum surface roughness was observed were labelled as optimized machining parameters. Experimental design parameters of experiment no. 6 had highest grey relation grade refer Figure 1. Thus, the optimal machining parameter were found to be as cutting speed of 770 rpm and a feed rate of 0.2 mm/rev and depth of cut as 0.7mm. The further steps of experimentation included evaluation of machining performance with different modes of lubrication (dry, wet/flooded, MQL with conventional cutting fluid, MQL with nano cutting fluid) on optimized parameters.

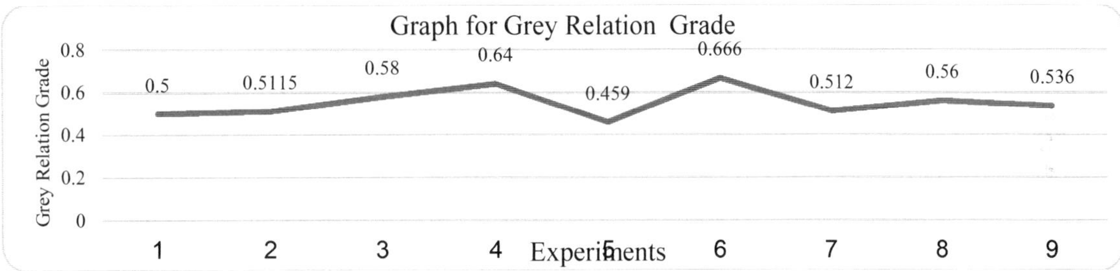

Figure 1. Average grey elation grade for factors

3.2. Chip thickness

Cutting Thickness of chip measures competence of machining process. Lower chip thickness is supposed to be given for good lubrication at machining area and improved machining efficiency [8]. Various chips were collected during the experiment and thicknesses of chips were measured using electronic micrometer. The chips for dry machining were much darker in colour owing to higher temperature along with twisted shape whereas, the chips for MQL with BN nano cutting fluid were much regular along with smooth surface which specified reduced machining temperature. The average chip thickness variation graph for modes of lubrication was plotted as seen in figure 2. which shows chip thickness is comparatively low for MQL with BN nano cutting fluid as compared with dry mode of lubrication.

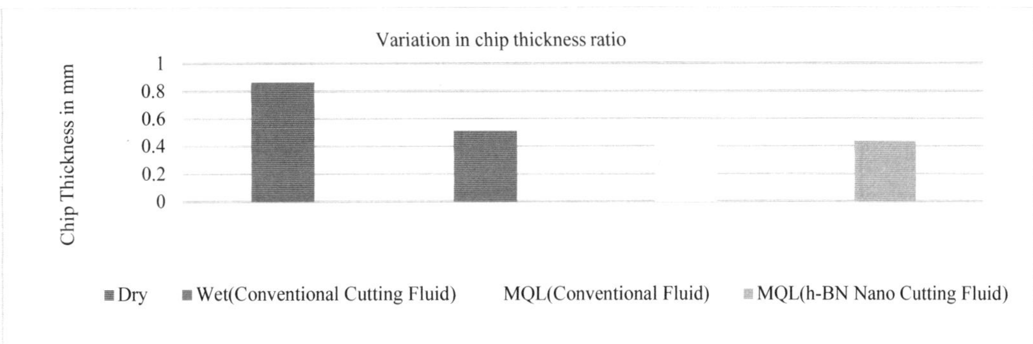

Figure 2. Variation in Chip Thickness Ratio with mode of Lubrication

3.3. Surface Roughness

Surface roughness is another parameter for evaluating machining performance. Good surface roughness results in better quality surface strength, resistance to corrosion and enhanced creep life of a workpiece. The subsequent benefits from good surface finish increases productivity with reduced re-processing of workpiece giving satisfied technical specifications. The surface roughness was recorded by Mitutoyo surface roughness tester. Results are recorded in figure 3, which shows that MQL with h-BN nano cutting fluid gives minimum surface roughness value and hence improved the surface finish.

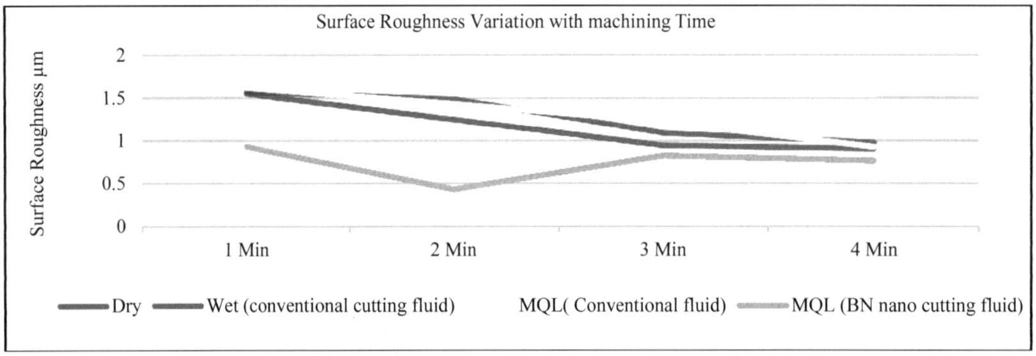

Figure 3. Variation in Surface Roughness with machining time

3.4. Tool wear

Tool wear is a crucial aspect to predict tool life and effective machining process. The tool wear is measured by profile projector at varying interval of time for all lubricating conditions and is observed to increase with machining time and also adds up to high chip-tool interface temperature. High tool wear is attained for dry machining shadowed by wet, MQL with conventional cutting and minimal by MQL with h-BN nano cutting fluid due to better heat dissipation. The figure 4, shows the comparison in tool wear with various modes of lubrication.

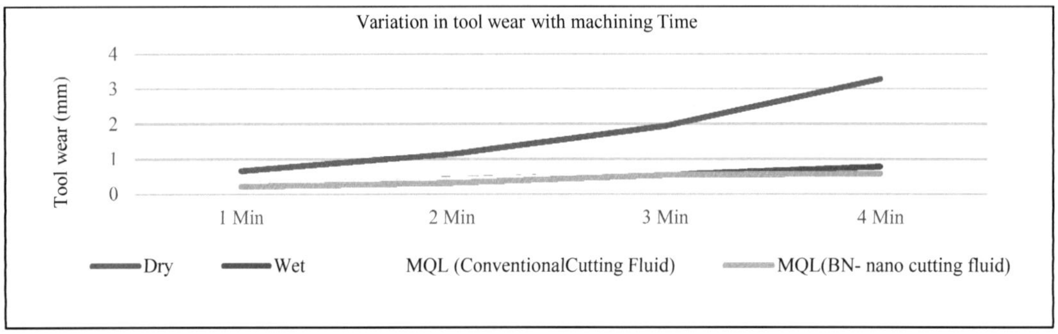

Figure 4. Variation in Tool Wear with Machining Time

3.5. Cutting forces

Cutting forces in machining are directly affected by lubrication during machining. An appropriate cushion of lubricant formed by MQL with h-BN nano cutting fluid at tool workpiece interface reduces the machining temperature along with cutting forces. The cutting forces are measured by lathe tool dynamometer and is seen that cutting force get reduced with h-BN nano cutting fluids refer figure 5.

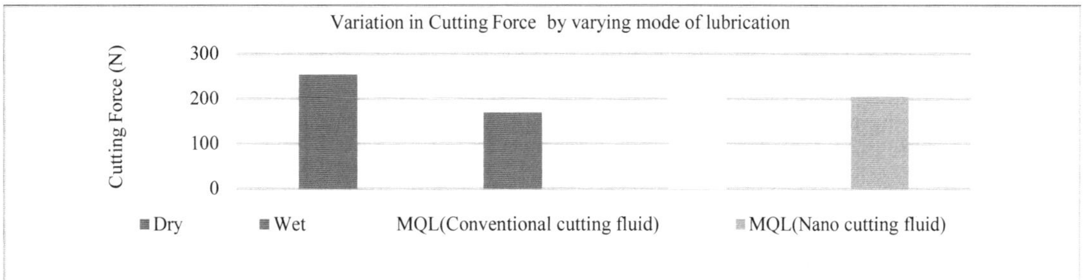

Figure 5. Variation in cutting force with different lubrication condition

3.6. Chip tool Interface Temperature

Cutting zone temperature is another important parameter affected by lubrication method. It is evident by study of thermo physical properties of BN nano cutting fluid that it shows improved thermal conductivity in comparison with conventional cutting fluid. This factor adds on to its lubricious property. The Extech dual laser infrared thermometer is used for measuring the temperature which showed that interface temperature for Dry machining is much higher when compared to the cutting temperature for wet, MQL (conventional cutting fluid) and MQL with nano cutting fluid. It is also seen that relative difference between wet, MQL (conventional fluid) and MQL (BN cutting fluid) is also less. It is inferred from figure 6 that the MQL with BN nano cutting fluid gives minimum cutting interface temperature.

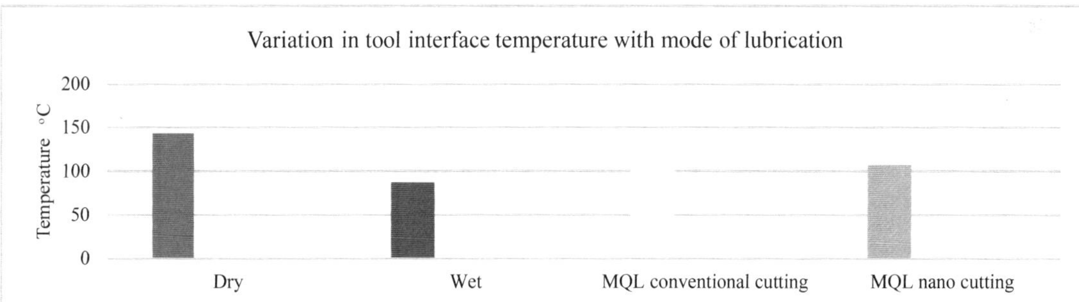

Figure 6. Variation in Tool cutting interface temperature with various lubrication modes

4. Conclusion

Various aspects of machining performance and green technology have been taken into consideration to reach at the best sustainable lubrication in machining. The effect of h-BN nano-fluids in enhancing the MQL lubrication capabilities during machining of Inconel-625 has been investigated and results pertaining to machining of Inconel - 625 under dry, wet, MQL (conventional fluid) and MQL (h-BN nano cutting fluid) are listed below to arrive at inference regarding preeminence of MQL with h-BN nano cutting fluid.

- Surface of chip achieved for MQL conventional is much smooth comparison to tangled chips of dry machining. Also, chip thickness for MQL with h-BN nano cutting fluid was much less in comparison to other lubricating techniques.
- During machining of Inconel-625 surface finish is significantly affected due to intense stress and temperature at tool workpiece interface. The surface roughness was found to be much lower for MQL with BN nano cutting fluid application due to effective enhancement of lubricating properties in comparison to other methods.
- Application of MQL with h-BN nano cutting fluid resulted into lower tool wear. The h-BN nano cutting fluid with lower viscosity and improved thermal conductivity gives better heat dissipation on the interface and hence reduces the tool wear to greater extent.

- Application of MQL with h-BN nano cutting fluid helps in better lubrication and adds up to lower down cutting forces compared to dry and MQL conventional lubricating technique.
- MQL with BN nano cutting fluid helps in lowering machining temp. (due to increase in K value) and better lubrication adds up to reduced cutting forces as compared to other techniques.

Based on the above inferences, it is evident that the employed h-BN nano cutting fluid with MQL application showed significant improvement in machining performance and thus MQL integrated with h-BN nano cutting fluid is a sustainable and greener lubricating option in machining.

5. References

[1] R. Autret and S. Y. Liang. (2003) "Minimum quantity lubrication in finish hard turning," in Proc. International Conference on Humanoid, Nanotechnology, Information Technology, Communication and Control, Environment, and Management, 26-28.

[2] Y. Kamata, T. Obikawa. (2007) "High speed MQL finish-turning of Inconel 718 with different coated tools" Journal of Materials Processing Technology 192–193.

[3] D. G. Thakur, B. Ramamoorthy, and L. Vijayaraghavan. (2009) "Optimization of Minimum Quantity Lubrication Parameters in High Speed Turning of Superalloy Inconel 718 for Sustainable Development", World Academy of Science, Engineering and Technology,54,224-226.

[4] Hwang Y, Park HS, Lee JK, Jung WH. (2006) "Thermal conductivity and lubrication characteristics of nanofluids". Curr Appl Phys 6:67–71.

[5] Shen, B., Shih, A.J., Tung., S.C., (2008) "Application of nanofluids in minimum quantity lubrication grinding" Tribology Transactions, 51, 730–737.

[6] M Amrita, RR Srikant, AV Sitaramaraju, MMS Prasad, P Vamsi Krishna.(2014)"Preparation and characterization of properties of nano graphite-based cutting fluid for machining operations" in Materials Manufacturing Process;29, 600-605,.

[7] Bikash Chandra Behera, Chetan, Sudarsan Ghosh, P Venkateswara Rao. (2014), "Effects on forces and surface roughness during machining Inconel 718 alloying using Minimum Quantity Lubrication", 5th International & 26th All India Manufacturing Technology, Design and Research Conference.

[8] L.B Abhang, Hameedullah M. (2010), "Experimental Investigation of Minimum Quantity Lubricants in alloy steel turning". International Journal of Engineering and Technology", Vol.2(7),3045-3053

Journal of Physics: Conference Series **1355** (2019) 012034 doi:10.1088/1742-6596/1355/1/012034

Dielectric Properties of Eco-friendly Nanofluids

Bijuna Kunju K[1] and Shemim S.S.[2]
[1] Professor and Head, Department of Electrical and Electronics Engineering, T.K.M.C.E, Kollam
[2] Head, Department of Physics, T.K.M.C.E, Kollam

E-mail: kbijuna@gmail.com, shemimsoorj@gmail.com

Abstract. Power and energy sector is adopting the modern technology in all science streams for improving the efficiency after the inclusion of distributed generation and smart grids. The key component in power transmission and distribution is a Transformer. Oil filled transformers are used in High Voltage Transmission and Low Voltage Distribution. To improve the operational efficiency of these huge machines, researches are carried out to prepare insulating oils with high dielectric strength. The present work focuses on the enhancement of dielectric properties of natural oil based insulating fluid by modifying with suitable nanoparticles. The available edible oils such as coconut oil, rice bran oil, sesame oil and sunflower oil are chosen for this purpose. Nanoparticles of TiO_2, Fe_3O_4, SiO_2, h-BN and Al_2O_3 are selected as additives. The properties such as Break Down Voltage, Tan delta and Viscosity are compared for different samples. The values obtained has proved the opportunity to develop natural vegetable base oil with nanomaterial like white graphene having superior qualities compared to the mineral based transformer oil.

1. Introduction

Power transformers are integral part of AC power transmission. Without them the concept of power transfer is almost meaningless. Transformer oil serves the dual purpose of acting as an insulation and coolant for which mineral oil was used. But it has adverse impacts on the environment. With the extent of ultra-high voltage ac power transmission, it is desirable to develop a new kind of oil insulation system, with higher dielectric strength compared to conventional insulation systems, in order to improve the long term operational reliability of power transformers and thus power systems. According to the technical specifications of transformer oils and IEC standards, the important properties to classify transformer oils are the physical properties, chemical properties, and most importantly, the electrical properties. These are examined in several publications [1-6] . Certain processes are implemented in the [4-6] to improve the electrical properties as well as other properties of base oils.

Researches regarding nanofluid were always active. But the concept of nanofluid was developed very recently to replace transformer oil. Vegetable oils are environmental friendly, but they lack properties to compete with the existing standards needed for insulating oil in transformers. Each geographical zone has its own typical and abundant vegetable oil. In crude form it is not suitable for use in transformers. So an extensive study is conducted in various samples obtained by incorporating the base oil with various nano particles and analyze the variations in their properties.

2. Preparation Of Insulating Oil

Solid nanoparticles are dispersed in the host oil using ultrasonic treatment. There are several important properties such as dielectric strength, flash point, viscosity, specific gravity and pour point and all of them have to be considered when qualifying certain oil as transformer oil. Nano particles are selected which remain electrically neutral with increase in voltage level. When the voltage level across the oil is increased, these nanoparticles become polarized and will be able to trap the moving electrons. Thus the net electron flow decreases which results in the increase of Breakdown Voltage (BDV). Also, the nanofluid should have high pour point and stability to be used as transformer oil.

An electronic weigh balance was calibrated and accurately measured the amount of nano particles. About 550 ml of base oil is taken and the amount of nano particle was chosen satisfying the weight to volume ratio 0.1g/100ml. The ultrasonic agitator was filled with water till the prescribed level and the tightly sealed container was immersed in it. The agitation process was carried out for 20 minutes. After 20 minutes the agitator automatically stops and a delay of 10 minutes was given to allow the base oil to cool. Then the process was repeated till a 3 hour agitation was completed. By this the nanofluid is prepared from the base oil and it was stored in an airtight container.

3. Tests Conducted On The Insulating Oil

3.1 Breakdown Voltage Test

The breakdown voltage of oil is the minimum voltage that causes a portion of oil to lose its dielectric strength and transform it from an insulator to conductor. Breakdown voltage is a characteristic of an insulator that defines the maximum voltage difference that can be applied across the material before the insulator collapses and conducts. Oil test kits are used to find out the breakdown strength of transformer oil. The equipment are complete in themselves and supplied with oil cell as per IS: 335 and IS 6792. Its application however is restricted to oils having a kinematic viscosity of less than 50 centistokes at 27°C. The oil test kit consists of shell fitted with two standard-compliant test electrodes. The distance between these electrodes can be adjusted by using feeler gauge. The shell is properly cleaned and distance between them is adjusted to 2.5mm. The base oil is filled in the vessel of the testing device. A test voltage is applied to the electrodes and is continuously increased up to the breakdown voltage with a constant, standard-compliant slew rate of e.g. 2 kV/s. At a certain voltage level breakdown occurs in an electric arc, leading to a collapse of the test voltage. An instant after ignition of the arc, the test voltage is switched off automatically by the testing device. Ultra-fast switch off is highly desirable, as the carbonization due to the electric arc must be limited to keep the additional pollution as low as possible. The transformer oil testing device measures and reports the root mean square value of the breakdown voltage. After the transformer oil test is completed, the insulation oil is stirred automatically and the test sequence is performed repeatedly. As a result the breakdown voltage is calculated as mean value of the individual measurements.

3.2 Tan-delta Test

A very low frequency test voltage is applied across the equipment whose insulation to be tested. First the normal voltage is applied. If the value of tan delta appears good enough, the applied voltage is raised to 1.5 to 2 times of normal voltage, of the equipment. The tan delta controller unit takes measurement of tan delta values. A loss angle analyzer is connected with tan delta measuring unit to compare the tan delta values at normal voltage and higher voltages, and analyze the results. During test it is essential to apply test voltage at very low frequency. If frequency of applied voltage is high, then capacitive reactance of the insulator becomes low, hence capacitive component of electric current is high. The resistive component is nearly fixed; it depends upon applied voltage and conductivity of the insulator. At high frequency as capacitive current, is large, hence, the amplitude of vector sum of capacitive and resistive components of electric current becomes large too. Therefore, required apparent power for tan delta test would become high enough which is not practical. So to keep the power requirement for this dissipation factor test, very low frequency test voltage is required. The frequency range for tan delta test is generally from 0.1 to 0.01 Hz depending upon size and nature of insulation.

3.3 Viscosity Test

Viscosity is a property arising from collisions between neighbouring particles in a fluid that are moving at different velocities. When the fluid is forced through a tube, the particles which compose the fluid generally move more quickly near the tube's axis and more slowly near its walls: therefore some stress, (such as a pressure difference between the two ends of the tube), is needed to overcome the friction between particle layers to keep the fluid moving. For the same velocity pattern, the stress required is proportional to the fluid's viscosity. [3]. A good oil should have low viscosity so that it offers less resistance to the convectional flow of oil thereby not affecting the cooling of transformer. Low viscosity of transformer oil is essential, but it is equally important that, the viscosity of oil should increase as less as possible with decrease in temperature. The most common method of determining kinematic viscosity in the lab is the capillary tube viscometer. In this method, the oil sample is placed into a glass capillary U-tube and the sample is drawn through the tube using suction until it reaches the start position indicated on the tube's side. The suction is then released, allowing the sample to flow back through the tube under gravity. The narrow capillary section of the tube controls the oil's flow rate; more viscous grades of oil take longer to flow than thinner grades of oil. This procedure is described in ASTM D445 and IS 3104.

4. Test Results And Analysis

This Section gives the details of all the test results as described in the previous Section.

4.1 BDV Results

The results of BDV test done on plain base oils are given in Table 1.

Table 1. Breakdown Voltage of pure samples.

	Sample Oil	BDV (kV/cm)
1	Mineral Oil	88
2	Coconut Oil	70
3	Sunflower Oil	60
4	Rice Bran Oil	104
5	Sesame Oil	62

It can be noted that the value of BDV is much high for a pure Rice Bran oil compared to all other samples.

The base oils are then mixed with the nanoparticles in the ratio of 0.1g/100ml. The samples were then tested at a span of several days as the particles were settling with time. The testing frequency is 1 day, 5 days, 15 days and 30 days. The complete results are given in Figure 1.

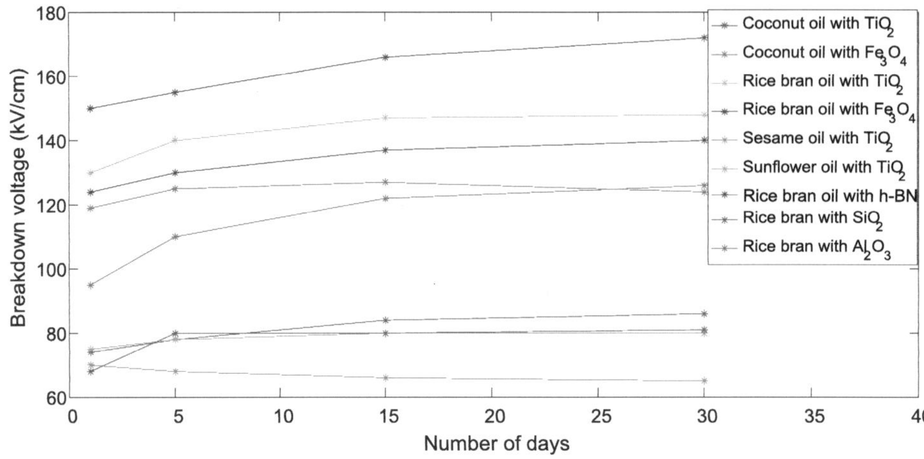

Figure 1: Breakdown Voltage of various samples of Nanofuids checked at intervals.

As per IS 335, the minimum value of BDV for any transformer insulating oil is 65 KV/cm. Oils like rice bran oil and coconut oil satisfied this condition in its untreated form itself. Oils like sunflower as well as sesame oil satisfied the condition after it was agitated with nanoparticles.

The values of BDV given in Table 1 show that pure rice bran oil has the greatest BDV. The results given in Figure 1 indicates an increase in BDV of Coconut Oil for about 5.7% to 22.8% with increase in number of settling days from 1 to 30 with the addition of nanoparticles of TiO_2 whereas an increase in BDV of about 35.7% to 80% is noted with increase in number of settling days from 1 to 30 with the addition of nanoparticles of Fe_3O_4.

The results for Sesame oil shows a change of about 12.9% to 6.45 % with increase in number of settling days with the addition of nanoparticles of TiO_2. It shows a decrease in BDV from its maximum value with increase in number of settling days. The BDV of Sunflower oil shows a change of 25.% to 29.5 % with the addition of nanoparticles of TiO_2.

An increase in BDV of Rice bran oil of about 25% to 42.3% is observed with increase in number of settling days with the addition of nanoparticles of TiO_2. 19.2% to 34.6 % increase is noted with the addition of nanoparticles of Fe_3O_4. An interesting feature is noted in the rice bran oil with Al_2O_3 where the BDV swings between 14% and 19% where it rises from first day to 15th day and then decreases till 30th day. The change in BDV with the addition of SiO_2 is seen to go down with days.

The most promising result is observed when Rice Bran Oil is mixed with h-BN (White grapheme) nanoparticles. The results show a rise in BDV from 150kV/cm to 172 kV/cm as the days increase for settling.

It was also observed that, as the agitation time increases, the stability of the nanofluid increases. The total agitation time was fixed as five hours from previous results. As the settling time increases, the BDV increases to a limit and then remains constant in most of the cases.

4.2 Tan Delta Results

The results of Tan Delta and Loss angle test done on different samples are given in Table 2.

Table 2. Tan Delta Results for Samples.

	Sample Oil	Tan Delta
1	Transformer Oil	0.054
2	Transformer Oil + TiO_2	0.010
3	Transformer Oil + Fe_3O_4	0.061
4	Coconut Oil Plain	2.22
5	Coconut Oil + TiO_2	0.26
6	Rice Bran Oil Plain	0.132
7	Rice Bran + TiO_2	0.128
8	Rice Bran + Fe_3O_4	0.09
9	Rice Bran + h-BN	0.18

On the basis of tests performed, it was found that the tan δ value of mineral oil was 0.054. It was found that the tan δ value of coconut oil was 2.22, which is a significantly higher value. The value of tan δ increases as the charge carriers increases. The tan δ value of the rice bran oil is around the specified value. As per IEC, the preferred value of tan δ is 0.2/mm .On addition of nanoparticles, the tan δ value increases further. Almost all combinations done except coconut oil + TiO2 with value 0.26, had a tan δ value below the preferred value. By using suitable substances, it may be possible to reduce tan δ further.

4.3 Viscosity Results

The results of the tests done on the same samples for viscosity are given in Table 3.

Table 3. Viscosity Results for Samples.

	Sample Oil	Viscosity (cSt)
1	Transformer Oil	10.48
2	Transformer Oil + TiO_2	21.06
3	Transformer Oil + Fe_3O_4	11.335
4	Coconut Oil Plain	40.388
5	Coconut Oil + TiO_2	57.65
6	Rice Bran Oil Plain	62.844
7	Rice Bran + TiO_2	60.244
8	Rice Bran + Fe_3O_4	53.43
9	Rice Bran + h-BN	52.55

On the basis of tests performed, it was found that mineral oil has a viscosity of 10.48cSt. For samples with TiO_2 and Fe_3O_4 nanoparticles, the values are 21.06 cSt and 11.335 cSt respectively at room temperature. As per IEC, the preferred values of viscosity are 27cSt at 27degree Celsius and 12 cSt at 40degree Celsius.

For rice bran oil, viscosity value is 62.844cSt. With the addition of TiO_2 and Fe_3O_4 nano particles, viscosity reduces to 60.244cSt and 53.43cSt respectively. Even with the addition of nano particles the values are quite larger than those preferred.

In the case of Coconut oil, viscosity value is 40.388cSt and with the addition of TiO2, the value increased to 57.65cSt. Combination of white graphene and rice bran oil gave a viscosity of 52.55cSt, which is lower than the value of pure rice bran oil. From the results, it was found that none of nanofluid made from natural oil have viscosity below 27cSt.

Generally, the natural oils are more viscous than mineral oils due to high intermolecular forces of attraction between the large sized oil particles. The viscosity can be decreased by reducing the size of the particles. This can be achieved by the emulsification of oil by using suitable agents. Moreover a combination of thin oil with regular oil can also reduce viscosity.

5. Conclusion

Samples of Nanofluids are prepared from Transformer oil (mineral base) and other natural vegetable environment friendly oils. The eligibility tests of each sample for using it as insulating medium in transformers revealed a very high chance of manufacturing a superior quality nanofluid with Rice Bran Oil as base and h-BN as an additive, since this sample recorded the highest BDV. The other properties of Tan Delta and viscosity can be brought to the desirable limits only with additional surfactants or binders. Further research on identifying suitable binding agents is under process.

Acknowledgments

The authors would like to acknowledge the support and funds provided by the TEQIP Phase II implemented in TKM College of Engineering, Kollam, Kerala.

References

[1] Y. Z. Lv, Y. Zhou, C. R. Li, Q. Wang, and B. Qi,‖, Recent Progress in Nano fluids Based on Transformer Oil: Preparation and Electrical Insulation Properties, *IEEE Electrical Insulation Magazine, September/October* — Vol. 30, No. 5,2014.

[2] Bin Du, Jian Li, Feipeng Wang, Wei Yao, and Shuhan Yao, Influence of Monodisperse Fe3O4 Nanoparticle Size on Electrical Properties of Vegetable Oil-Based Nanofluids, *Article ID 560352, Journal of Nanomaterials* Volume 2015.

[3] W.M.L.B. Naranpanawe, M.A.R.M. Fernando, J.R.S.S. Kumara, E.M.S.N. Naramapanawa, and C.S. Kalpage, Performance Analysis of Natural Esters as Transformer Liquid Insulation – Coconut, Castor and Sesame oils, *2013 IEEE 8th International Conference on Industrial and Information Systems, ICIIS 2013*, Aug. 18-20, 2013.

[4] N. A. Mohamad, R. Yunus ,M. Z. A. AbKadir, N. Azis, J. Jasni, A Study on the Dielectric Properties of Palm Oil and Coconut Oil‖, *IEEE International Conference Power & Energy (PECON)* 2014.

[5] Jian Li, Zhaotao Zhang, and Ping Zou, Stanislaw Grzybowski, Markus Zahn, Preparation of a Vegetable Oil -Based Nanofluid and Investigation of Its Breakdown and Dielectric Properties‖, *IEEE Electrical Insulation Magazine, September/October,* Vol. 28, No. 543,2012.

[6] B. S. H. M. S. Y. Matharage, M. A. R. M. Fernando, Performance of Coconut Oil as an Alternative Transformer Liquid Insulation, *IEEE Transactions on Dielectrics and Electrical Insulation,* Vol. 20, No. 3; June 2013.

Effect of graphene, SiC and graphite addition on hardness, microstructure and electrical conductivity of microwave sintered copper MMCs fabricated by powder metallurgy route

C Ayyappadas[1], O S Shanu[2], Abhilash Vijayan[2], Mohammed Irfan A[2], V Vishnu[2], M Shamnadh[3]

[1] Assistant professor, Department of Mechanical Engineering, NSS College of Engineering, Palakkad, Kerala-678008, India

[2] M.Tech student, Computer Integrated Manufacturing, Department of Mechanical Engineering, TKM College of Engineering, Kerala-691005, India

[3] Assistant professor, Computer Integrated Manufacturing, Department of Mechanical Engineering, TKM College of Engineering, Kollam, Kerala-691005, India

E-mail: ayyappadas755@gmail.com

Abstract. Copper MMCs reinforced with graphene, SiC and graphite flakes with varying mass fractions of 2.5, 5, and 7.5 wt% SiC, 2.5, 5, 7.5, and 10 wt% graphite and 0.2, 0.4, 0.6 and 0.8wt% graphene particles were fabricated through powder metallurgy route. The powder mixtures were blended and then compacted under a uniaxial pressure of 400MPa for both Cu-SiC and Cu-Gr composites and 600Mpa for Cu-Gn composites. Green compacts were sintered by microwave (MW) sintering process. MW sintering was performed at 950°C in open atmosphere with a heating rate of 25°C/min for all composites. Densification parameter (DP) was determined for all MMCs and a reduction in DP values were observed with an increase in weight percentage of reinforcement particles as a result of increase in porosity. Higher mass fractions of reinforcement particles in the Cu matrix tend to offer a diffusion barrier to copper atoms. Micro-structural analysis was performed using Optical and scanning electron microscope which revealed a homogeneous microstructure of all composites at low mass fractions of Sic, graphite and graphene. In order to detect the presence of any oxides of Cu and other elements, EDS analysis was carried out. Hardness data showed an increasing trend for Cu-SiC and Cu-Gn composites and a declining trend for Cu-Gr composites. Porosity has a significant effect on the electrical conductivity values of the composites.

1. Introduction

Copper MMCs have superior mechanical and thermal properties compared to that of pure copper [1]. Adding dispersed particles into the copper matrix improves the mechanical properties of copper [1,2]. Carbide and oxide phases impart sufficient thermal stability even at elevated temperatures. Cu-SiC metal matrix composites possess the properties of both copper and Silicon Carbide, ie high thermal conductivity of copper and low coefficient of thermal expansion of SiC [1,3]. This composite is majorly used for heat sinking or heat spreader applications. A Cu-SiC heat spreader offers high thermal conductivity and an adjustable coefficient of thermal expansion (CTE) over a desired range

[4,5]. Copper–graphite MMCs have excellent thermal and electrical conductivities, and the properties of graphite, i.e. solid lubricating and small thermal expansion coefficient [6]. These composites find wide applications in brushes, and bearing materials due to the excellent mechanical and thermal properties. Copper-graphite MMCs with low weight percentages of graphite is also used in slip rings, switches, relays, connectors, plugs and low voltage DC machines with very high current densities [7].They are excellent in terms of machining and brazing, also a good candidate for thermal management applications for high power density packaging of advanced semi-conductor devices. These MMCs exhibits superior mechanical, electrical and thermal characteristics. Cu-Gr composites are widely used in bearing materials and brushes because of its excellent thermal and self-lubricating properties [7,8].

It was reported that graphene is the strongest material ever tested [9]. It has been preferred as an excellent reinforcement because of its superior tensile strength (130 GPa), Young's modulus (1TPa), thermal conductivity (5000W/mK), electrical conductivity, and large specific surface area (SSA-2630m^2/g) [10]. Cu-Gn composites are reported to have superior electrical, mechanical and thermal properties as compared to pure copper [11]. Graphene is the building block of graphite, a one atom thick 2D allotrope of carbon with a honeycomb structure. Production of graphene in bulk quantity is an expensive process. Researchers around the globe have been working on graphene based devices like batteries and solid state devices which are predicted to have superior performance and life. Dispersion of graphene in copper phase is challenging due to large difference in relative density of the materials.

Sintering involves heating and reducing the porosity of a green compact. A conventional sintering process uses some external sources and heat energy is transferred by conduction convection and radiation. Generation of heat takes place in the external heating element and it is transferred to the core of the material by conduction heat transfer. Initially the surface of the material is heated and as the temperature rises the heat is transferred to the core, resulting in the development of internal stress and a thermal gradient in the radial direction. In order to achieve thermal equilibrium conventional sintering process requires considerable time. On the other hand microwave sintering process is a distinctive process which enables us to reduce sintering cycle times. Microwave radiations are electromagnetic radiations with frequency ranging from 300 MHz to 300 GHz, with region of wavelengths between 1mm to 1m in Electromagnetic spectrum. Microwave spectrum usually includes 3 bands, Extremely High Frequency (EHF) 30GHz to 300GHz, Super High Frequency (SHF)-3GHz to 30GHz, Ultra High Frequency (UHF) 300MHz to 3GHz. Microwave radiations of frequency of the order of 2.5 GHz are commonly used in commercial microwave furnaces. Unlike a conventional sintering process, microwave heating results in rapid consolidation of the composites by volumetric heating. In this process heat is transferred from the core to the surface of the material which results in low surface temperature of the samples. Thus thermal degradation of the material can be considerably reduced. As compared to conventional sintering process, high heating rates can be enabled in microwave process which reduces the sintering time. Rustum Roy et al. investigated the microwave sintering characteristics of transparent alumina. Sintering was performed at 1750^{0}C for 15 minutes, and it was reported that the samples exhibited full densification and good transparency [12]. Manoj Gupta et al. fabricated Al-SiC nano composites by microwave sintering followed by hot extrusion process. They summarized microwave process as an eco-friendly consolidation method which offers higher heating rates and shorter processing cycle times without compromising on the properties of the material [13].

In this study attempts were made to fabricate copper matrix composites reinforced with varying weight percentage of SiC, graphite and graphene particles through powder metallurgy route. The effect of reinforcements on the microstructural and electrical properties of copper composites was studied.

2. Experimental procedure

2.1 Materials

For the production of composites, gas atomized copper powder with a purity of 99.5% and an average particle size of 100μm (Manufacturer: Sigma Aldrich, USA) has been used. Fine quality graphene powder (Angstron materials, USA) of average particle size 100nm, SiC powder of 99.9% purity (Manufacturer: Sigma Aldrich, USA) having an average particle size of 20 μm and 99.9% pure graphite flakes of average particle size 50μm (Manufacturer: Angstron materials, USA) were used as the reinforcement phases. Each composition with a varying percentage of 2.5, 5, and 7.5 wt% SiC particles, 2.5, 5, 7.5, and 10 wt% graphite flakes and 0.2, 0.4, 0.6 and 0.8wt% graphene particles and pure copper powder were mixed in a screw mixer for 30 minutes each. Spheroidal form of copper particles are shown in figure 1a), which is the result of gas atomization process. The morphology of graphene and graphite are flaky in nature and that of SiC is angular as seen in figure 1b), 1d) and 1c). Powder characteristics of as received powders are shown in table 1.

Figure 1. Powder morphology of as received powders a) Pure copper b) graphene c) SiC and d) graphite

Table 1. Powder characteristics of as received powders

Characteristics	Copper	Graphite	Graphene	SiC
Apparent density (g/cm^3)	4.46	0.16	0.016	1.27
Tapped density (g/cm^3)	5.32	0.34	0.14	0.26
Flow rate in sec/50g	32	No flow	No flow	14
Average particle size (μm)	100	50	100nm	20
Theoretical density (g/cm^3)	8.96	2.26	2.2	3.21
Shape	Spherical	Flaky	Flaky	Angular
Purity (%)	>99.5	99.99	99.99	99.99

2.2 *Fabrication of Cu-SiC, Cu-Gn and Cu-Gr composites*

Blended powders were compacted in a closed cylindrical die of 16 mm diameter under a uniaxial pressure of 400MPa for both Cu-SiC and Cu-Gr composites and 600Mpa for Cu-Gn composites. Zinc stearate powder was applied to the die walls and rod in order to facilitate easy ejection. The green compacts thus produced were sintered by microwave sintering process at a temperature of 950° C with a heating rate of 25°C/min and a holding time of 1 hour in all cases. A 6 kW, 2.45GHz multimode cavity commercial furnace was employed for the sintering process. An insulation package with multilayer wall was incorporated to ensure the uniform heating of the samples. Samples were allowed to cool in the furnace itself.

2.3 *Metallographic sample preparation*

For capturing the optical microimages, metallographic preparations were performed on the samples as per ASTM E407 standard. First the samples were polished by traditional process by employing different grades of emery sheets ranging from 200-2000 grit sizes. Then the samples were polished in a disc polishing machine with diamond paste and Al_2O_3 media. A mirror finish was produced on all the MMC surfaces. Samples were etched with a solution containing 5g $Fecl_3$, 10ml HCl, 50ml glycerol and 30ml water as per the standard. A controlled etching process was performed where the sample was rubbed against a cotton cloth soaked with the etching media for a time period of 15-60 seconds.

2.4 *Characterization*

Sintered densities were determined by a method based on Archimedes principle. Micro structural analysis was performed using (Zeiss Axioscop A40, Germany) Optical microscope and (Zeiss EVO MA 10) scanning electron microscope. SEM-EDS analysis was performed to detect the presence of Cu, SiC, graphite, graphene and any presence of oxides of these. Vickers micro hardness test was performed on all the samples using a Vickers micro hardness tester (Matsuzawa MMT-X7 model, Japan) based on optical measurement. A load of 100g was applied and a dwell period of 10s was maintained for testing all samples. Five readings were taken on each of the samples by carefully picking a point which includes both the matrix and the reinforcement homogeneously. Pores present in the samples were carefully avoided to obtain better hardness data. Electrical conductivity of the samples was determined by an instrument which works on the principle of 'four point collinear probe'. Resistivity of the samples was measured, which is the reciprocal of electrical conductivity. ASTM B193 standard test procedure was followed in the measurement process.

$$\text{Bulk resistivity} = 2\pi S(V/I) \qquad \{1\}$$

Where S is space between probes, I is the test current and V is the measured voltage

3. Results and discussion

3.1. *Relative density*

Relative density plots of the composites sintered at 950^{0}C are shown in figure1. Higher relative density was observed for Cu-Gn composites with a maximum value of 90% as shown in figure 1. Relative density of the Cu-SiC composites varied between 80-86%, Cu-Gr between 76-87% and for Cu-Gn in the range 87-90%. Relative density of pure copper sintered by microwave sintering process was determined as 86%. It was observed that the relative density of the composites decreases with the increase in addition of the reinforcement particles [14]. Lowest relative density was observed for Cu-10wt% Gr MMC. This is due to the solid lubricating nature of graphite which separate the copper grains and inhibits the development of a continuous copper network. The relative density values of Cu-Gn composites were much closer as the weight percentage of graphene particles were below 1%. Density of the graphene (2.21 cm^3)is much lower than that of copper (8.89g/cm^3). The density of SiC powder was 3.21 g/cm^3, 2.26 g/cm^3 for graphite flakes and 2.21 g/cm^3 for graphene powder. Due to this the relative density of the composites becomes low. Another possible reason may be the presence of porosity in the composites [15]. For lower weight percentage of SiC, Cu-SiC interface was found to be less and hence less copper atom diffusion barrier. Under such conditions copper atoms can easily diffuse and result in higher densification values of the composite. Similar behavior was predicted for Cu-Gr and Cu-Gn composites [5,15]. The decrease in relative density of composites can be attributed to the formation of oxides which was confirmed by SEM-EDS analysis. Densification parameter was relatively high for all composites with low weight fractions of reinforcements. With the increase in wt% of SiC, porosity observed in Cu-SiC composites were considerably high. Pores present in Cu-Gn composite are a result of large mismatch in relative density of the particles. This has significant effect on properties, and still redeemed by adding graphene with outstanding properties. Agglomeration tendency of graphite was observed in the entire process of mixing, blending and while taking microstructures. Relative density of Cu-Gr composites reduces as a result of the same. Addition of graphite after an optimum value thus impedes the properties of the Cu-Gr MMCs.

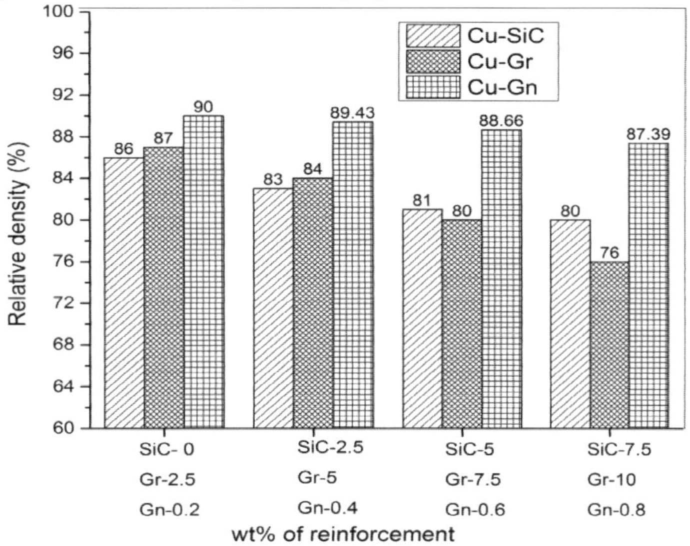

Figure 2. Relative density of Cu-SiC, Cu-Gr and Cu-Gn composites

3.2. Microstructure

Optical micrographs of the Cu-SiC composite sintered by microwave process at 950°C are shown in figure 3. The bright areas represents the copper matrix as shown in microstructure of pure copper in figure 3(a) ,while the dark spots indicate the uniformly distributed SiC particles in the copper matrix as in figure 3(d). Microstructure of the composite plays a vital role in predicting the mechanical, thermal and electrical behavior of the composites. When the reinforcement is uniformly distributed in the matrix it enhances the mechanical properties of the composite [16]. As the weight percentage of SiC was increases it was observed that the SiC particles get embedded into the copper grains due to the ductile nature of copper [17]. Large grains of copper of the order of 100μm were observed in the micrographs of pure copper, figure 3(a). The distinction between the particle size of the matrix and the SiC particles can be predicted from the micrographs. Compared to Cu-Gr composites, Cu-SiC composites exhibited a homogeneous microstructure even at higher weight fractions of SiC. Hard SiC ceramic phase does not agglomerate easily unlike graphite as observed in Cu-Gr composites. Presence of alumina particles were observed in the microstructures of all composites. Hard alumina particles get embedded into the copper matrix during metallographic sample preparations. A slight variation in color in Cu-SiC composites is observed due to the presence of alumina particles.

Graphite particles have a natural tendency to form agglomerates as observed from the optical micrographs of Cu-Gr composites. Presence of graphite agglomerate in the copper matrix cannot be avoided completely due to the inherent nature of graphite as shown in figure 4(c). Graphite particles due to its light nature sometimes floated on the surface which could be observed while taking the micrographs [7,8]. The mixing of copper and graphite powder was a challenge due to large difference in relative density values. For low mass fractions of graphite upto 5 wt%, homogeneous microstructure

of Cu-Gr was observed. Many researchers have reported that by adding 5-7 wt% of graphite into copper matrix, the mechanical and thermal properties of Cu-Gr MMCs can be enhanced to a greater extent [18].
Microimages of the Cu-Gn composites exhibited a homogeneous distribution of graphene particles within the copper matrix. In the micrographs the dark area indicates the uniformly distributed graphene particles as seen in figure 5(a). As the weight percentage of graphene increased a tendency to form agglomerates was observed as shown in figure 5(c) and (d). Some of the graphene particles were embedded in the copper matrix while the rest of the particles were observed in the grain boundary of the copper matrix, figure 4(a). Embedded particles do not have a re-arranging ability. With the increase in addition of more graphene the number of particles present in grain boundary of copper increases and leads to agglomeration [9]. High specific surface area and low density of graphene poses great challenge while mixing the powders. Homogeneity can be maintained by increasing the mixing time and by controlling the size of Cu grains.

Figures 6-8 shows the SEM micrographs of the composites. From the SEM micrograph of Cu-SiC composite it was observed that the SiC particles were present in the grain boundaries of the copper grains. Dark areas indicate the SiC particles as illustrated in figure 6(d). Clustering behavior of Sic particles were observed in the SEM micrographs, which was not evident in optical images as seen in figure 6(d). Uniform distribution was much higher for composites containing lower weight percentage of SiC particles. Apart from the Cu-SiC, interface no other phases were seen in the micrographs of Cu-SiC composites. In the micrographs of Cu-Gr composites, agglomerates of graphite were observed due to the inherent nature of graphite to form clusters as shown in figure 7 (a) and (b). Dark areas indicate the graphite agglomerates in the micrographs. The variation observed in the hardness value can be justified by the SEM micrograph. Presence of any interfacial products were not observed from the microstructure which confirms that no reactions takes place between copper and graphite during the sintering process [7,8]. Apart from the Cu-Gr interface no other phases were observed in the micrographs. Graphite induces into the matrix due to its soft nature which was evident from the micrographs [18]. Due to the low solubility of carbon in copper only mechanical bonding between two phases occurred.

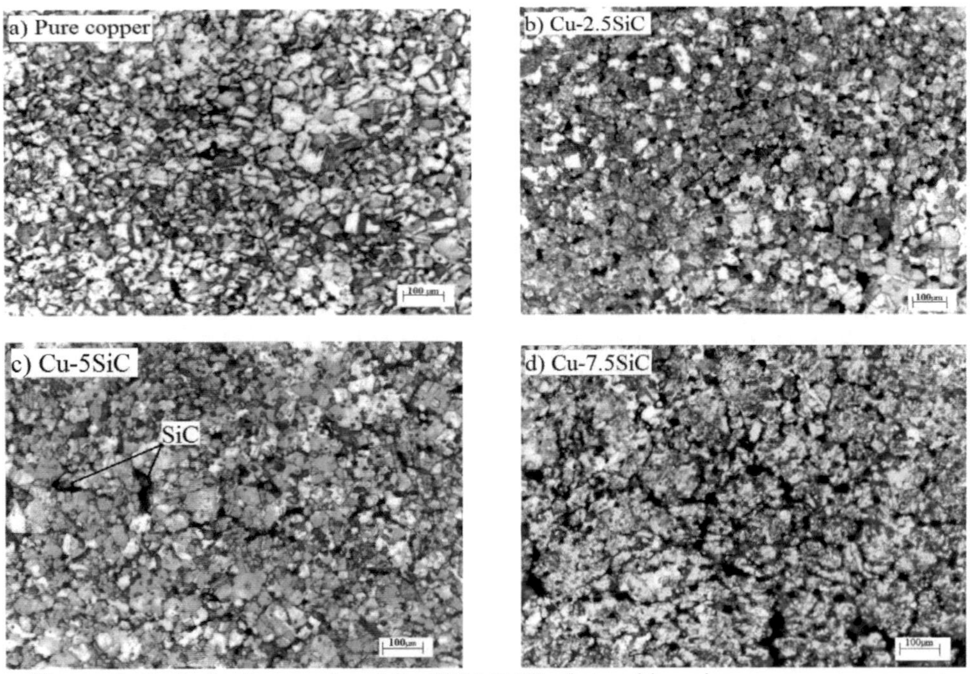

Figure 3. Optical micrographs of Cu-SiC MMCs sintered by microwave process

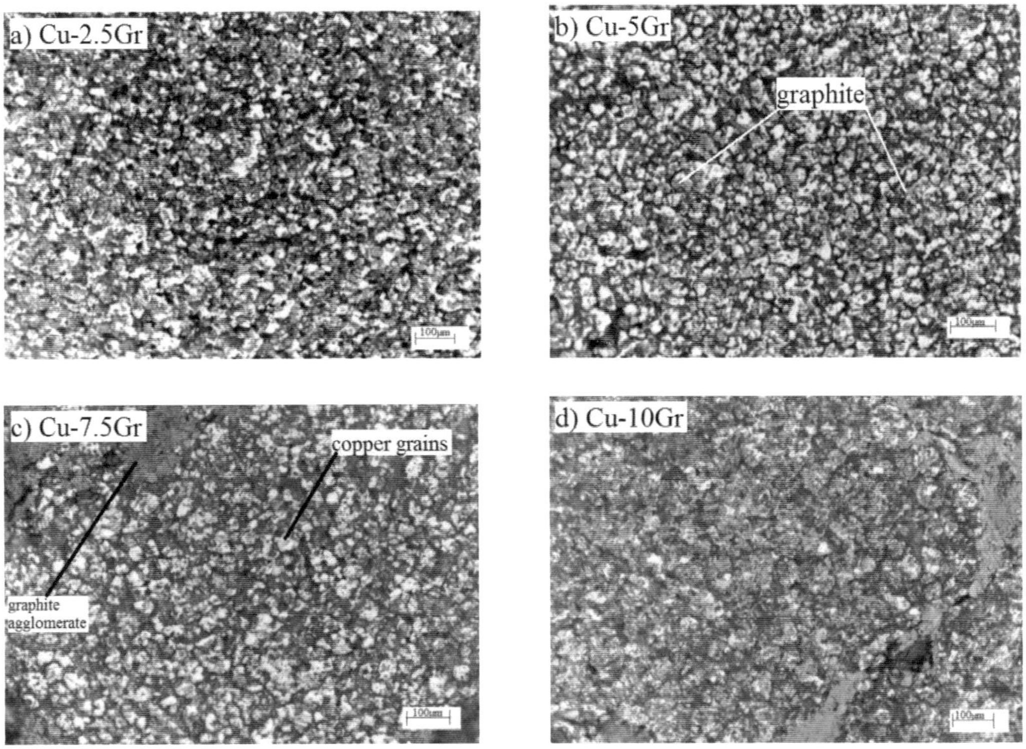

Figure 4. Optical micrographs of Cu-Gr MMCs sintered by microwave process

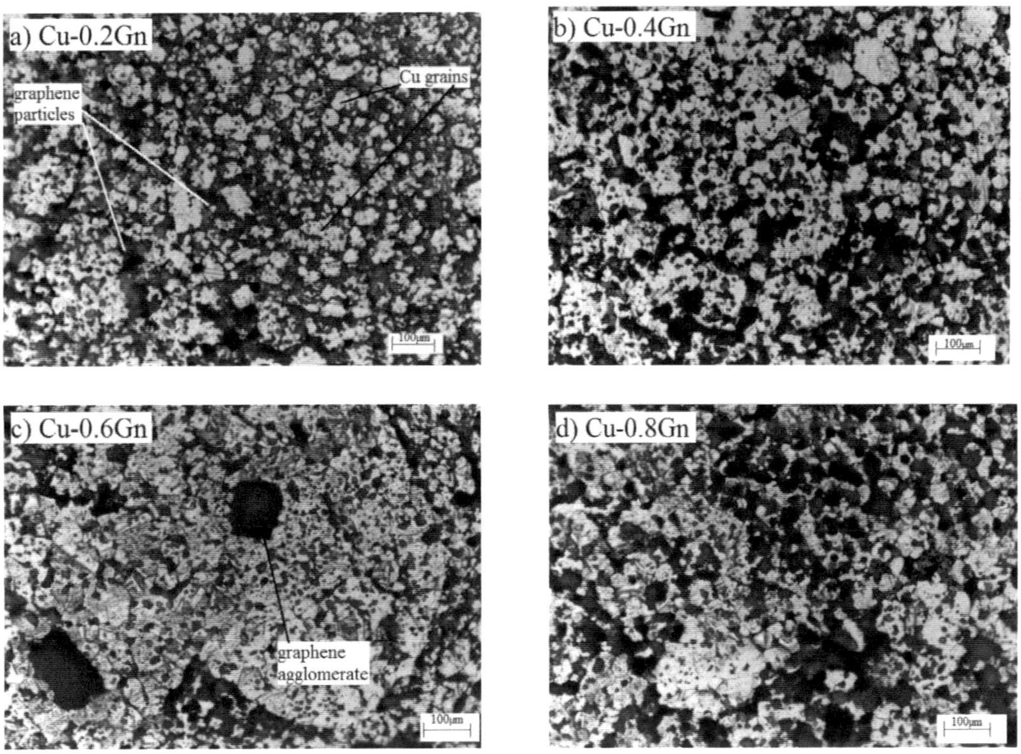

Figure 5. Optical micrographs of Cu-Gn MMCs sintered by microwave proces

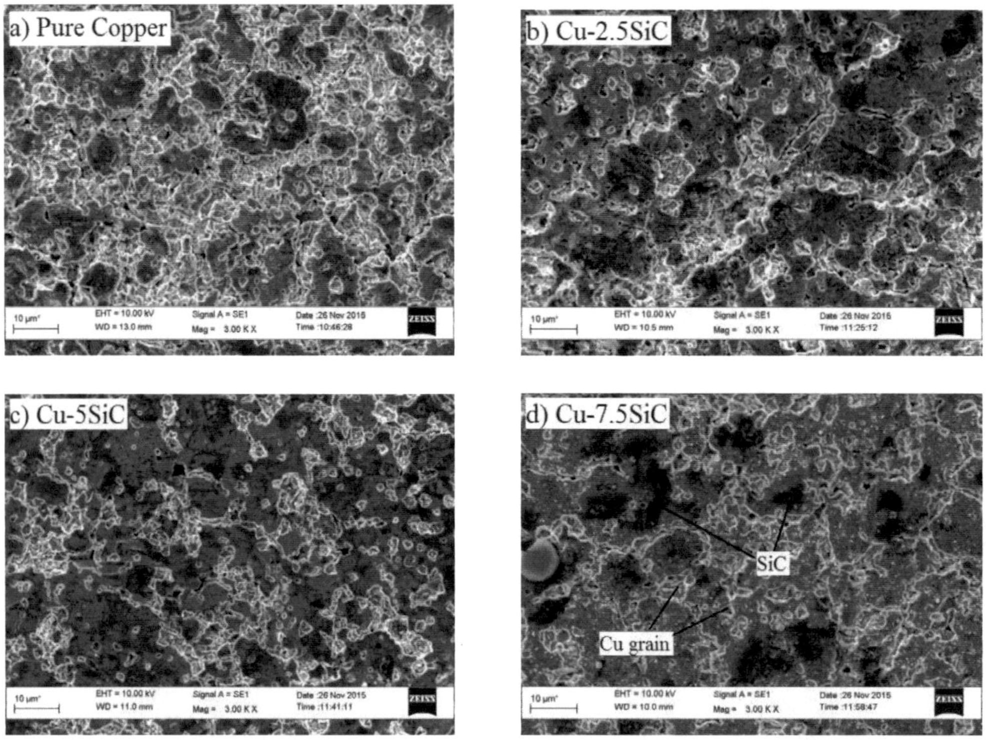

Figure 6. SEM micrographs of Cu-SiC MMCs sintered by microwave process

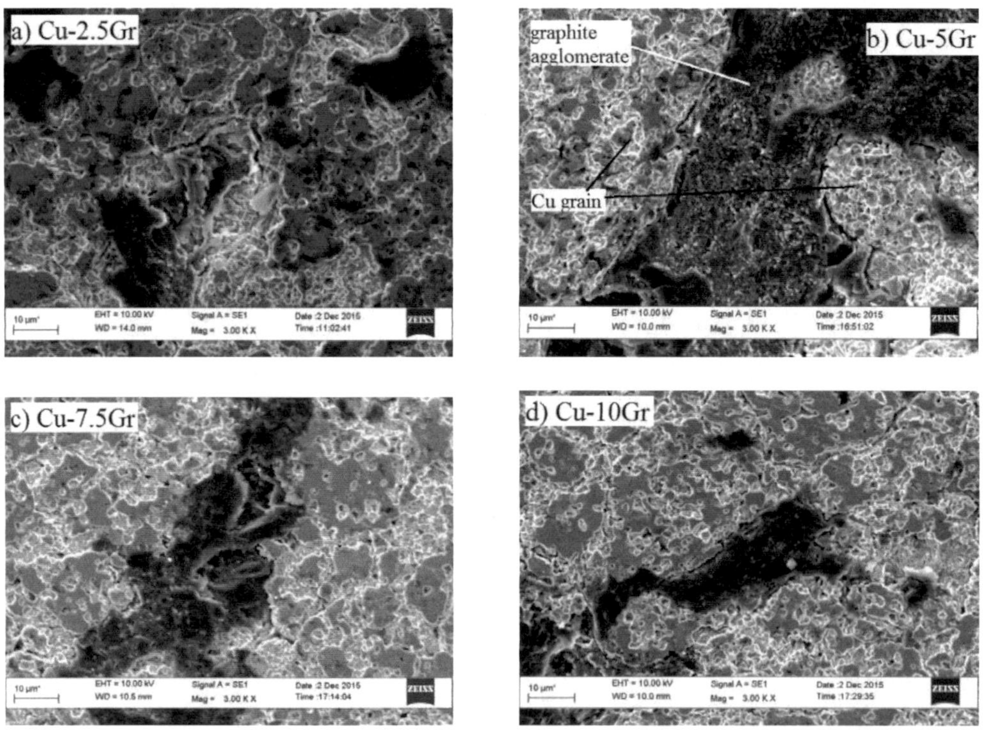

Figure 7. SEM micrographs of Cu-Gr MMCs sintered by microwave process

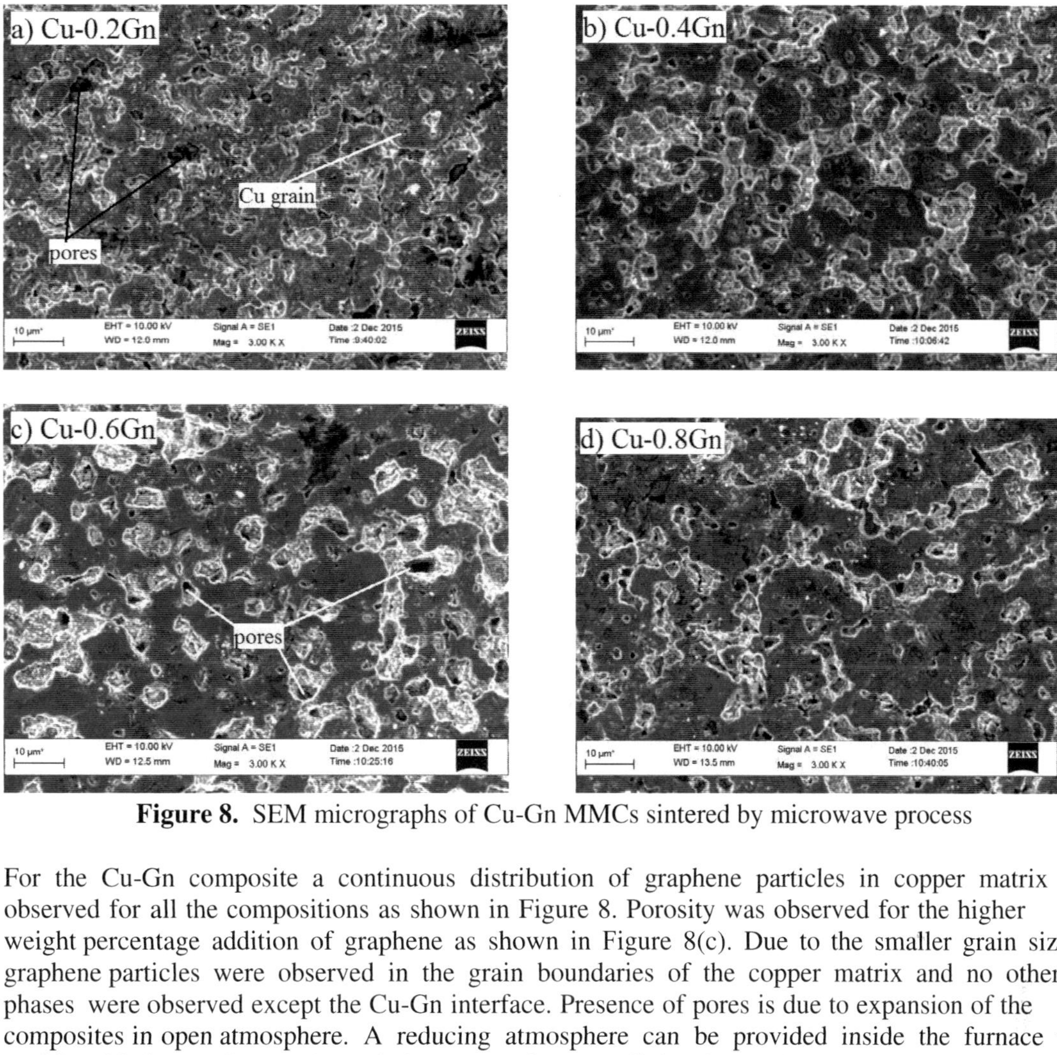

Figure 8. SEM micrographs of Cu-Gn MMCs sintered by microwave process

For the Cu-Gn composite a continuous distribution of graphene particles in copper matrix was observed for all the compositions as shown in Figure 8. Porosity was observed for the higher weight percentage addition of graphene as shown in Figure 8(c). Due to the smaller grain size, graphene particles were observed in the grain boundaries of the copper matrix and no other phases were observed except the Cu-Gn interface. Presence of pores is due to expansion of the composites in open atmosphere. A reducing atmosphere can be provided inside the furnace to avoid oxidation and expansion of the composites. A slight change in density was observed due to the presence of atmosphere in furnace.

SEM-EDS analysis was performed to analyze the presence, morphology and distribution of Cu, SiC, graphite, Graphene and any oxides of these in the composite, shown in figure 9-12. As the green compacts were sintered at 950°C oxides of copper might have formed which was confirmed by the EDS analysis, and the presence of oxygen was detected [5,12,19]. Presence of chlorine and aluminium was detected in EDS analysis of pure copper and Cu-SiC composites. During etching and polishing processes of the composites, etching media and alumina particles get entrapped into minute pores present in MMCs [1]. EDS analysis only confirmed the presence of Si and C which in turn can be predicted as SiC present in the composite. It was reported that at high temperatures SiC decomposes into Si and C. But presence of any carbides of copper was not detected. Copper is not a carbide former and presence of any such phases is thus ruled out at present. Oxygen detected in the Cu-SiC by EDS analysis can be supported by figure 9 [15]. Due to the ductile nature of copper alumina particles get embedded into the matrix, which was difficult to remove and was detected in the EDS analysis. Presence of oxygen detected might be from alumina media used for etching.

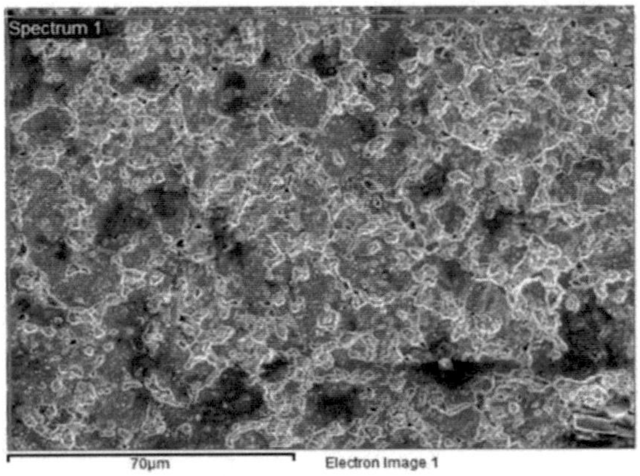

Figure 9. Field under consideration for SEM-EDS analysis of Cu-7.5SiC MMC

Table 2. SEM-EDS analysis of Cu-7.5SiC composite sintered by microwave sintering process at 950°C

Element	Weight%	Atomic%
C K	5.20	17.99
O K	9.66	25.10
Si K	0.34	0.50
Cl K	1.71	2.01
Cu L	83.10	54.39
Totals	100.00	100.00

Figure 10. Field under consideration for SEM-EDS analysis of Cu-10Gr MMC

Table 3. SEM-EDS analysis of Cu-10Gr composite sintered by microwave sintering process at 950°C

Element	Weight%	Atomic%
C K	28.28	64.07
O K	4.10	6.97
Cu L	67.63	28.96
Totals	100.00	100.00

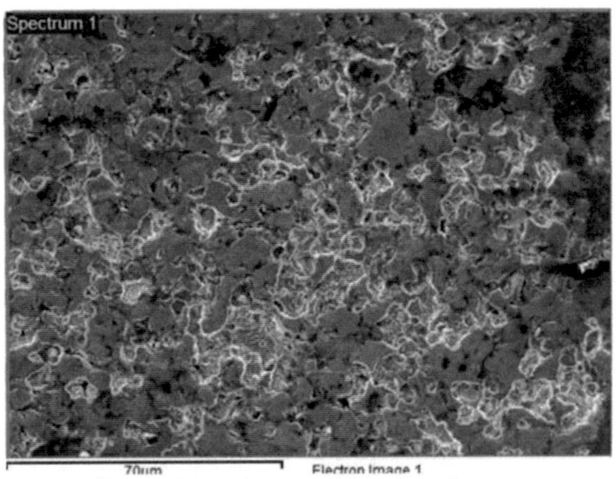

Figure 11. Field under consideration for SEM-EDS analysis of Cu-0.8Gn MMC

Table 4. SEM-EDS analysis of Cu-0.8Gn composite sintered by microwave sintering process at 950°C

Element	Weight%	Atomic%
C K	4.68	17.26
O K	7.90	21.86
Cu L	87.41	60.88
Totals	100.00	100.00

3.3. Hardness

Figure 12 shows the hardness values of the composites sintered by microwave process at 950°C. Indenter of the hardness tester was carefully placed on the composites so as to pick a point which constitutes both the matrix and reinforcement for getting accurate hardness values. It was observed that the hardness value increases with the increase in content of the reinforcement particles except for Cu-Gr composite which was observed low for the range in the study. SiC particles are excellent barriers to the dislocation movement in the copper matrix. Increasing the SiC content, strongly impeded the plastic flow, resulting the hardness of Cu–SiC composite to increase as shown in the figure 12 [1,20]. Higher hardness value was observed for Cu-Gn composite for the entire range due to the superior mechanical properties of graphene despite the presence of pores [21]. Areas with pores were carefully avoided while selecting points. It was reported that graphene has a tensile strength of 130Gpa and youngs modulus of 1Tpa [10]. Highest hardness value of 82 ± 2.4HV_{100} was observed for Cu-0.8Gn. Due to the soft nature of graphite the hardness values of the Cu-Gr composites were observed to be the lowest. As the weight percentage of graphite increased, hardness values were found to be significantly low. The hardness results showed an increasing trend in the beginning and started to decline with more addition of graphite. Presence of adequate mechanical mixed layer (MML) is reported to have significant effect in the tribological and mechanical properties of graphite. Cu-5Gr shows better hardness values compared to other compositions of Cu-Gr. This can be attributed to homogeneous microstructure of the composite. Lowest hardness value of 35.3±2.2 HV_{100} was observed for Cu-10Gr. Presence of large pores was detected in these composites. It was observed that a very small weight percentage of graphene particles added to the matrix could improve the hardness values to a bigger extend [9,10]. Dispersion strengthening is the main strengthening mechanism which was responsible for the improvement in the hardness values of the composites [8]. The hardness results mainly projected the mechanical properties of the reinforcement particles where a high hardness value was observed for Cu-SiC and Cu-Gn composites. Lower values of Cu-Gr composites were observed due to the soft nature of graphite, SiC being hard and graphene with superior mechanical properties contributed to improvement in hardness values.

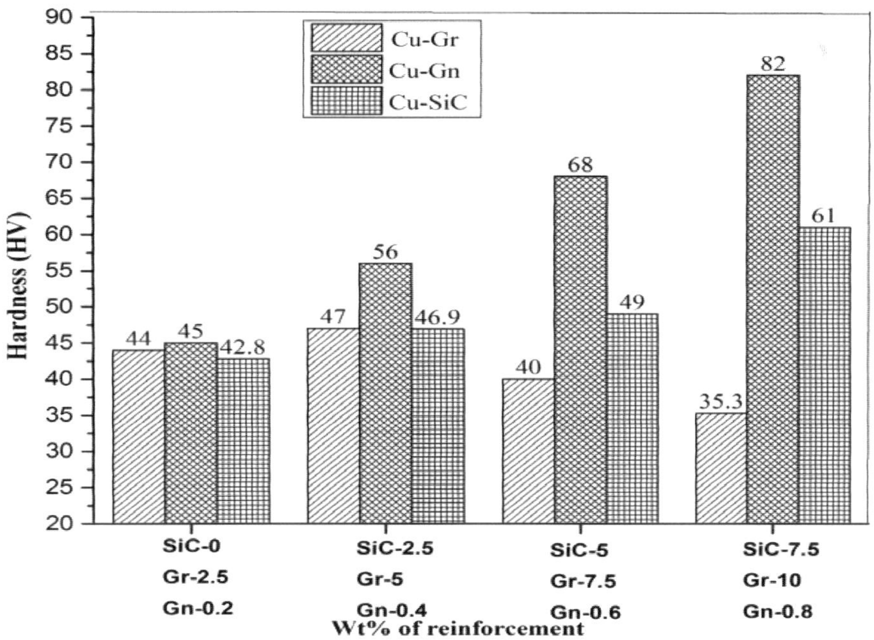

Figure 12. The variation of Vickers micro-hardness for the composites sintered by microwave sintering processes at 950° C

3.4. Electrical conductivity

Electrical conductivity of all sintered composites are shown in table 5. Pure copper exhibited excellent electrical conductivity. Annealed copper has an electrical conductivity of 59 MS/m (100%) which is set as the standard.6%IACS variation from that of annealed copper is due to presence of porosity and slight variations in purity of composites by the encapsulation of oxides. Cu-0.2Gn MMC with highest relative density value exhibited the highest electrical conductivity of 95% IACS. Due to presence of porosity in the composites conductivity values started to decline with more addition of graphene. Lowest electrical conductivity reported for Cu-Gn MMC was 86% IACS for Cu-0.8Gn composite. For most of the application which demands for high hardness and significantly high conductivity values Cu-0.8Gn composites can be employed. A tradeoff between electrical conductivity and hardness is preferred for use in brushes, slip rings etc. Presence of a conductive network is inevitable in attaining high electrical conductivity values. With the addition of ceramic SiC particles where there is a great mismatch between the conductivity values, a drastic decline in electrical conductivity of Cu-SiC composites was observed. Addition of 2.5wt% SiC results in a significant reduction in conductivity of copper. A maximum of 80% IACS and a minimum of 54% IACS was observed for Cu-2.5SiC and Cu-7.5SiC composites. Adding more SiC into copper increases hardness but reduces electrical conductivity to a greater extent. Almost 75% reduction in conductivity was observed for Cu-7.5SiC composite as compared to pure copper sample. Conductive network of copper is hindered by the addition of large fraction of less conductive SiC particles. Cu-2.5Gr composite exhibited an electrical conductivity value of 88% IACS. Graphite is a good conductor of electricity, so addition of small weight percentages does not bring much change in the observed values. Relative density values decline with addition of large weight percentages and which in turn reduces the conductivity values of MMCs. Lowest value among the Cu-Gr composites was observed for Cu-10Gr MMC with an electrical conductivity of 55% IACS. It was observed that an optimum addition of graphite (2.5-5 wt%) enhances the properties without large variations in electrical conductivity values.

Table 5. Variation of electrical conductivity of Cu-SiC, Cu-Gr and Cu-Gn MMCs sintered by microwave process at 950^{0}C

Composition	Electrical conductivity (%IACS)	Composition	Electrical conductivity (%IACS)	Composition	Electrical conductivity (%IACS)
Pure Copper	94	Cu-2.5Gr	88	Cu-0.2Gn	95
Cu-2.5SiC	80	Cu-5Gr	81	Cu-0.4Gn	92
Cu-5SiC	65	Cu-7.5Gr	60	Cu-0.6Gn	89
Cu-7.5SiC	54	Cu-10Gr	55	Cu-0.8Gn	86

4. Conclusion

Microwave sintering process was employed to successfully fabricate copper matrix composites with a varying addition of SiC, graphite and graphene. The effect on hardness, microstructure and electrical conductivity of these MMCs were investigated.

- Microwave sintering resulted in better consolidation of the composites.
- Relative density of Cu-Gn composites exhibited better values as compared to its counter parts. Densification response of all composites found to reduce with an increase in addition of the reinforcement particles.
- Optical and SEM micrographs revealed the uniform distribution of the reinforcement particles in the copper matrix for Cu-SiC, Cu-Gn and Cu-Gr composites for low weight percentages of reinforcement particles.
- No other second phases were observed in the micrographs except the copper reinforcement interface
- SEM-EDS analysis revealed the presence of oxygen in all the composites due to oxidation of copper at sintering temperature.
- From Vickers micro hardness results it was observed that a higher content of SiC and graphene could improve the hardness of the composite drastically. Due to the soft nature of graphite, hardness values of Cu-Gr MMCs were seen to decline with addition graphite.
- A maximum hardness value of 82 ± 2.4HV_{100} was observed for Cu-0.8Gn composite. Lowest value observed was 35.3±2.2 HV_{100} for Cu-10Gr composite.
- For obtaining superior electrical properties, an optimum quantity of graphene and graphite is preferred. For higher hardness combined with tailorable electrical conductivity of Cu-SiC composites, 2.5-5 wt% SiC is suggested.

Scope for future work

- Thermal conductivity and wear behavior of the composites can be studied.
- Coefficient of thermal expansion (CTE) of the composites can be investigated.

Acknowledgment

The authors would like to acknowledge the facilities and equipment available in VIT, Vellore provided by Department of Science and technology, India, under various projects, which was utilized for the successful completion of the work.

References

[1] Celebi Efe G, Zeytin S and Bindal C 2012 The effect of SiC particle size on the properties of Cu–*SiC composites Mater. Des. (1980-2015)* **36** pp 633–639

[2] Tjong S C and Lau K C 2000 Tribological behaviour of SiC particle-reinforced copper matrix Composites *Mater. Letrs.* **43** pp 274–280

[3] Zhu J, Liu L, Zhao H and Shen B 2007 Microstructure and performance of electroformed Cu/nano-SiC composite. *Mater. Des.* **28** pp 1958–1962

[4] Kuen-Ming Shu and Tu G C 2003 The microstructure and the thermal expansion characteristics of Cu/SiCp composites *Mater. Sci. Eng.* **349** pp 236–247

[5] Celebi Efe G, Ipek M, Zeytin S and Bindal C 2012 An investigation of the effect of SiC particle size on Cu–SiC composites *Compos. Part B eng.* **43** pp 1813–1822

[6] Kovacik J, Emmer S and Bielek J 2015 Thermal conductivity of Cu-graphite composites *Int. J. Therm. Sci.* **90** pp 298–302

[7] Rajkumar K, Aravindan S and Kulkarni M .5 2012 Wear and Life Characteristics of Microwave-Sintered Copper-Graphite Composite *J. Mater. Eng. Perform.* ASM InternationalDOI:10.1007/s11665-012-0161-z

[8] Chandana priyadarshini samal 2012 *Microstructure and mechanical property study of Cu-graphite metal matrix composite prepared by powder metallurgy route* (Thesis,NIT Rourkela)

[9] Varol T and Canakci A 2015 Microstructure, Electrical Conductivity and Hardness of Multilayer Graphene/Copper Nanocomposites Synthesized by Flake Powder Metallurgy *Met. Mater. Int.* **21** pp 704–712

[10] H Prashantha Kumar and Anthony M Xavior 2014 Graphene Reinforced Metal Matrix Composite (GRMMC): A Review *Procedia Eng.* **97** pp 1033–1040

[11] Ayyappadas C, Muthuchamy A, Raja Annamalai A and Dinesh K. Agrawal 2017 An investigation on the effect of sintering mode on various properties of 5 copper-graphene metal matrix composite, http://dx.doi.org/10.1016/j.apt.2017.04.013

[12] Roy R ,Agrawal D, Cheng J and Gedevanishvili S 1999 Full sintering of powdered metal bodies in a microwave field *Nature* **399** pp 668–670.

[13] Penchal Reddy M, Shakoora R A, Gururaj Parande, Vyasaraj Manakari, Ubaid F, Mohamed A M A and Manoj Gupta 2017 Enhanced performance of nano-sized SiC reinforced Al metal matrix nanocomposites synthesized through microwave sintering and hot extrusion techniques *Progress in Natural Science: Materials International* **27** pp 606–614

[14] Celebi Efe G, Yener T, Altinsoy I, Ipek M, Zeytin S and Bindal C 2011 The effect of sintering temperature on some properties of Cu–SiC composite *J. Alloys Compd.* **509** pp 6036–6042

[15] Celebi Efe G F, Altinsoy I, Ipek M, Zeytin S, Bindal C 2011 Some properties of Cu–SiC composites produced by powder metallurgy method *Kovove Mater* **49** pp 131–136

[16] Ovidko I A 2014 Metal graphene nano composites with enhanced mechanical properties; a review, *Rev.Adv. Mater. Sci.* **38** pp 190-200

[17] Celebi Efe G F 2010 *Development of Conductive Copper Composites Reinforced with SiC* (Ph.D. thesis, Sakarya University, Institute of Science and Technology)

[18] Samal C P, Parihar J S and Chaira D 2013 The effect of milling and sintering techniques on mechanical properties of Cu–graphite metal matrix composite prepared by powder metallurgy route J. Alloys Compd. **569** pp 95–101

[19] Kang H K 2005 Microstructure and electrical conductivity of high volume Al_2O_3-reinforced copper matrix composites produced by plasma spray *Surf. Coat. Technol.* **190** pp 448–452

[20] Ramesh C S, Ahmed R N, Mujeebu M A and Abdullah MZ 2009 Fabrication and study on tribological characteristics of cast copper-TiO2-boric acid hybrid composites *Mater. Des.* **30** pp 1632–1637

[21] Jagannadham K 2012 Thermal Conductivity of Copper-Graphene Composite Films Synthesized by Electrochemical Deposition with Exfoliated Graphene Platelets *Metall. Mater. Trans. B* **43** pp 316–24

Journal of Physics: Conference Series **1355** (2019) 012036 doi:10.1088/1742-6596/1355/1/012036

Investigation of Magnetostriction in Second Generation High Temperature Superconductors

K.B. Ashok, R. J. Thomas, M. Jose Prakash.
Department of Mechanical Engineering T K M College of Engineering, India

E-mail: ashokbalan189@gmail.com

Abstract. Superconductors help to drastically improve the efficiency in every field by making the transmission losses of electrical energy to minimum. Superconductors show special properties only when their parameters like temperature, magnetic field, current density, mechanical strain are within their critical limits. Effect of magnetic field on second generation high temperature superconducting compounds have been studied in past few years. In this paper, the numerical simulation of the magnetostriction effect of second generation superconductor REBCO is discussed. The results obtained from numerical studies is compared with the experimental results reported in literature. The findings show that there is good agreement and all the three samples are showing rapid change in magnetostriction when the magnetic field is above 10^6 A/m.

1. Introduction

Magnetostriction is one of the peculiar phenomena shown by superconductors. It describes the change in dimensions of a material due to a change in its magnetization. This phenomena was discovered by James Joule in 1840. Magnetostriction depends on the strength of the magnetic field and the material. Magnetostriction of a single crystal first generation superconductor BSCCO ($Bi_2Sr_2CaCu_2O_8$) material is in the order of 10^{-4} [1-2]. This falls in the category of giant magnetostriction as the value is greater than the magnetostriction shown by other superconductors. Superconducting crystal of Magnesium Dibromide (Mg_2B) has the magnetostriction in the order of 10^{-5} [3]. Most of the high temperature superconductors (HTS) show this effect in the order of 10^{-6} [4]. But the type II superconductor Niobium Titanium (NbTi) shows in the order of 10^{-7} [5]. Strong ferro and antiferromagnetic materials attains magnetostriction value in the order of 10^{-5} to 10^{-2}; while that for diamagnetic and paramagnetic materials they are in the order of 10^{-7} to 10^{-5} [6].

Compared to other superconducting materials, the rare earth Barium Cupric oxide (REBCO) is found to be a promising one. Maruyamaa et al. reported the development of REBCO HTS power cables and succeeded in reducing current loss to one-sixth the loss of that of a BSCCO cable [7]. REBCO coated conductors are more attractive because of their high critical temperature of 92 K. They also have high mechanical strength [8]. Considering the positives of REBCO, it may be expected that superconducting wires and cables will be the next generation ones. However, there is limited numerical studies on REBCO especially in bringing out the effects of magnetostriction on their performance while in application.

The effect of magnetostriction in REBCO materials has an influence when they are used to make superconducting cables. The type-II superconductors being brittle in nature and the changes in their dimensions due to magnetostriction will result in increased strain inside the cables made out of them.

Journal of Physics: Conference Series **1355** (2019) 012036 doi:10.1088/1742-6596/1355/1/012036

Consequently, the performance of these superconducting cables will be degraded. Through experimental investigations it is difficult to identify the individual influence of each parameter on the performance due to the intertwined effects. Mathematical modelling has an important role in such situations. Studies using the mathematical model will not only help to save the money and time, but also in improving the theoretical knowledge of these effects through various parametric investigations.

This paper provides the details of the process by mathematically modelling the effects of magnetostriction in various superconducting materials. The approach will be to construct and analyse the mathematical model representing various physical situations. Being a multi-physics problem, modelling is a complex process here and it requires modification and customization of existing routines in the commercially available software. Literature shows that COMSOL Multiphysics was used to successfully model the Surface Acoustic Wave (SAW) sensors made out of magnetostrictive material; the model was generated using the AC/DC and the structural mechanics modules of COMSOL Multiphysics [9]. Numerical model to examine the magnetostriction of bulk high temperature superconductor under uniform and non-uniform magnetic field is reported by Li X et al. [10].

2. Basics Concepts for the calculation of Magnetostriction

The magnetostriction component along any direction can be calculated as a nonlinear function of the magnetization.

$$\lambda_i = \frac{3}{2}\lambda_s\left[\left(\frac{Mi}{Ms}\right)^2 - \frac{1}{3}\right] \quad (1)$$

Where λ_s is the saturation of magnetostriction and direction cosine is the ratio of magnetization along the required direction M_i and the saturation magnetization M_s of the material. In this work, it is assumed that the material is sufficiently pre-stressed such that all magnetic moment are perpendicular to the direction of magnetization at the beginning of magnetization process.

The free strain due to magnetostriction is incorporated using an extension of the Hooke's Law. COMSOL uses a generalized Hooke's Law to describe the constitutive relationship between the stress and strain tensors in a solid mechanics problem. This general form of the linear constitutive relation allows the incorporation of initial strain in a material which does not contribute to mechanical stress unless the material is mechanically constrained. Because, by definition, the magnetostriction or magnetization-induced free strain in the material is equivalent to an initial strain, this idea is used to incorporate the magnetostriction calculated from equation 1.

3. Modelling

The size of superconducting crystal was taken as 1.05 x 3.3 mm as shown in the figure 1. An air domain is created around the superconductors to realistically model the magnetic flux path. The boundaries of this air domain are magnetically insulated which ensure that flux does not diverge out of the modelling domain.

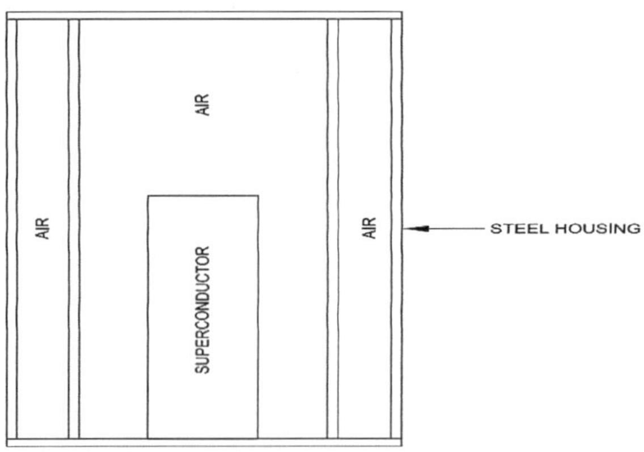

Figure 1: Modelling arrangement

The current is applied externally by creating a coil around the superconductors. The external current density is the total current through the coil divided by the longitudinal cross sectional area. The steel housing used in the model is designed to create a closed magnetic flux path thereby minimizing flux leakage. The non-linear magnetic behaviour of the steel is modelled by using a hysteresis curve to specify the magnetic constitutive relation in the steel housing. Bottom side of the superconductor is fixed to the steel housing. Due to the external current around the superconductors create a magnetic field and magnetize the superconductor. The current density is varied from 0 to 10^{10}A/m^2.

Table 1: Properties of $YBa_2Cu_3O_{7-x}$, $DyBa_2Cu_3O_{7-x}$ and $HoBa_2Cu_3O_{7-x}$ [4, 13, 15]

Compound	Young's Modulus (GPa)	λ_tx 10^{-6} (Anisotropic magnetostriction Constant)	Poisson's ratio
$YBa_2Cu_3O_{7-x}$	148	5	0.3
$DyBa_2Cu_3O_{7-x}$	95	-84	0.3
$HoBa_2Cu_3O_{7-x}$	95	-66	0.3

Incorporation of a nonlinear hysteresis curve help in modelling magnetic saturation at sufficiently high magnetic field. COMSOL Multiphysics have good collection of magnetic field vs magnetisation (HB) properties of different materials. In the same manner as mentioned above, a non-linear HB relation is also used for superconducting material [11-12].Physics controlled mesh is used to solve the model. It may be noted that in this work, the magnetostriction effect parallel to the magnetic field is only considered as the Poisson's ratio is 0.3 and the effect will be more significant in the longitudinal direction. This has also been experimentally shown by Lu J et al. [11], where the magnetization and hysteresis loss of a 1x1 mm^2 REBCO sample was measured in different field and field orientation. They have found that hysteresis losses are dependent on the magnetic field and its field angle for a *Super Power SCS 4050* REBCO conductor.

The various issues faced during the process of numerically simulating is discussed. The HB properties of superconducting compound is not available at higher magnetic fields and therefore, had to be extrapolated and assumed. Another challenge is that COMSOL Multiphysics can't take magnetization and demagnetization values corresponding to the single magnetic field. So magnetostriction effect during the magnetization is considered in this study. The demagnetisation studies need to be conducted separately. At the end, for validation, the results obtained through numerical investigation are compared with the experimental data available in the literature.

4. Results and Discussion

Numerical studies have been conducted on the 3 type of samples selected is given separately. The dimensional variation in the longitudinal direction is captured with the increase in magnetic field and is plotted accordingly. Figure 2 shows the effect of magnetic field on magnetostriction for Yttrium Barium Cupric Oxide (YBCO) $YBa_2Cu_3O_{7-x}$ specimen. The results obtained from experimental studies are also shown in the graph for comparison. It may be observed that the numerical results are in good agreement wherever the experimental results are available. It may be also inferred that when the magnetic field reaches the range of 10^6 A/m, there is a sudden increase in the effect of magnetostriction on the material.

The area where experimental results are available is zoom and plotted as inset in Figure 2. It may be seen that experimentally also there is a trend for a sudden increase in magnetostriction effect when increasing the field beyond certain limit.

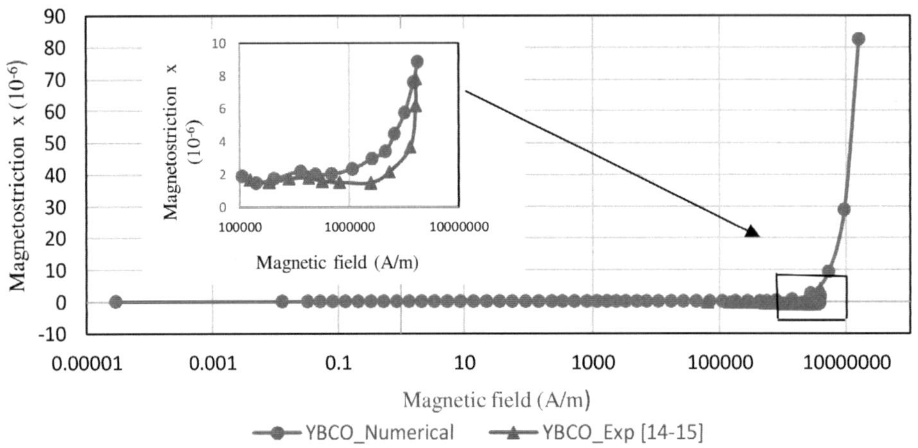

Figure 2: a) Magnetostriction vs Magnetic Field (YBCO) b) Magnified View of specified area

Figure 3 illustrates the magnetostriction of $DyBa_2Cu_3O_{7-x}$ compound under varying magnetic field. $DyBa_2Cu_3O_{7-x}$ shows the magnetostriction in the negative direction in the order greater than that of $YBa_2Cu_3O_{7-x}$ compound shown in Figure 2. Negative value indicates that there is contraction in the material in the longitudinal direction parallel to the magnetic field. Comparison of experimental and numerical results depicts good agreement. In this sample also it may be observed that when the magnetic field is in the range of 10^6 A/m, there is a sudden increase in magnetostriction effect. It shown that experimentally also there is a trend for a sudden increase in magnetostriction effect when increasing the field beyond certain limit.

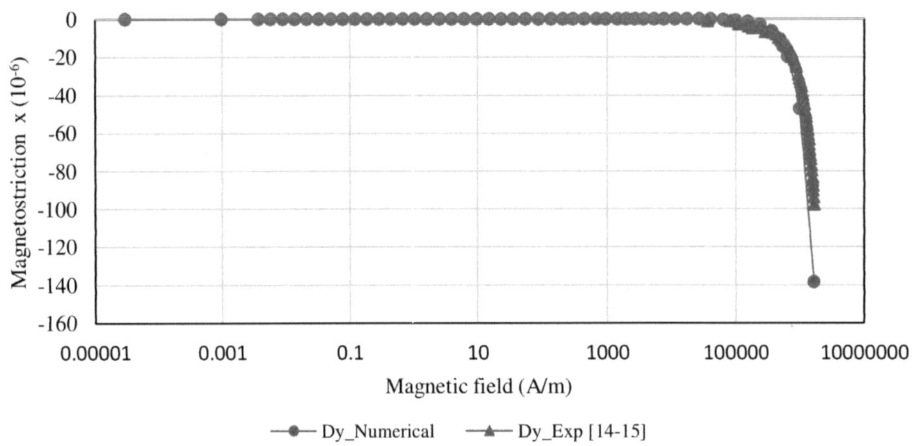

Figure 3: a) Magnetostriction vs Magnetic Field ($DyBa_2Cu_3O_{7-x}$)

Figure 4 shows the effect of magnetic field on magnetostriction of $HoBa2Cu3O7_{-x}$ compound. This compound shows a similar characteristics (negative values) as that YBCO compound along the direction parallel to the magnetic field. It may be observed that when the magnetic field is increased beyond the order of 10^5 A/m, there is a sudden increase in magnetostriction as similarly observed for other samples. Magnetostriction at the field greater than 10^7 A/m is not reported in the open literature due to the practical difficulty in applying such high magnetic fields.

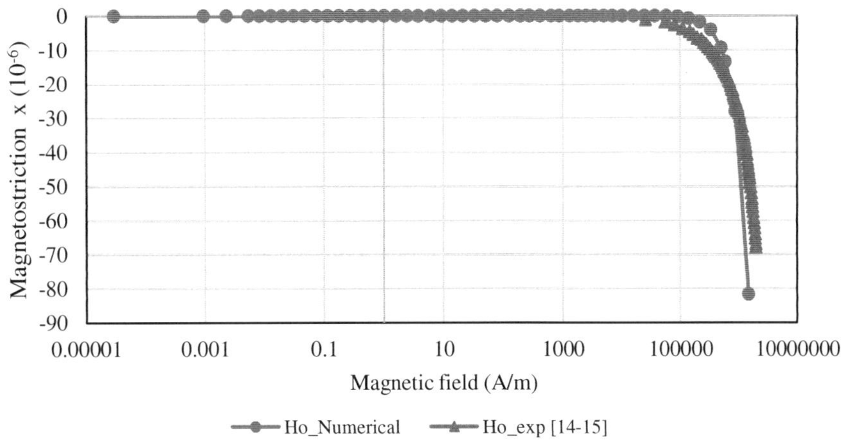

Figure 4: a) Magnetostriction vs Magnetic Field ($HoBa_2Cu_3O_{7-x}$)

$DyBa_2Cu_3O_{7-x}$ and $HoBa_2Cu_3O_{7-x}$ shows similar results of negative variation and have good agreement with the experimental results. They have magnetostriction effect in the order of 10^{-5}, when the magnetic field is in the range of 10^6 A/m. It is also seen that the order of magnetostriction is higher in these compounds compared with YBCO superconducting compound. In YBCO, the expansion is observed as the effect of magnetostriction and is in the order of 10^{-6} and the sudden increase is at 10^6 A/m.

The dimension of superconductor vary due to magnetostriction effect because, when the magnetic field increases above the lower critical magnetic field, magnetic lines penetrate through the superconductor and decreases the condensation energy. This effect is associated with change in volume and leading to magnetostrictive strain. When the magnetic field is loaded to the expansive (positive) magnetostrictive materials, the tensile stress occurs parallel to applied magnetic field. On the other hand, the opposite phenomena will occur for compressive (negative) magnetostriction [16].

Out of the 3 samples, 2 of them had a negative effect (contraction) due to magnetostriction. It may be due to the anisotropic magnetostriction properties of materials. The anisotropic magnetostriction constant is given in the Table 1. It indicates that Dy and Ho compounds have negative values and the 2 samples with their presence, showed negative magnetostriction in the direction parallel to the applied field.

All the 3 compounds showed rapid rise in magnetostriction when the magnetic field crosses a particular value. It may be due to the fact that the magnetostriction is directly proportional to the square of magnetization as understood from Equation 1.

5. Conclusion

The numerical studies have been conducted to analyse the effect of magnetic field on different high temperature superconducting samples and the following conclusions are obtained. When the magnetic field attain a value of 10^6 A/m, it is found that magnetostriction effect in YBCO compound is in the order of 10^{-6} and other two compounds have the order of 10^{-5}. In comparison with the experimental results available in the literature, it may be concluded that numerical simulation will be very helpful in calculating magnetostriction effect. It is also understood that the magnetostriction effect shown by $DyBa_2Cu_3O_{7-x}$, $HoBa_2Cu_3O_{7-x}$ are negative and the order of magnetostriction greater than the YBCO compound.

6. References

[1] H Ikuta, Y Nakayama, N Hirota, K Kishio and K Kitazawa,1993*Giant Magnetostriction of Cuprates in Superconducting State*Applied Superconductivity **Vol.1**, Nos 3-6. Pp 635 – 643.

[2] H Ikuta, NHirota, K. Kishio and K. Kitazawa 1994*Magnetostriction in High-To Cuprate Single Crystals,Physica***C 235-240** 237-240.

[3] A Nabiaeka, BKundysa, Yu Bukhantseva, S Vasilieva, A Wi sniewskia, J Junb, S M Kazakovb, J Karpinskib, and HSzymczak 2002 *Magnetostriction in superconducting MgB2*Physica**B 319** 286–292.

[4] M Braden, P Bfhm, F Seidler, H Kalenborn, and D Wohlleben, 1990 *Magnetostriction of several high-Tc superconductors*, Z. Phys. B - Condensed Matter **79**, 173-176.

[5] A Gerbera, R Schleserb, P J E M van der Lindenb, and PWyderb, 2002, *Magnetostriction of the model type-II superconductor NbTi*Physica**B 319** 293–302.

[6] Eremenko V V, SirenkoV A, Szymczak H, and Nabialek A, 1999*Magnetostriction of superconductors (a review) Low Temperature Physics*, **25(4)** 225–241.

[7] Maruyama O, Ohkuma T, Masuda T, Ohya M, Mukoyama S, Yagi M and Hayakawa N 2012 *Development of REBCO HTS Power Cables* Physics Procedia**36** 1153–1158.

[8] Van der Laan D C, Noyes P D, Miller G E, WeijersH W and Willering G P 2013*Characterization of a high-temperature superconducting conductor on round core cables in magnetic fields up to 20 T* Superconductor Science and Technology**26(4)** 045005.

[9] Elhosni M, Elmazria O, Petit-Watelot S, Bouvot L, ZhgoonS,Talbi A and Boumatar O, 2016*Magnetic field SAW sensors based on magnetostrictive-piezoelectric layered structures: FEM modeling and experimental validation*. Sensors and Actuators A: Physical, **240**, 41–49.

[10] Li X, Jiang L, Wu H, and Gao Z 2017*The magnetostriction in a superconductor-magnet system under non-uniform magnetic field*. Physica C: Superconductivity and Its Applications, **534**, 50–54.

[11] Lu J, Choi E S, Kandel H, Abraimov D V and Markiewicz W D 2014 *Hysteresis Loss of REBCO Conductor Up to 35 T* IEEE Transactions on Applied Superconductivity **24(3)** 1–4.

[12] Semenov S V, and Balaev D A 2018 *Temperature behavior of the magnetoresistance hysteresis in a granular high-temperature superconductor: Magnetic flux compression in the intergrain medium*. Physica C: Superconductivity and Its Applications, **550**, 19–26.

[13] Lei M, Sarrao J L, Visscher W M, Bell T M, Thompson J D, Migliori A, 1993 *Elastic constants of a monocrystal of superconductingYBa2Cu3O7−δ*. Physical Review **B 47(10)** 6154–6156.

[14] A Del Moral, M R Ibarra, J I Arnaudas, P AAlgarabel, Marquina Moran and M AAlario1989*MagnetostrictionAnd Thermal Expansion Of High T Magnetic Superconductors* Rebazcu307_ ~ (Re = Eu, Sm, Gd, Dy, Ho, Er, Tm And Y)Physica**C 161** 48-58.

[15] A Del Moral , M R Ibarra , J I Arnaudas , PAAlgarabelMarquina Moran And M AAlario,1988 *Magnetostriction And Thermal Expansion Of High T,. Magnetic Superconductors* Rebazcu307_ ~ (Re = Eu, Sm, Gd, Dy, Ho, Er, Tm and Y) Journal of Magnetism and Magnetic Materials **76** & **77**612-614.

[16] Yoshitake Nishi, Yoshihito Matsumura, Atsushi Kadowaki and Shingo Masuda, 2005 *Composition Dependence on Compressive Magnetostriction of Fe–Sm Binary Alloy Thin Films* Materials Transactions, Vol. **46**, No. 12 pp. 3063 to 3066.

Synthesis and Characterization of Ultrasonically Treated Al/Al_2O_3 Nanocomposites

Syed Muhammed Fahd[1], Krishnakumar T. S.[1], Aju kumar V. N.[1] and Agesh Markose[1]
[1] Assistant Professor in Mechanical Engineering, TKM College of Engineering, Kollam

E-mail: muhammed.fahd@gmail.com

Abstract. Aluminium-Alumina (Al/ Al_2O_3) nanocomposites are synthesized for different amounts of reinforcement through liquid route. Ultrasonic treatment (UST) is carried out during the solidification of the melt. It is expected that the addition of nanoparticles to the melt and UST provides grain refinement for the specimens due to increased number of nucleation sites and cavitation effects respectively. The effects of UST on these nanocomposites are studied using the results obtained from optical microscopy, hardness and XRD. Reduction of porosity and grain refinement is observed in the synthesized nanocomposites.

Keywords: Ultrasonic treatment, grain refinement, nanocomposites

1. Introduction

Aluminium and aluminium alloys owing to their low density and high strength to weight ratio finds numerous applications in automobile and aerospace industries. Mechanical properties of these alloys can be further improved by reinforcing them with nanoparticles, called nanocomposites. Various methods have been developed for synthesizing aluminium matrix nanocomposites out of which the liquid route using stir casting is considered economical [1]. Ultrasonic treatment (UST) in aluminium alloy melts is known to yield degassing effect and grain refinement. These effects are due to cavitation generated by the high frequency ultrasound wave [2].
Ultrasound treatment in the synthesis of nanocomposites can disperse the particulate reinforcements uniformly. Introduction of nanoparticles to the melt increases heterogeneous nucleation sites leading to grain refinement [3][4]. The objective of this study is to investigate the effect of alumina nanoparticle addition in pure aluminium for ultrasonically treated melts.

2. Experimental Methods

Specimens of Al/Al_2O_3 nanocomposites are synthesised with aluminium as the matrix and alumina nanoparticle as the reinforcement. Alumina nanoparticle used for this study is purchased from Sigma Aldrich, Bangalore and is of 50 nm average size. Specimens prepared are i) aluminium, ii) aluminium with 2% alumina nanoparticle and iii) aluminium with 3% alumina nanoparticle, with ultrasonic treatment for all the three specimens during solidification of the melt. These are then characterized using optical microscopy, Vicker's hardness and X-ray diffraction. Details of the specimens used for characterization are given in table 1.

Table 1. Details of specimens used for characterization

Specimen No	Details
S1	As received
S2	With UST and no reinforcement
S3	With UST and 2% reinforcement
S4	With UST and 3% reinforcement

To synthesize the nanocomposite specimens following method is used. Pure aluminium is melted on a table top furnace in a clay bonded graphite crucible. Melt is heated up to 740°C and alumina nanoparticle is added. The mixture is mechanically stirred for 10 minutes and is poured into a steel mould. The steel mould is then placed on the ultrasonic agitator and ultrasonic wave is propagated through the mould till the composite melt solidifies. Experiments are conducted with ultrasound frequency of 40 KHz. The schematic diagram of the experimental setup is shown in figure 1.

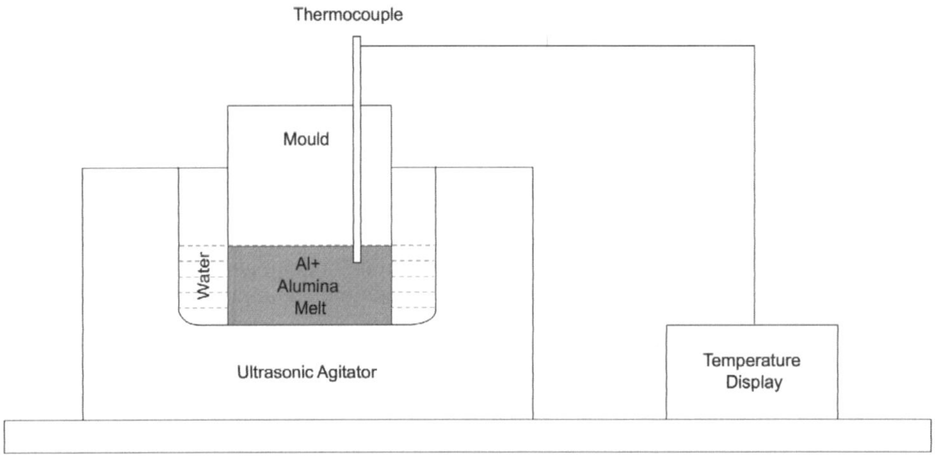

Figure 1. Experimental Setup for the synthesis of nanocomposite specimens

3. Results and Discussion

3.1 Optical Microscopy Analysis

The microstructures of the specimens prepared are given in figures 2 and 3. As seen from the microstructures it is evident that significant grain refinement has happened for specimens which are ultrasonically treated.

The average grain sizes of the specimens are determined using the line intercept method and is given in figure 4. The average grain size of the as-received specimen is greater than 70 μm. For specimen S2 the average grain size is found to be < 50 μm. For specimen S3 also the average grain size is found to be < 50 μm. And for specimen S4 the average grain size is around 20 μm. It is also observed that in comparison with the as received specimen there is significant reduction of porosity. This can be attributed to the degassing effect that the UST has [1].

Journal of Physics: Conference Series **1355** (2019) 012037 doi:10.1088/1742-6596/1355/1/012037

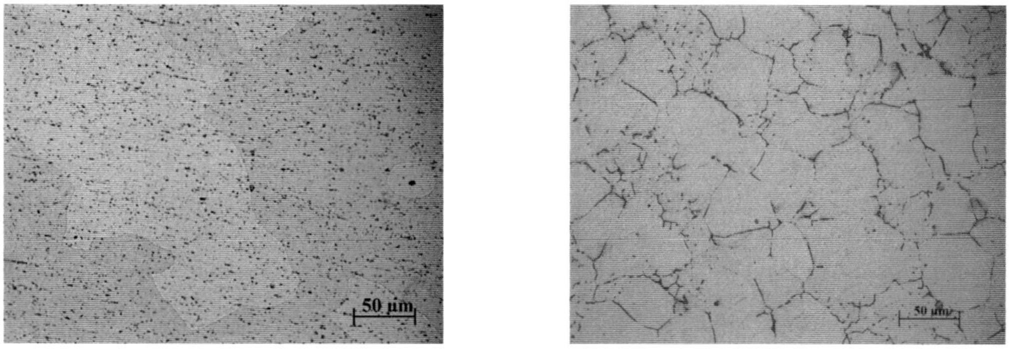

Figure 2. Microstructure of the a) as-received specimen, b) specimen with UST and no reinforcement

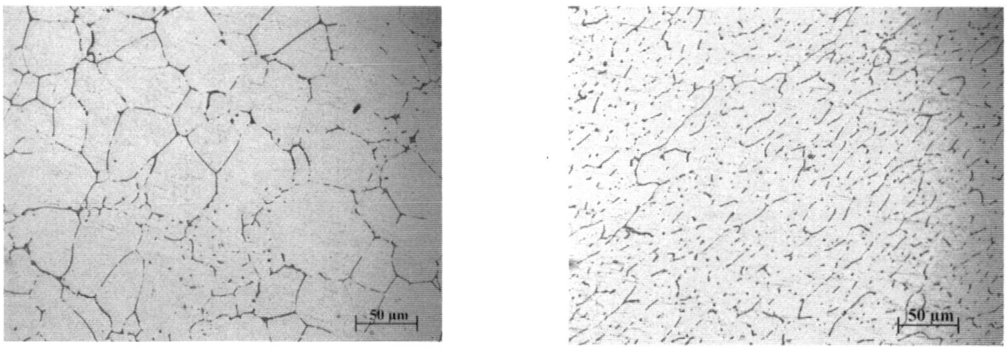

Figure 3. Microstructure of the specimen a) with UST and 2% alumina b) with UST and 3% alumina

All ultrasonically treated specimens are found to have non-dendritic equiaxed grains, which is consistent with the results of earlier studies [4]. It is also observed that a bimodal distribution of grain sizes exists in the UST specimens, which improves the ductility of the material [5].

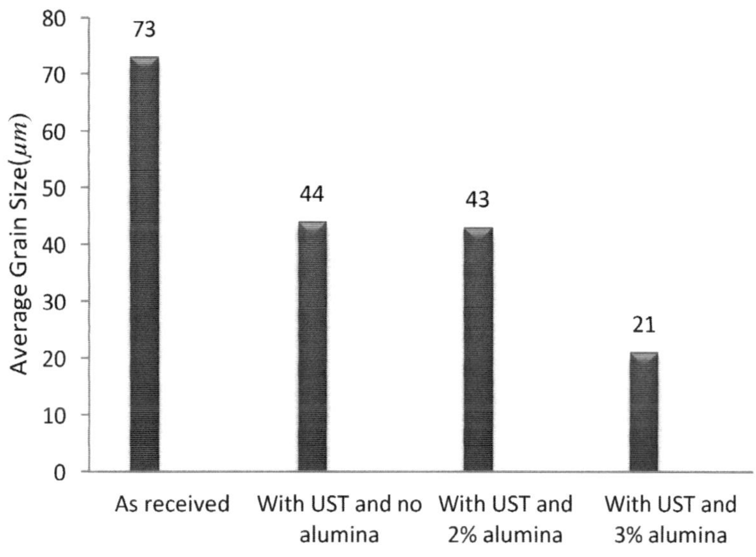

Figure 4. Comparison of grain size for specimens

A significant reduction in grain size is observed in the specimen S4. This is due to the fact that adding 3% by weight alumina nanoparticle to the melt and ultrasonic treatment has increased the number of nucleation sites there by decreasing the grain size of the material.

3.3 X-ray Diffraction Studies

It has been observed from our XRD studies that the crystallographic texture strongly depends on the percentage of Al_2O_3 in the composite. For specimen S4 where the percentage of Al_2O_3 is 3%, a double texture of (111) and (311) is observed whereas the other samples with no alumina and 2% alumina shows a strong (111) texture [6] as shown in figure 5.

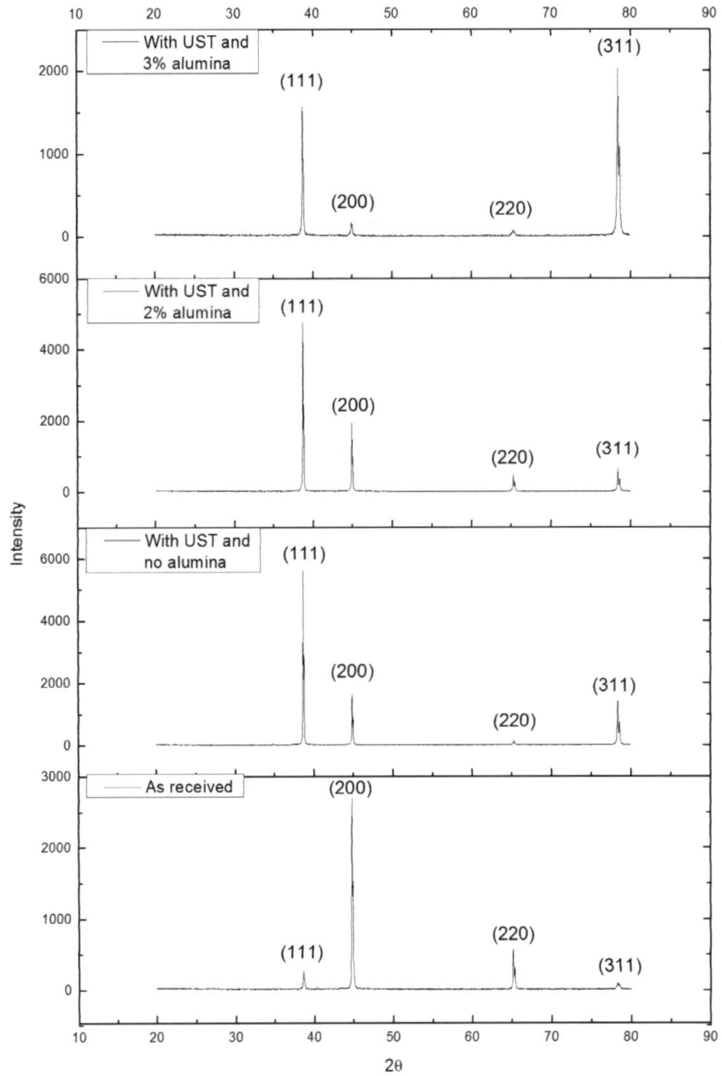

Figure 5. XRD Plots for the specimens

3.3 Hardness Testing

The hardness values of the ultrasonically treated specimens are given in figure 6. It is observed that the hardness values increase as the percentage of reinforcement in the composite increases. The increase in hardness values can be attributed to two main factors; one is the strengthening effect due to reinforcements and the second is the decrease in grain size. Both act as hindrances to dislocation motion there by strengthening the composite [7].

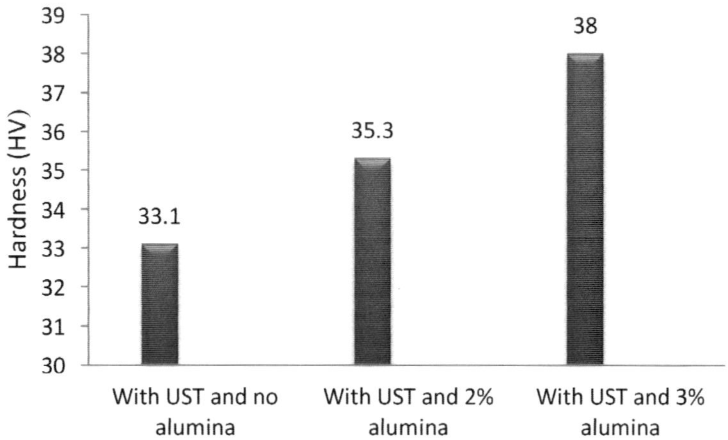

Figure 6. Comparison of hardness for specimens

4. Conclusions

UST aluminium and Al/Al_2O_3 nanocomposites were successfully synthesised. It was found that UST decreases the grain size and gives a equiaxed non-dendritic structure. Significant grain refinement was observed in Al /3% Al_2O_3 nanocomposite with ultrasonic treatment during solidification. Another observation was the reduction in porosity of the ultrasonically treated specimen. The crystallographic texture was found to be dependent on the percentage of alumina nanoparticles in the composite. It is also observed that the addition of nanoparticles and ultrasonic treatment resulted in the increase of hardness values.

References

[1] Casati R and Vedani M 2014 *Metals* **4** 65
[2] Eskin D G 2016 *Material Science and Technology* **33** 636
[3] Xuan Y and Nastac L 2018 *Ultrasonics* **83** 94
[4] Wang G, Wang Q, Easton M A, Dargusch M S, Qian M, Eskin D G and StJohn D H 2017 *Scientific Reports* **7** 9729
[5] Oskooie M S and Asgharzadeh H 2014 *IOP Conf. Series: Mat. Sci. and Engineering* **63** 012022
[6] JCPDF *Database International Centre for Diffraction Data* 1998 PA,Card No. PDF/851327
[7] Jia S Zhang D Xuan Y Natsac L 2015 *Applied Acoustics* **103** 226

Modelling and Analysis of a Progressive Cam Tool for Blade Fuse Holder: A Case Study

Asif Afzal, Mohammed Sirajuddin, Abdul Razak R.K, Chethan K.M, Avinash
P. A. College of Engineering (Affiliated to Visvesvaraya Technological University, Belagavi), Mangaluru, India

E-mail: asif.afzal86@gmail.com

Abstract. Sheet metal segments delivered from press working find wide applications in the territories going from telephonic gadgets to space stations. To design a progressive tool for sheet metal components according to the given details is a testing assignment. This article covers design and modelling of a progressive cam press tool for the component blade fuse holder, which is used in the Printed circuit boards of cars. Few very essential basics of press tools are briefly mentioned. The design and modelling process begins with the careful examination of the component geometry and material specifications. The tool development and modelling of the tool carried out by modelling software Solid Works are provided in detail. Points of interest and detriments of progressive tools over stage tools are additionally spotlighted.

1. Introduction

Sheet metal is a metal formed into flat and thin pieces. It is one of the fundamental forms used in metalworking and can be cut and bent into a numerous different shapes. Countless day to day objects are constructed from the sheet metals [1]. Thicknesses can vary significantly, even though extremely thin thicknesses are well thought-out for foil or leaf, and sheet metal components that are thicker than 0.25 inches (6 mm) are usually considered for plates. Sheet metal is accessible as flat pieces or as a coiled strip. The coils are produced by running a continuous sheet of metal through a roll slitter. The thickness of the sheet metal components is known as its gauge. The sheet metal gauge typically varies from 8 gauges to about 30 gauges. The higher the sheet metal gauge, the thinner the metal will be. There are several different metals that can be prepared into sheet metal, such as brass, aluminium, copper, nickel, steel, titanium and tin. For decorative uses, important sheet metals include platinum, gold, and silver [2].

Sheet metal has applications in car bodies, airplane wings, and roofs for building, medical tables, and several other things. The sheet metal of iron and of materials having high magnetic permeability, which is also known as laminated steel cores, has many applications in electric machines and transformers [3]. Historically, a vital use of sheet metal was in plate armor worn by horse cavalry, and further sheet metal continues to have various decorative uses, together with, in horse tack [4], [5]. The prime objective of this work was to design a progressive cam tool to produce the component that exceeds the expectations of the consumer in aspects like cost, quality and on time delivery. It holds out the strong probability of achieving, reduced overall cost of manufacturing a product by producing acceptable parts at the lower cost, raise the production speed by designing tools that will generate parts and rapidly possible and uphold quality by designing tools, which will, over and over again produce parts with the obligatory precision [6-14].

The rest of the paper is composed as follows: Section II presents the component study and its drawing. Section III gives the 3D modeling details of tool and analysis details of the component. Section VI presents the manufacturing of the component. In section V we conclude with a brief report on challenges and hereafter work in progressive tools.

2. Progressive tool

The major parts of standard press tool die and progressive tool are mentioned in this section. Important press cutting operations generally employed are also mentioned.

2.1. Press Tool Elements

In general press tool consists of the following two half's:

- Top half, which consists mainly of: Punches, Punch plate, Punch back plate, Top plate/ top bolster, Shank, Guide bush. All these parts are shown in Figure 1 and 2.
- Bottom half, which consists of: Stripper plate, Guide plate, Die plate, Bottom plate/ bottom bolster, Guide pillar. All these parts are clearly shown in Figure 2 and 3.

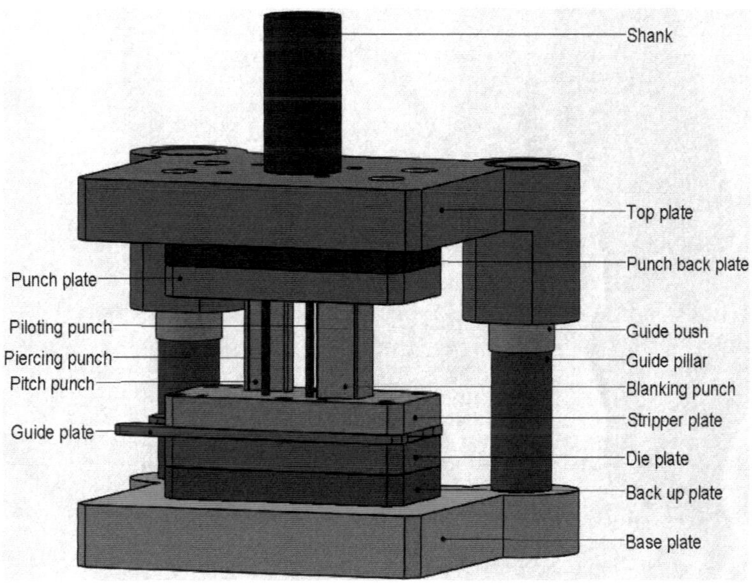

Figure 1. Press tool elements

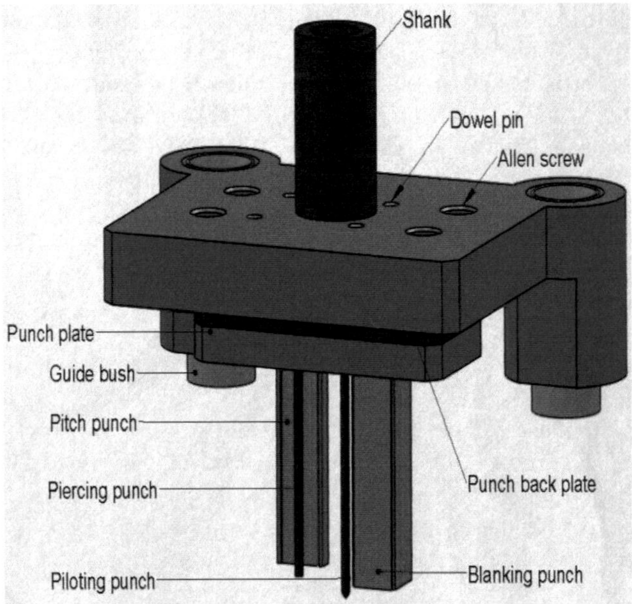

Figure 2. Top half of the tool

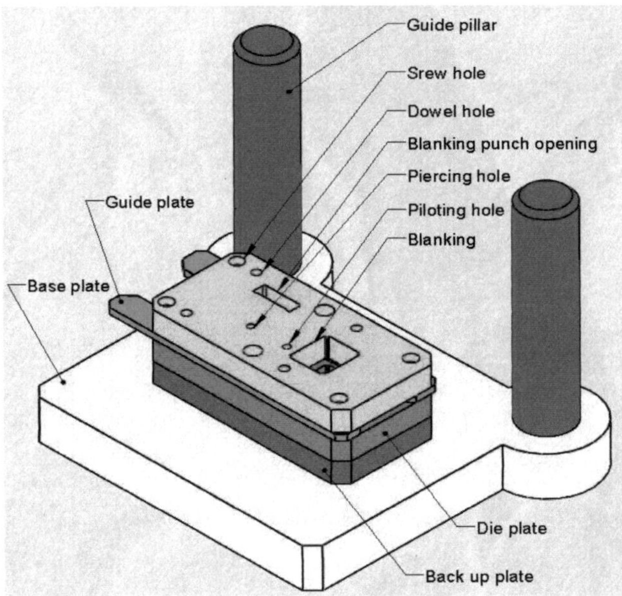

Figure 3. Bottom half of the tool

2.2. General Press Tool Operations
General operations done carried out using press tools are mainly:
- Cutting operations: Piercing Notching, Trimming, Cut off and Part off.
- Non cutting operations: Bending, Forming, Drawing, Curling and Dimpling.

3. Design of progressive cam tool

Journal of Physics: Conference Series **1355** (2019) 012038 doi:10.1088/1742-6596/1355/1/012038

3.1. Component: Blade Fuse Holder

The study of the component is the most important and the first step for the designer. The component drawing is studied to know the important features like its geometry, special and important features of the component. The component drawings are carefully scrutinized to extract the maximum possible amount of information as shown in Figure 4. The solid models of the component are as shown in Figure 1-3. By using these models the entire tool is built. Table 1 gives the details of the component.

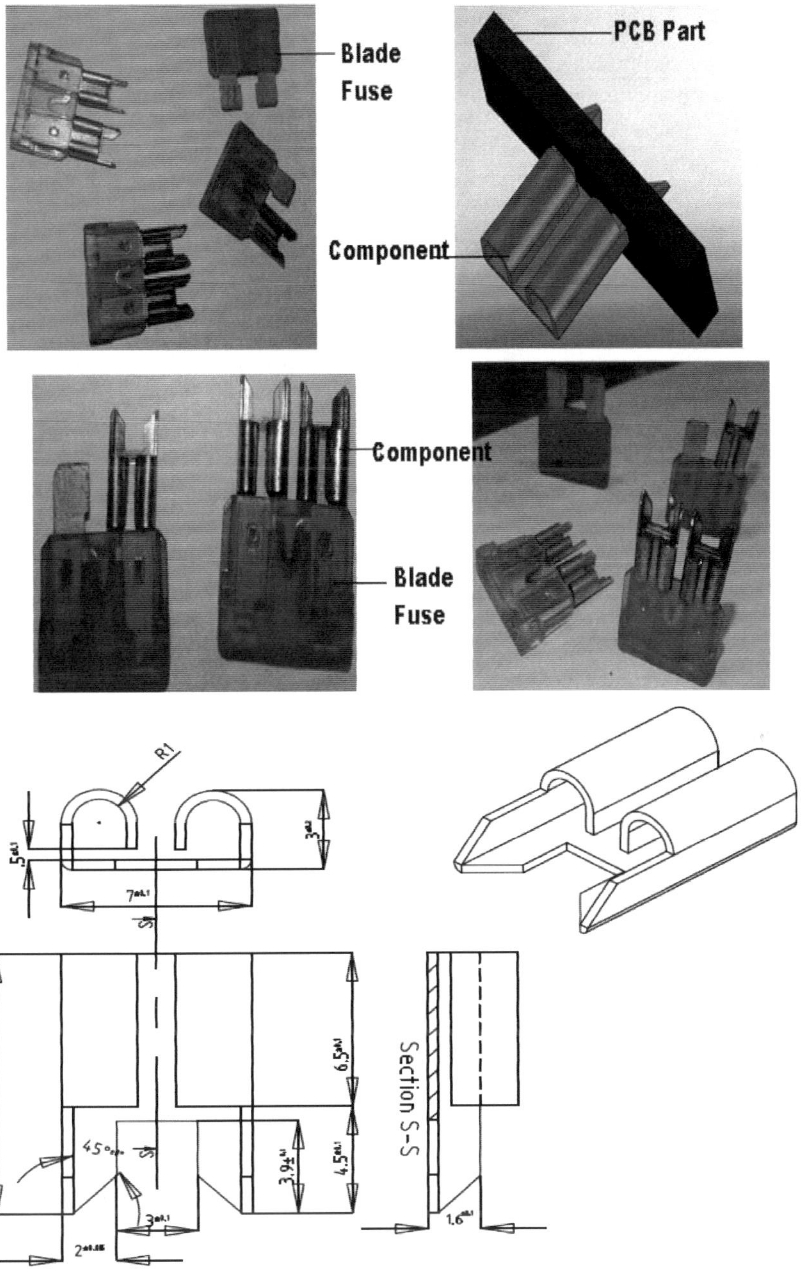

Figure 4. Component drawing and its application

Table1. Details of the component

Name of the component	Blade fuse holder (Clip for PTF200)
Material	CuSn5 (Phosphor bronze), Typical cold
Mass of the component	0.48 gms(approx)
Weight of the component	4.7 N
Thickness of the component	0.4±0.05mm
Density of the component	8.85 gm/cm3
Length of component	11.0±0.2
Width of component	7±0.1 mm (approx)
Number of components required	1,75,000/ week
Maximum height	3.2 mm
Minimum height	2.8 mm
Volume of component	54.08 mm³
Surface area of component	135.2 mm².

3.2. Calculation

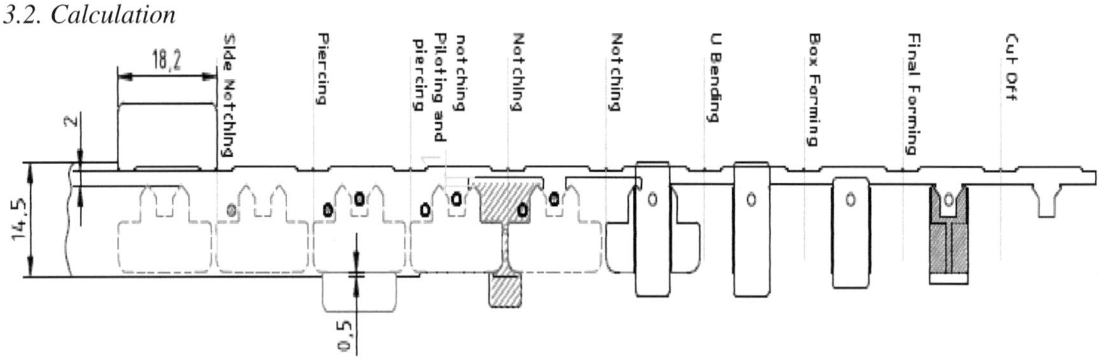

Figure 5. Wide Row Strip Layout

- Strip Width = 14.5 mm
- Pitch = 18.2 mm
- No. of Stages = 9

Figure 5 shows the strip layout designed for the progressive tool having nine stages wide strip layout. The following shows the values of different parts of the tool calculated using suitable appropriate formulas mentioned in Table 2.

Table 2. Details of the parts of tool

Part of the tool	Formula	Values approximated
Scrap for the Strip	A = 1.0 x t (For Single row)	1.0mm
% Area of Utilization	Area of Blank x No. of Rows / Strip Width x Pitch	51.23 %
Press tonnage for press selection	total shear force required / 70 - 80 %	25T press
Cutting Clearance	C = 0.005 x t x √fs	0.015 mm per side
Punch plate	Thickness= 0.75 x Td	19mm

Punch back plate	Thickness= 0.5 x Td	9mm
Stripper plate	Thickness= 0.75 x Td	19mm

3.3. 3D modelling of tool and analysis of the component

The finite element method (FEM), which is sometimes mentioned as finite element analysis (FEA), is a technique which performs computations to produce approximate solutions of boundary value problems in almost all the fields of engineering. Depending on the nature of physical problem being analysed, the field variables may include temperature, fluid velocity, physical displacement, heat flux and to name only a few. This section gives the complete 3D model on the progressive cam tool in different views. This tool was modelled using Solid Works modelling software package. Analysis carried out using the same software is also presented for the Blade Fuse Holder. Number of nodes created for FEA analysis are 2396 with number of elements = 1461. Force applied on the top curl of the holder is equal to 1 KN/ m2. Tetrahedral element mesh was created applying fixed i.e. no deformation at the bottom of the holder as it is in contact with the fuse extension part as shown in Figure 6. Displacement and stress concentration analysis of the component was carried out for the applied force.

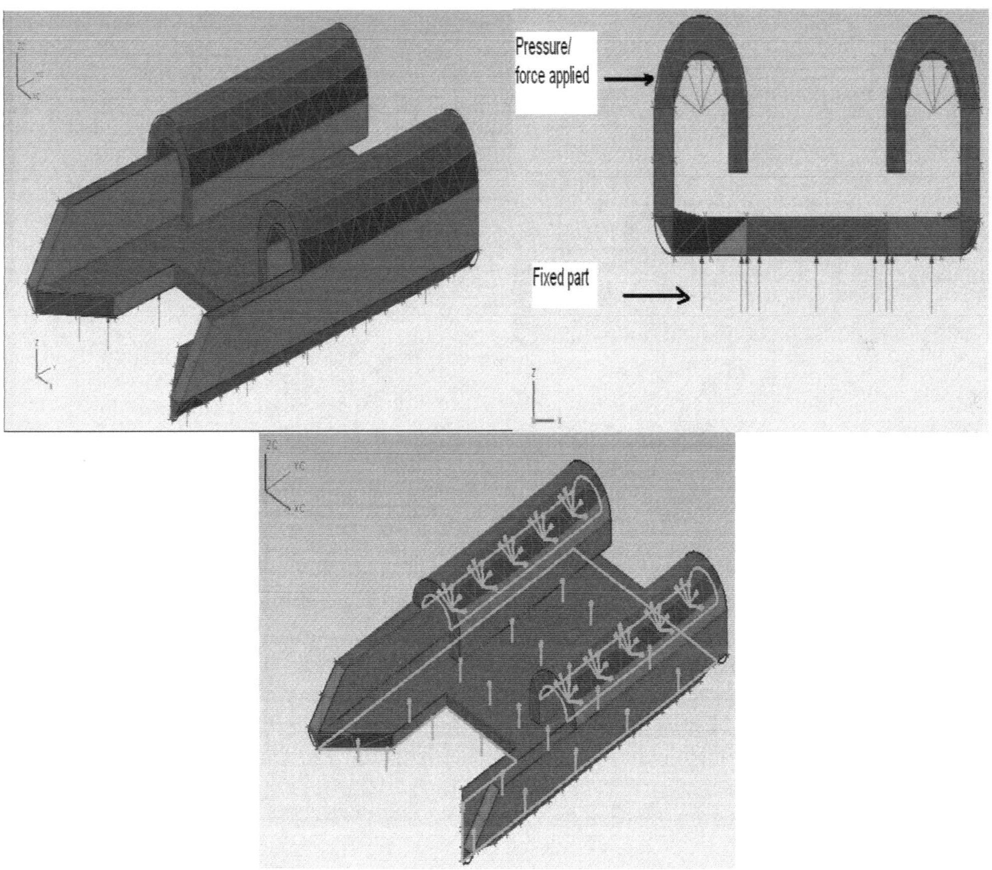

Figure 6. Isometric view of component with meshes, fixed and deformed parts

4. Results and Discussions

4.1. 3D Model of Tool

Journal of Physics: Conference Series **1355** (2019) 012038 doi:10.1088/1742-6596/1355/1/012038

The 3D model of the tool designed according to values obtained in Table 2 is modelled accordingly as shown in Figure 8. The designed and modelled tool in its closed form, top half, bottom half and explode view showing the entire parts of the tools are provided below.

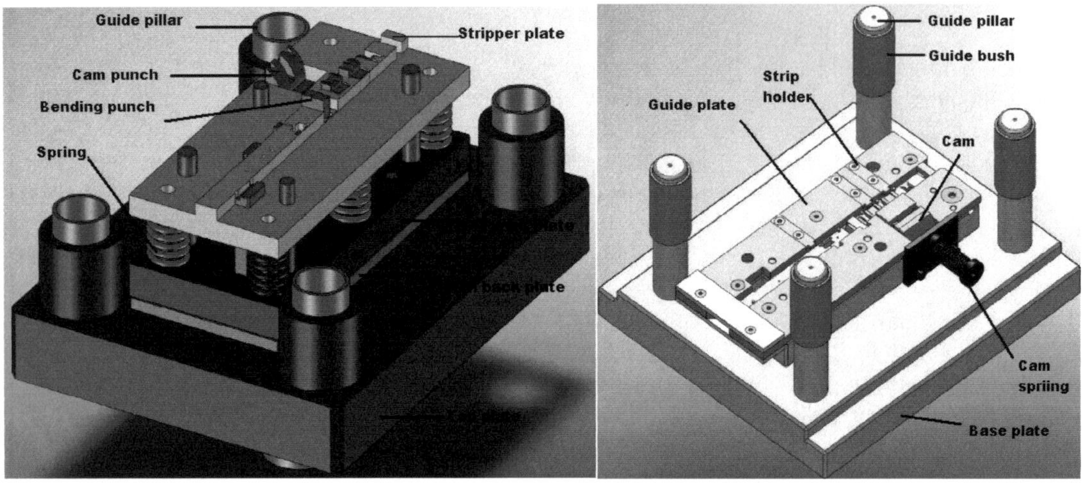

Figure 7. Top half and bottom half of the tool

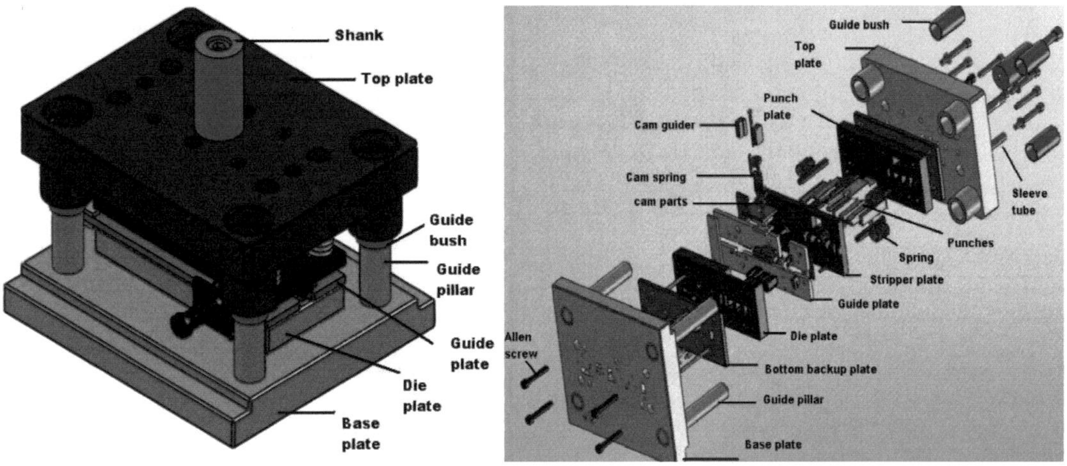

Figure 8. Assembled half of the tool and exploded view of the tool

4.2. Analysis for the progressive component

The finite element analysis of the blade fuse holder component using Solid Works software is shown in Figure 9-10. From Figure 9 it can be seen that the maximum deformation is occurring at the end of the curled component. This occurrence is due to expansion of the component during its assembly with the actual blade fuse. The component is considered to be fixed at its flat end, hence there is nil deformation. However a closer look reveals that the stress concentration is highest at the junction of curled component and the U shaped part as shown in Figure 10. It is worth emphasizing that there has to be a further deeper investigation using coupled thermal, mechanical, and electrical consideration as during the operation electric current continuously flows through the component leading several structural changes. From the above analysis it is clear that the component does not have any effect on the physical behaviour. As load applied and deformation is very less leading to very less stress concentration, this component is safe to be used in electrical parts. Force applied to the part is almost nil hence it does not affect the material.

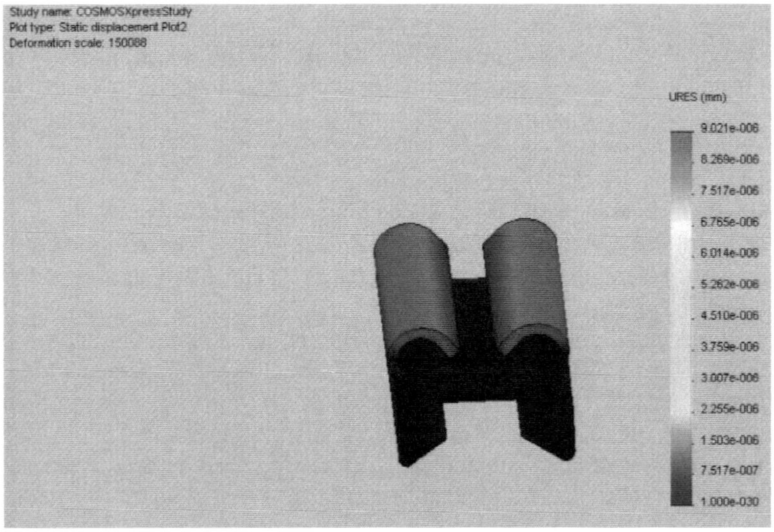

Figure 9: Displacement to the component

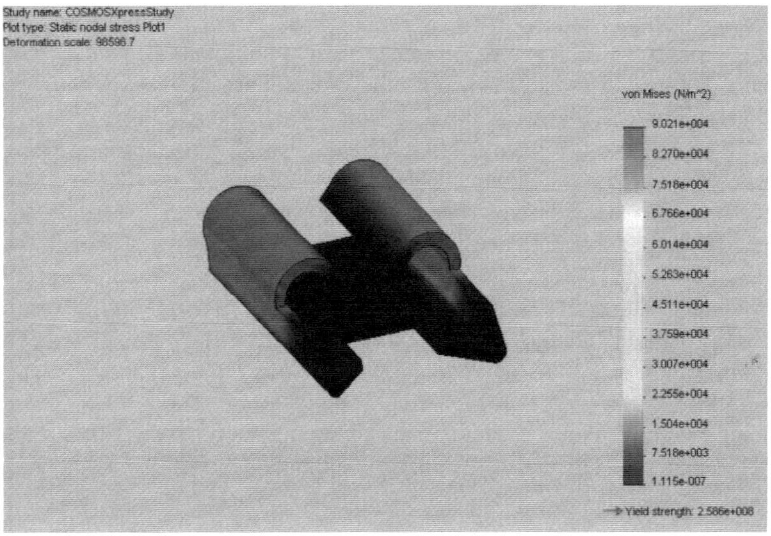

Figure 10. Deformed part of the component during assembly and stress concentration

5. Conclusion

In this work, modelling, study and analysis of the progressive tool for blade fuse holder was carried out successfully. The drawing of the component was studied carefully using Auto Cad, as it was more critical to do its development. The three dimensional modelling of the progressive tool using cams employing solid works is carried out successfully. Also, the analysis of blade fuse holder component was performed using the same software. While modelling the tool it was found that the component needs cam for its curling operation hence increasing the complexity of modelling and number of press operations. Employing cams further can be avoided for designing of progressive tools as it leads to increased manufacturing time and cost. Stage tools to be preferred for complex press operations. Progressive tools anyway reduce labour cost and provide increased production. Analysis, carried out using Solid Works suggests, having no effect of force and displacement on the component.

6. References

[1] WANG, Zhe-ying and WANG, Ke and JIANG, Jie, Design of progressive die for stamping smart card plate based on Pro/E [J], Forging \& Stamping Technology, 2010,v-6,pp-029

[2] Ismail, HS and Chen, ST and Hon, KKB, Feature-based design of progressive press tools, International Journal of Machine Tools and Manufacture, 1996, v-36,no-3.pp-367—378

[3] Kumar, S and Singh, R, A low cost knowledge base system framework for progressive die design, Journal of Materials Processing Technology, 2004, 153, 958—964.

[4] Tor, SB and Britton, GA and Zhang, WY, A knowledge-based blackboard framework for stamping process planning in progressive die design, The International Journal of Advanced Manufacturing Technology, 2005, 26, 7-8, 774—783.

[5] Kumar, Shailendra and Singh, Rajender, An intelligent system for automatic modeling of progressive die, Journal of Materials Processing Technology, 2007,194,1, 176—18

[6] Jiang, RD and Lauw, BT and Nee, AYC, Insert design automation for progressive dies, The International Journal of Advanced Manufacturing Technology,28, 3-4, 279—285 , 2006.

[7] Sim, Sung-Bo and Lee, Sung-Taeg and Jang, Chan Ho, A study on the development of center carrier type progressive die for U-bending part process, Journal of materials processing technology, 28,3-4, 279—285, 2004.

[8] Kumar, S and Singh, R, A knowledge-based system for selection of progressive die components, Journal of Achievements in Materials and Manufacturing Engineering, 20,1-2,475—478, 2007.

[9] Kumar, S and Singh, R, A knowledge-based system to automate the selection of progressive die components, International Journal of Computational Materials Science and Surface Engineering,1,1, 85—96, 2007.

[10] Zhan-jun, LIU, Design of multi-position progressive die for contact-plug parts [J], Forging \& Stamping Technology, 1, 030, 2008.

[11] Yang, Ming and Manabe, Ken-ichi and Ito, Kuniyoshi, Micro press forming and assembling of micro parts in a progressive die, Journal of Mechanical Science and Technology, 21, 10, 1452—1455 , 2007.

[12] Neuenschwander, Thomas R,Method and progressive die assembly with separate blanking and stacking for the manufacture of laminated parts, US Patent 5,349,740, Google Patents. 1994

[13] Kumar, S and Singh, R, An automated design system for progressive die, Expert Systems with Applications,38,4, 4482—4489, 2011.

[14] Kumar, S and Singh, R, Automation of strip-layout design for sheet metal work on progressive die, Journal of materials processing technology, 195,1, 94—100, , 2008.

Comparison of Mechanical Properties of Zn-27Al Alloy Using Various Casting Techniques

Sreejith J[1], Nishad K[2], Krishnadas M[3], Krishnaprasad C N[4], Mamatha Muralidharan[5], Krishnaprasad V[6] and Rajesh Menon B[7].

[1] Research Scholar, Department of Mechanical Engineering, Amrita Vishwa Vidyapeetam, Coimbatore, India.
[2,3,4,5,6] B-Tech Student, Department of Mechanical Engineering, NSS College of Engineering, Palakkad, India.
[7] Associate Professor, Department of Mechanical Engineering, NSS College of Engineering, Palakkad, India.

E-mail: sree8333@gmail.com

Abstract. Zinc-aluminium alloys are cost effective substitute for wide range of applications in the industrial purpose mainly as bearing materials. Zn- 27 Al is the material selected due to its good mechanical and tribological properties. Casting is one of the methods which can be used to fabricate the alloys. Sand casting and stir casting are used to develop Zn- 27 Al alloy. Both mechanical and wear properties were studied. Hardness, tensile strength and impact strength is highest for the stir casted alloy when compared to sand casted Zn-27 Al.. The composition of the prepared alloys justifies the better properties of stir casted alloy and the compositions are analysed by Energy Dispersive X-Ray (EDX) analysis.

1. Introduction

The commercial usage of Zinc based cast alloys, commonly referred to as “ZA” alloys, and is now increasing day by day. The common fabricating methods of ZA alloys are sand casting, permanent mould, shell mould, stir casting and high-pressure die casting. These ZA alloys possess better mechanical properties when compared with the conventional zinc die casting alloys and those of cast iron, copper alloys and aluminium. In addition, they have excellent wear resistance, bearing properties and machinability. Advantage of cast properties is low melting energy consumption because of low melting temperatures and hence increased mould stability and die life.

In 1990, Jim Birch found out the possibilities of different casting methods to fabricate the zinc-aluminium alloys like sand casting, gravity die casting and pressure die casting. Zinc-aluminium alloys possess better mechanical properties than aluminium alloys, plastics, copper alloys, cast iron and steel fabrications [1]. The ZA-27 is a high-strength alloy but ZA-12 and ZA-8 alloys are moderate to high strength materials. All the above mentioned alloys can be fabricated using sand cast, pressure die cast and permanent molding [2]. The studies were conducted on the effect of rare earth elements on the Zn-12Al alloy fabricated using sand casting. The rare earth metal used for this purpose is mischmetal. And the impact toughness of the alloy is directly proportional to the mischmetal addition due to the grain size refinement [3]. Manganese also can be added into the Zn-Al alloy. Researchers were conducted on the microstructure and mechanical properties of manganese added Zn-8Al alloy and finally the formation of complex intermetallic compound ($MnAl_6$) [4]. The 2% addition of copper content obtains the highest wear resistance and tensile strength for the monotectoid-based zinc-

aluminium alloys but the hardness is directly proportional to the percentage addition of copper [5]. The garnet reinforced zinc-aluminium alloys shows reduced wear rate than the unreinforced alloys and it exhibits abrasion wear at low loads and delamination wear at high loads [6]. Both fatigue strength and fatigue life increases for the binary zinc-aluminium alloys as the percentage of aluminium increases. But copper addition upto 2% can enhance the fatigue strength of binary zinc-aluminium alloys [7]. The heat treated Zn-27Al has better tribological properties when compared to the as-casted alloys and it is due to the breaking of dendritic structure [8]. Also, studies reveals that the zinc-aluminium alloys without addition of copper and silicon exhibits lower mechanical properties [9]. The wear and electrochemical behaviour of hypereutectic Zn-Al alloys and role of grain microstructure and macrostructure was also studied for the different ranges of Zn-Al alloys (Zn-1Al, Zn-2Al, Zn-3Al and Zn-4Al) [10].
Hence, the present study focus on the mechanical properties (hardness, tensile strength, yield stress and impact strength) of the Zn-27 Al alloy fabricated using both stir casting and sand casting techniques.

2. Experimental Procedure

The following section deals with the material selection, fabrication of the alloys, microstructure, mechanical tests and wear tests.

2.1 Selection of materials

Zinc-aluminium alloys possess wide range of applications in the industrial area due to its good mechanical properties such as hardness and tensile strength. Zn- 8 Al, Zn- 12 Al, Zn- 27 Al are the most commonly used binary Zn-Al alloys. The material selected for the present research work is Zn-27 Al due to its better mechanical and wear properties.

2.2 Selection of fabrication techniques

Fusion method, electro-deposition method, reduction method and powder metallurgy are the most common methods used for the fabrication of alloys. Fusion method is selected for developing Zn- 27 Al binary alloy. Sand casting is the conventional method usually uses two piece moulds. Stir casting is an advanced method which uses mechanical stirrer for the uniform distribution of elements..

2.3 Fabrication of alloys

Pure zinc and aluminium are used for developing Zn- 27 Al using both sand casting and stir casting process.

2.3.1 Sand casting

Figure 1 represents the schematic diagram of sand casting technique. The metals are melted in a furnace at 650 °C. The moulds are prepared using silica sand. Cylindrical die of size Ø 30 mm × 120 mm and impact testing die of size 15 mm × 15 mm × 140 mm are made. The molten metal is allowed to fill the mould and then the alloys are removed after cooling.

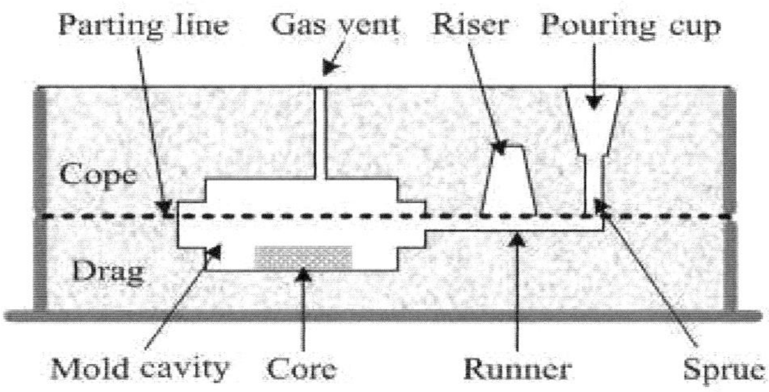

Figure .1. Sand casting process

Figure .2. Hardness testing specimen by sand casting process

Figure .3. Tensile testing specimen by sand casting process

Figure .4. Impact testing specimen by sand casting process

Figures 2, 3 and 4 shows the hardness, tensile and impact testing specimens fabricated using sand casting process.

2.3.2 Stir casting

In the stir casting process, the molten metals are distributed by mechanical stirring, and then the molten matrix is poured into the mould. Mechanical stirring is the key element of the stir casting process.

In preparing metal matrix composites by this method, there are some factors need to be considered:

1. The difficulty of achieving a uniform distribution of the metals.

2. Wettability between the two main substances.

3. Porosity in the cast metal matrix.

4. Chemical reactions between the materials.

5. Rate of solidification and relative density.

6. Geometry of the mechanical stirrer.

7. Position of stirrer in the melt, melt temperature

Pure aluminium was bought as ingot and pure zinc as thin sheets. Materials were cut into suitable size to be put into the furnace. The stirring speed and stirring time was set at 350 rpm and 10 minutes respectively. The alloy was melted to 650°C in a graphite crucible. Molten metal was poured into dies of tensile specimen (Ø 14 mm × 140 mm) (figure 7), impact test specimen (15mm x 15mm x 140) (figure 8), hardness testing specimen (Ø 30mm x 140) (figure 6). All these specimens were casted out with a solidification tome of 2hrs. Figure 5 represents the stir casting apparatus.

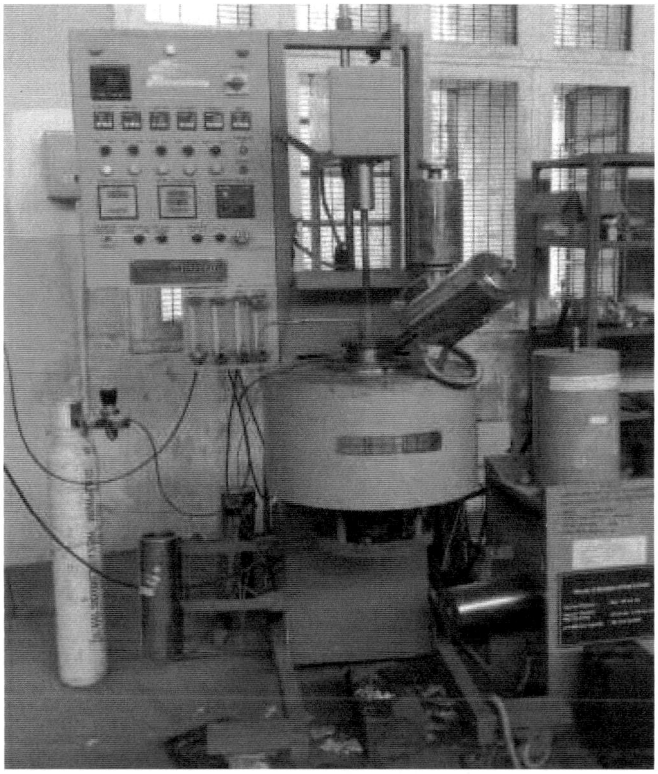

Figure .5. Stir casting equipment

Figure .6. Hardness testing specimen by stir casting process

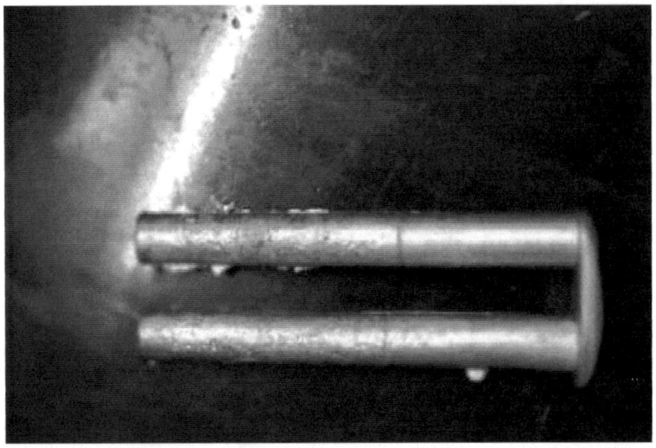

Figure .7. Tensile testing specimen by stir casting process

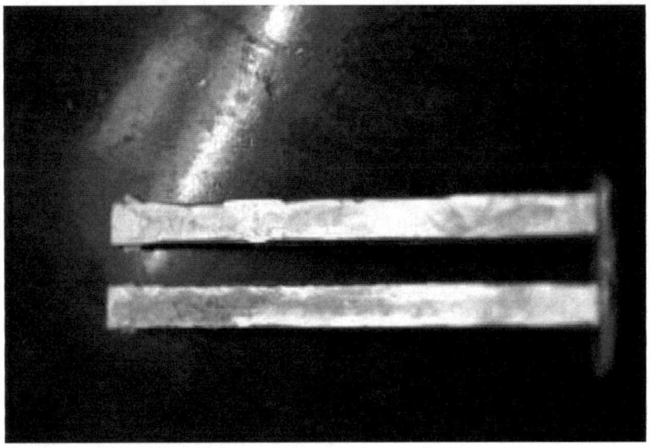

Figure .8. Impact testing specimen by stir casting process

2.4. Hardness testing

Hardness of the material is found out using the Brinell indentation hardness principle .This test is conducted according to ASTM E10:2015 standards. The model of the machine is AKB 3000 .loading capacity of the machine is 3000 kgf. The diameter of the indenter chosen was 5mm .casted material were machined to suitable size (25mm x 27mm) . load of 750 kgf was applied for 10 sec. Hardness was taken on 3 different regions of the specimen . Figure 9 represents the Brinell Hardness Testing Machine.

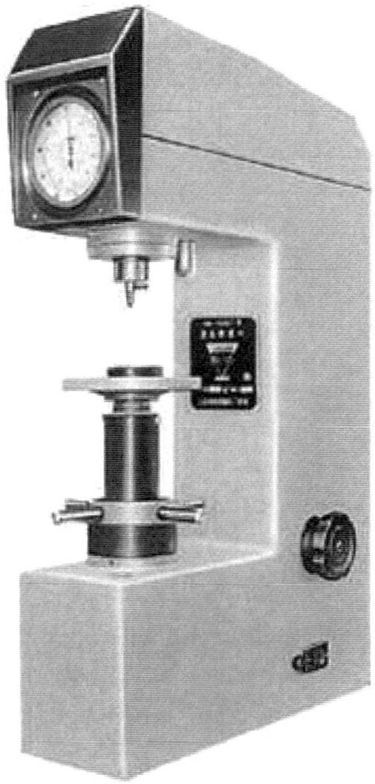

Figure. 9. Brinell Hardness Testing Machine

2.5. Tensile testing

The specimen for tensile test were prepared from the casted alloy having gauge dia 9mm and gauge length of 36mm with overall length of 150mm and is shown in the fig. The specimen was made according to the ASTM E8 standards. The specimen were cleaned before the experiment using emery paper of grit size 200. The machine used for the experiment was UTM M30, where maximum load was limited to be 30kN and maximum elongation to 1000mm. Fillet diameter of the specimen was 15.89 mm. The grips in the testing machine were properly serrated before the commencement of the experiment. Figure 10 shows the tensile testing apparatus.

Journal of Physics: Conference Series **1355** (2019) 012039 doi:10.1088/1742-6596/1355/1/012039

Figure. 10. Tensile Testing Apparatus

2.6. Impact testing

Charpy pendulum impact tests were carried out on the cast samples to have comparative impact properties of the stir and sand castings. The impact test is done by placing a square shaped V-notched specimen in the machine (placed as cantilever beam). The specimens have V-shaped notch of 45°. The notch is located on tension side of specimen during impact loading. Depth of notch is generally taken as t.5 to t/3 where't' is thickness of the specimen. The V-notched impact test standard specimens with dimensions of (10mm X 10mm X 55 mm) were made according to ASTM-A370. Model of the machine used was AIT-300-EN .Capacity of the machine was 300 J. The test was conducted at room temperature (25.3 °C). Schematic diagram of the impact testing apparatus is shown in figure 11.

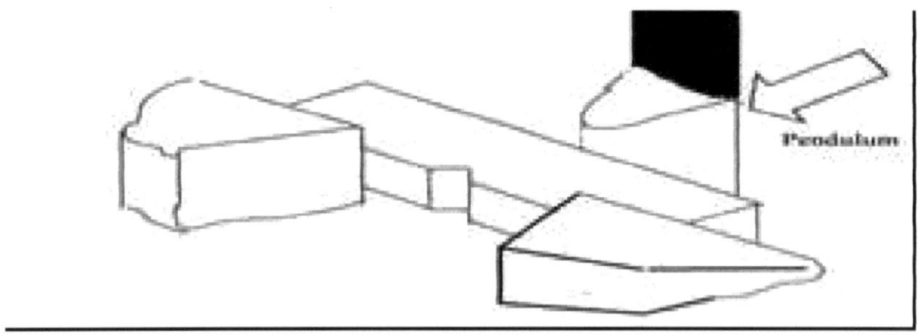

Figure. 11. Schematic diagram of impact test

3. Results and Discussions

This section deals with the comparison of mechanical properties (hardness, tensile and impact) and microstructure of the alloy prepared using both sand casting and stir casting process. Also, the Energy Dispersive X- Ray Analysis (EDX) comparison is discussed in this section.

3.1. Brinell hardness testing

Table 1 depicts the comparison of Brinell hardness of the alloy prepared using both stir casting and sand casting process. Figure 12 depicts the graphical representation of the comparison of the Brinell hardness values.

Table 1. Comparison of Brinell hardness

Process	Alloy	Observed Value HBW 5/750			Average value HBW 5/750
		1	2	3	
Sand casting	Zn-27Al	114	114	116	115
Stir casting	Zn-27Al	156	143	143	147

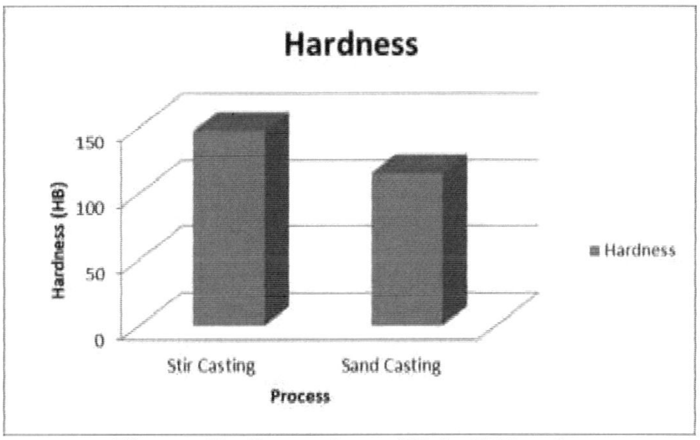

Figure. 12. Comparison of Brinell Harndess

From the graph, it is clear that the stir casted alloy possess better hardness property when compared to sand casted alloy. This is because of the better distribution of materials when using stir casting process. The aluminium particles are properly mixed with the zinc matrix because of mechanical stirring process. The better mixing property of stir casted alloy is validated using EDX analysis.

3.2. Tensile testing

The comparison of tensile strength and yield stress is shown using table 2. Figure 13 and figure 14 represents the tensile strength plot and yield strength plot respectively.

Table 2. Comparison of tensile and yield strength

Process	Yield strength (N/mm^2)	Tensile strength (N/mm^2)
Sand casting	195.8	225.0
Stir casting	143.2	194.2

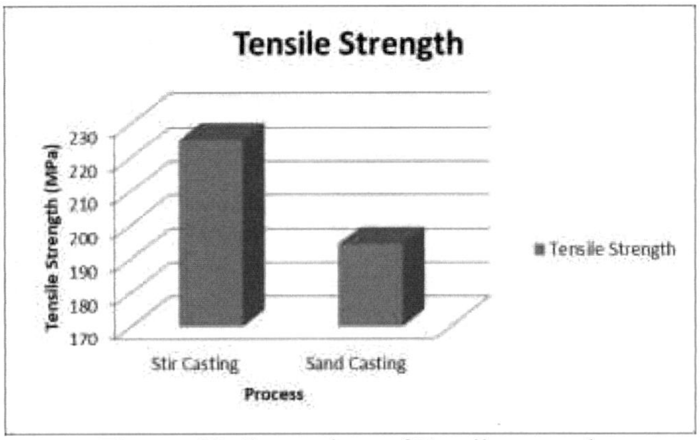

Figure. 13. Comparison of Tensile strength

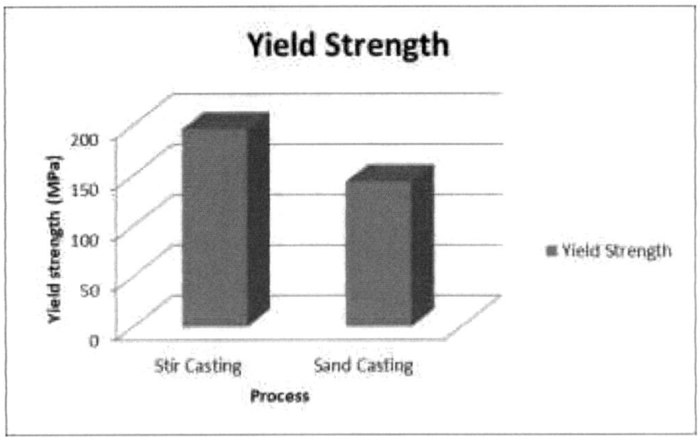

Figure. 14. Comparison of Yield strength

3.3. Impact testing

The impact strength in terms of energy (J) of the specimen was found out by Charpy test and the values were recorded of both the castings (Table 3 and Figure 15).

Table 3. Impact strength variation comparison

Process	Specimen Temperature (^0C)	Thick (mm)	Width (mm)	Length (mm)	Charpy Impact Engergy (Joules)
Sand casting	25.3	10.01	10.00	55	1.68
Stir casting	25.3	10.01	10.00	55	2.0

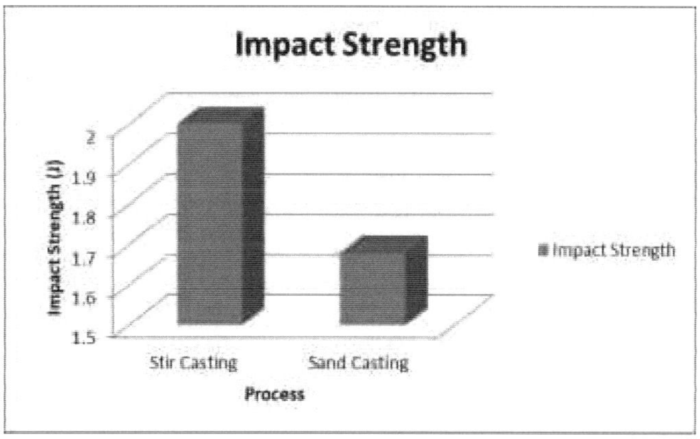

Figure. 15. Comparison of Impact strength

3.4. Microstructure analysis

Figures 16 and 17 represent the microstructure analysis of sand casted specimen and stir casted specimen, respectively. From figure 16, the stir casted alloy specimen has the uniform distribution of zinc-aluminium matrix and the alloy distributions of elements are justified using EDX analysis.

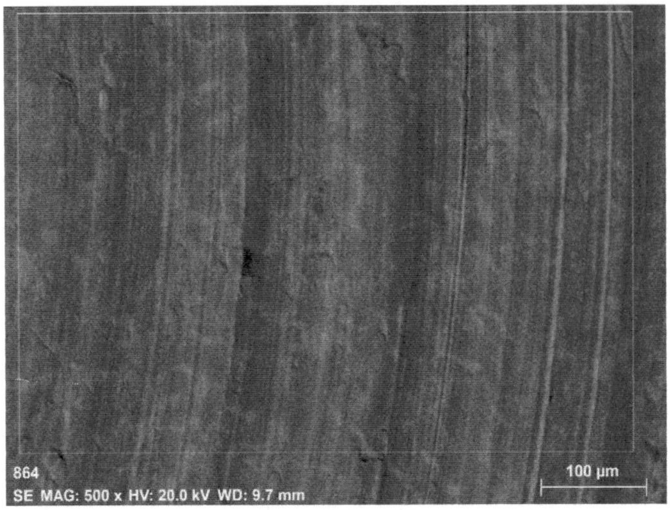

Figure. 16. Microstructure analysis of stir casted alloy

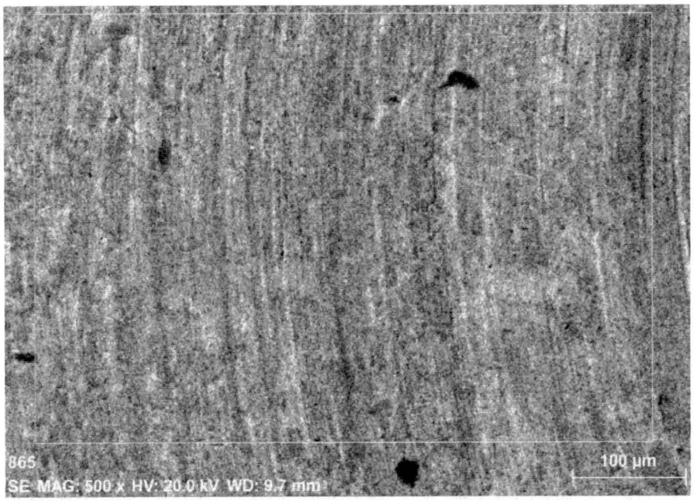

Figure. 17. Microstructure analysis of sand casted alloy

3.5. EDX analysis

Figure 18 and 19 shows the EDX analysis of both sand casted and stir casted alloy respectively.

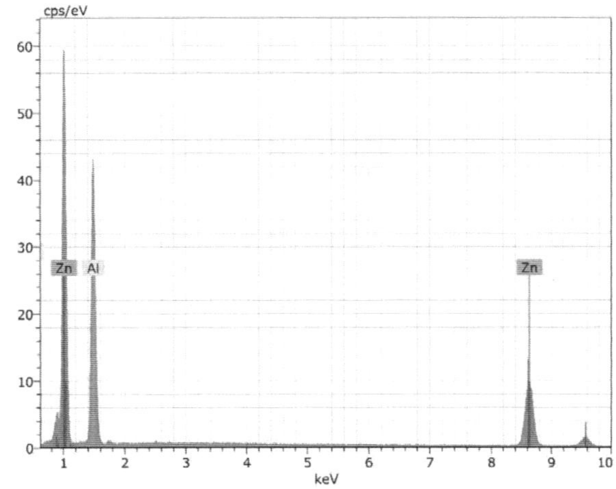

Spectrum: SAMPLE 1

Element	Series	unn. C [wt.%]	norm. C [wt.%]	Atom. C [at.%]	Error (3 Sigma) [wt.%]
Zinc	K-series	55.35	53.61	24.74	4.69
Aluminium	K-series	30.06	29.12	32.57	4.42
Carbon	K-series	16.68	16.16	40.60	9.06
Oxygen	K-series	1.15	1.11	2.09	1.02
	Total:	103.24	100.00	100.00	

Figure. 18. EDX analysis of sand casted alloy

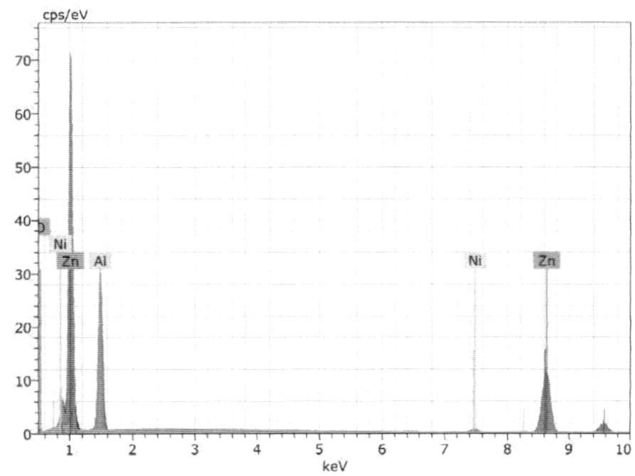

Spectrum: SAMPLE 2

Element	Series	unn. C [wt.%]	norm. C [wt.%]	Atom. C [at.%]	Error (3 Sigma) [wt.%]
Zinc	K-series	64.78	63.25	34.22	5.46
Aluminium	K-series	24.38	23.80	31.20	3.61
Carbon	K-series	11.09	10.82	31.88	6.73
Nickel	K-series	1.28	1.25	0.75	0.23
Oxygen	K-series	0.90	0.88	1.95	0.88
	Total:	102.42	100.00	100.00	

Figure. 19. EDX analysis of stir casted alloy

From the figures and tables, it is clear that the percentage of zinc (63.25%) an aluminium (31.20%) is more when the alloys are fabricated using the stir casting process. In sand casted alloys, the percentage quantity of zinc and aluminium is 53.61% and 29.12% respectively. Hence the mixing property or uniform distribution of elements in the stir casted alloy enhances the mechanical properties of alloys.

4. Conclusions

- Zn-27 Al alloy is fabricated using sand casting and stir casting techniques successfully.
- Stir casted Zn-27Al possess better hardness, tensile strength and impact strength when compared with sand casted alloy due to the better distribution of elements in stir casted specimen.
- The proper mixing of the elements in the stir casted alloy is justified using EDX analysis.

References

[1] J Birch 1990 *Casting Manager, Zinc Development Association* **11** 42.

[2] E Gervais, R J Barnhurst and C A Loong 1985 **37** *Appl. Technol.* 43.

[3] Y Chen 1994 *J. Alloys and Compounds* **206** 169.

[4] Ahmet Turk, Mehmet Durman and E Sabri Kayali 2007 *J. Mater. Sci.* **42** 8298.

[5] T Savas and A Pas 2004 *Tribol. Int.* **37** 45.

[6] G Ranganath, S C Sharma, and M Krishna 2001 *Wear* **251** 1408.

[7] T Savaşkan, M Aydın, H A Src, M Aydın, and H A Odabas 2013 *Mater. Sci, Technol.* **17** 631.

[8] M B Ã, S Mitrovic, and B Jeremic 2010 *Tribiology Int.* **43** 16.

[9] Z Balalan and M Kaplan 2007 *Int. J. Sci. Technol.* **2** 75.

[10] A E Ares, L M Gassa, C E Schvezov, and M R Rosenberger 2012 *Mater. Chem. Phys.* **136** 394

Homogenization of Mechanical Properties of Unidirectional Fibre Reinforced Composites with Matrix and Interface Defects: A Finite Element Approach

Siddharth S[1], Dr.Ajith Ramesh[2#]
Department of Mechanical Engineering, Amrita School of Engineering, Coimbatore, Amrita Vishwa Vidyapeetham, India

E-mail: [1]siddharth.sreeram@gmail.com
[2#]r_ajith@cb.amrita.edu (Corresponding Author)

Abstract. Fiber Reinforced Composites find increasing applications in the areas of Aerospace, Military Armours, Bullet-proof vests, etc. As the composites are composed of two different constituents, there arises a need to determine the effective properties of the homogenous composites. Experimental determination of the effective properties is very expensive considering the amount of experiments that are required to be conducted, the time and cost to be incurred for each experiment, and the permutations and combinations of the optimal fiber volume fraction. The effective properties are essential for modeling of composites with reference to real-time applications. The micro-mechanics approach reduces most of the above mentioned complexities and helps in accurately evaluating the effective properties. In the presented paper, the properties like Young's Modulus, Poisson Ratio, and Shear modulus of a healthy (defect free) composite is obtained by modeling a Representative Volume Element (RVE) using the commercial Finite Element Analysis (FEA) solver – Abaqus,with application of Periodic Boundary Conditions (PBC). The presented research focuses on Fiber-Reinforced Metal-Matrix composites like AA2024-Al_2O_3 and the Ceramic-Matrix composites like ZrB_2-SiC. In general, defects in composites arise during the manufacturing process. Matrix Crack, Interfacial De-bonding and Fiber Crack are the major defects which degrade the mechanical properties of composites. This paper presents the modeling of Interfacial de-bonding using the Cohesive-Zone Modelling (CZM) technique for every 90° variation in the fiber-matrix interface and the subsequent evaluation of the corresponding homogenous properties. Matrix Crack is modelled as a matrix defect with a 'V' notch for varying a/w ratios. For every variation in matrix crack, the corresponding properties are estimated. Numerical evaluation of the individual effects of interfacial de-bonding and fully grown matrix cracks are followed by the modelling of the coupled effects.

1. Introduction

Composite Materials have been used extensively due to their high strength along the fiber direction and high strength-to-weight ratio. Fiber-Reinforced Metal-Matrix and Ceramic Matrix Composites find tremendous applications specifically in automotive, aerospace, mineral processing, and armour applications [1]. UD Ceramic Matrix Composites are mostly used in high temperature blast proof applications like face sheets for blast proof sandwich panels. These kind of high temperature composites are either manufactured by Hot Isostatic Pressing (HIP) or Powder Metallurgy, i.e., Compaction followed by Plasma arc sintering. Finding the homogeneous properties of the composite experimentally is an expensive process as these high temperature materials and its corresponding manufacturing process are expensive. So theoretical and numerical methods are employed to determine the homogenous properties (i.e.) Young's Modulus, Poisson Ratio and Shear Modulus of these materials. These properties are essential to carry out numerical modelling of these materials. Micromechanics is the most efficient tool to find the homogenous properties of the composite analytically and numerically [2]. Most of the available analytical and numerical models do not incorporate combined effects of matrix defects, interface defects and high temperature effects in a single model. Sadik L. Omairey [3] has developed a Python code, integrated with ABAQUS which calculates the effective properties of the unidirectional composite. ABAQUS is a popular finite element tool used for modelling highly non-linear problems with high accuracy [4]. These CMCs are basically used at very high temperature regimes like blast. So, estimation of these homogenous properties at elevated temperature. Hashin [5] has performed thermo-elastic analysis on composites at elevated temperatures .Eric W. Neuman et al [6] have experimentally determined the mechanical properties of ZrB_2 at ambient and elevated temperatures. Modelling the interface defect is generally done by cohesive surfaces. Wu-Gui Jiang et al [7] have modelled and extracted the homogenous properties of the composite with imperfect interface. Y. Benveniste [8] uses cohesive surfaces at a variation of angle along the fiber-matrix interface to model the interface defects. Matrix and Fiber crack are mostly modelled using user subroutines.

Our goal is to homogenize the effective properties of the composites with all the above mentioned criteria. As our intention is to observe the trend in degradation of properties due to defects in the matrix and the interface, we have simplified the matrix crack into a 'V' notch with varying a/w ratios and 90° de-bonding between the fiber-matrix interface. For these mentioned cases, the homogenous properties were extracted and the degradation in properties is noted.

2. Finite Element Procedure

2.1 Modelling Representative Volume Element

The Representative Volume Element is modelled with unit dimensions with fiber embedded in the matrix. The fibre is unidirectional in nature and runs along the x axis of the geometry of RVE is shown in the figure 1

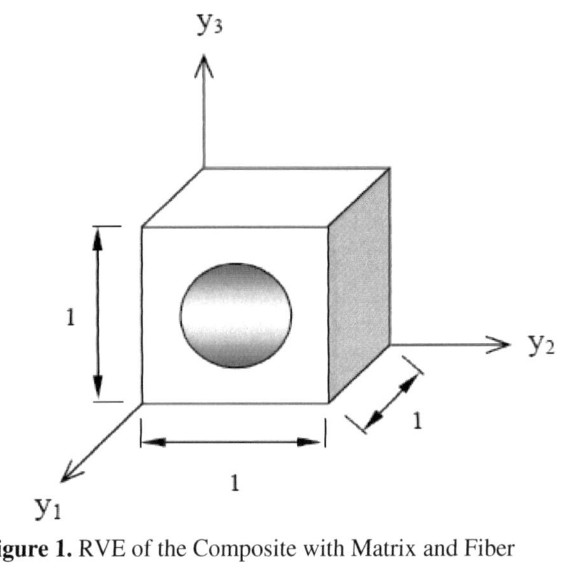

Figure 1. RVE of the Composite with Matrix and Fiber

For modelling the RVE, commercial finite element solver ABAQUS is used. The assumptions considered prior to the modelling are that the fibers are placed in equal distances and the radius is same throughout the composite. It is also ensured that volume fraction criterion is satisfied.

2.2 Materials

The materials used for the homogenization are UD Fiber Reinforced Metal Matrix Composite and a UD Fiber Reinforced Ceramic Matrix composite. The UD CMC used in the study is Ultra High Temperature Ceramic Composite used for blast proof applications. The material properties of the matrix and reinforcements are listed in Table 1 and 2 as given below.

.

Table 1. Matrix Properties of CMC and MMC.

Matrix	E(GPA)	Nu (POISSON RATIO)
AA2024	73.1	0.33
ZrB_2	500	0.11

Table 2. Fiber Properties of CMC and MMC.

Reinforcement	E(GPa)	Nu	G(GPa)
Silicon Carbide	E_1=403.2	Nu_{12}=0.148	G_{12}=93.1
	E_2=262.6	Nu_{13}=0.148	G_{13}=93.1
	E_3=262.6	Nu_{23}=0.194	G_{23}=109.5
Alumina (NEXTEL 610)	373	0.27	

In the case of matrix material, both MMC and CMC are considered to be isotropic in nature. However, the reinforcement Nextel 610 alumina fibre is considered as isotropic and Silicon Carbide fiber is considered as transversely isotropic. It is to be noted that after homogenization, the end result is a transversely isotropic material.

2.3 Periodic Boundary Condition

When it comes to homogenization of UD fiber reinforced composites, Periodic Boundary Conditions would fetch us exact results. Materials like fiber reinforced composites are generally represented as a repeated array of structures which is periodic. Periodic Boundary Condition implies that each RVE has same deformation and there is no separation between the neighbouring RVE's. For a cubical RVE the displacement on the pair of opposite surfaces are given by equations 1 and 2.

$$u_i^{j+}=\varepsilon_{ik}x_k^{j+}+u_i^* \quad (1)$$

$$u_i^{j-}=\varepsilon_{ik}x_k^{j-}+u_i^* \quad (2)$$

where j+ is along positive X direction and j- is along negative X direction. The difference of equations 1 and 2 is equation 3.

$$u_i^{j+}-u_i^{j-}=\varepsilon_{ik}+\Delta x_k^j \quad (3)$$

For an RVE, Δx is a constant. So, the above equation reduces to equation 4.

$$u_i^{j+}(x,y,z)-u_i^{j-}(x,y,z)=C_i^{\,j}(i,j=1,2,3) \quad (4)$$

The constants, c^{11}, c^{22} and c^{33}, represent the tensile or compressive behaviour on the RVE model due to the reaction of three normal traction components , whereas the other three pairs of constants , $c^{21} = c^{12}$, $c^{31} = c^{13}$and $c^{32} = c^{23}$ correspond to the shear deformations due to three shear traction

Journal of Physics: Conference Series **1355** (2019) 012040 doi:10.1088/1742-6596/1355/1/012040

components . This form of boundary conditions satisfies displacement periodicity. It can be seen from the above equation that the difference of the dis placements for the corresponding points for a pair of opposite boundary surfaces are specified, but the individual displacement components are still functions of the coordinates.

The PBC is applied as a set of three boundary conditions for variation in the value of C
SET 1: $C^1_1 = C^2_2 = C^3_3 = 0.012$ mm

SET 2: $C^1_1 = C^2_2 = 0.018$ mm

SET 3: $C^1_2 = C^2_1 = 0.016$ mm

SET 4: $C^2_3 = C^3_2 = 0.016$ mm where i represents phase and j represents direction.

Figure 2.1-2.4 represents output for 4 sets of boundary conditions for Aluminium-alumina composite and 3.1-3.4 represents ZrB2-SiC Composite

Figure 2.1. Set 1 Boundary Condition

Figure 2.2. Set 2 Boundary Condition

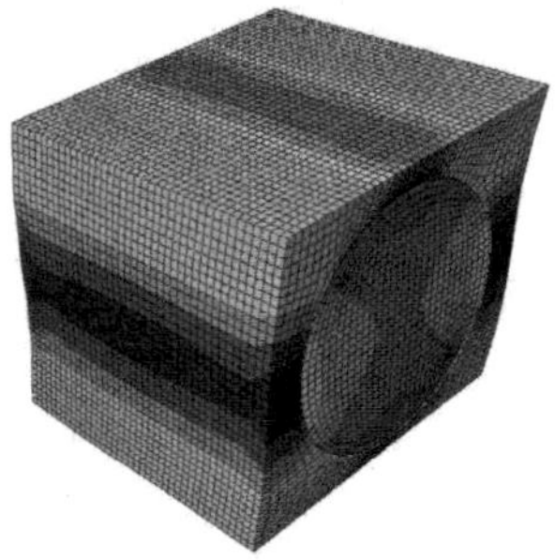

Figure 2.3. Set 3 Boundary Condition

Figure 2.4. Set 4 Boundary Condition

Figure 2. Aluminium-Alumina Composite Outputs

Figure 3.1. Set 1 Boundary Condition

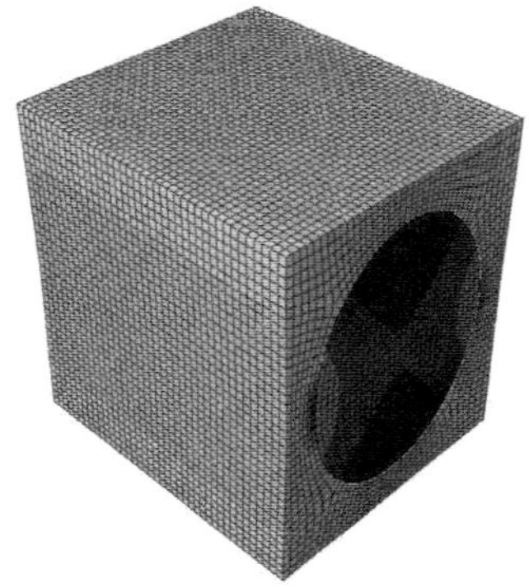

Figure 3.2. Set 2 Boundary Condition

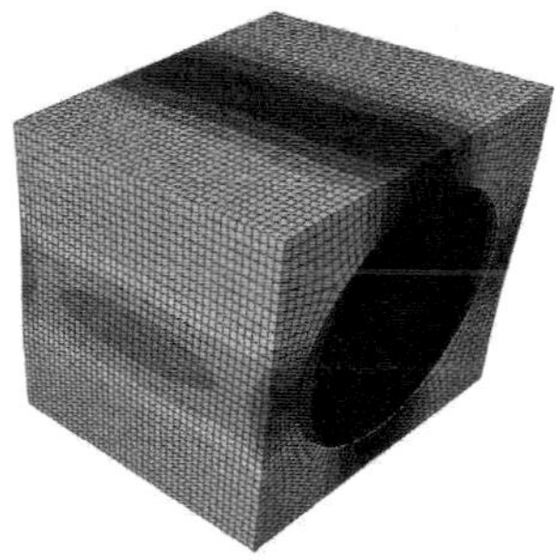

Figure 3.3. Set 3 Boundary Condition

Figure 3.4. Set 4 Boundary Condition

Figure 3. ZrB_2-SiC Composite Outputs

2.4 Prediction of Elastic Constants

$$[\varepsilon] = [S][\sigma] \tag{4}$$

where [S] is the compliance matrix in equation (4),

$$[S] = \begin{bmatrix} S11 & S12 & S13 & 0 & 0 & 0 \\ S12 & S22 & S23 & 0 & 0 & 0 \\ S13 & S23 & S33 & 0 & 0 & 0 \\ 0 & 0 & 0 & S44 & 0 & 0 \\ 0 & 0 & 0 & 0 & S55 & 0 \\ 0 & 0 & 0 & 0 & 0 & S66 \end{bmatrix} \tag{5}$$

From the above matrix (5) the engineering elastic constants can be determined by following relations

$$E1 = \frac{1}{S11} \qquad Nu12 = -\frac{S12}{S11} \qquad G12 = \frac{1}{2S44}$$

$$E2 = \frac{1}{S22} \qquad Nu13 = -\frac{S13}{S11} \qquad G13 = \frac{1}{2S55}$$

$$E3 = \frac{1}{S33} \qquad Nu23 = -\frac{S23}{S22} \qquad G23 = \frac{1}{2S66}$$

2.5 Results from Healthy Case Simulations

After applying the above described periodic boundary conditions and calculating the compliance matrix constants, the Young's modulus, Poisson Ratio and Shear Modulus of the MMC and CMC with varying volume fraction from 20-47% is as shown in Figure 4.1-4.6 below. The corresponding figures below represent the variation of properties with respect to different debonding angles. Figures 4.1-4.6 represents variation of E_1, E_2, Nu_{12}, Nu_{23}, G_{12}, G_{23} vs volume fraction

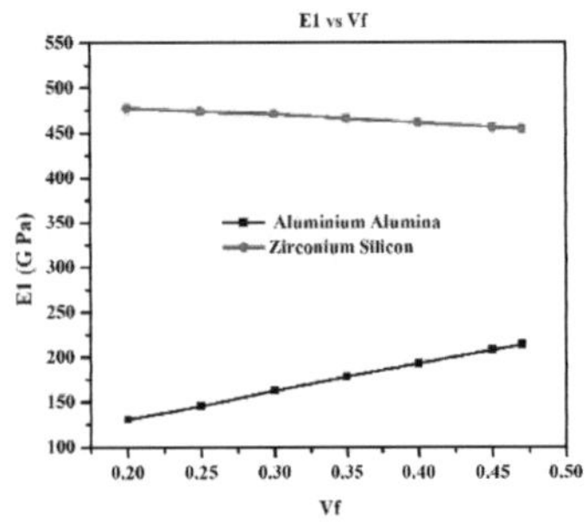

Figure 4.1. Vf vs E_1

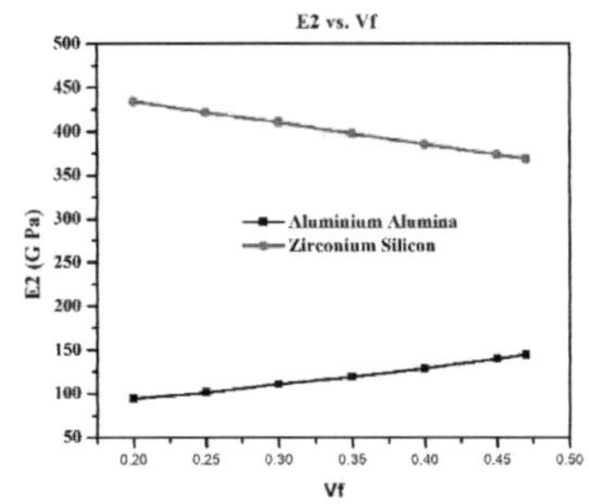

Figure 4.2. Vf vs E_2

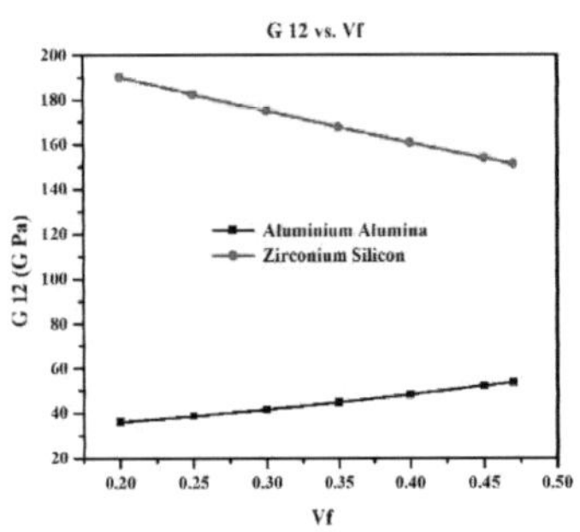

Figure 4.3. Vf vs G_{12}

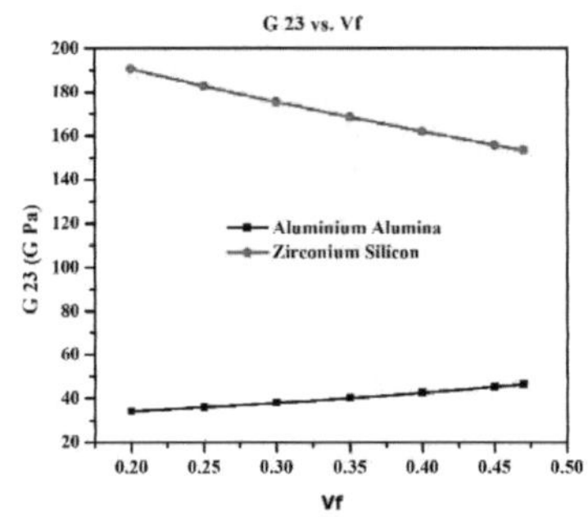

Figure 4.4. Vf vs G_{23}

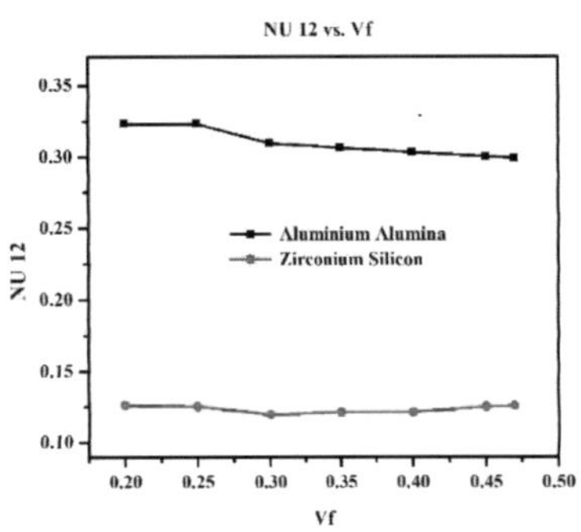

Figure 4.5. Vf vs Nu_{12}

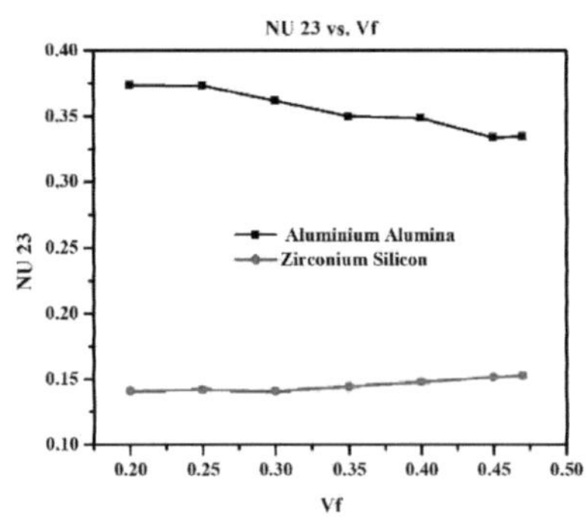

Figure 4.6 . Vf vs Nu_{23}

Figure 4: Homogenous properties versus volume fraction

2.6 Inference from Healthy Case Simulation

With the outputs from Figure 4.1-4.6, the following inferences are made for variation of properties with respect to change in fibre volume fraction

- . As the MMC is Aluminium-Alumina, the properties are dominated by the fiber, the Young's modulus along the fibre direction (i.e.) E_1 increases with increases by 46% at 47% fibre volume fraction

- But the CMC is ZrB2-SiC where the matrix property completely dominates the fiber properties. So E_1reduces by 5% as fibre volume fraction increases
- E_2 of the MMC increases by 53% and for CMC it reduces by 15%
- As both composites are transversely isotropic in nature E_2 and E_3 are same
- G_{12} of MMC increases by 49 % and decreases by 21 % for CMC
- G_{23} increases by 36 % for MMC and reduces by 20% for CMC
- There is a decrease in Poison's Ratio as fibre volume fraction goes high for Aluminium-Alumina MMC, but for ZrB2-SiC CMC it almost remains a constant

3. Matrix crack:

The damage scenario in composite materials is peculiar when compared to metals. Failure in metals is defined by slow propagation of cracks, i.e., crack initiation and propagation in a series of notch blunting, while the failure in composites is a continuous process of damage growth, including various complicated failure mechanisms and damage modes. As our interest is in homogenizing the effective properties in the presence of defects in the matrix, crack initiation and propagation are not modelled. Instead, a 'V' notch with varying a/w ratios is modelled and the homogenization is performed. The 'V' notch model is give in the Figure 5.

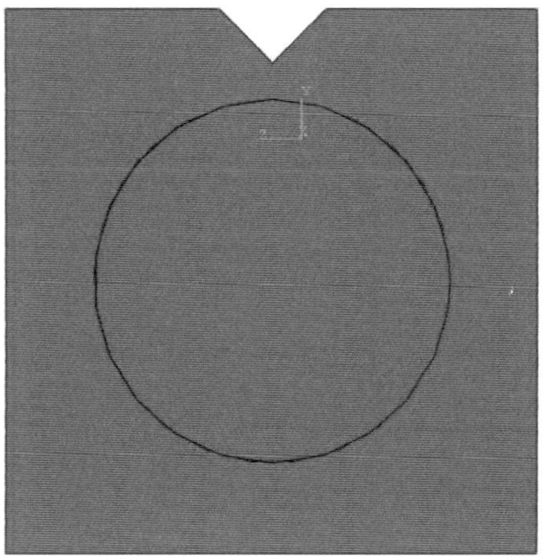

Figure 5. Composite with v notch

The a/w ratio is varied from 0.02 mm to 0.1mm, and for all corresponding 'V' notches the effective properties are homogenized. Table 1 and Table 2 displays the properties of the matrix and fibre materials.The corresponding figures below represent the variation of properties with respect to different a/w ratio. Figures 6.1-6.6 represents variation of E_1, E_2, Nu_{12}, Nu $_{23}$, G_{12}, G_{23} vs a/w ratio

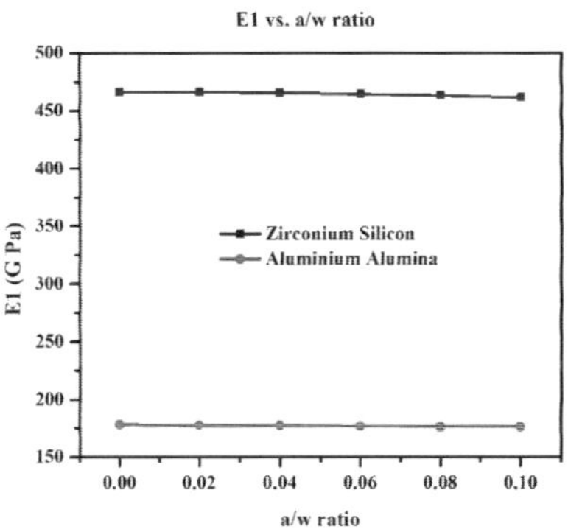

Figure 6.1. a/w ratio vs E_1

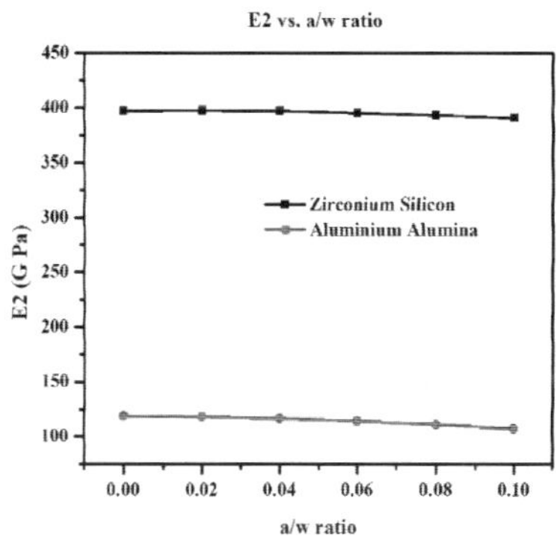

Figure 6.2. a/w ratio vs E_2

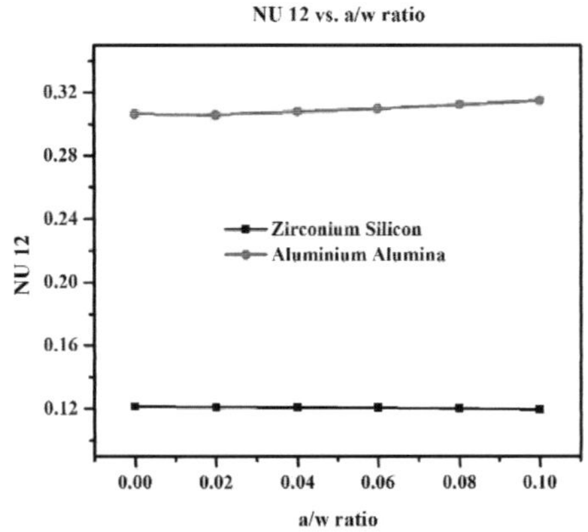

Figure 6.3. a/w ratio vs Nu_{12}

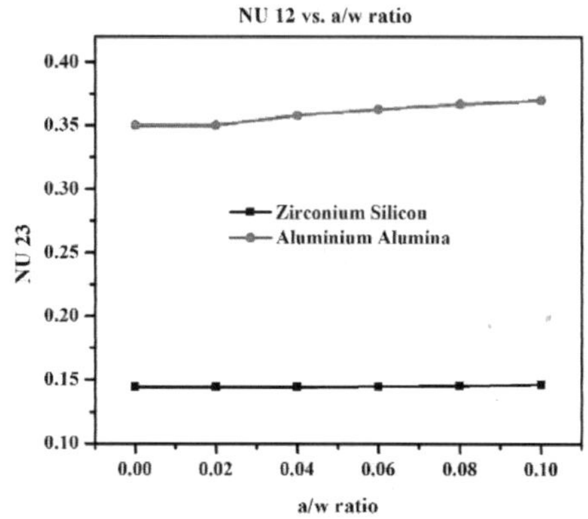

Figure 6.4. a/w ratio vs Nu_{23}

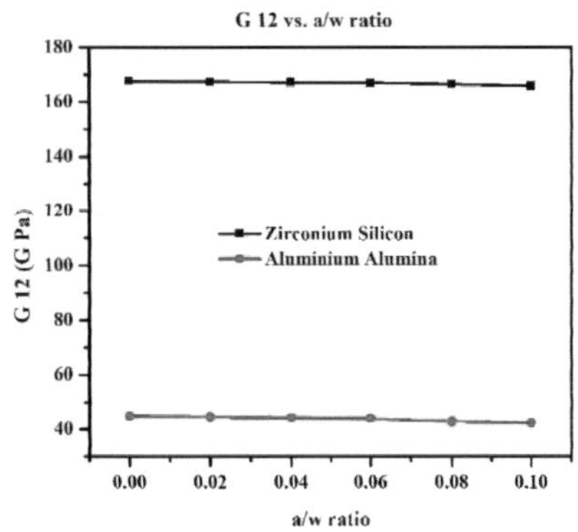

Figure 6.5. a/w ratio vs G_{12}

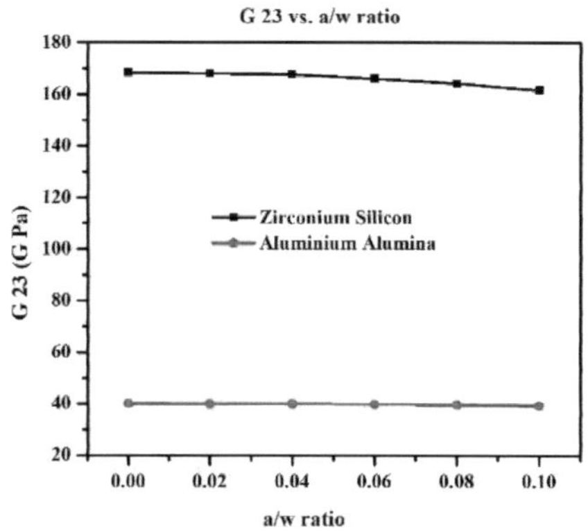

Figure 6.6. a/w ratio vs G_{23}

Figure 6: Homogenous properties versus a/w ratio

3.1 Inference from Matrix Defect

For 35% Fibre Volume fraction, for the variation of a/w ratio for 0.02-0.1 from the figure 6.1-6.6, following observations are made

- Matrix Crack doesn't affect E_1 for both MMC and CMC, as the reduction is only 1% along the fibre direction.
- It doesn't affect Poisson ratio's of both the materials as they almost remain constant for variation of a/w ratios
- E_2 for ZrB2-SiC composite reduces by 2% for variation in a/w ratio and for Aluminium-Alumina Composite it reduces by 10%
- G_{12} for ZrB2-SiC Composite reduces by just 1% and for Aluminium-Alumina composite it reduces by 5%
- G_{23} for CMC reduces by 4% and for MMC it reduces by 2%
- Nu $_{12}$ and Nu $_{23}$ remain almost a constant for the CMCs but increases by 4% and 7% for MMCs

4. Interfacial Debonding:

Interfacial debonding is one major contributor for degradation of property of the composite. UD MMC and CMC are manufactured by either Hot Isostatic Pressing (HIP) of Powder Metallurgy (Compaction followed by Plasma Sintering). The issue occurs with debonding when the applied pressure in HIP process or Compaction process is not an optimum value. Debonding occurs when the applied pressure exceeds the optimum pressure value. Debonding generally degrades the transverse tensile strength of the composite and it doesn't affect the strength of the composite in the fibre direction. Experimental studies have revealed that the interfacial properties play a very significant role in affecting the effective orthotropic properties of the composite. Such imperfect interface in composite materials may arise from the presence of impurities at phase boundaries or sometimes due to wettability issues. These impurities are due to formation of oxide films or bonding agents at fibre/matrix interface. The existence of these imperfect interfaces will affect the overall effective property of the composite material significantly and thus leading to debonding of the fiber from the matrix and ultimately failure of the material.

4.1 Modelling Debonding

A 3D RVE is modelled with imperfect interface to study the effect of degradation of properties of the composite. The imperfect interfaces between the fiber and the matrix are taken into account by introducing cohesive contact surfaces. The influences of the interface on the elastic constants and the tensile strengths are examined through these interface models . The effect of the variation of the imperfect interface for every 90° of the fiber and matrix surface is studied. For every 90° the reduced cohesive stiffness is calculated and is assigned to the surface and the rest of the surface is tied to the matrix surface. The cohesive stiffness is calculated using given formula below

Interface Stiffness Constants

$$K_{nn}=K_{ss}=K_{tt}=K_{interface}$$
$$E_{interface}=(E_{matrix}+E_{fibre})/2$$
$$H_{interface}=0.1*\text{Fiber radius, fibre radius}=0.3338\text{mm}$$
$$K_{interface}=E_{interface}/H_{interface}$$

For Aluminium- Alumina Composite E matrix= 73.1GPa and E fibre=373GPa. For ZrB_2-SiC CMC E matrix =500GPa and E fibre=403.2GPa.

$K_{interface}$ for Zirconium – Silicon Carbide Composite is $1.3562*10^7$ and $K_{interface}$ for Aluminium-Alumina Composite is $6.6982*10^6$. These corresponding stiffness values are assigned at every 90° variation in angle and the rest of the surface are tied. The rest of the homogenization procedure and the boundary conditions remain the same.

4.2 Output from corresponding boundary conditions:

The respective 4 sets of boundary conditions are applied on the RVE with induced defects and corresponding outputs for respective boundary conditions are shown in the figures given below. Figure 7.1-7.4 represents output with respect to 4 set of boundary condition for 0°-90° debonding. Figure 8.1-8.4 represents output with respect to 4 set of boundary condition for 0°-180° debonding. Figure 9.1-9.4 represents output with respect to 4 set of boundary condition for 0°-270° debonding. Figure 10.1-10.4 represents output with respect to 4 set of boundary condition for 0°-360° debonding.

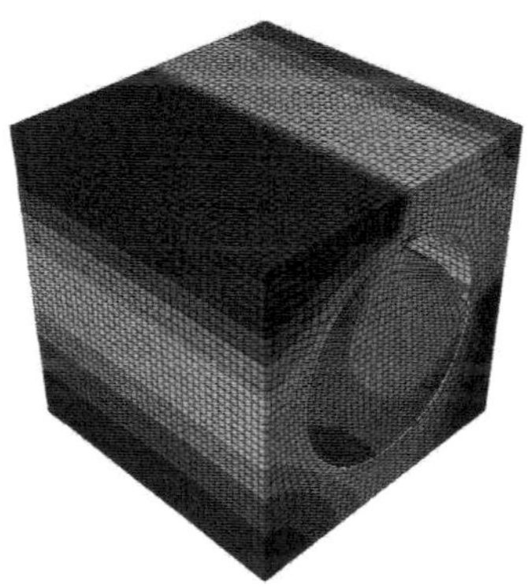

Figure 7.1. Set 1 BC

Figure 7.2. Set 2 BC

Figure 7.3. Set 3 BC

Figure 7.4. Set 4 BC

Figure 7. Output for 0°-90° Debonding

Figure 8.1. Set 1 BC

Figure 8.2. Set 2 BC

Figure 8.3. Set 3 BC

Figure 8.4. Set 4 BC

Figure 8. Output for 0°-180° Debonding

Figure 9.1. Set 1 BC

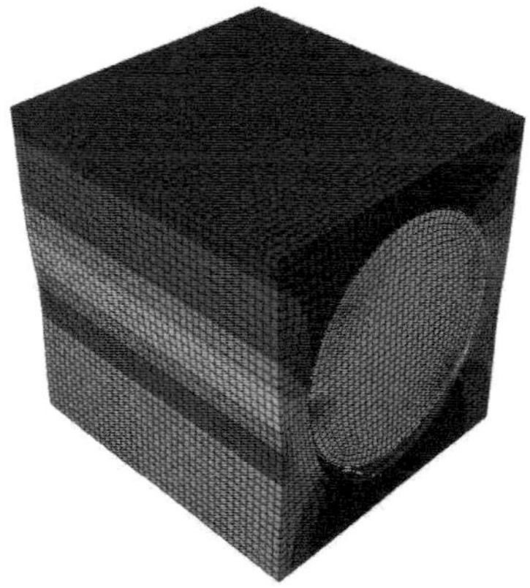

Figure 9.2. Set 2 BC

Figure 9.3. Set 3 BC

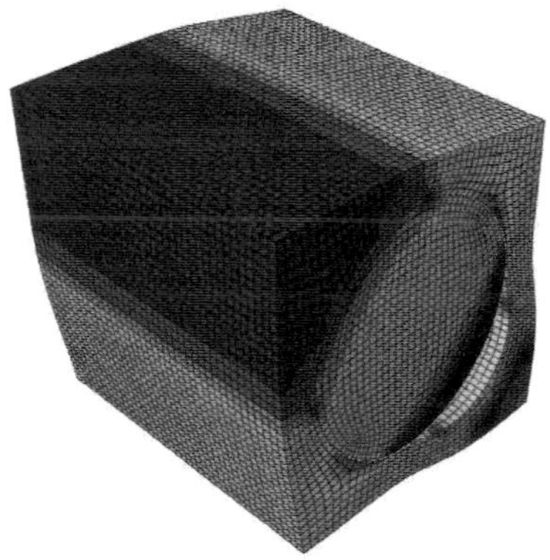

Figure 9.4. Set 4 BC

Figure 9. Output for 0°-270° Debonding

Figure 10.1. Set 1 BC

Figure 10.2. Set 2 BC

Figure 10.3. Set 3 BC

Figure 10.4. Set 4 BC

Figure 10. Output for0°- 360° Debonding

4.3 Results:

The corresponding figures below represent the variation of properties with respect to different debonding angles. Figures 11.1-11.6 represents variation of E_1, E_2, Nu_{12}, Nu_{23}, G_{12}, G_{23} vs debonding angle

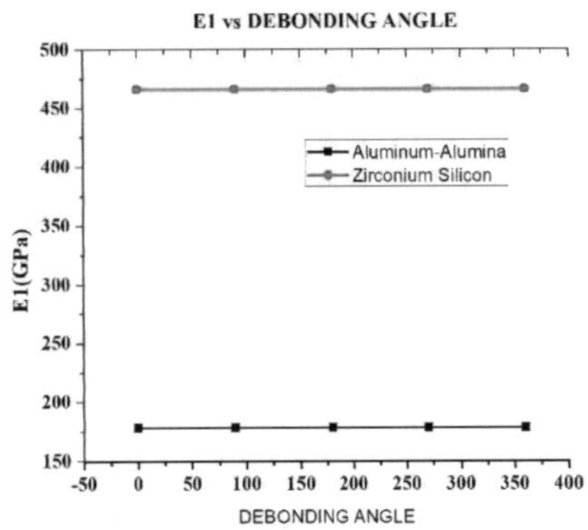

Figure 11.1. Debonding angle vs E_1

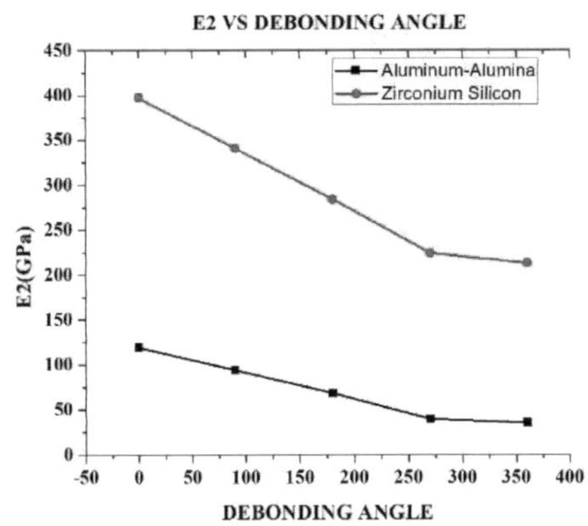

Figure 11.2. Debonding angle vs E_2

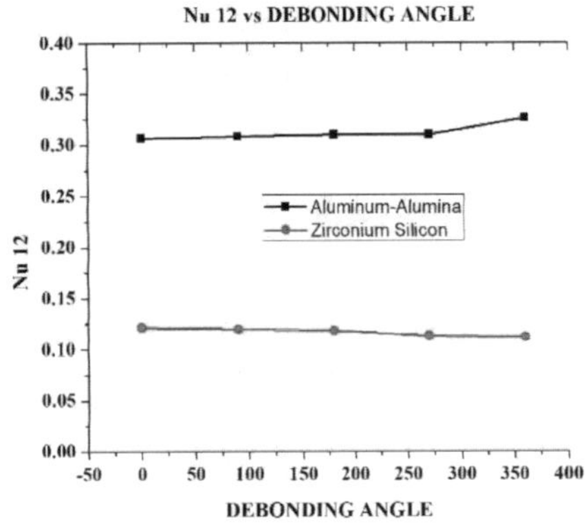

Figure 11.3. Debonding angle vs Nu_{12}

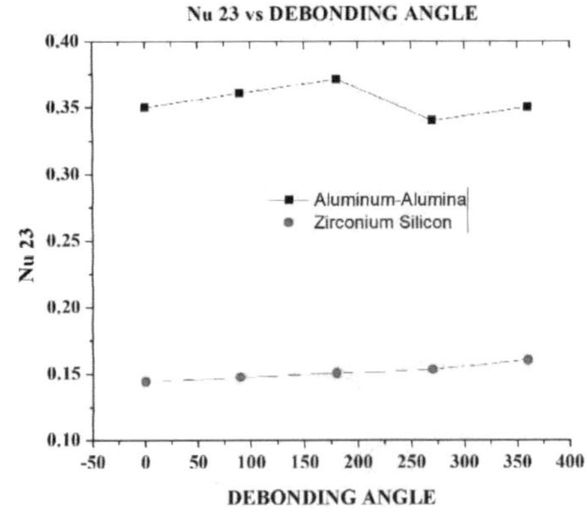

Figure 11.4. Debonding angle vs Nu_{23}

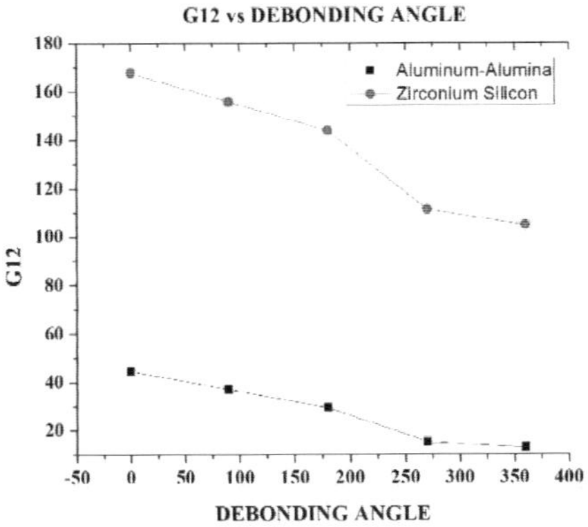

Figure 11.5. Debonding angle vs G_{12}

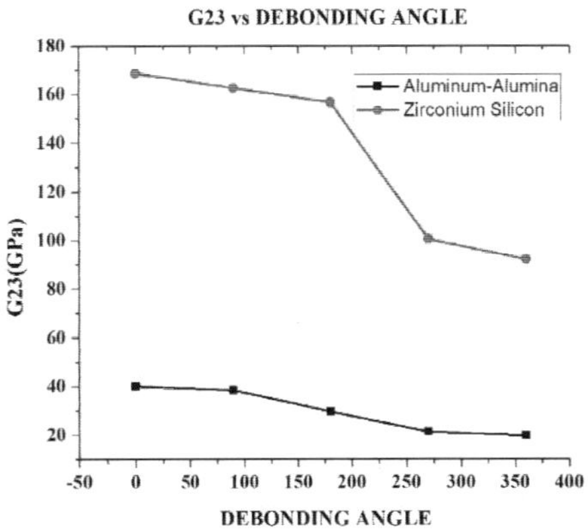

Figure 11.6. Debonding angle vs G_{23}

Figure 11: Homogenous properties versus debonding angle

4.4 Inference from Debonding Results:

For 35% Fibre Volume fraction, for the variation of debonding angle from ratio from 90°-360 following observations are made from figure 11.1-11.6

- Debonding doesn't affect E_1 for both MMC and CMC
- Poisson ratio's of both material remains a constant for variation of debonding angle
- E_2 for ZrB2-SiC Composite reduces by 46% from 0°-360° variation in debonding angle and for Aluminium-Alumina Composite it drastically reduces by 70%.
- G_{12} for ZrB_2-SiC Composite reduces by 37% and for Aluminium-Alumina composite it almost doubles about 71% reduction in G_{12}
- G23 for CMC reduces by 45% and for MMC it reduces by 51%.

5. Conclusion:

The following conclusions were drawn from the above analysis between healthy case and the damage induced case

- E_1 for MMC increases by 46% from 20%-47% Fibre Volume fraction.
- Analysis for the crack and Debonding has been conducted for 35% Volume fraction where E_1 almost remains same for both matrix crack and debonding, as these phenomena doesn't affect young's modulus along the fibre direction.
- E_1 decreases by 5% as Fibre volume fraction increases for a CMC.
- The reason behind the reduction in E_1 is Young's Modulus of the Matrix dominates young's modulus of the fibre
- Transverse Young's Modulus E_2 increases by 53% for MMC and reduces by 15% for the CMC
- With matrix crack the degradation of E2 is only 2% for the CMC and 10% for the MMC
- .Debonding degrades E2 for both MMC and CMC in a large scale by 46% for CMC and 70% for MMC
- Poisson Ratio almost remains the same for all 3 cases for both CMC and MMC
- G_{12} increases by 49% in MMC and decreases by 21% in CMC. Matrix crack only has very less effect of the degradation on CMC by 1% and % in MMC
- Debonding has a very high impact on G_{12} reducing by 37% for CMC and has a very drastic effect on MMC by degrading it by 71%. G_{23} goes down by 45% for CMC and 51% for MMC.

The above study shows that defects in the matrix doesn't affect the properties much. On the other hand defects in the fibre matrix interface degrades the material to a larger scale by reducing its properties.

6. References

[1] Kumar G V, Rao, C, and Selvaraj, N, "Mechanical and dry sliding wear behaviour of Al7075 alloy-reinforced with SiC particles", Journal of Composite Materials, vol. 46, pp. 1201-1209, 2011
[2] Zihui Xia, Yunfa Zhang, Fernand Ellyin., "A unified periodical boundary conditions for representative volume elements of composites and applications.," International Journal of Solids and Structures 40 (2003), 1907–1921.
[3] Sadik L Omairey, Peter Dunning, Srinivas Sriramula, "Development of an ABAQUS plugin tool for periodic RVE homogenisation, "Engineering with Computers (2018).
[4] Ramesh A, Sumesh, C S, Abhilash, P M, and Rakesh, S, "Finite Element Modelling of Orthogonal Machining of Hard to Machine Materials", International Journal of Machining and Machinability of Materials, vol. 17, pp. 543-568, 2015.
[5] Z Hashin, "Thermoelastic properties of fiber composites with imperfect interface" Mechanics of Materials 8 (1990), 333-348.
[6] Eric W Neumann, Georgy E Hilmas, William G Fahrenheit," Strength of Zirconium Diboride to 2300°C", " Journal of the American Ceramic Society", Vol 40(2012), 47-50.
[7] Wu-Gui Jiang, Ren-Zhi Zhong, Qing H. Qin and Yong-Gang Tong, "Homogenized Finite Element Analysis on Effective Elastoplastic Mechanical Behaviours of Composite with Imperfect Interfaces .," International Journal of Molecular Sciences 15 (2014), 23389-23407.
[8] Y Benveniste., "The effective mechanical behaviour of composite materials with imperfect contact between the constituents .," Mechanics of Materials 4 (1985), 197-208.

Enhancing the thermophysical properties of PAG lubricant using graphene nano-sheets.

S S Sanukrishna, Ajin V Raju, Anantha Krishnan, G H Harikrishnan, A Amal, T S Krishna Kumar and M Jose Prakash

Department of Mechanical Engineering, TKM College of Engineering, Kerala, India

E-mail: sanukrishna@sctce.ac.in

Abstract. One of the most attractive methods to improve the thermo physical and heat transfer properties of conventional coolants and lubricants is the use of nanoparticles in it. Nanofluids are stable colloidal suspensions of nanoparticles and base fluids. The nanofluids exhibit superior thermal performance compared to conventional heat transfer fluids and this leads to the best hope for lubricants based nanofluids as well. Nanolubricant is a new class of lubricant produced by dispersing nano-sized particles of metals, metal oxides, carbon and its allotropes into conventional lubricating oils. In the present study, experiments were conducted to measure the thermophysical and rheological properties of polyalkyleneglycol (PAG) refrigerant compressor oil suspended with graphene nano-sheets. The effect of particle concentration and temperature on thermal conductivity and viscosity of nanolubricant were elucidated. The volume fraction is varied from 0.05 to 0.8% and the temperature range is from 15^0C to 85^0C. The rheological properties have been investigated at various shear rates, particle concentrations, and temperatures. Furthermore, a correlation is proposed to predict the thermal conductivity of Graphene-PAG nanolubricant. The results revealed that the thermal conductivity and viscosity of the nanolubricant is strongly depends on the volume fraction and temperature. The nanolubricant exhibits higher thermal conductivity and viscosity compared to the PAG oil. Unlike in the case of pure lubricant, the shear rate has a crucial impact on the behavior of nanolubricant. Interestingly, the pure lubricant, which is a Newtonian fluid, is transformed into non-Newtonian when nanoparticles are appended with it. From the studies,, it was observed that the nanolubricant having graphene particle concentration of 0.05% is optimum for refrigeration application.

1. Introduction

Exponentially growing energy demands necessitate high efficient heat transfer fluids and lubricants in diverse thermal systems. The performance of lubricants directly affects global energy consumption. For this reason, enhancing the thermo-physical and tribological properties lubricants is imperative. The addition of chemical compounds is a well-known practice to enhance the thermal and tribological capabilities of coolants and lubricants. The development of various nanoparticles has opened up new avenues in various fields including cooling, lubrication, medicine, composite etc. Metallic and metal oxide nanoparticles, carbon and its derivatives (fullerenes, onions, nanotubes, and graphene), have been used now a days to develop nano-coolants and nanolubricants.

In the year 1995, Choi coined the term nanofluids and found that nanoparticles can offer anomalous enhancement of thermal conductivity to the base fluids [1]. Since 1995, research in this area has diversified tremendously into many disciplines. A series of breakthroughs were witnessed when Choi and Eastman et al. and Das et al. observed more than 40% enhancement in thermal conductivity by using 0.3% Cu particles of 10 nm size and 150% increase by using multi-walled carbon nanotubes (MWCNT) in engine oil [2,3]. In 2008 Baltez et al. studied anti-friction properties of CuO, ZrO_2 and ZnO nanoparticles. Interestingly, another wave of serious interest evolved among researchers when single-atom-thick graphene was successfully extracted from bulk graphite in 2004. Few reports have been published on graphene based nanolubricants and nanofluids by Rasheed et al., and Senatore et al. in 2013 and 2015 [4, 5]. However, comprehensive studies focusing on graphene based nanolubricants are scarce. It is imperative to understand the anti-friction behavior, enhanced thermal transport phenomenon, rheological and tribological characteristics of graphene based nanofluids and nanolubricants owing to its different nature from other nanoparticles.

This study mainly focuses on the thermal and rheological characterization of graphene-Polyalkylene glycol (PAG) nanolubricant. Specifically, to determine the thermo-physical properties like thermal conductivity and viscosity of graphene based nanolubricant at different composition and temperatures. Furthermore, efforts are taken to propose a regression model to predict the thermal conductivity of graphene based nanolubricant.

2. Methodology

2.1. Preparation of nanolubricant

Preparation of nanolubricant is the key step in the experimental studies. Nanofluids are not mere solid to fluid suspensions. Graphene-PAG nanolubricants were prepared at different volume fractions (0.05, 0.1, 0.15, 0.3, 0.5% and 0.8vol %). The stable nanolubricants were prepared thorough magnetic agitation and ultra-sonication. No evidence of sedimentation or coagulation was noticed after 5 days of preparation. The properties of graphene nano-sheet are summarized in table1.

Table 1. Properties of graphene nano-sheet.

Properties	Values
Purity	>99%
Thickness	5-10nm
Length	4-8 micron
Surface area	$190 m^2/g$
Thermal conductivity	≈3000 W/mK
Tensile modulus	>1000GPa

2.2. Determination of thermal conductivity

The thermal conductivity of base lubricant and nanolubricants were measured with KD2-Pro thermal property analyzer (Decagon devices, Inc., USA). The principle of operation of the device is transient hot wire method and the maximum deviation is ±5.0%. The probe of the instrument consists of a needle with a heater and temperature sensor inside it. Electric current is supplied through the heater and the temperature of the probe is measured over time. The apparatus meets the standards of ASTM D5334. A refrigerated and heating circulator (F-25, Julabo, Germany) with an accuracy of 0.01°C was used to maintain the temperature of the sample constant. The thermal conductivity of nanolubricant at different particle volume concentrations was measured in the temperature range of 15°C to 45°C.

2.3. Determination of rheological behaviour

Rheological behaviour was evaluated in the temperature range of 25-85°C at varying shear rates and particle dosing levels. A Brookfield LVDV-II +Pro, plate-and-cone rheometer together with a constant temperature circulator (JULABO F-25, Germany) was used for the measurements. All the measurements were performed under steady state conditions within the torque range of 10-100%.

3. Results and discussion

3.1. Thermal conductivity

Figure 1 shows the variation of thermal conductivity with respect to volume fraction and temperature. From the figure it is evident that, as graphene concentration in the PAG lubricant increases, the thermal conductivity increases first then slightly decreases. The maximum value of thermal conductivity is obtained as 0.167 W/mK at a particle concentration of 0.05vol%, which is 9.86% greater than that of pure lubricant. The major contributing factor for this enhancement is increased Brownian motion and subsequent thermal interaction and higher thermal conductivity of nano-sheets.

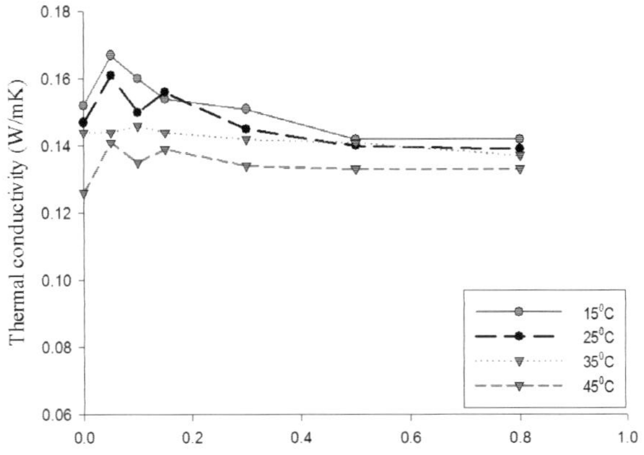

Figure 1. Variation of thermal conductivity with volume fraction.

The temperature dependent thermal conductivity of graphene-PAG nanolubricant is shown in figure 2. As temperature increases thermal conductivity decreases. However, it is observed that for 0.05% particle concentration, the value of thermal conductivity is always greater than that of pure lubricant. According to the study, the most suitable concentration as an alternative lubricant for the refrigerant compressor is 0.05vol%.

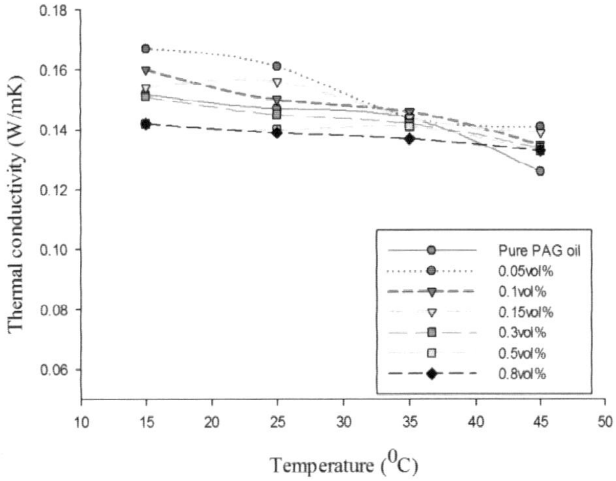

Figure 2. Temperature dependent thermal conductivity variation.

3.2. Viscosity

The viscosity of the nanolubricant is measured at different shear rates and temperatures. Figures 3-8 represent the variation in viscosity with respect to shear stresses at different temperatures. The figure shows that the viscosity of nanolubricant at all particle concentrations and temperatures displays non-Newtonian behaviour, especially at lower shear rates and temperatures. That is the viscosity varies with respect to shear rates. Interestingly, as the temperature and shear rate increases, the nanolubricant displays Newtonian behaviour.

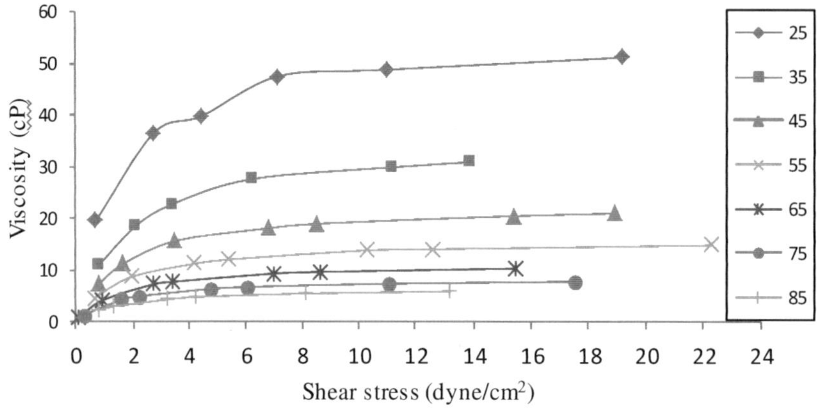

Figure 3. Shear stress vs viscosity at a volume fraction of 0.05% (at different temperatures).

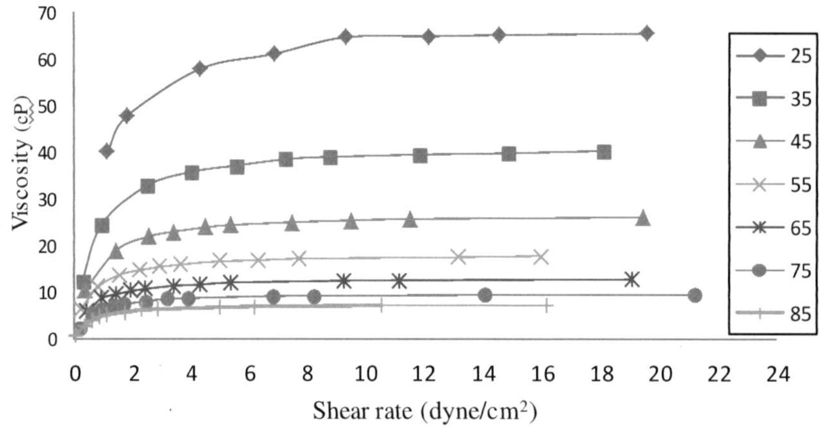

Figure 4. Shear stress vs viscosity at a volume fraction of 0.1vol% (at different temperatures).

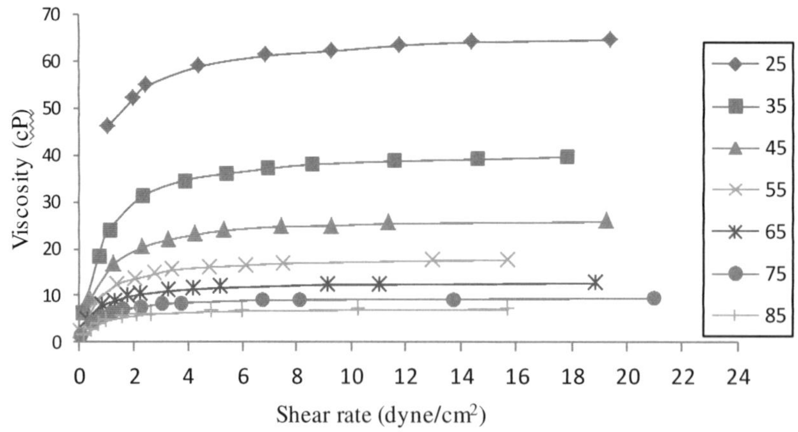

Figure 5. Shear stress vs viscosity at a volume fraction of 0.15vol% (at different temperatures).

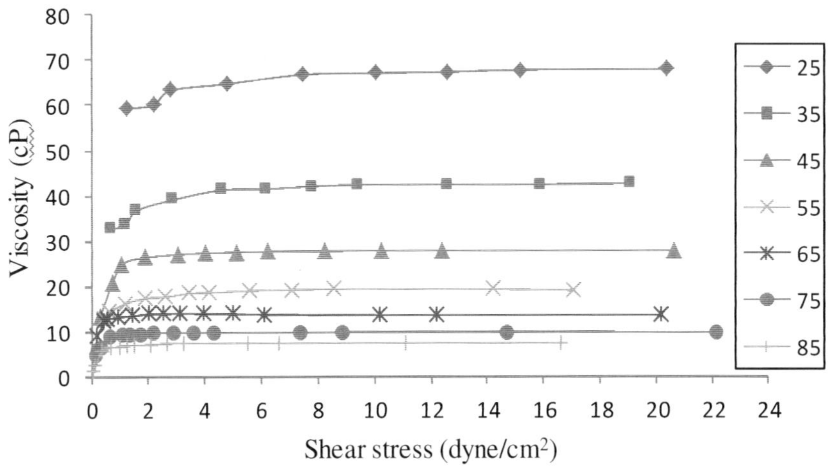

Figure 6. Shear stress vs viscosity at a volume fraction of 0.3vol% (at different temperatures).

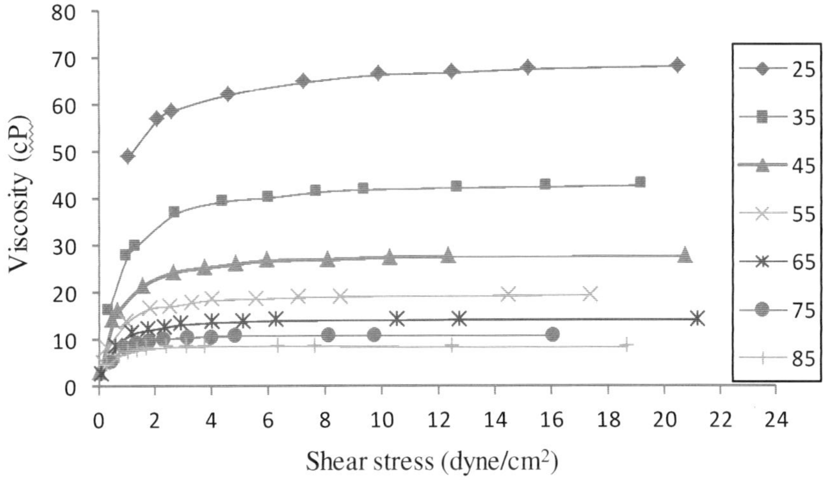

Figure 7. Shear stress vs viscosity at a volume fraction of 0.5vol% (at different temperatures).

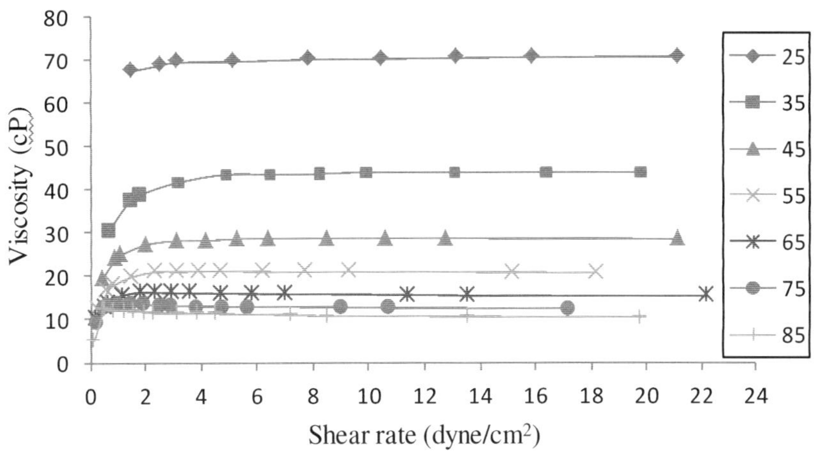

Figure 8. Shear stress vs viscosity at a volume fraction of 0.8vol% (at different temperatures).

Figure 9 shows the variation in viscosity of graphene based nanolubricant with respect to temperature and volume fraction.

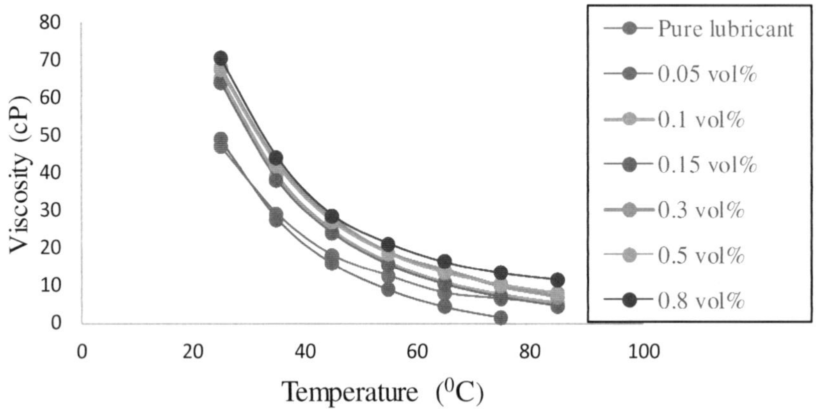

Figure 9. Temperature dependent viscosity.

The viscosity of pure lubricant and nanolubricant decreases with the intensification of temperatures at all particle concentrations. This is due to the weak van-der Waals force of attraction between the molecules. However, graphene-PAG nanolubricant at lower particle concentration (0.05vol %) exhibits a slight decrease in viscosity than that of pure PAG oil, especially at elevated temperature. This may be an added advantage since reduced viscosity significantly saves the pumping power.

3.3. Correlation to predict the thermal conductivity

From the experimental data, a correlation is proposed for predicting the thermal conductivity of graphene-PAG nanolubricant at different concentrations and temperatures, which is shown in equation 1.

$$k = -0.001T - 0.014\varphi + 0.166 \quad (1)$$

where T is the temperature in °C, φ is the particle concentration and k is the thermal conductivity in W/m K. The regression model with an R-squared value higher than 0.98, well predicts the thermal conductivity of nanolubricant in the temperature region of 15-80°C and volume fraction between 0.05 to 0.8vol%, as shown in Figure 10.

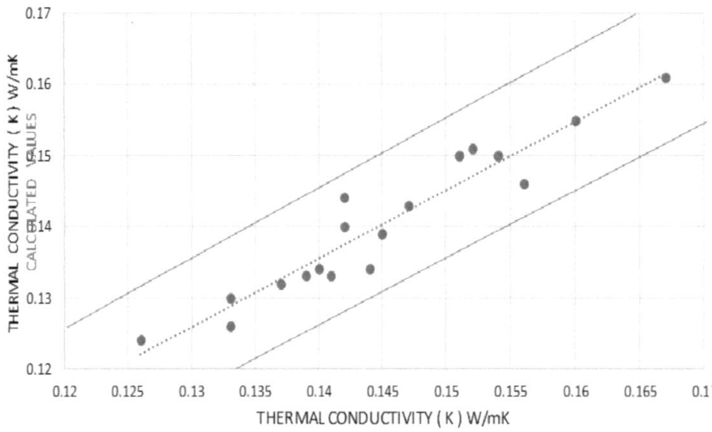

Figure 10. Experimental data vs model predictions.

4. Conclusions

Thermal conductivity and viscosity of the graphene-PAG nanolubricant have been experimentally determined . The experimental results show that the addition of nanoparticle to the lubricant oil enhances the thermal conductivity. The thermal conductivity was 9.86% higher than that of pure PAG lubricant at a particle concentration of 0.05vol%. Similar to thermal conductivity; the volume concentration and temperature have significant effects on the viscosity of nanolubricants as well. As the particle concentration increases, the viscosity also increases. In the case of particle concentration 0.05vol%, the nanolubricant displays slight detrimental effect on viscosity than that of pure PAG lubricant. The nanolubricant exhibits non-Newtonian behaviour especially at lower shear rates. The proposed correlation well predicts the thermal conductivity of nanolubricant. It is concluded that, among the particle concentration considered for the investigation, the optimum concentration of graphene nanoparticle is 0.05%.

References

[1] Eastman J A, Choi S U S, Li S, Yu W and Thompson L J 2001 Anomalously increased effective thermal conductivities of ethylene glycol-based nanofluids containing copper nanoparticles *Appl. Phys. Lett.* **78** 718–720.

[2] Kumar P C M, Kumar J, Tamilarasan R, Sendhil Nathan S and Suresh S 2014 Heat transfer enhancement and pressure drop analysis in a helically coiled tube using Al_2O_3 / water nanofluid *J. Mech. Sci. Technol* **28** 1841–1847.

[3] Sharif M Z, Azmi W H,. Redhwan A A M, and Mamat R 2016 Investigation of thermal conductivity and viscosity of Al_2O_3/PAG nanolubricant for application in automotive air conditioning system *Int. J. Refrig.* **70** 93-102.

[4] Rasheed A K, Khalid M, Rashmi W, Gupta T C S M and Chan A 2016 Graphene based nanofluids and nanolubricants - Review of recent developments *Renew. Sustain. Energy Rev.* **63** 346–362.

[5] Berman D, Erdemir A, and Sumant A V 2014 Graphene: A new emerging lubricant *Mater. Today* **17** 31–42.

OPTIMAL LAYUP SCHEMES WITH SELECTIVE DISPERSION OF CORE/SHELL MICROPARTICLES IN PLY INTERFACES OF GLASS/EPOXY COMPOSITE LAMINATES FOR LOW VELOCITY IMPACT

Mubarak Ali M[1], Sunil C Joshi[2]

[1]Assistant Professor, Department of Mechanical Engineering, TKM College of Engineering, Kollam, Kerala, India 691005

[2]Associate Professor, Department of Mechanical and Aerospace Engineering, Nanyang Technological University, Singapore, 639798

Email: muba0001@e.ntu.edu.sg, mubarakali@tkmce.ac.in

Abstract: The overall objective of the investigation presented in this paper was to study the effect of selective dispersion core/shell microparticles (CSP) within the different ply interfaces on impact damage behaviour of glass/epoxy laminates. These laminates were fabricated with the 30 g/m^2 CSP particle areal weight per interface. A series of impact experiments were done with instrumental drop tower device at impact energy of 7.8 J and impactor velocity of 2.4 m/s, which is within a practical low velocity impact range. The analysis of the impact load, contact duration indicates that the presence of CSP particles in the bottom layers has a more pronounced effect on the impact performance improvement as compared to the top layers. In the case of selective dispersion, the presence of particles in the middle plies slightly degrades the performance when compared to the selective dispersion of particles only in top or bottom plies.

Keywords: Glass/epoxy composite laminates, Impact behaviour, Optimal layup scheme, Core/shell particles, Selective Dispersion

Introduction

Composite materials are being used extensively in aerospace, automotive, and marine industries to build lightweight and cost-effective structures. Their usage is also gaining momentum in orthopaedics and spacecraft applications. Composites are well established for its high specific strength and high specific stiffness, which make them a desirable choice for weight critical applications [1-4].

One of the major drawbacks of laminated composites is their vulnerability to impact [5-6]. Aircraft composite structures are often exposed to impact damages from hailstones, tools dropped during maintenance or manufacturing, collisions by service vehicles, foreign objects thrown up from the ground or debris stones from the runway propelled by the tires [7]. These low velocity impacts can create internal damages within the laminated composites which often cannot be discovered by visual inspection and eventually can grow under the applied loading. These damages result in drastic reduction in the residual strength in tension, compression, shear and bending, and to some extent, affect the fatigue and vibration characteristics of the laminates [8-9].

A simple and inexpensive technique for the toughening of composite materials consists of selectively toughening the matrix resin between the plies by using thermoplastic or rubber particles [10]. This technique appears to be the best practical method so far for improving the impact toughness of composites [11]. A new type of particles called core/shell microparticles has been successfully used to improve the impact behaviour of prepreg laminates [12-14]. In this study, selective dispersion of core/shell microparticles has been investigated to understand and arrive at an optimum layup schemes.

Experimental Details

The laminates investigated in this research were fabricated using GFE8 L530-7781 prepreg (refer Table 1 for its details) and about 6 to 9 samples were tested in each layup scheme.

Table 1 Specifications of E-glass fabric of GFE8 prepreg

Glass fabric name		HexForce 7781
Nominal warp count/cm		57/2.54
Nominal fill count/cm		54/2.54
Weave		8 Harness Satin
Fabric weight		301 g/m^2
Fabric thickness		0.22 mm
Breaking strength	Warp	98.82 kN/m
	Fill	78.80 kN/m
Filament diameter		6 μm
TEX		66
No of strands in a yarn		1

The chosen CSP particles have a soft rubber (PBA) core and a transparent shell of poly (methyl methacrylate) (PMMA) with an epoxy functional group grafted to the shell with the average particle size of about 160 μm. The structure and particle size distribution of the CSP particles has been explained elsewhere [7]. In order to distinguish the effect of particle dispersion in top layers and bottom layers, various layup schemes were worked out. Table 2 shows the various sample codes and its descriptions. Figure 1 illustrates the schematic of different layup schemes used, indicating the point of application of the impact load with tup. The Original laminates (ORG) has 8 layers and are fabricated without CSP particles. Altered laminates (ALT) were fabricated with CSP particles dispersed in 3 and 5 layers with two different configurations; particles dispersed only in top layers and particle dispersed only in bottom layers. All the impact tests were conducted on the GFE8 laminates at the incident impact energy of 7.8 J, which corresponds to the impactor velocity of 2.4 m/s. Dynatub Drop Tower Impact testing machine is used for the present experiments and the details of the impact-testing machine is explained elsewhere [6].

Table 2 Sample codes for various layup schemes

Sample code	Description	Figure
ORG0	Original Laminate without particle dispersion	1a
ALT-7ALL	Particle dispersed in all 7 interfaces	1b
ALT-3BTM	Particle dispersed only in bottom 3 interfaces	1c
ALT-3TOP	Particle dispersed only in top 3 interfaces	1d
ALT-5BTM	Particle dispersed only in bottom 5 interfaces	1e
ALT-5TOP	Particle dispersed only in top 5 interfaces	1f

Results and Discussions

Figure 2 shows the maximal load values that the laminate of the various layup schemes withstand during the impact test. It is apparent from Figure 2 that, all the altered laminates irrespective of particle dispersion in 3 or 5 interfaces and top or bottom interfaces, exhibited higher maximal load compared to that of the ORG0 laminate. It can also be observed that the ALT-7ALL and the ALT-3BTM laminates show same maximal load. The other layup schemes such as the ALT-3TOP, ALT-5BTM and ALT-5TOP showed lower maximal load when compared to the ALT-3BTM. Hence, in terms of maximal load, the ALT-3BTM is optimal layup scheme as it withstands the same maximal load as that of the ALT-7ALL.

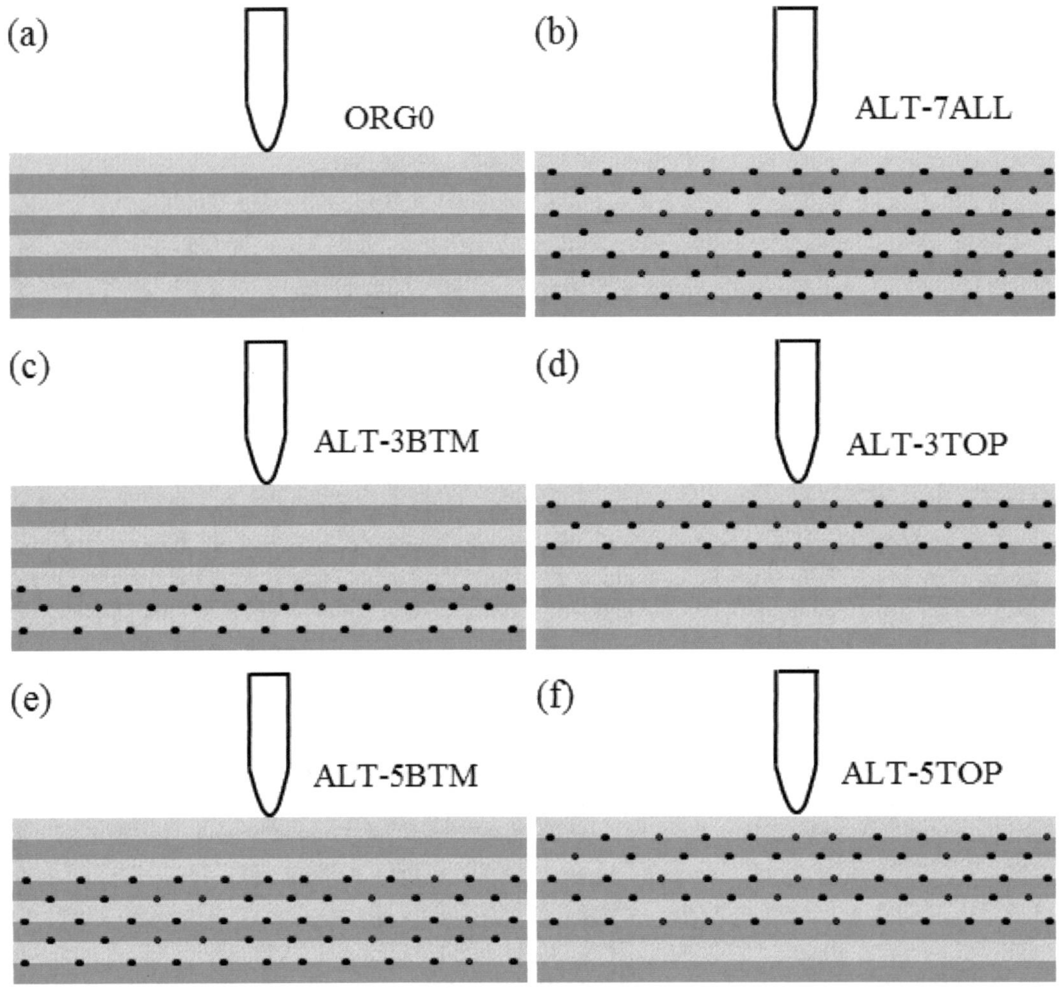

Figure 1 Schematic representations of different layup schemes

Figure 3 depicts the contact duration of various laminates with different layup schemes. In terms of the contact duration, the ALT-7ALL shows the lowest, followed by the ALT-3BTM. The contact duration of all the altered laminate with particles was better than the ORG0 laminates irrespective of the layup schemes.

Figure 4 represents the damage index (DI) for various layup schemes. This also indicates the same trend as the maximal load. The ALT-3BTM exhibited the least damage index of all the layup schemes, even better than the ALT-7ALL layup. This is primarily due the high elastic stored energy of the ALT-3BTM layup when compared to all other layup schemes. The damage index of the all layup schemes with particles indicated improvement when compared to that of the ORG0 laminates. It can be clearly deduced that in terms of maximal load, contact duration and damage index, that all the layup schemes with particles incorporation (whether it is 3 layers or 5 layers; top layers or bottom layers) showed better performance than the ORG0 laminates. Hence, the presence of CSP particles had definite improvement on the GFE8 laminates irrespective of layup schemes.

The ALT-3BTM and the ALT-5BTM shows better maximal load, contact duration and DI than the ALT-3TOP and the ALT-5TOP respectively. Hence it can be deduced, that the presence of CSP particles in the bottom layers has a more pronounced effect on the impact performance improvement as compared to the top layers.

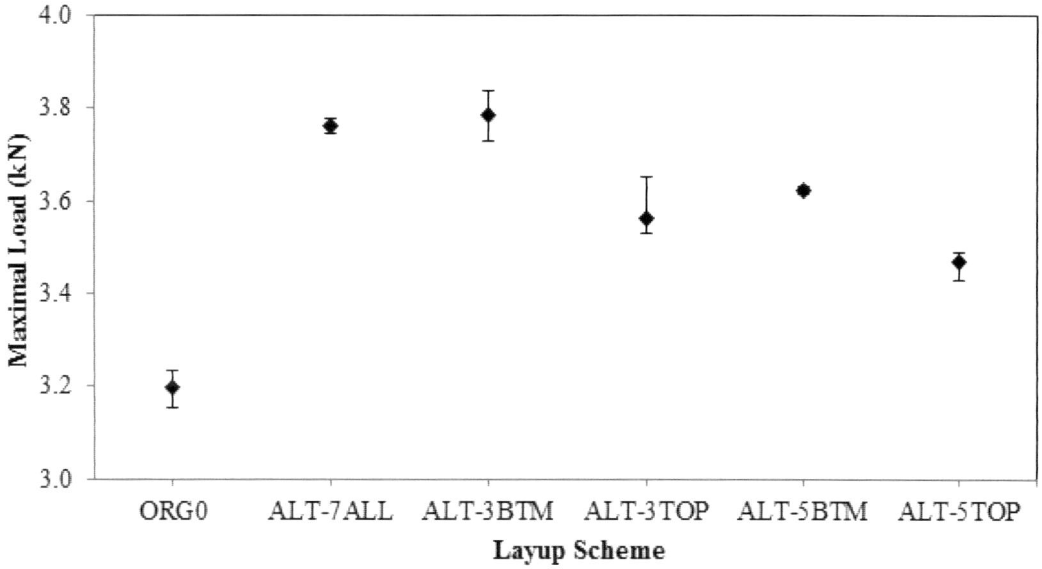

Figure 2 Maximum load for various lay up schemes

It can be understood that ALT-3BTM shows better impact response than ALT-5BTM in terms of maximal load, contact duration and damage index. Similarly ALT-3TOP shows better impact response than the ALT-5TOP samples. Hence, it can be said in the case of selective dispersion that the presence of particles in the middle plies slightly degrades the performance when compared to the selective dispersion of particles only in top or bottom plies.

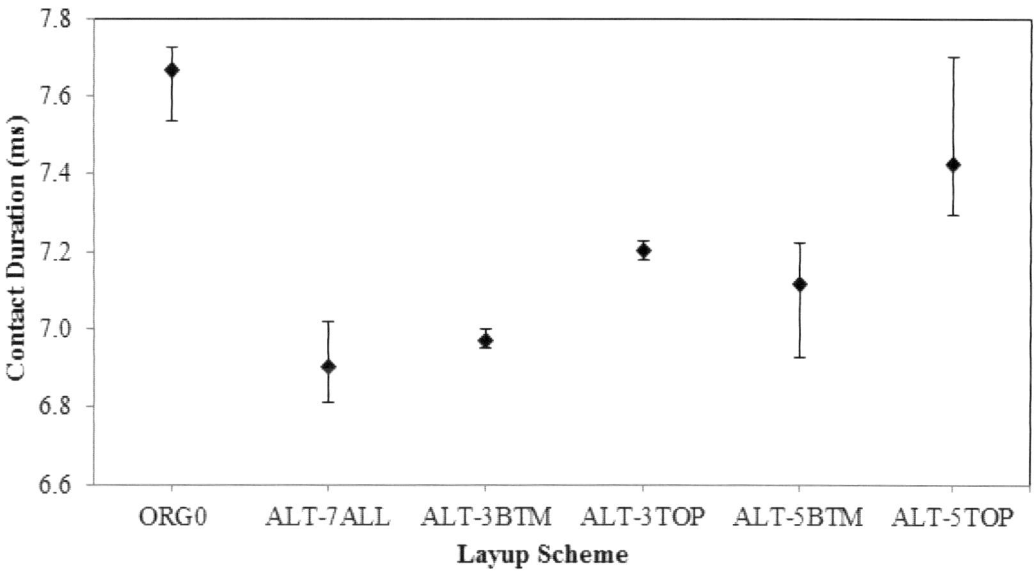

Figure 3 Contact duration for various layup schemes

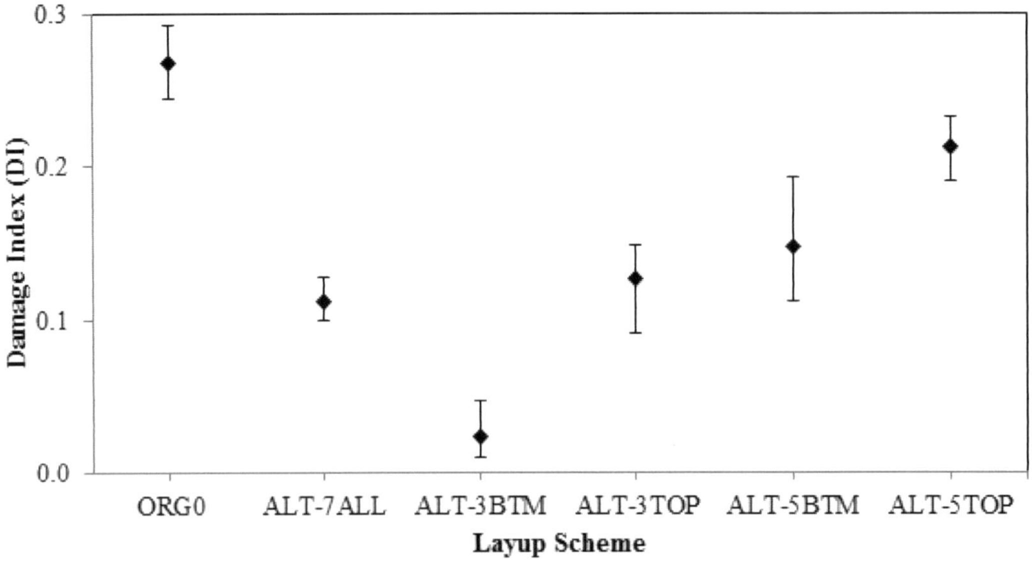

Figure 4 Damage index (DI) for various layup schemes

It has been explained in previous study [6], that the first damage in the laminate precipitated as the matrix cracks in the bottom face of the laminate. When the impact load is applied, the bottom layers will be in tension and the top layers will be under compression. This means incorporation of particle within the interlaminar regions between the layers subjected to tension exhibit superior resistance to impact load and damage.

Conclusions

The present study deals with identifying the optimal layup scheme and effect of selective dispersion of Core/shell particles in the interlaminar interlayers of the glass/epoxy composite laminates. The laminates were fabricated with Core/shell particles dispersed in all seven layers, top and bottom 5 layers and top and bottom 3 layers. The presence of CSP particles had definite improvement on the GFE8 laminates irrespective of layup schemes. In case of selective dispersion, the presence of particles in the middle plies slightly degrades the performance when compared to the selective dispersion of particles only in top or bottom plies. Incorporation of particle within the interlaminar regions between the layers subjected to tension exhibit superior resistance to impact load and damage.

Acknowledgement

The authors thank School of Mechanical and Aerospace Engineering, Nanyang Technological University, Singapore for the kind support and material facilitation for this research.

References

[1] Mubarak Ali, S. C. Joshi and M. T. H. Sultan (2017), Palliatives for Low Velocity Impact Damage in Composite Laminates, *Adv Mater Sci Engg*, **2017** (2017) Article ID 8761479.

[2] Vishnu Prasad, Deepak Suresh, MA Joseph, K Sekar, Mubarak Ali (2018), Development Of Flax Fibre Reinforced Epoxy Composite With Nano Tio2 Addition Into Matrix To Enhance Mechanical Properties, *Mater Today-Proc*, 5 (5) Part 2, 11569-11575.

[3] V Prasad, CVM Hunize, RI Abhiraj, MA Jospeh, K Sekar, M Ali (2019), Mechanical Properties of Flax Fiber Reinforced Composites Manufactured Using Hand Layup and Compression Molding—A Comparison, In: Shanker K., Shankar R., Sindhwani R. (eds), *Adv Ind Prod Engg*, Lecture Notes in Mechanical Engineering. Springer, Singapore 781-789.

[4] Vishnu Prasad, MA Joseph, K Sekar, Mubarak Ali (2018), Flexural and impact properties of flax 3ibre reinforced epoxy composite with nano TiO_2 addition, *Mater Today-Proc*, 5 (11) Part 3, 24862-24870

[5] Mubarak Ali and C. J. Sunil (2013), Energy Absorption Characteristics of Interface Modified GFRP Laminates under Low Velocity Impact", *Adv Mater Res*, (626), 589-593.

[6] Ali Mubarak, C Joshi Sunil (2012), Impact perforation characteristics of glass/epoxy composite laminates with modified ply interfaces, Proceedings of the 2nd Annual International Conference on Materials Science, Metal and Manufacturing (M3 2012), 22-27.

[7] Mangalgiri, P.D. (1999). Composite Materials for Aerospace Applications, *B Mater Sci,* **22** 657.

[8] Cantwell, W.J. and Morton, J. (1991). The Impact Resistance of Composite Materials - A Review, *Compos,* **22** 347.

[9] Richardson, M.O.W. and Wisheart, M.J. (1996). Review of Low-Velocity Impact Properties of Composite Materials, *Compos Part A: Appl S,* **27** 1123.

[10] Mubarak Ali and C. J. Sunil (2013), Fractography of Particle Strengthening Mechanisms at Interfaces in Prepreg Composites, *Adv Mater Res*, (816-817), 196-200.

[11] Gilbert, E.N., Hayes, B.S. and Seferis, J.C. (2003). Interlayer Toughened Unidirectional Carbon Prepreg Systems: Effect of Preformed Particle Morphology, *Compos Part A Appl S,* **34** 245.

[12] Mubarak Ali and Sunil C Joshi, Damage Evolution in Glass/Epoxy Composites Engineered using Core Shell Microparticles under Impact Loading, *J Mater Sci,* **48** (2013) 8354.

[13] Mubarak Ali and Sunil C Joshi, Impact damage resistance of CFRP prepreg laminates with dispersed CSP particles into ply interfaces, *Inter J Damage Mech*, **21** (2012) 1106.

[14] Mubarak Ali, Sunil C Joshi and Adrian Tan, Tailoring of Bonded Composite Scarf Joint Interface for Impact Damage Mitigation and Stiffness Compatibility, *Plast Rubber Compos* **45** (2016) 43.

Journal of Physics: Conference Series **1355** (2019) 012043 doi:10.1088/1742-6596/1355/1/012043

Effect of laser power on microhardness of NiCrBSi laser clads deposited on AISI410 stainless steel

Gaurav Durge[a], Aayush Chandak[b], Abhimanyu Kumar Jaiswal[a], Ravikumar Dumpala[a]

[a] Department of Mechanical Engineering, Visvesvaraya National Institute of Technology, Nagpur 440010, India
[b] Shreenath Engineering Industries,K-39/1, MIDC Indl.Area, Hingna Rd, Nagpur-440016, Maharashtra, India

E-mail: flowers.ravi@gmail.com

Abstract: In the present experimental study, NiCrBSi alloy was deposited on martensitic stainless steel (AISI 410 grade) using laser cladding technique. High power diode laser (HPDL) was used to obtain good fusion, optimum dilution & crack free single track laser clad by varying laser input power (3200 W, 3600 W & 4000 W) while keeping other processing parameters constant. Powders of NiCrBSi alloy with different wt. % of Ni, Cr, B and Si (50HRC & 60HRC) were used in the current study for the deposition. Pre-heating of the substrate was carried out at around 300°C using oxyacetylene flame before deposition in order to reduce the cracks and porosity. Effect of power on microhardness of laser clad was not prominent in the range of power 3200 to 4000 W). In the case of alloy 1 maximum microhardness 692.2 $HV_{0.3}$ was observed for power 3600 W, while for alloy 2 with power 3600 W maximum microhardness of as-deposited was 755.1 $HV_{0.3}$.

1. Introduction

With the rapid development taking across all over the world, a vast development in laser technology has taken place and laser cladding is a widely used technique for repairing and remanufacturing work in industries such as aerospace, automotive etc. Laser cladding technologies have led to a significant increase in reliability of the product developed by it. Laser cladding is a process in which a high power laser beam is directed to melt the cladding material and lead to the fabrication of very thin layer on the substrate. The coatings produced by laser cladding have many advantages over other process such as low dilution, metallurgical bonding with the substrate, minimum thermal distortion etc [1].

The AISI 410 grade martensitic stainless steel is used as a substrate. The AISI 410 stainless steel have many applications in manufacturing of bolts, bushings, gas turbine parts, mine equipment, steam turbine parts, valve parts etc. Material is magnetic in all conditions.
NiCrBSi is one of the most frequently used material for laser cladding, as it has many advantages such as low price, high hardness and more corrosion resistance in comparison with other similar alloys [2, 3].The corrosion and wear-resistant NiCrBSi coating is found to be the most reliable solution to protect the low-alloyed steel substrate having a significantly lower melting temperature than the deposited alloy [4].

The laser powers used for analysis were 1500, 2000, 2500, 3000, 3500 W and with the increase of the laser power, the average microhardness of the cladding layer was found to be reduced. This was because the increase of laser power led to entrance of more Fe element into the cladding layer, which enhanced the dilution effect of the cladding layer. Therefore, the average hardness of cladding layers decreased with the increase of laser power [5].

Preheating of the substrate material was done at 450 °C to avoid cracking. The laser power was varied from 1100 to 1900 W in two equal steps of 400 W and keeping other parameters constant. Slower cooling rate and high dilution at higher laser powers was supposed as the reason for the decrease in microhardness. Also, the dilution rate was found minimum at the laser power 1100 W and 1500 W and it also showed higher hardness [6].

As we know the melted volume of the substrate is an important parameter which measures the quality of the laser cladding process. This is due to the fact that the melted volume of the substrate material together with the deposed material volume determines the dilution (%). The best cladding quality is achieved when the base material surface just melts enough to obtain a good metallurgical bonding between the clad layer and the substrate material. Among the various varying parameters used for laser cladding such as laser power, cladding speed, and powder feed. The most significant factors for the substrate melted volume is the laser power [7].

The aim of present work is to study the effect of laser power on the microhardness of cladded layer of NiCrBSi alloy of 50HRC & 60HRC on AISI410 Stainless steel plate. Laser cladding is done using high power diode laser by varying the power by 3200W, 3600W & 4000W.High power diode laser is used for cladding process because the high power direct diode laser is an excellent heat source for many cladding applications. A combination of the large beam profile and the short wavelength make it ideal for the rapid deposition of relatively large areas of clad material. The simplicity, ease of use, high efficiency and compact nature of the direct diode laser makes it appropriate for many industrial applications [8]. The substrate material is preheated at temperature 300°C before performing laser cladding operation as maximum microhardness is achieved by preheating for nickel alloy [9].

2. Experimental Details

Laser cladding is performed on AISI 410 Stainless steel plate, for fabricating NiCrBSi alloy onto it and argon gas is used as the carrier and shielding gas. During the process, a shielding gas is always used to protect the molten material from the atmosphere [10]. Two different alloys of NiCrBSi powder having microhardness 50 HRC and 60 HRC are being used. Laser cladding is performed for 3 powers i.e 3200W, 3600W and 4000W using high power diode laser. Pre-heat treatment of the substrate material is carried out at 300°C using oxyacetylene flame before deposition in order to reduce the cracks and porosity. Chemical composition (Wt. %) of stainless steel 410 is given in table 1:

Table 1: Nominal chemical composition (Wt. %) of stainless steel 410

Material	C	Cr	Si	Ni	Mn	Fe
Stainless Steel 410	2.86	18.52	0.6	6.88	1.90	Bal

NiCrBSi powder is fed using co-axially powder feeder in a controlled rate. NiCrBSi powder of 50HRC and 60HRC is used having the chemical composition as mentioned in table 2:

Table 2: Nominal chemical composition of 50HRC & 60HRC NiCrBSi powder

Hardness	**Ni**	**Cr**	**B**	**Si**	**C**	**Fe**
50 HRC	Balance	12.5	2.3	3.3	0.5	3.9
60 HRC	Balance	15	3.1	4.3	0.7	4.2

To deposit laser clad, Laserline LDF 4000-100 Fibre coupled diode laser integrated with CNC is used. The powder is delivered by a Sulzer Metco Twin-10 C powder feeder and injected into the processing area through a co-axial nozzle. All the optics and nozzles are integrated with a Kuka KR16 robot arm having 6 DOF.

The processing parameter used during laser cladding operation are given in Table 3

Table 3: Processing parameter used during laser cladding operation:

Sample No.	Power (W)	Parameters values
S 1	3200 W	
S 2	3600 W	Scanning speed - 20mm/s Powder feed rate - 60g/min Spot size - 5mm Stand-off distance - 20mm
S 3	4000 W	

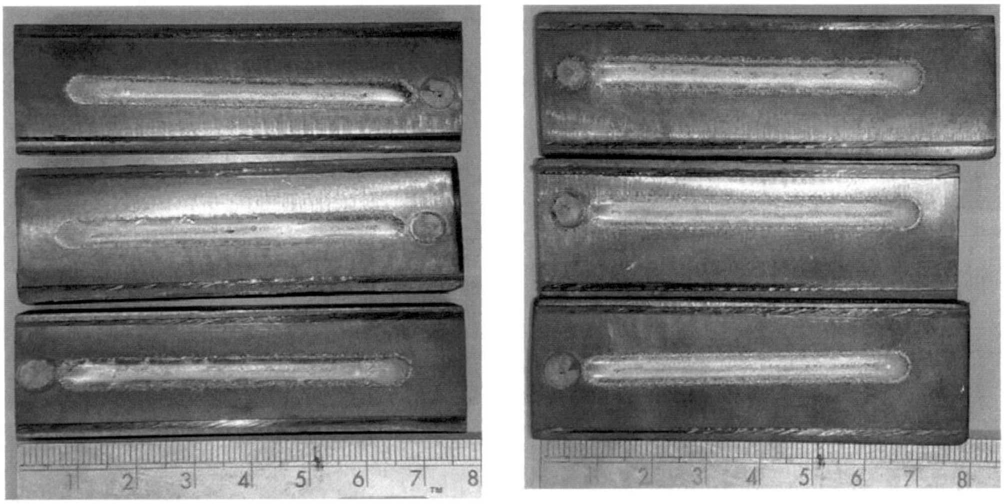

50HRC NiCrBSi Laser cladded samples 60HRC NiCrBSi Laser cladded samples

Figure 1- ***Laser cladded sample (50HRC & 60HRC) at powers 3200W, 3600W & 4000W***

Figure 1 shows the laser cladded samples of 50HRC and 60HRC alloys by varying laser power as 3200W, 3600W and 4000W.

2.1 Microhardness Testing

Microhardness of the transversely sectioned surface of the sample is measured using Vickers hardness tester [Mitutoyo (Model- HM 112)] at a load of 300g with the loading time of 15sec. A minimum of 8 indentations has been executed for each sample.

3. Results & Discussions

Microhardness of the as-deposited NiCrBSi alloy of 50HRC and 60HRC deposited using the laser at different powers are measured using Vicker hardness tester and the graph is plotted for showing the variation in microhardness for both the alloys.

Microhardness is measured on the tranverse section at eight different points. Results shows that the microhardness of the clad layer is fairly uniform except near the top surface. The hardness value decreased in the dilution zone till reaching the base metal. Figure 2 and Figure 3 represents the comparison of microhardness of three samples 50HRC NiCrBSi alloy. Effect of laser power on microhardness of the cladding layer is studied which shows that as the laser power is increased from 3200W to 3600W, the microhardness of as-deposited coating is found to be increased from 752.7HV to

Journal of Physics: Conference Series **1355** (2019) 012043 doi:10.1088/1742-6596/1355/1/012043

755.1 HV for 60HRC alloy and 691.9HV to 692.2HV for 50HRC alloy, which doesn't show much significant change in microhardness on raising power. When the laser power is further increased to 4000W the microhardness decreased to a value of 724.6HV for 60HRC alloy and 688.8HV for 50HRC alloy. This is because more Fe element enters into the cladding layer, with the increase of laser power, which enhances the dilution effect of the cladding layer [7]. Therefore, the average hardness of cladding layers decreases with the further increase of laser power above 3600W.

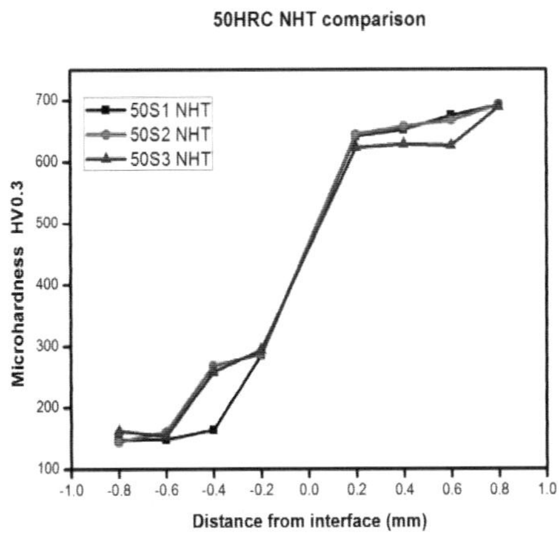

Figure 2- Micro-hardness plot of50HRC as-deposited NiCrBSi alloy for different power

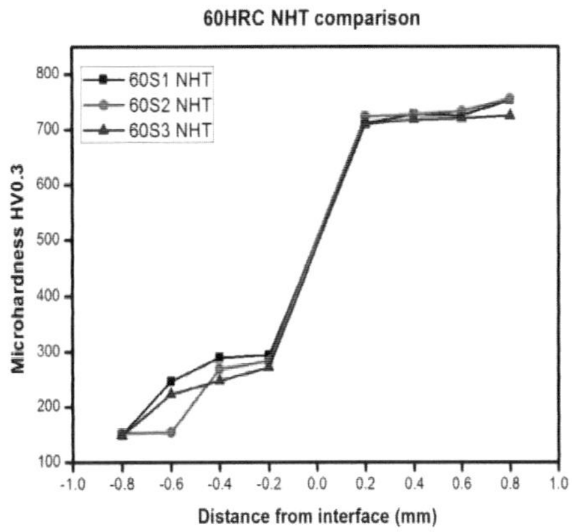

Figure 3- Micro-hardness plot of 60HRC as-deposited NiCrBSi alloy for different power

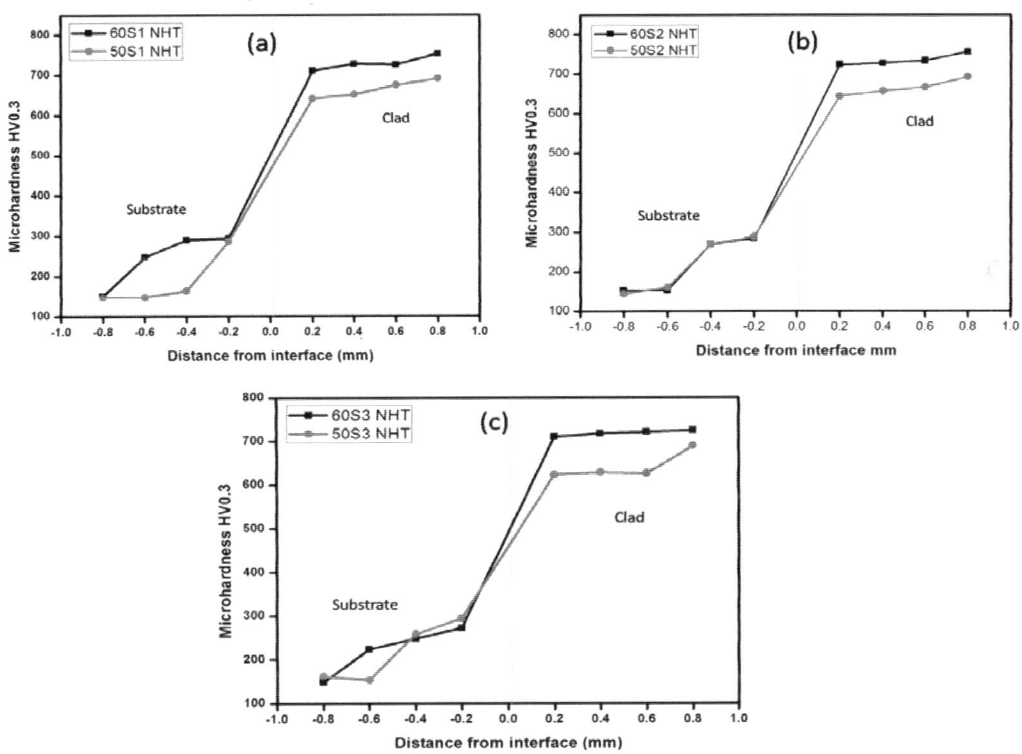

Figure 4- Microhardness plot of single track clad of 50HRC & 60HRC NiCrBSi alloy on

a) S1-3200W b) S2-3600W c) S3-4000W

For laser power of 3200W (S1), the maximum microhardness of 60HRC alloy is about 752.7 HV in the top part of cladding layer for as-deposited coating, and the microhardness of clad layer ranges from 710.5 to 752.7 HV and that for 50HRC alloy the microhardness value ranges from 641.3 to 691.1 HV. Similarly, for laser power of 3600W (S2), the maximum value of microhardness of 60HRC alloy is about 755.1 HV in the top part of cladding layer for as-deposited coating and the microhardness of clad layer ranges from 723.1 to 755.1 HV and that for 50HRC alloy the value ranges from 643.9 to 692.2 HV. For laser power of 4000W (S3), the maximum microhardness value of 60HRC is about 724.6 HV in the top part of cladding layer for as-deposited coating, and the microhardness of clad layer ranges from 709.9 to 724.6 HV and for 50HRC alloy the microhardness ranges from 622.6 to 688.8 HV.

Effect of power on microhardness of laser clad was not prominent in the range of power 3200 W to 4000 W and for all laser powers, the microhardness of 60HRC alloys are higher than the 50 HRC alloys.

4. Conclusions

NiCrBSi alloys of 50HRC and 60HRC is fabricated on AISI410 stainless steel using high power diode laser by varying the laser power as 3200W, 3600W and 4000W by the laser cladding process. From this research work we came to conclusions as mentioned below :

4.1 Effect of power on microhardness of laser clad is not prominent in the range of power (3200 to 4000 W).

4.2 Microhardness is measured on the transverse section of cladded specimen by making 8 indentation at different points including the top portion as well as the dilution zone of cladded layer. Maximum hardness is obtained at top most point, which continuously decreased with increasing depth and dilution zone is having the least microhardness.

4.3 For all three samples cladded at powers (3200W, 3600W & 4000W), microhardness of 60 HRC alloy clad is higher than that of 50HRC alloy clad.

5. References

[1] P. Monson, W. Steen, Comparison of laser hardfacing with conventional processes, *Surface Engineering*, 6 (1990) 185-193.
[2] I. Hemmati, V. Ocelík, J.T.M. De Hosson, Effects of the alloy composition on phase constitution and properties of laser deposited Ni-Cr-B-Si coatings, *Physics Procedia*, 41 (2013) 302-311.
[3] F. Weng, H. Yu, C. Chen, K. Wan, Influence of Nb and Y on hot corrosion behavior of Ni–Cr-based superalloys, *Materials and Manufacturing Processes*, 30 (2015) 677-684.
[4] Z. Bergant, U. Trdan, J. Grum, Effect of high-temperature furnace treatment on the microstructure and corrosion behavior of NiCrBSi flame-sprayed coatings, *Corrosion Science*, 88 (2014) 372-386.
[5] W. Kai-ming, F. Han-guang, L. Yu-long, L. Yong-ping, W. Shi-zhong, S. Zhen-qing, Effect of power on microstructure and properties of laser cladding NiCrBSi composite coating, *Transactions of the IMF*, 95 (2017) 328-336.
[6] S. Gnanasekaran, G. Padmanaban, V. Balasubramanian, Effect of Laser Power on Metallurgical, Mechanical and Tribological Characteristics of Hardfaced Surfaces of Nickel-Based Alloy, *Lasers in Manufacturing and Materials Processing*, 4 (2017) 178-192.
[7] D.M. Goodarzi, J. Pekkarinen, A. Salminen, Effect of process parameters in laser cladding on substrate melted areas and the substrate melted shape, *Journal of Laser Applications*, 27 (2015) S29201.
[8] S. Barnes, N. Timms, B. Bryden, I. Pashby, High power diode laser cladding, *Journal of Materials Processing Technology*, 138 (2003) 411-416.
[9] J. Jiang, G. Lian, M. Xu, C. Li, B. Chen, B. Li, Influence of Preheating Temperature on Mechanical Properties of Laser Cladding Layer, in: *ASME 2016 11th International Manufacturing Science and Engineering Conference, American Society of Mechanical Engineers*, 2016, pp. V001T002A031-V001T002A031.

[10] S. Zanzarin, S. Bengtsson, A. Molinari, Study of carbide dissolution into the matrix during laser cladding of carbon steel plate with tungsten carbides-stellite powders, *Journal of Laser Applications*, 27 (2015) S29209.

Determining the formability of AA5052 sheets in annealed and H32 condition

R Anand Sekhar

Assistant Professor, TKM College of Engineering, Kollam-691005

E-mail: anand@tkmce.ac.in

Abstract

Forming Limit Diagram (FLD) is an easy and efficient way of determining the safe region for formability in sheet metals. In this work formability of AA5052 H32 sheets of thickness 1 mm is studied by developing a FLD. Since traditional generation of FLDs requires sophisticated equipment, an attempt has been made to generate FLDs using tensile testing and Erichsen cupping tests. Reduced section tensile samples were prepared with varying notch radius of 10 mm and 12 mm for the test. Tensile testing was carried out on a Universal Testing Machine (UTM) with a maximum capacity of 120 kN. Erichsen cupping tests were carried out on rectangular specimens with length of 90mm and widths of 60, 40, 25 and 10 mm respectively. The process (tensile and Erichsen testing) was repeated for this alloy in the annealed conditions with annealing temperature of 345°C. FLDs were plotted for both as-received and annealed conditions. The annealed samples shows a better formability compared to as received samples.

Keywords : Forming Limit Diagram, Erichsen test, Annealing, Formability

1. Introduction

Many materials fracture before appreciable deformation. This phenomenon has limited the use of materials in many applications. Determining the amount of deformation a material can withstand before failure is very much needed for its effective use. Aluminium alloys find a range of application in many fields like aerospace, automobiles, household appliances. Also aluminium alloys finds applications in light weight structures. Therefore determining the formability of aluminium alloys is gaining importance in the present scenario.

Forming Limit Diagram (FLD) is a graphical method for determining the limit strains in a material subjected to deformation. A number of test are available to determine the limit strains in a material [1], [2]. Limit strains are the strains which a material can withstand just before fracture. Thus FLD plays a crucial role in the design of many forming processes. Moshksar et al [3] studied the FLD for Al305 sheets of thickness 1mm. FLD was plotted for inplane and outplane conditions. He concluded that in outplane FLD the amount of strain was higher. Through the test he determined the formability regimes for Al305 sheets. Naka et al [4] studied the effect of parameters like temperature and forming speed on FLD of Al-Mg alloy. He concluded that FLD depends on forming speed at high temperatures only. Also strain rate plays a major role in increasing formability regimes of FLD at high temperatures. A number of studies have been made on determination of formability of the alloys [1], [5]–[8]and also on FLDs [2], [3], [9]–[13].

In the present study AA5052 H32 sheets with thickness 1mm was used for the tests. Forming Limit Diagram was plotted using tensile test and Erichsen cupping test data. FLDs were plotted for both as received and annealed conditions. The main aim of the present work is to determine a safe formable region for the alloy under consideration.

Journal of Physics: Conference Series **1355** (2019) 012044 doi:10.1088/1742-6596/1355/1/012044

2. Experimental details

2.1. Material and Equipment: AA5052 H32 sheet of thickness 1mm was used in this experiment. The main elements in the alloy apart from Al are Mg (2.7 %), Fe (0.21 %) and Cr (0.20 %)

FLD was plotted by conducting two tests. Negative minor strain of FLD was determined by using uniaxial tensile test. Tensile testing was done on a Universal Testing Machine (UTM). Maximum capacity of the UTM was 120 kN. The cross head speed used was 2.5 mm/min. Deep drawing experiments were done with the help of Erichsen cup testing machine Samples were deformed up to tearing by using a hemispherical punch (20mm diameter) and a die in Erichsen Cupping machine.

2.2. Test Procedure : Negative minor strain region of FLS was determined by using uniaxial tensile test. For this tensile sample with gauge length of 70 mm and notch radius of 10 mm and 12 mm were cut from AA5052 H32 sheets of thickness 1 mm using CNC machining. Tensile samples are cut 45 degree to the rolling direction.

Positive major strain region of FLD was determined using Erichsen Cupping test. A set of specimens with a fixed length and different widths were prepared. Rectangular specimens with varying widths of 60 mm, 40 mm, 25 mm and 10 mm were also cut from AA5052 H32 sheets of 1 mm thickness using CNC machining. Nature of strain changes from balanced biaxial tension to uniaxial tension by varying width from 60 mm to 10 mm respectively. Punch used for Erichsen cup test has a diameter of 20 mm. All the samples were subjected to cupping till onset of tearing.

Circles of diameter 2.5mm were engraved on all the specimens by using laser etching. The tests were conducted on all the samples up to fracture. After deformation the circles got stretched into ellipses with major axis and minor axis indicating deformation The two diameters were measured on three distinct regions namely safe region, necked region and fractured region. Major and minor diameters of the ellipses were photographed using a stereomicroscope and measured using Image J software. Major strain (ε_1) and minor strain (ε_2) were measured at two distinct regions namely safe region and teared region. Graphs were plotted with major strain (=ln(major dia./original dia.)) versus minor strain (=ln(minor dia./original dia.)). A groove appears in the deformed region indicating localized necking. In order to improve formability the experiments were repeated after annealing the samples at 345 °C. FLD was plotted for both as received and annealed conditions.
Specimen geometry used for uniaxial tension and Erichsen cupping test are shown in figure1 and 2. Table 1 shows the shape and dimensions of specimen used for both tests.

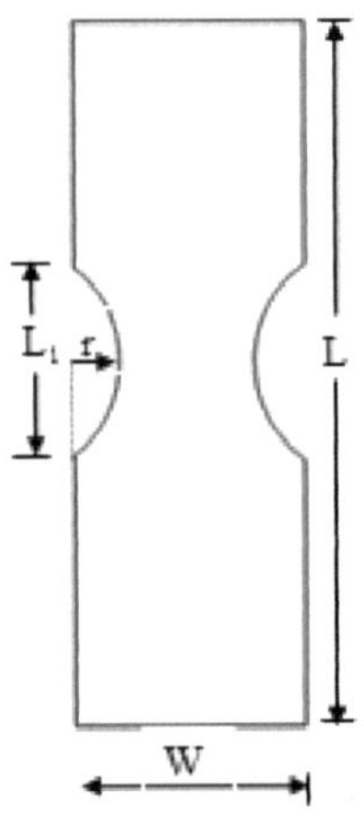

Figure1. Specimen Geometry for Uniaxial tension test

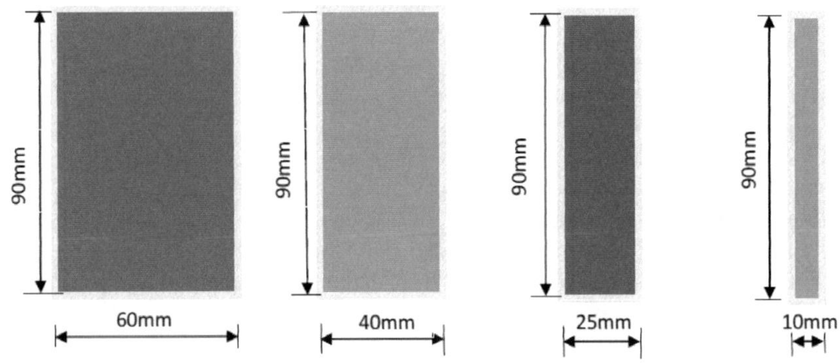

Figure2. Specimen Geometry for Erichsen Cupping test

Table1. Dimensions of specimens used for both tests.

Specimen type	**Specimen Width, W (mm)**	**Specimen length, L (mm)**	**Notch radius , r (mm)**	**Length of notch, L_1 (mm)**
Notched Specimen (Uniaxial tension test)	a) 50 b) 50	a) 150 b) 150	a) 10 b) 12	a) 70 b) 70
Rectangular Specimen (Erichsen cupping test)	a) 60 b) 40 c) 25 d) 10	a) 90 b) 90 c) 90 d) 90	No Notch	

3. Result and discussion

From the uniaxial deformation of the notched specimens with circular grids, we will get the limit strains. These limit strain represents the negative minor strain region of the FLD .Similarly for plotting the positive strain region of FLD Erichsen cupping test was done on rectangular specimens with circular grids.

3.1. Formability in the As Received Condition: Uniaxial deformation test was conducted on notched specimens with notch radius 10mm and 12 mm respectively. Major Strain versus Minor Strain graphs were plotted for both samples. figure3 (a) and (b) shows the limit strain values for both samples. These are the strain values just before failure.

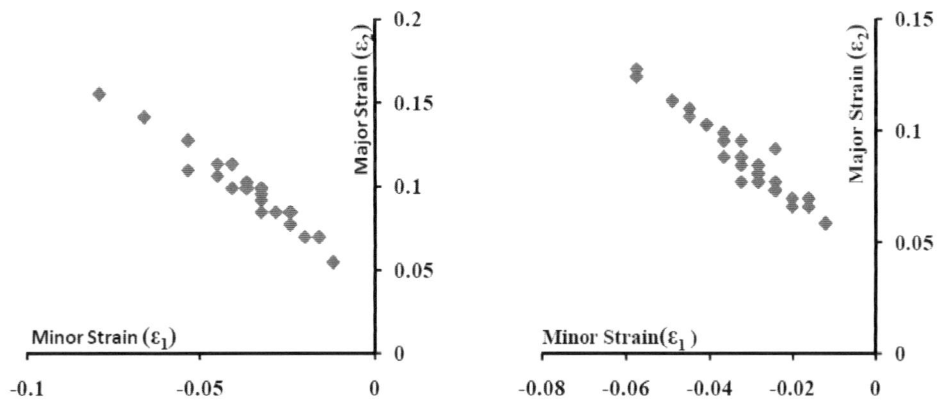

Figure3. Major and Minor strain values measured after Uniaxial deformation (as received) on specimens with varying notch radius (r) **a)** r=10mm **b)** r=12mm

Erichsen cupping test was conducted on rectangular specimens of width 60mm, 40mm, 25mm and 10mm respectively. Here also major strain versus minor strain value is plotted and the results are shown in figure4. (a-d). These are the strain values just before failure

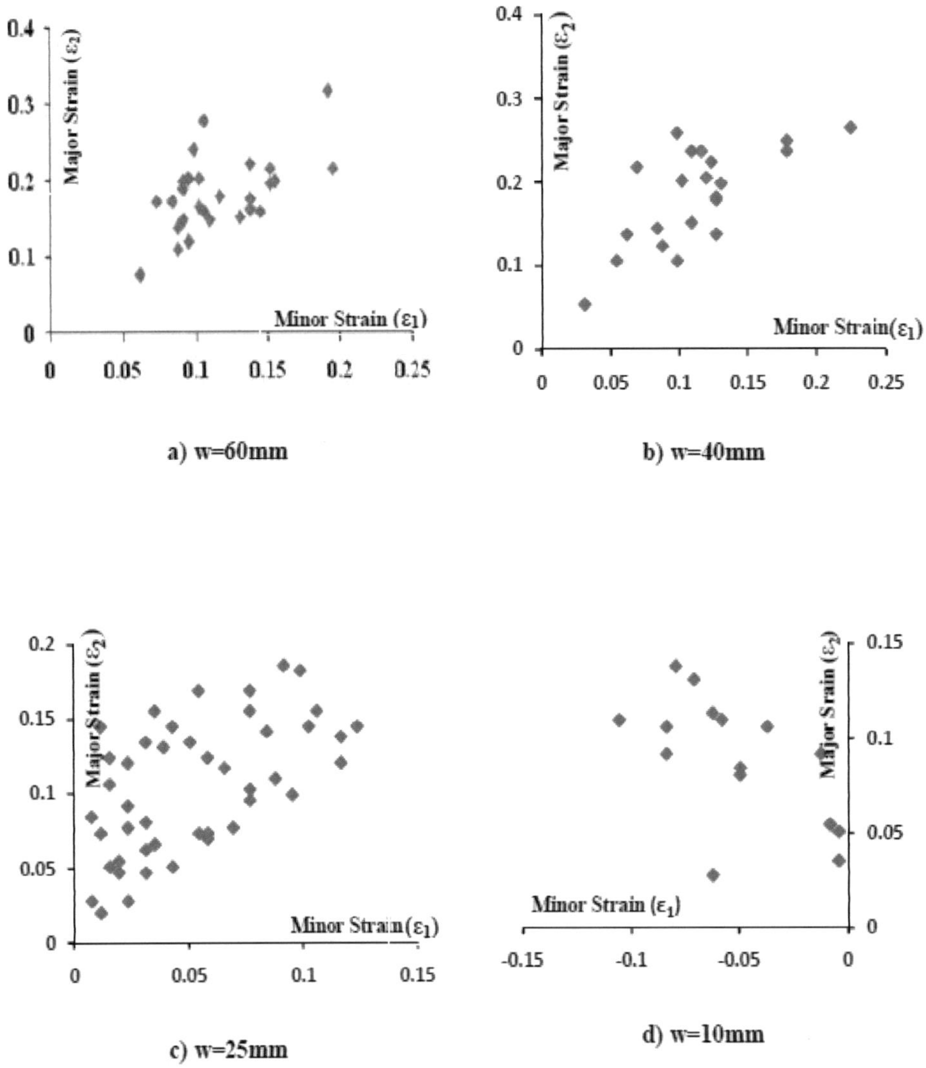

Figure4. Major and Minor strain values measured from Erichsen Cupping test (as received) on specimens with varying width (w) **a)** w=60mm **b)** w=40mm **c)** w=25mm **d)** w=10mm

Considering all data points from uniaxial deformation test and Erichsen test a graph was plotted. In the graph ordinate represents major strain and abscissa represents minor strain. It is as shown in figure.5. On the negative minor strain region (second quadrant) of the graph a line was drawn demarcating safe and unsafe regions. Similarly a demarcating line was drawn on the positive minor strain region (first quadrant). Joining the two (demarcating line and curve) an FLD is drawn which indicates the limit of the formability region diagram for the material AA5052 H32 in the as received condition.

.

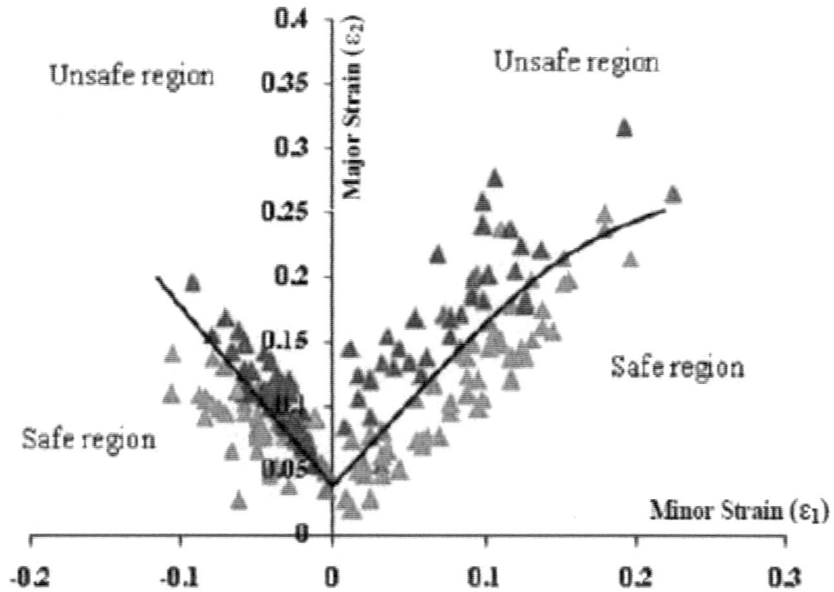

Figure5. FLD (as received condition)

3.1. Formability in annealed condition

As a further improvement in formability AA5052 H32 sheets were subjected to annealing at a temperature of 345ºC for a holding time of 10 min. Uniaxial deformation test and Erichsen test were repeated as for the as received case and graphs were plotted. Comparing the tensile test data for annealed condition shown in figure6 with the corresponding data's for as received condition (figure3) it can be concluded that ductile regime is more for the former case. The reason for this is that the material in the as received condition is cold rolled one and so severely strain hardened.

Journal of Physics: Conference Series **1355** (2019) 012044 doi:10.1088/1742-6596/1355/1/012044

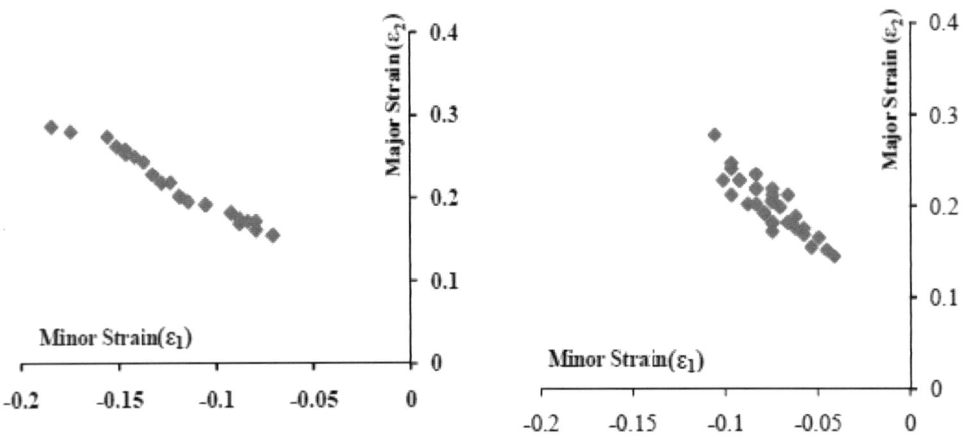

Figure.6. Major and Minor strain values measured after Uniaxial deformation (annealed) on specimens with varying notch radius (r) **a)** r=10mm **b)** r=12mm

Erichsen cupping data of the rectangular specimens are shown in figure7 (a-d). These are the strain values just before failure. Here also we can see that when when the width of the samples were varied the state of strain changes from biaxial to uniaxial stretching. Also from the figure7 (a-d) it is evident that the strain values for the rectangular specimens are more which indicated better formability for the annealed samples.

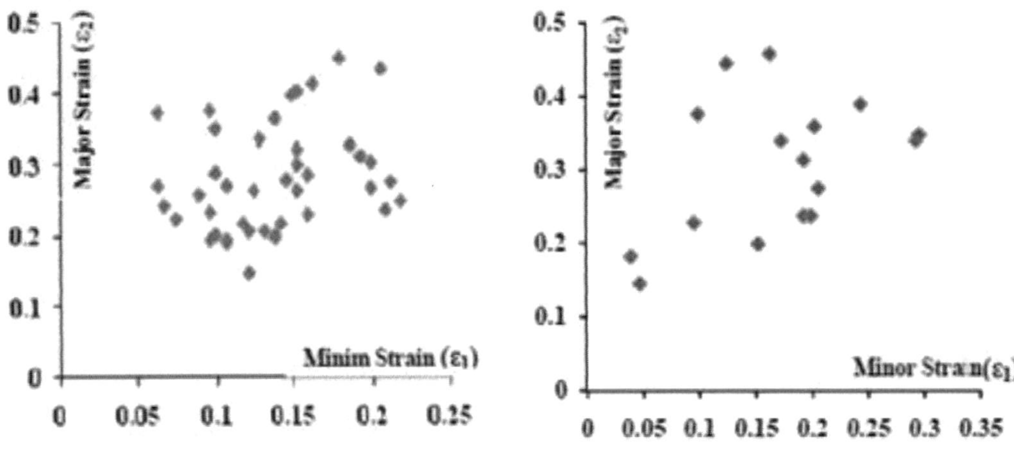

b) w=40mm

a) w=60mm

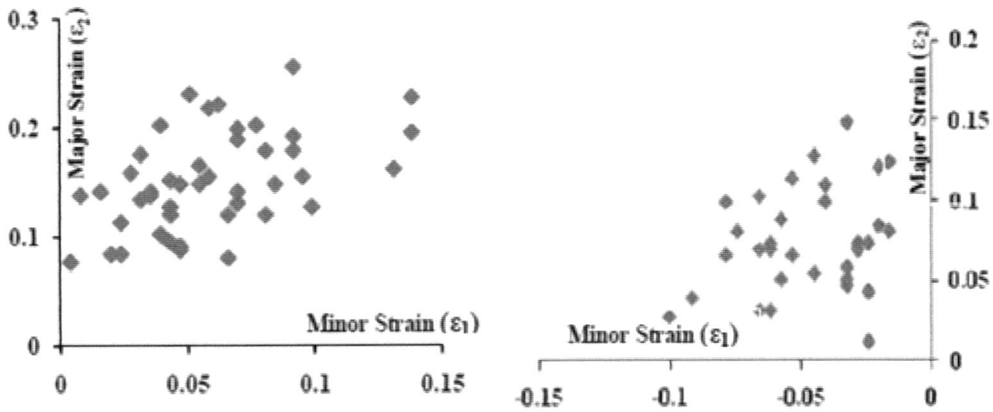

Figure7. Major and Minor strain values measured from Erichsen Cupping test (annealed) on specimens with varying width (w) **a)** w=60mm **b)** w=40mm **c)** w=25mm **d)** w=10mm

FLD for the annealed condition was plotted taking into consideration all the data points from uniaxial deformation test and Erichsen test. (figure8).

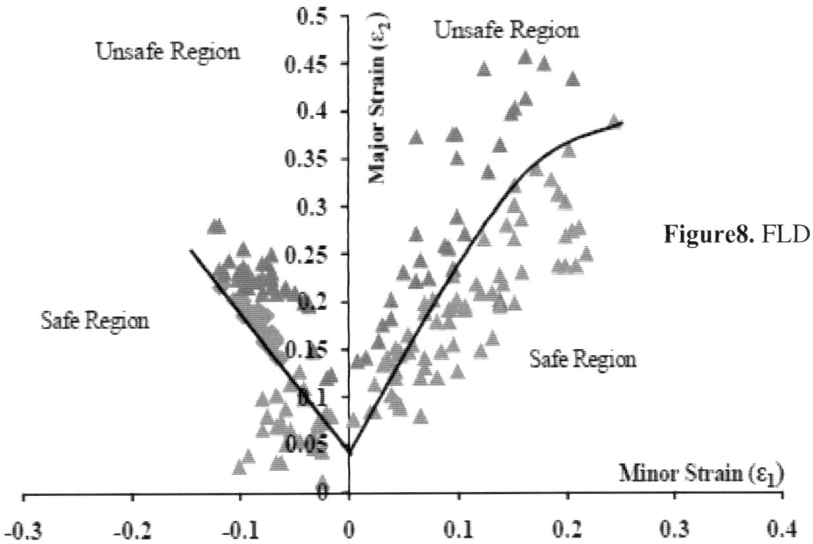

Figure8. FLD (annealed condition)

For comparison both FLDs were plotted a single graph (figure 9). From the figure it can be inferred that the FLD for annealed case is above the FLD for as received one. Thus a part of unsafe region in the as received case becomes the safe region for annealed case. This indicates that the limit of formability has been increased by annealing.

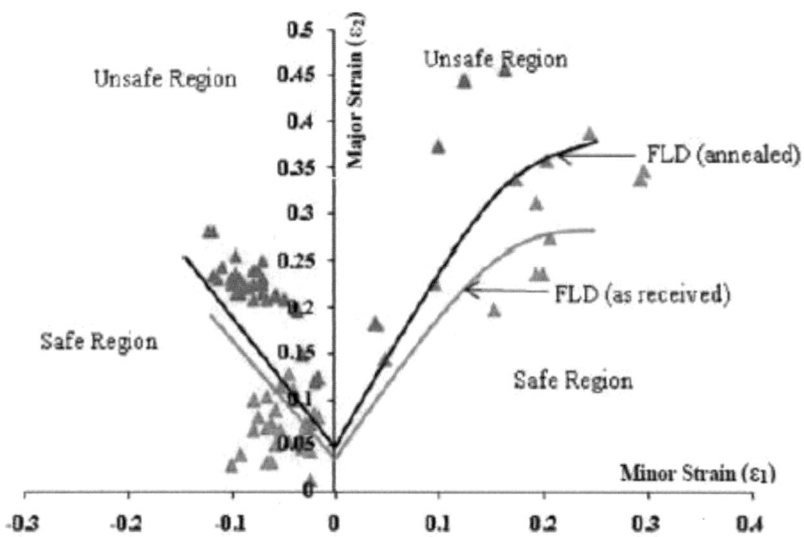

Figure9. Comparison of FLDs for annealed and as received conditions.

figure10 shows the samples after deep drawing test. As the sample width decreases the failure region got shifted from center to the edges.

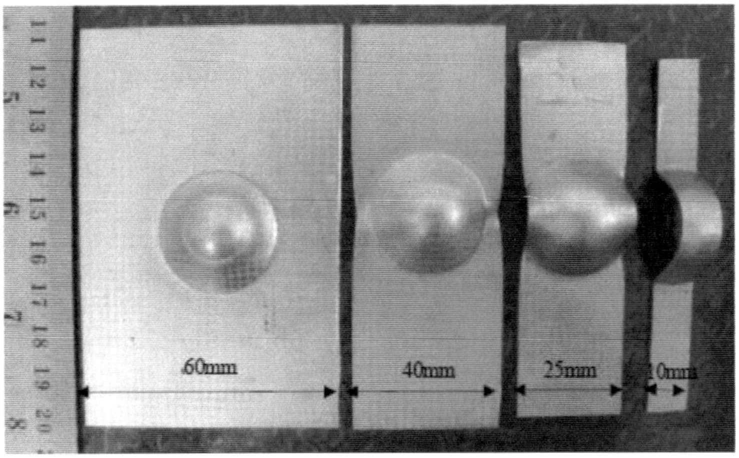

Figure10. Rectangular specimens subjected to Erichsen Cupping test

Conclusion

Forming Limit Diagram (FLD) for AA5052 sheets were determined using uniaxial tensile deformation and Erichsen cupping tests on grid marked specimens in the as received (H32) condition and annealed condition (345°C temperature for 10 min). As width decreases the failure was concentrated near the edges rather than in the center of the Erichsen test specimens. FLD for the annealed condition lies above the FLD for as received material which implies better formability in the annealed condition.

References

[1] K. Chandra Sekhar, R. Narayanasamy, and K. Velmanirajan 2014 *Mater. Des.*, vol. 53, pp. 1064–1070, 2014.
[2] Z. Marciniak, K. Kuczyński, and T. Pokora 1973 *Int. J. Mech. Sci.*, vol. 15, no. 10, pp. 789–800.
[3] M. M. Moshksar and S. Mansorzadeh 2003 *J. Mater. Process. Technol.*, vol. 141, no. 1, pp. 138–142.
[4] T. Naka, G. Torikai, R. Hino, and F. Yoshida, *J. 2001 Mater. Process. Technol.*, vol. 113, no. 1–3, pp. 648–653.
[5] R. Narayanasamy, R. Ravindran, K. Manonmani, and J. Satheesh 2009*Mater. Des.*, vol. 30, no. 5, pp. 1804–1817.
[6] K. Velmanirajan, A. S. A. Thaheer, R. Narayanasamy, R. Madhavan, and S. Suwas, *J. 2013 Mater. Eng. Perform.*, vol. 22, no. 4, pp. 1091–1107.
[7] S. Wagner 1994 *Talat*, pp. 1–30.
[8] H. J. Kleemola and J. O. Kumpulainen, *J. 1980 Mech. Work. Technol.*, vol. 3, no. 3–4, pp. 289–302.
[9] R. Narayanasamy and C. Sathiya Narayanan 2007 *Mater. Des.*, vol. 28, no. 1, pp. 16–35.
[10] A. F. Ávila and E. L. S. Vieira, *J. 2003 Mater. Process. Technol.*, vol. 141, no. 1, pp. 101–108.
[11] H. Keife, 1980 Journal of Mechanical Working Technology, 4 133--143 vol. 4, pp. 133–143.
[12] A. Parmar and P. B. Mellor, 1980 *Int. J. Mech. Sci.*, vol. 22, no. 3, pp. 133–150.

Polyurea Coated Steel Plates For Blast Mitigation In Armoured Vehicles

Agesh Markose[1], Rakesh Pillai. R[2]
[1,2]Assistantt Professor, TKM College of Engineering, Kollam.
E-mail: ageshmarkose@gmail.com

Abstract. Increasing terror attacks and development of Improvised explosive devices (IEDs) has resulted in the development of protective structures for armored personnel vehicles (APV). These shaped plates either deflect or absorb the energy imparted to them. A well designed plate should absorb the energy as much as possible and transmit the least impulse to the vehicle structure. In this context the response of polyurea coated steel plates have recently been explored with the purpose of increasing the effectiveness of these shaped plates. Experimental studies have shown the capability of polyurea coated steel plates in mitigating the blast load. Hence there is scope for a numerical study on the polyurea coated V-shaped steel plates for finding its response. The objective of this study is to conduct FE based analysis of V-shaped steel plates coated with polyurea. The important parameters which affect the response of the plate are the mass of the explosive, the parameters of the V shaped plate and thickness of polyurea layer. The response variables under consideration are the deflection of the plate and imparted impulse. A numerical procedure validated for flat plates coated with polyurea, has been used for the above study. The results are helpful in identifying an optimal structure design for resisting blast loads.

1. Introduction

New and improvised explosive devices in the form of land mines are widely used in war affected zones. The armoured personal vehicles (APV) are the only means of transportation under such conditions. These APVs always uses shaped metallic plates as blast load resisting structures for effectively resisting blast loads. These coated metallic plates can deflect away and absorb a major part of the blast load. Hence specific impulse transferred to the vehicle crew is significantly reduced thus reducing the possibility of damage. APV like South African CASSPIR are known to use 'V' shaped plates as blast mitigation structures. Traditionally the plates are made from the armoured steel. Latest research has also highlighted the use of sandwich structures, using hyperplastic materials like polyurea for the above purpose as given in Ref. [4,6]. Polyurea is an elastic polymer formed by the rapid chemical reaction between isocyanates and amines. Polyurea coatings exhibit resistance to fragmentation and impact loads along with blast energy absorbing capability. These materials when applied as an external coating on the, metals plate are seen to improve the blast resistant property of the shaped plates.

The mechanical response of flat plates and beams has been extensively investigated by several researchers [1-3]. In these studies experimental and numerical procedures have been used for the analysis of circular and quadrangular plates under varying magnitudes of loads and with various boundary conditions. On the contrary few studies are available related to the response of the V shaped plates coated with Polyurea subjected to blast loading.

Investigators have conducted an experimental and numerical modelling of the response of the bilayer plate made of metal plate and polyurea, under shock loading as presented in Ackland et al. [4] as well as in Amini et al. [6]. Yuen et al. [5] has conducted exclusive studies on the response of the V shapes plates. The V shaped plates mounted on ballistic pendulum were subjected to explosive loading. Agesh et al. [7] has conducted numerical simulations on the response of full scale V plates and flat plates under blast loading.

In the present paper the objectives are as follows,

1. To develop a numerical model for simulating the response of metallic plates with polyurea coating and validate the published results.
2. To use the validated model to assess the effectiveness of polyurea coated test plates.

The response parameters simulated are the midpoint displacement and the force or impulse transmitted to the boundaries for varying coating thickness.

2. Problem description and numerical modelling.

APV like Casspir use V shaped metallic hull made from armored steel as protection against blast loading. These plates provide adequate protection to the vehicle and crew from land mines. A schematic diagram of the shaped plate used for the numerical simulation is shown in Figure 1.

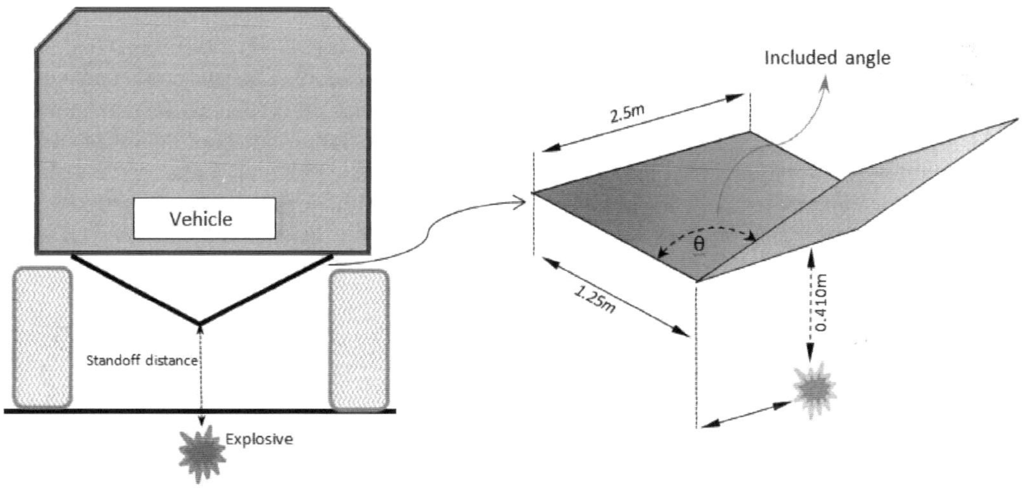

Figure 1. Schematics illustrating V-shaped hull and the terminologies

Here we are considering the response of flat plate and scaled down model of hull plate under shock loading. The validation is done with the flat plates under explosive loading. The analysis is carried out for scaled down V plate undergoing shock loading. The dimensions of the models are given in figure 1.

There are broadly two sets of parameters which affect the response of the APV, the ones related to the blast conditions and which relate to the structure. The focus of our analysis is on the structural part and hence a simplified blast model is assumed. The parameter under consideration is the thickness of the coating. The 3d numerical study is conducted with Abaqus version 6.12.

Table 1. Model dimensions

	Parameters of Flat plate for validation	Parameters of scaled down Plate
Dimensions of projected area	1000mmx1000mm	300x300mm^2
Standoff distance	10mm	----
Plate thickness	4mm-6mm	1mm
Polyurea thickness	7.7mm, 15.7mm	1mm,2mm, 4mm
Plate material	Bluescope XLERPLATE 350 grade	Mild steel

2.1 Geometric and material modelling

The plate angle of 145^0 has been selected based on recommendation from the numerical study conducted by the authors as given in Ref. [7]. A plate of 1000mm x1000mm has been selected for validation. The 3D model is created with Abaqus and discretized using lagrangian elements. The plate has been modelled with shell elements while the polyurea layer has been modelled with solid elements. Shell elements S4R (4 noded, reduced integration elements with hour glass control) has been used since the thickness is far lesser than the other dimensions and also due to its computational efficiency. First order solid element C3D8R (8noded, 3 dimensional reduced integration element) has been used for modelling the hyperelastic material, polyurea. Similar elements have been used previously in Ackland et al. [4], for modelling these materials under impulsive loading. A schematic cross section of the coated V plate is shown in Figure 2 and FE mesh used in numerical analyses is shown in Figure 3.

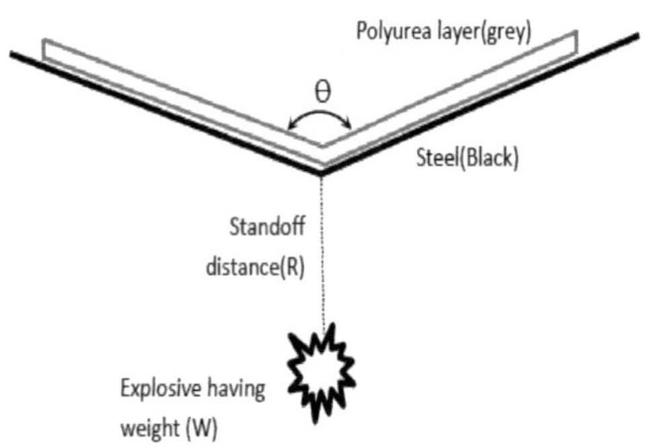

Figure 2. Schematic cross section of a coated V plate.

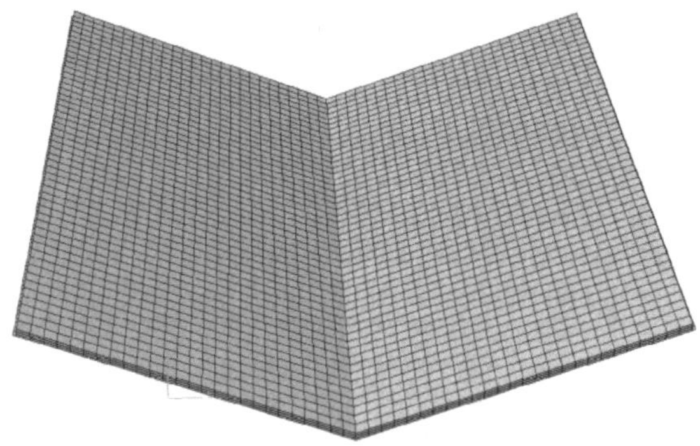

Figure 3. FE model of coated V plate.

The Johnson Cook material model which is dependent on the strain rate has been used in calculation of finite deformations. The Von Mises flow stress given by Johnnson et al. [8] is expressed as,

$$\sigma = [A + B\varepsilon^n][1 + Cln\dot{\varepsilon}^*][1 - T^{*m}] \quad (1)$$

Here ε is the equivalent plastic strain, $\dot{\varepsilon}^* = \frac{\dot{\varepsilon}}{\dot{\varepsilon}_0}$ is the dimensionless plastic strain rate and $\dot{\varepsilon}_0$ is the reference strain rate. $T^*$is the homologus temperature . The constants A, B,n, C and m in the equation may be determined by fitting the flow stress data, based on static and dynamic tests. All the material parameters for steel in the Johnson cook elastoplastic model have been evaluated by Iqbal et al. [9]. The hyper elastic strain energy function can be expressed as a two parameter Mooney Rivlin model as

$$\psi = C_{10}(I_1 - 3) + C_{01}(I_2 - 3) + \frac{1}{d}1(J - 1)^2 \quad (2)$$

Here C_{01} and C_{10} are material constants, $I_1$and I_2 are the first and second invariant of the deformation tensor, J is the determinant of the elastic deformation gradient and d is the material incompressibility factor. The value ofC_{10}= 875.2 kPa, C_{01}=6321.3 kPa and d=4x10^{-7} (kPa)$^{-1}$

2.2 Blast load modelling

The validation part of the analysis deals with the simulation of the plate which undergoes shock loading from a centrally placed explosive. The explosive charge upon detonation generates a shock wave propagating towards outer direction with a sudden jump in the field parameters (like particle velocity, density, pressure and internal energy) across the shock front. The difference of shock pressure over the atmospheric pressure is denoted by blast over pressure, (P_s-P_a). The shock wave can be represented as shown in figure 4.

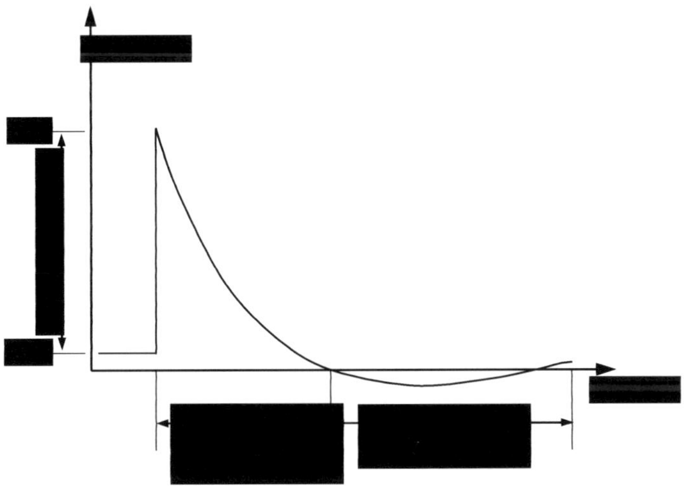

Figure 4. Typical pressure time history in blast wave.

The pressure time history of the blast wave can be represented by the modified Friedlander equation, Ref [7].

$$p(t) = (p_s - p_a)\left[1 - \frac{t-t_a}{t_d}\right]e^{\frac{-(t-t_a)}{\theta}} \quad (3)$$

The second part of the analysis deals with a plate which is loaded centrally. Parameters for governing blast loading on V-shaped Plates are given below.

- Radius r of Shock loaded area is 17.5mm.
- Loading region on V-Plate taken as circular region with radius r

- Peak pressure at centre of the plate is 250bar .

The equation suggested by Brode [10] was used to calculate the blast overpressure ,in the near field as a function of scaled distance and is given as,

$$P_{pos}= (\frac{6.7}{Z\hat{}3}+1)\text{bar for } (P_{pos}>10\text{bar}) \quad (4)$$

where R, is the radial distance between the target and charge, W is weight of detonator charge in kg & Z is scaled distance expressed as $Z=\frac{R}{W^{1/3}}$ .P_{pos} represents the blast overpressure acting on the plate. The peak pressure values of 150bar, 200 bar and 350bar has been used in the simulation of the V plate . Another convenient option for the analysis of this fluid structure interaction problem is to utilize the built in blast generation function Conwep. The time history of the loading and its variation with the angle of incidence θ is given in Ref. [7]. The total pressure is given as

$$P(t) = P_i(t)[1 + \cos\theta_i - 2\,cos^2\theta_i] + P_r(t)cos^2\theta_i, for\ cos\theta_i \geq 0,$$

$$P(t) = P_i(t),\ for \cos\theta_i \leq 0 \quad (5)$$

Here $P_i(t)$ is a function of incient pressure, $P_r(t)$, is afunction of refelcted pressure, θ_i is the angle of incidence measured between point of intersection on the target surface and the straight line connecting the point with the explosive location.

The imparted impulse (I) is be obtained by integrating the pressure acting with respect to the time duration of the loading.

$$I = \int_0^{t_d} Pdt \quad (6)$$

Here P represents the applied pressure due to blast loading function Conwep and t_d is the duration of loading. The time evolution of pressure and calculation of impulses due to blast loading due to conwep loading has been demonstrated by Agesh et al.[7].

3. Results

In the first part of the study the parameters like deformation of the plate and polyurea layer was simulated and compared with the experimental result given in Ackland et al [4]. A mesh convergence study was conducted on the plates subjected to Conwep loading. The mesh size for the scaled down model was varied as , .006, 0.003 and.001m. The mesh size for the scaled down model was selected as .002m based on the results.

Figure 5 shows the comparison between the deformation contours of 6mm flat plate where the numerical results are compared with experiments conducted on the same plates by Ackland et al. [4]. From the comparison, it is seen that the numerical model is able to capture the experimental trends fairly accurately for the uncoated plate. A comparison of the total force transmitted to the boundaries for the flat plate and V shaped plate with 4mm thick coating are presented in Figure 6. It can be seen that the flat plate transmits the maximum force while the V shaped profiling with a polyurea layer reduces the force significantly.

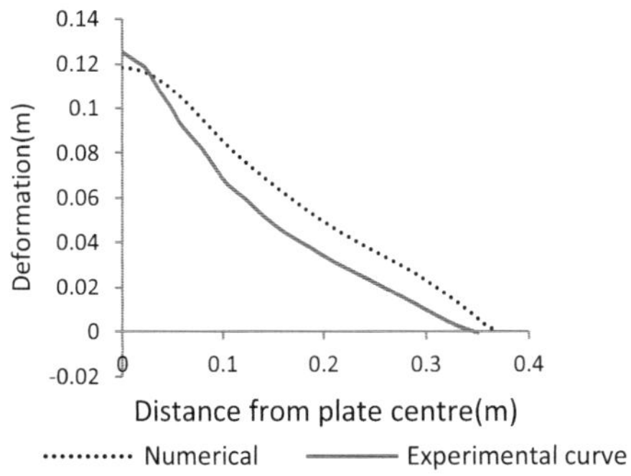

Figure 5. Deformation contour of 5mm thick coated steel plate.

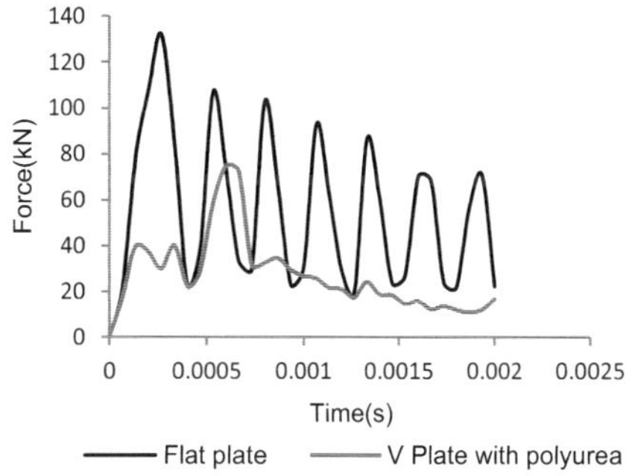

Figure 6. Response of the flat plate and V plate with polyurea layer

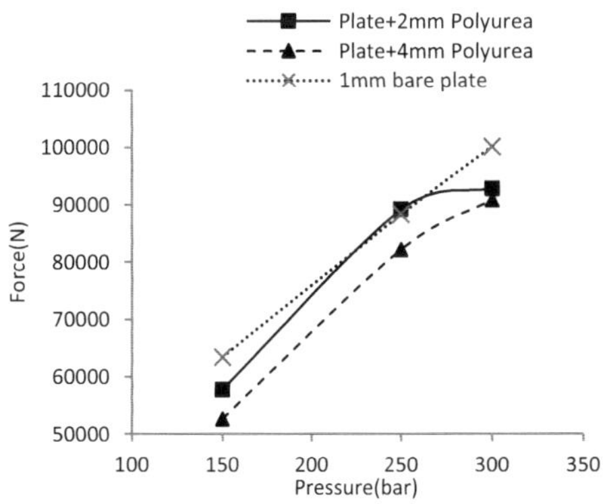

Figure 7. Force transmitted for 1mm plates with polyurea coating

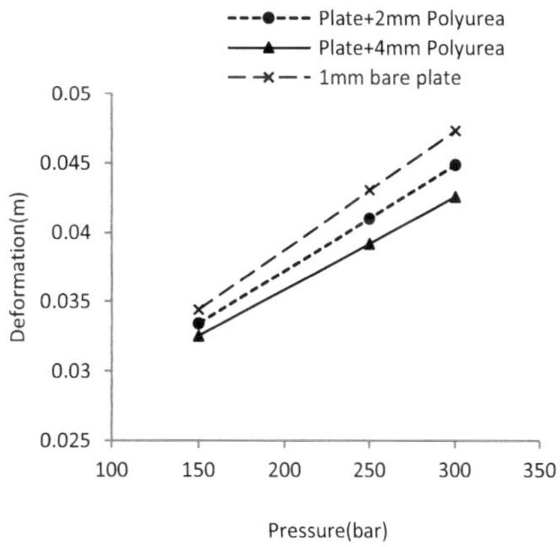

Figure 8. Deformation for the 1mm plates with polyurea coating

The Figure 7 shows the variation of maximum force transmitted on the V shaped plates with the variation of coating thickness. Figure 8 shows a similar variation of maximum displacement with input pressure. It is seen that there is a reduction in both deformation and transmitted force with increasing value of coating thickness, for the pressure values under consideration. This reduction could be partly due to the increasing stiffness of the structure as a result of coating and also due to the viscoelastic energy dissipation in the polyurea layer.

4. Conclusion

The main findings were

- The V shaped plate offers substantial reduction in the transferred reaction forces to the base structure.
- The polyurea layer is capable of reducing the deflection imparted to the structure.
- From the results it also seems that the thickness of the polyurea layer affects the final force transmitted and needs further investigation for proper quantification of results.

References

[1] Teeling-Smith RG and Nurick GN1991The deformation and tearing of thin circular plates subjected to impulsive loads *Int. J. Impact. Eng.* 11(1)77-91.

[2] Rudrapatna NS, Vaziri R and Oslon MD 1999 Deformation and failure of blast loaded square plates *Int. J. Impact Eng.* 22(4), 449-67.

[3] Jacob N, Chum Kim Yuen S, Nurick GN, Bonorchis D, Desai SA and Tait D, Scaling aspects of quadrangular plates subjected to localized blast loads – experiments and predictions 2004 *Int.J. Impact Eng.* 30(8,9), 1179-208.

[4] Ackland K, Anderson C and TuanDuc Ngo 2013 Deformation of polyurea-coated steel plates under localised blast loading *Int.J. Impact Eng.* 51, 13-22.

[5] Chung Kim Yuen S, Langdon GS, Nurick GN, Pickering EG, Balden VH 2012 Response of V-shape plates to localised blast load: Experiments and numerical simulation *In. J. Impact Eng* 46,97-109.

[6] Amini MR Simon J Nemat-Nasser S 2010 Numerical modeling of effect of polyurea on response of steel plates to impulsive loads in direct pressure-pulse experiments *Mechanics of Materials* 42, 615–627.

[7] Agesh M Rao CL 2017 Mechanical Response Of Vshaped plates under blast loading *Thin walled structures* 115, 12-20.

[8] Johnson GR Cook WH 1983 A constitutive model and data for metals subjected to large strain, high strain rate and high temperature In: *Proc. of the 7th symposium on ballistics*, Hauge , Netherlands 541-7.

[9] Iqbal MA Senthil K Bhargava P Gupta NK 2015 The characterization and ballistic evaluation of mild steel *Int. J. Impact Eng* 78, 98-113.

[10] Brode HL 1955 Numerical solutions of spherical blast waves *J. App. Phys* **26**(6), 766-75.

AUTHOR INDEX